NHTI Library
Concord's Community College
Concord, NH 03301

Cryogenics
A TEXTBOOK

Cryogenics
A TEXTBOOK

S.S. Thipse

Alpha Science International Ltd.
Oxford, U.K.

Cryogenics
A Textbook
552 pgs. | 291 figs. | 11 tbls.

TP
482
.T55
2013

S.S. Thipse
Senior Assistant Director
Power Train Engineering
The Automotive Research Association of India
Pune – 411 038

Copyright © 2013

ALPHA SCIENCE INTERNATIONAL LTD.
7200 The Quorum, Oxford Business Park North
Garsington Road, Oxford OX4 2JZ, U.K.

www.alphasci.com

All rights reserved. No part of this publication may be reproduced, stored in a retrieval system, or transmitted in any form or by any means, electronic, mechanical, photocopying, recording or otherwise, without prior written permission of the publisher.

ISBN 978-1-84265-729-4

Printed in India

Dedicated to My Brother
Yogesh S. Thipse

Preface

A common question bothering students and teachers alike is "why can't we find a single textbook which fulfills our need? This is true for Cryogenics as there are many standard books available on RAC, however very few on Cryogenics. The students continue searching for textbooks with relevant study material. Keeping in mind the requirement of students, a need was felt to write a textbook, which would cover all relevant topics from syllabi of prominent universities as well as include modern developments in the field of Cryogenics. In particular my experience of teaching this subject at universities has indeed shaped up this book nicely.

The subject of Cryogenics now is included in the syllabus of most universities at undergraduate and graduate levels. This textbook thus aims to provide a broad coverage of theory, derivations, solved problems repeatedly appearing in university examinations across the country. Care has been taken to make this book as student friendly as possible along with an exam-oriented approach. Advanced topics like cryocoolers; cryogenic alternative fuels have been discussed in greater detail, which will benefit all students.

It gives me great pleasure in presenting this book to the student and academic community. Please feel free to communicate any errors, omissions, suggestions, and criticisms of the text to me. I shall acknowledge them and improve the forthcoming editions. Time and again I receive E-mails from satisfied students who express their heartfelt appreciation for their good scores as a result of referring my text books. What more could an author ask for? I sincerely thank my students and colleagues for encouraging me to write this textbook as well as for their overwhelming response to previous textbooks. I also thank my publishing company "Narosa Publishing House" for their continued support and excellent management to bring this book out on time. Particularly, I would like to thank Mr. N.K. Mehra for his contribution in bringing this book to you. Last but not the least I dedicate this book to my beloved brother Yogesh who has been a source of unflinching support and care.

S.S. Thipse

Preface

Contents

Preface *vii*

Review of Basic Thermodynamics and Heat Transfer — 1.1
Introduction — 1.1
Thermodynamics — 1.1
Thermodynamic System — 1.1
Ideal Gas and Equation of State — 1.3
Modes of Heat Transfer — 1.6

Introduction to Refrigeration and Basic Cycles — 2.1
Introduction — 2.1
History of Refrigeration — 2.1
Important Theory Questions — 2.6
Solved Problems — 2.8
Unsolved Theory Questions — 2.11
Unsolved Problems — 2.11

Introduction to Cryogenics — 3.1
Introduction — 3.1
Need for Cryogenics — 3.2
History of Cryogenics — 3.2
Chronology of Cryogenics — 3.5
Cryogenic Terminology — 3.5
Applications of Cryogenics — 3.6
Worldwide Scenario for Cryogenic Industry — 3.8
Indian Cryogenics Scenario — 3.9
References — *3.10*
Questions — *3.10*

Basics of Cryogenics — 4.1
Introduction — 4.1
Cryogenic System Performance Parameters — 4.1
Joule-Thomson Experiment — 4.2
Joule-Thomson (JT) Coefficient — 4.3
Joule-Thomson Coefficient for Real gases — 4.4
Inversion Curve — 4.6
Thermodynamically Ideal Cryogenic System — 4.7
Adiabatic Expansion of an Ideal Gas — 4.8

Reverse Carnot Cycle	4.10
Reverse Brayton Cycle or Bell Coleman Cycle	4.11
Methods of Producing Low Temperature	4.14
Using Enthalpy of Mixing (Mixing of Salt with Water)	4.14
Expansion in a Turbine	4.15
Throttling (Isenthalpic Expansion)	4.16
Thermoelectric Cooling	4.17
Adiabatic Demagnetization Refrigeration	4.18
Solved Problem	4.20
References	*4.21*
Questions	*4.21*

Cryogenic Material Properties — 5.1

Introduction	5.1
Ultimate and Yield Strength	5.2
Fatigue Strength	5.2
Impact Strength	5.3
Elastic Moduli	5.4
Hardness and Ductility	5.4
Thermal Conductivity	5.5
Specific Heat of Solids	5.7
Specific Heat of Liquids and Gases	5.8
Coefficient of Thermal Expansion	5.9
Electrical Conductivity	5.11
Super Conductivity	5.11
Theories of Superconductivity	5.14
Applications of Superconductivity	5.17
References	*5.18*
Questions	*5.18*

Cryogenic Fluids (Cryogens) — 6.1

Introduction	6.1
Properties of Cryogens	6.1
Cryogen 1: Liquid Nitrogen (LN_2)	6.2
Cryogen 2: Liquid Oxygen (LO_2)	6.4
Cryogen 3: Liquid Hydrogen (LH_2)	6.6
Cryogen 4: Liquid Argon (LAr)	6.9
Cryogen 5: Liquid Helium (LHe)	6.11
Cryogen 6: Liquid Neon (LNe)	6.12
Cryogen 7: Liquid Fluorine (LF)	6.13
Cryogen 8: Liquid Carbon-dioxide (LCO_2)	6.14
Superfluidity	6.15
References	*6.18*
Questions	*6.18*

Gas Liquefaction Systems — 7.1

Introduction	7.1
Liquefaction System Performance Parameters	7.3
Thermodynamically Ideal Liquefaction System	7.3
Joule-Thomson (J-T) Effect	7.4

Adiabatic Expansion in Turbine	7.6
Simple Linde-Hampson (L-H) System	7.7
Pre-cooled Linde-Hampson System	7.9
Linde Dual Pressure System	7.10
Cascade System	7.11
Claude System	7.13
Claude Dual Pressure System	7.14
Kapitza Liquefaction System	7.15
Heylandt Liquefaction System	7.16
Comparison of Liquefaction Systems	7.16
Problems	*7.17*
References	*7.20*
Questions	*7.20*

Liquefaction Systems for Inert Gases Helium, Neon and Argon — 8.1

Introduction	8.1
Liquefaction of Argon	8.2
Argon Manufacture from Purge Gas Recovery of Ammonia Plants	8.2
Liquefaction of Neon	8.4
Pre-cooled Linde Hampson (LH) Cycle for Neon Liquefaction	8.4
Liquefaction of Helium	8.8
Pulse Tube (PT) Helium Liquefaction System	8.8
Collins Helium Liquefaction System	8.9
Simon Helium Liquefaction System	8.11
References	*8.12*
Questions	*8.13*

Liquefaction Systems for Natural Gas and Hydrogen — 9.1

Introduction	9.1
Liquefied Natural Gas Plants	9.4
History of Liquefaction of Hydrogen	9.11
References	*9.19*
Questions	*9.19*

Cryogenic Equipment — 10.1

Introduction	10.1
Cryogenic Heat Exchangers	10.1
Plate and Fin Heat Exchanger	10.2
Coil Wound Heat Exchanger	10.3
Collins Heat Exchanger	10.6
Classification of Heat Exchangers	10.6
Cryogenic Compressors	10.13
Cryogenic Turbo Expanders	10.14
Cryogenic Pumps	10.18
Cryogenic Turbo Alternators	10.20
Cryogenic System Optimization	10.20
Problems	*10.22*
References	*10.22*
Question	*10.23*

Gas Separation and Purification Technologies — 11.1
Introduction — 11.1
Thermodynamically Ideal Gas Separation System — 11.2
General Characteristics of Mixtures — 11.3
Principles of Rectification — 11.8
Types of Rectification Columns — 11.17
Helium Separation from Natural Gas — 11.29
Refrigeration Purification — 11.30
Solved Problems — *11.32*
References — *11.35*
Questions — *11.35*

Air Separation Technologies — 12.1
Introduction — 12.1
Thermodynamics of Air Separation — 12.2
Introduction to Cryogenic Gas Separation Plants — 12.3
Cryogenic Air Separation Units (ASU) — 12.4
Cryogenic ASU Cycles — 12.7
Non-Cryogenic Air Separation Units (ASU) — 12.8
Air Separation By Rectification — 12.10
Expansion Engine for Cryogenic ASU Plants — 12.11
Tonnage Air Separation Plants — 12.12
Membrane Separation Units for Air — 12.16
Chemical Separation Units for Air — 12.19
Recent Trends in Air Separation Units — 12.20
References — *12.22*
Questions — *12.22*

Cryogenic Refrigeration Systems — 13.1
Introduction — 13.1
Refrigerator Effectiveness — 13.2
Thermodynamically Ideal Refrigeration System – Isothermal Source — 13.4
Thermodynamically Ideal Refrigeration System – Isobaric Source — 13.5
Joule-Thomson (JT) Refrigeration — 13.6
Cascade or Pre-cooled Joule-Thomson Refrigerator — 13.8
Refrigerator Optimization — 13.9
Expansion Engine Refrigeration or Claude Refrigeration — 13.10
Regenerators — 13.12
Solved Problems — 13.14
References — *13.17*
Questions — *13.17*

Cryogenic Refrigerators — 14.1
Introduction — 14.1
Phillips Refrigerator — 14.2
Importance of Regenerator for Phillips Refrigerator — 14.4
Vuilleumier (VM) Refrigerator — 14.4
Solvay Refrigerator — 14.6
Gifford Mc-Mohan (G-M) Refrigerator — 14.8
Magnetic Cooling or Adiabatic Demagnetisation — 14.11

Magnetic or Magneto-Caloric Refrigerator	14.14
Thermal Valves	14.18
Helium Dilution Refrigerator	14.18
Solved Problem	14.20
References	*14.20*
Questions	*14.21*

Cryocoolers — 15.1

Introduction	15.1
Classification of Cryocoolers by Walker's Chart	15.2
Mechanical Cryocoolers	15.3
Cryocooler Requirements	15.4
Cryocooler vs Cryostat	15.4
Joule-Thomson (JT) Cryocooler	15.5
Brayton Cryocooler	15.8
Plank Sorption Cryocooler	15.10
Gifford Mc-Mahon (G-M) Cryocooler	15.13
Pulse Tube Cryocooler	15.15
Stirling Cryocooler	15.18
Comparison of Cryocoolers	15.22
Applications of Cryocoolers in Space	15.23
Solved Problem	15.24
References	*15.25*
Questions	*15.25*

Cryogenic Insulations — 16.1

Introduction	16.1
Insulation Performance Regarding Heat Transfer	16.1
Expanded Foam Insulation	16.3
Solid Foam Insulation	16.4
Gas Filled Powders and Fibre Insulation	16.5
Vacuum Insulation	16.6
Evacuated Porus and Fibre Insulation	16.8
Opacified Powder Insulation	16.9
Multilayer Insulation (MLI)–Super Insulation	16.10
Composite Layer Insulation (CLI)	16.12
Glass Microsphere Insulation	16.13
Vapor Shielding	16.13
Comparison of Insulations	16.14
References	*16.16*
Questions	*16.16*

Cryogenic Instrumentation and Measurement — 17.1

Introduction	17.1
Cryogenic Liquid Level Measurement	17.1
Hydrostatic Gauges	17.1
Electric Resistance Gauges	17.3
Capacitance Liquid Level Probes	17.3
Magnetic Bond Type Level Measurement	17.5
Thermodynamic Liquid Level Gauge	17.5

Cryogenic Flow Measurement	17.6
Orifice Meter	17.6
Venturi Meter	17.7
Turbine Flowmeter	17.8
Capacitance Flowmeter	17.9
Electromagnetic Flowmeter	17.10
Ultrasonic Flowmeter	17.10
Cryogenic Pressure Measurement	17.10
Bourdon Gauge	17.11
Electronic Scanned Pressure (ESP) Modules	17.11
Cryogenic Pressure Transducers	17.12
Cryogenic Strain Gauge Sensors	17.12
Cryogenic Pressure Sensitive Paint (PSP)	17.13
Cryogenic Temperature Measurement	17.13
Thermodynamic Temperature Scale	17.13
Thermocouples	17.16
Metallic Resistance Thermometer	17.17
Semiconductor Resistance Thermometer	17.19
Magnetic Thermometer	17.20
Constant Volume Gas Thermometer	17.21
Vapor Pressure Thermometer	17.23
Cryostat	17.25
References	*17.27*
Questions	*17.27*
Cryogenic Storage and Transfer Systems	**18.1**
Introduction	18.1
Basic Cryogenic Storage or Dewar Vessel	18.1
Dewar Inner Vessel Design	18.4
Dewar Outer Vessel Design	18.5
Dewar Vessel Support Design	18.6
Cryogenic Industrial Storage Tanks	18.7
Cryogenic Storage in Space	18.10
Cryogenic Piping	18.12
Draining of Cryogenic Tanks	18.14
Cryogenic Safety Devices for Tanks	18.16
Cryogenic Transfer Systems	18.17
Uninsulated Transfer Line	18.18
Expandable Transfer Line	18.18
Porous Insulated Transfer Line	18.18
Vacuum Insulated Transfer Line	18.19
Cryogenic Transfer Line Joints	18.20
Cryogenic Valves	18.21
Cryogenic Two Phase Flow	18.22
Pressure Drop across two Phase Flow	18.24
Cool Down of Transfer Lines	18.24
References	*18.28*
Questions	*18.28*

Cryogenic Vacuum Technology | 19.1
Introduction | 19.1
Flow Regimes in Vacuum Systems | 19.1
Conductance in Vacuum Systems | 19.2
Calculation of Pump Down Time for a Vacuum System | 19.3
Components of Vacuum Systems | 19.4
Mechanical Vacuum Pumps | 19.6
Diffusion Pumps | 19.7
Ion Pumps | 19.8
Cryopumping | 19.10
Getters and Sorption Pumps | 19.12
Baffles and Cold Traps | 19.13
Vacuum Gauges | 19.15
Vacuum Valves | 19.20
Problem | *19.22*
References | *19.22*
Questions | *19.23*

Cryogenic Plant Design | 20.1
Introduction | 20.1
Cryogenic Plant Design : General Principles | 20.2
Codes and Standards for Cryogenic Plants | 20.3
Safety Considerations in Design of Cryogenic Plants | 20.3
Design Assessments for Cryogenic Plants | 20.4
Thermal Design for Cryogenic Plants | 20.4
Temperature Considerations for Cryogenic Plant Equipment | 20.5
Pressure Considerations for Cryogenic Plant Equipment | 20.6
Materials of Construction for Cryogenic Plants | 20.7
Mechanical Design of Cryogenic Plant Equipment | 20.8
Design Considerations for Cryogenic Plant Specific Equipment | 20.8
Maintenance, Inspection and Monitoring of Cryogenic plants | 20.11
Construction of Cryogenic Plants | 20.11
References | *20.12*
Questions | *20.12*

Safety for Cryogenic Systems | 21.1
Introduction | 21.1
Cryogenic Health Hazards | 21.1
Personal Safety in Cryogenic Plants | 21.3
General Safety Practices for Cryogenic Plants | 21.4
Cryogenic Plant Safety Considerations | 21.5
Cryogenic Plant Critical Safety Systems | 21.8
References | *21.9*
Questions | *21.9*

Advances in Cryogenics | 22.1
Introduction | 22.1
Advances in Superconductivity | 22.2
Advances in Cryogenic Propulsion | 22.3
Advances in Cryocoolers | 22.3

Adiabatic Demagnetization Refrigerators (ADR) — 22.5
Advances in Cryosurgery — 22.6
References — *22.7*
Questions — *22.7*

Applications of Cryogenics - I Industrial Liquefied Gases — 23.1
Introduction — 23.1
History of Industrial Liquefied Gas Industry — 23.3
Industrial Liquefied Gases: Industry Scenario — 23.4
Types of Industrial Gases — 23.6
Industrial Gas: Liquid Oxygen — 23.8
Industrial Gas: Liquid Nitrogen — 23.9
Industrial Gas: Liquid Argon — 23.10
Industrial Gas: Liquid Acetylene — 23.10
Industrial Gas: Solid Carbon Dioxide — 23.10
Industrial Gas: Liquid Hydrogen — 23.11
Industrial Gas: Liquid Helium — 23.11
References — *23.12*
Questions — *23.12*

Applications of Cryogenics - II Biological and Medical Applications — 24.1
Introduction — 24.1
History of Cryobiology — 24.2
Human Cryobiology — 24.2
Cryopreservation — 24.2
Cryonics — 24.3
Cryosurgery — 24.4
Cryogenic Medical Gases — 24.5
Liquefied Medical Oxygen — 24.6
Liquefied Medical Oxygen Plant — 24.7
References — *24.9*
Questions — *24.9*

Applications of Cryogenics - III Propulsion and Space Technology — 25.1
Introduction — 25.1
Cryogenic Aircraft Development — 25.1
Cryogenic Propulsion — 25.2
Disadvantages of Cryogenic Propulsion — 25.3
Cryogenic Propellants — 25.4
Cryogenic Injectors — 25.4
Cryogenic Engine — 25.5
Cryogenics for Space Applications — 25.6
Cryogenic Tanks — 25.7
Solid Cryogens — 25.8
Cryosorption — 25.8
References — *25.9*
Questions — *25.9*

Applications of Cryogenics - IV Food Freezing and Storage — 26.1
Introduction — 26.1

Need for Food Freezing	26.2
History of Food Freezing	26.3
Advantages of Food Freezing	26.3
Freezing of Vegetables	26.4
Freezing of Fish	26.4
Hazards of Improper Food Freezing	26.6
Food Freezing Techniques	26.7
Blast Freezer	26.8
Benefits of Blast Freezing	26.9
Immersion Freezers	26.12
Benefits of Immersion Freezing	26.12
Benefits of Flash Freezing	26.15
References	*26.16*
Questions	*26.16*

Applications of Cryogenics - V Electronics and Computers — 27.1

Introduction	27.1
Semiconductor Devices for Cryogenic Power Electronics	27.2
Cryogenic Electronics for Space Operation	27.3
Cryogenic Electronics for Superconductivity	27.5
Cryogenic Electronics for Transport	27.6
Cryogenic Electronics in Advanced Sensor Systems	27.8
Research Trends in Cryo-Electronics	27.8
Introduction to CryoComputing	27.9
Cryogenic Processor	27.9
Cooling of Computers by Cryocoolers or Cryogenic Liquids	27.10
Cryogenic Cooling of Supercomputers	27.11
Cryogenic Cooling of CPU Chips	27.12
Cryogenics for Telecommunication Applications	27.13
References	*27.14*
Questions	*27.14*

Applications of Cryogenics - VI Transport — 28.1

Introduction	28.1
Liquefied Natural Gas (LNG) as a Transportation Fuel	28.2
Production of LNG	28.3
Properties of LNG	28.3
Economics of LNG	28.4
Advantages of LNG	28.4
Disadvantages of LNG	28.4
Hazards of LNG	28.5
Material Compatibility of LNG	28.5
Transportation of LNG	28.5
Storage of LNG	28.6
Piping for LNG	28.7
LNG Dispensers	28.7
LNG Vehicles	28.8
Liquefied Hydrogen (LH_2) as a Transportation Fuel	28.9
History of Liquid Hydrogen	28.10

Production of Liquid Hydrogen	28.10
Properties of Liquid Hydrogen	28.11
Hazards of Liquid Hydrogen	28.11
Advantages of Liquid Hydrogen	28.12
Disadvantages of Liquid Hydrogen	28.12
Storage of Liquid Hydrogen	28.12
Transportation of Liquid Hydrogen	28.13
Piping for Liquid Hydrogen	28.14
Dispensers for Liquid Hydrogen	28.14
Vehicle Emissions from Liquid Hydrogen	28.15
BMW Liquid Hydrogen Cars	28.15
References	*28.17*
Questions	*28.17*

Applications of Cryogenics - VII Metallurgy and Manufacturing — 29.1

Introduction	29.1
Deep Cryogenic Treatment (DCT)	29.2
Cryogenic Tempering of Tools	29.4
Introduction to Cryogenic Manufacturing	29.5
Cryogenic Deburring	29.5
Cryogenic Milling	29.5
Cryogenic Shrink Fitting	29.6
Cryogenic Deflashing	29.6
Cryogenic Grinding	29.7
References	*29.8*
Questions	*29.8*

Applications of Cryogenics - VIII Power Generation and Waste water Treatment — 30.1

Introduction	30.1
IGCC Power Generation Plant	30.1
Introduction to Cryogenics for Waste Water Treatment	30.3
Cryogenic Waste Water Treatment Plant	30.4
References	*30.5*
Questions	*30.5*

Appendix - I	*A.1*
Appendix - II	*A.4*
Appendix - III	*A.5*
Index	*I.1*

CHAPTER 1

Review of Basic Thermodynamics and Heat Transfer

INTRODUCTION

In order to understand the subject of Refrigeration and Air-Conditioning (RAC), it is necessary to review the basic principles of Thermodynamics and Heat transfer. Both these subjects are the foundation of the theory of RAC. They are essential in explaining the various cycles, processes and also in analyzing the functioning of RAC equipment. We shall briefly review some principles in this chapter.

Part-I Review of Thermodynamics

Thermodynamics

It is the science of energy and entropy. The word is derived from Latin words thermos and dynamics i.e. movement of heat. In common terms, it is a science that deals with heat, work and their inter-conversion.

Thermodynamic System

It is an area upon which attention has been focused for a thermodynamic study. Systems are of three types namely 1) open system, 2) closed system and 3) isolated system. Surrounding is the area enclosing a system. System is separated from its surrounding by a boundary. Universe is the system and surrounding taken together. In an open system energy and mass can cross the boundary. In a closed system only energy can cross the boundary. In an isolated system neither energy nor mass can cross the boundary.

Macroscopic and Microscopic Approach

Analysis of a thermodynamic system on a molecular level considering the behavior of each individual molecule is the microscopic approach. Analysis of a thermodynamic system ignoring molecular effects and assuming that the working substance is continuous (without discrete particles) is the macroscopic approach.

Thermodynamic State

A state of the system is its configuration described in detail. Usually a state is defined by set of properties.

Thermodynamic Property

A thermodynamic property is an entity whose value depends solely on the initial and final state of the system and not on the path. If a property is dependent on the mass of a system, then it is known as an extensive property. If a property is independent on the mass of a system, then it is known as an intensive property. Properties per unit mass or specific properties are intensive properties. (Example: Specific Volume)

Thermodynamic Path

If a thermodynamic system passes through a series of states it is said to describe a thermodynamic path.

Thermodynamic Process

When a thermodynamic system changes its state it is said to undergo a thermodynamic process.

Thermodynamic Cycle

When a thermodynamic system undergoes a series of processes such that the starting point of the first process is same as the end point of the last process then it is said to have completed a thermodynamic cycle.

Thermodynamic Equilibrium

A system is said to be in thermodynamic equilibrium when it is in mechanical, chemical and thermal equilibrium. A system is said to be in mechanical equilibrium when its properties return to their original state value if disturbed. A system is said to be in thermal equilibrium when its temperature returns to its original value if varied. A system is said to be in chemical equilibrium when its chemical composition is invariant.

Heat

It is a transient energy flowing across a system boundary due to temperature difference between two systems or between a system and its surrounding. It is a path function and mathematically an inexact differential. Heat flowing into a system is denoted by a positive sign and heat flowing out of a system is denoted by a negative sign.

Work

Work is done by a system during an operation if the sole effect external to the system can be reduced to the rise of a weight against gravity. It is also a path function like heat and mathematically an inexact differential. Work done by a system is considered positive and the work done on the system is considered negative.

Energy

It is defined as the capacity of a body or a system to do work. Internal energy, kinetic energy, potential energy are examples of energy.

Zeroth Law of Thermodynamics

If two thermodynamic systems are in thermal equilibrium with a third system, then they are in thermodynamic equilibrium with each other.

First Law of Thermodynamics

Energy can neither be created, nor destroyed. It can only be converted from one form to another. Heat and work are equivalent. In an equation form, $Q = \Delta U + W$.

Steady Flow Energy Equation

It is the application of first law of thermodynamics to a steady flow process.

$$\overset{0}{m}\left[h_1 + \frac{V_1^2}{2} + gZ_1\right] + Q = \overset{0}{m}\left[h_2 + \frac{V_2^2}{2} + gZ_2\right] + W$$

In word form, the sum of internal energy (h), kinetic energy ($v^2/2$), potential energy (Z) and heat energy (Q) at inlet to a control volume is equal to the sum of internal energy, kinetic energy, potential energy and work energy at outlet.

Ideal Gas and Equation of State

An ideal gas obeys all gas laws under all conditions of pressure and temperature. Air is assumed to be an ideal gas. Ideal gas law or equation of state is $PV = mRT$.

Charle's Law

Volume of a given mass of gas varies directly with its absolute temperature if its pressure is constant. $V \propto T$ if $P = C$. Thus $V/T = C$.

Boyle's Law

Volume of a given mass of gas varies inversely with its absolute pressure if its temperature is constant.

$$V \propto 1/P \text{ if } T = C. \text{ Thus } PV = C.$$

1.4 Cryogenics

Avagadro's Law

Molecular weights of all perfect gases occupy the same volume under the same conditions of temperature and pressure.

Types of Thermodynamic Processes

Isochoric process – constant volume process ($V = C$).
Isobaric process – constant pressure process ($P = C$).
Isothermal process – constant temperature process ($T = C$ or $PV = C$).
Isentropic process – constant entropy process ($S = C$ or $PV^\gamma = C$).
Isenthalpic process – constant enthalpy process ($H = C$).
Polytropic process – ($PV^n = C$).

Specific Heat at Constant Volume (C_v)

It is defined as the heat required by a unit mass of a substance to raise its temperature by one degree at constant volume.

Specific Heat at Constant Pressure (C_p)

It is defined as the heat required by unit mass of a substance to raise its temperature by one degree at constant pressure.

Thermodynamic Relations

$$C_p - C_v = R \quad \text{and} \quad C_p/C_v = \gamma$$

For constant volume, $\delta Q = dU = m \times C_v \times dT$,
where $\quad dT = $ temperature difference.
For constant pressure, $\delta Q = dH = m \times C_p \times dT$

Enthalpy

It is defined as a property of convenience. It was coined to combine frequently occurring terms U and PV in thermodynamic expressions. It has no physical meaning and is purely hypothetical in nature. $H = U + PV$.

Entropy

Entropy is defined as a tendency of a system towards disorder. Change in entropy is given by $dS = \delta Q/T$. Entropy of the universe always goes on increasing.

Clausius Inequality

$\oint \delta Q/T < 0$ for an irreversible process. $\Phi = $ Cyclic Integral
$\oint \delta Q/T = 0$ for a reversible process.

Reversible Process

A process which when completed can be brought back to its original state without affecting the surrounding is a reversible process.

Irreversible Process

A process which when completed cannot be brought back to its original state without affecting the surrounding is an irreversible process.

Availability

It is a measure of the actual energy available for use out of the total energy of the system.

Irreversibility

It is a measure of how much a system is irreversible in nature.

Kelvin-Planck Statement of Second Law of Thermodynamics

It is impossible to construct a heat engine operating in a thermodynamic cycle, which will produce no other effect than raising of a weight while exchanging heat with a single reservoir.

Clausius Statement of Second Law of Thermodynamics

It is impossible to construct a heat engine operating in a thermodynamic cycle, which will produce no other effect than transfer of heat from a cooler to a hotter body while exchanging heat with a single reservoir.

Third Law of Thermodynamics

Entropy of a system is zero at absolute zero of temperature.

Comparison of First Law and Second Law of Thermodynamics

First law is silent about direction and extent of energy conversion, assuming all forms of energy as equal. Second law differentiates between forms of energy and provides the direction and extent of energy conversion. Heat is a low-grade energy as compared to work.

Thermal Efficiency

It is defined as the ratio of work output to heat input. $\eta = W/Q$. Note $0 < \eta < 1$.

1.6 Cryogenics

Coefficient of Performance (COP)

It is defined as the ratio of desired heat effect to work input.
  COP (Heat Pump) = Heat Rejected/Work Supplied.
  COP (Refrigerator) = Heat Absorbed/Work Supplied.

> ### Part-II Review of Heat Transfer

Modes of Heat Transfer

Conduction, convection and radiation are the three modes of heat transfer.

Conduction

It is that mode of heat transfer in which heat flows from one molecule to another without movement of individual molecules.

Fourier's Law of Heat Conduction

$Q = -kA\, dT/dx$, where Q = Quantity of heat flow in Watt, dT/dx = Thermal gradient in °K/m, A = Cross-section Area in m^2 and k = Thermal Conductivity of material in $W/m°K$.

Heat Conduction through Composite Wall

$Q = (T_i - T_o) / \Sigma x/Ak$, where T_i = Temperature at inlet surface of wall, T_o = Temperature at exit surface of wall, x = individual material wall thickness, A = Individual area of cross-section, k = Thermal conductivity of each material.

Heat Conduction through Pipe

$Q = 2\pi Lk\,(T_i - T_o) / \log_e(A_2/A_1)$, where T_i = Temperature at inlet surface of wall, T_o = Temperature at exit surface of wall, L = Length of Pipe, A_1 = Inner area of cross-section, A_2 = Outer area of cross-section, k = Thermal conductivity of pipe material.

Heat Conduction through Hollow Sphere

$Q = 4\pi k\,(T_i - T_o)(R_1 R_2 / R_2 - R_1)$ where T_i = Temperature at inlet surface of wall, T_o = Temperature at exit surface of wall, R_1 = Inner radius of cross-section, R_2 = Outer radius of cross-section, k = Thermal conductivity of sphere material.

Perfect Conductor

A material that allows heat to flow through it with minimum resistance. Example: Copper.

Perfect Insulator

A material that does not allow heat to flow through it. Example: Plastics

Critical radius of Insulation (R_c)

It corresponds to the value of outer radius of a pipe for which heat flow through the pipe is maximum.

$R_c = 2k/h$, where h = convective heat transfer coefficient in $W/m^2 °K$.

Overall Heat Transfer Coefficient for a Composite Slab of Two Materials

$Q = U \times A \times \Delta T$, where U = Overall heat transfer Coefficient.

$U = 1/(1/h_1 + b_1/k_1 + b_2/k_2 + 1/h_2)$, where b_1 and b_2 denote widths of the two slabs, k_1 and k_2 denote conductivities of the two slabs and h_1 and h_2 denote convective heat transfer coefficients of the two slabs.

Steady State

It is that condition of a body where temperature profiles over the body are uniform throughout and do not vary with time.

Differential Equation for Steady State Conduction

$\Delta^2 T = (1/\alpha) \, \partial T/\partial t$, where α = Thermal Diffusivity, Δ^2 = Laplace operator and $\partial T/\partial t$ = Time rate of change of temperature.

Unsteady State

It is that condition of a body where temperature profiles are not uniform throughout and vary with time.

Differential Equation for Unsteady State Conduction

$\partial^2 T/\partial x^2 = (1/\alpha) \, \partial T/\partial t$, where α = Thermal Diffusivity = $k/\rho C_p$, where ρ = density, C_p = specific heat at constant pressure, $\partial T/\partial t$ = Time rate of change of temperature.

Biot Number

$$B_i = hL/k.$$

Fourier Number

$$F_o = \alpha t / L^2.$$

1.8 Cryogenics

Fins

These are extended surfaces attached to any body to improve its heat transfer rate to the atmosphere by providing additional surface area.

Pin Fins

These are type of fins, which have a smaller cross-sectional area as compared to their length.

Differential Equation for Pin Fin

$d^2T/dx^2 = (hP/kA)(T_o - T_i)$, where P = perimeter of fin and A = cross-section area. T_o and T_i are the outlet and inlet temperatures respectively.

Fin Effectiveness

$$\eta = \tanh(mL)/(mL) \text{ where } m^2 = hP/kA.$$

Heat Transfer through Fin

$Q = \sqrt{hPkA}\,\theta_0 \tanh(mL)$ where θ_0 = Non dimensional temperature.

Convection

It is that mode of heat transfer where heat flows due to movement of molecules between a solid and a fluid or between two fluids.

Newton's Law of Convection

$Q = h \times A \times \Delta T$, where h = convective heat transfer coefficient.

Reynolds Number

$Re = \rho V d / \mu$, where ρ = density, V = velocity of flow, d = characteristic dimension, μ = absolute viscosity.

Natural Convection

The convection that takes place due to density difference caused by variation in temperature without any external driving force is known as free or natural convection.

Forced Convection

The convection that takes place due to an external driving force such as a pump is known as forced convection.

Dimensional Analysis for Natural Convection

The convective coefficient, h = Function of $(g, \beta, \Delta_t, L, \mu, \rho, C_p, k)$
Where g = gravitational acceleration, β = expansion coefficient, Δ_t = temperature difference between surface and fluid, L = characteristic length, μ = absolute viscosity, ρ = density, C_p = specific heat at constant pressure, k = thermal conductivity.

Prandtl number

$P_r = \mu C_p / k$, where μ = absolute viscosity, C_p = specific heat at constant pressure, k = thermal conductivity.

Nusselt number

$N_u = hL/k$, where h = convective heat transfer coefficient, L = Characteristic Dimension, k = thermal conductivity.

Grashoff Number

$G_r = g\beta \Delta t L^3 \rho^2 / \mu^2$, where μ = absolute viscosity, g = gravitational acceleration, β = expansion coefficient.

Friction Factor

$F = \Delta P/4 \, (L/D) \, (\rho V^2/2)$, where L/D = ratio of Length to Diameter, V = fluid velocity, ρ = fluid density, ΔP = pressure drop.

Dimensional Analysis for Forced Convection

The convective coefficient, h = Function of $(V, L, \mu, \rho, C_p, k)$
Where V = velocity of fluid, L = characteristic length, μ = absolute viscosity, ρ = density, C_p = specific heat at constant pressure, k = thermal conductivity.

Number Relations

N_u = function of (G_r, P_r) for natural convection.
N_u = function of (R_e, P_r) for forced convection.

Velocity Boundary Layer

It is the viscous layer of fluid adjoining any solid surface that contains a velocity gradient normal to that surface. Drag Coefficient = $C_d = 1.328 / (R_e)^{0.5}$.

Thermal Boundary Layer

It is the viscous layer of fluid adjoining any solid surface that contains a temperature gradient normal to that surface. For laminar flow, $N_u = 0.664 \, (R_e)^{0.5} (P_r)^{0.33}$.

Condensation

When a phase change from vapor to liquid occurs by giving out latent heat to a surface that is at a temperature lower than the saturation temperature of liquid, the process is called Condensation. Condensation is of two types: film and drop condensation.

Boiling

When a phase change from liquid to vapor occurs while extracting heat from a surface that is at a temperature higher than the saturation temperature of liquid, the process is called Boiling. Boiling is of three types: pool, film and nucleate boiling.

Heat Exchangers

These are equipments, which enable transfer of heat from one fluid to another. They are of three types 1) Parallel flow 2) Counter flow 3) Cross Flow.

Log Mean Temperature Difference (LMTD)

The temperature difference, which if assumed constant along the length of heat exchanger will give the same rate of heat transfer as will the actual temperature difference.

$$\text{LMTD} = \theta_m = (\theta_1 - \theta_2) / \ln(\theta_1/\theta_2)$$

Effectiveness (ε)

Denotes efficiency of Heat Exchanger, ε = Actual heat transfer/Maximum heat transfer.

Number of Transfer Units (NTU)

Denotes size of the heat exchanger. NTU = $U \times A / C_{min}$, where C_{min} = minimum specific heat of fluid.

Relation between Effectiveness and NTU

$\varepsilon = 1 - e^{-(1+C)\text{NTU}} / (1+C)$, where C = Capacity Ratio = C_{min}/C_{max}.

Factors for Design of Heat Exchanger

Fluid temperatures, Flow rates, Outer shape, Fouling factor, Mechanical strength.

Condenser

Equipment used for condensation of vapor through heat exchange with a cooling liquid. Types are Jet, Surface and Evaporative condenser.

Cooling Tower

A water spray chamber used to cool water by direct contact with air due to spraying.

Radiation

It is that mode of heat transfer where heat flows as electromagnetic waves within or without a medium.

Black Body

A black body absorbs 100% radiation falling on it.

Stefan Boltzman Law

Amount of energy emitted from a black body per unit area is given by $E = \sigma T^4$, where E = emissivity of the black body, σ = Stefan Boltzman constant = 5×10^{-8}.

Shape Factor (F)

It is a functional expression of the configuration of the two radiating surfaces.

Heat flow between two black bodies = $Q = F \times A \times \sigma \times (T_1^4 - T_2^4)$.

Absorptivity (α)

It is defined as the fraction of total amount of energy falling on a surface, which is absorbed. $\alpha = Q_a/Q$.

Reflectivity (γ)

It is defined as the fraction of total amount of energy falling on a surface, which is reflected. $\gamma = Q_r/Q$.

Transmittivity (τ)

It is defined as the fraction of total amount of energy falling on a surface, which is transmitted through it. $\tau = Q_t/Q$ and $\tau + \alpha + \gamma = 1$.

Planck's Law

$e\lambda = 2\pi C_1 / \lambda^5 (e^{C_2/\lambda T} - 1)$, where $e\lambda$ = monochromatic emissivity, λ = Wavelength of radiation, C_1 and C_2 are constants.

Wein's law

$\lambda_m \times T = 0.029 \, (m°K)$, where T = Temperature of body and λ_m = monochromatic wavelength.

Kirchoff's Law

For a black body, emissivity is equal to absorptivity. $\alpha = \varepsilon$.

Intensity of Radiation

It is defined as radiant flux emitted per unit solid angle per unit time.

Grey Body

A body, which absorbs only a fraction of energy falling on it.

Note: This concludes our review of basic thermodynamics and heat transfer.

CHAPTER 2

Introduction to Refrigeration and Basic Cycles

INTRODUCTION

Refrigeration literally means production of cooler temperatures than surroundings. It is more of an art than a science. Once considered as a luxury, it is now a necessity as it affects day-to-day functioning of human society. Refrigeration can be defined as per ASHRAE guidelines as follows: "Refrigeration is the artificial removal of heat from a substance or a space to produce a temperature lower than that which would exist under the influence of ambient conditions." ASHRAE stands for the American Society for Heating, Refrigeration and Air-Conditioning Engineers. This organization is a premier body, which develops standards for Refrigeration and Air-conditioning.

Refrigeration can also be defined as "The science of producing and maintaining temperatures below the surrounding temperature." In crude terms refrigeration is the science or art of cooling. In this text we shall study basic cycles of refrigeration, different types of refrigeration systems, their applications and equipment used.

HISTORY OF REFRIGERATION

Refrigeration has come a long way from the cooling of water in earthen pots to the sophisticated commercial refrigeration units of capacity of the order of thousand tons or more. In the B.C. era, the old civilizations of Egypt, India and China unknowingly utilized the science of refrigeration to cool water and produce ice. Earthen pots were used to store water and due to evaporation and radiant heat loss, cooling effect was achieved and ice was produced. The use of ice in preserving food and drinks was prevalent at that time. In the early A.D. era, the use of salt-water mixture in evaporative cooling was common in India. All these methods of early cooling were good enough for domestic applications but were unsuitable for commercial applications due to the lower amounts of ice production. This need for commercial refrigeration led to pioneering work in the field of machine based artificial refrigeration.

2.2 Cryogenics

In 1834, Perkins developed the first commercial hand operated refrigeration system which was popular in the United Kingdom for almost half a century. It had a hand-operated compressor, a water-cooled condenser, a throttle valve and an evaporator. The working refrigerant used was ether. The compressor would compress the ether vapor, which was then condensed in a condenser. Subsequent to the condensation, the liquid refrigerant was expanded by the throttle valve and then released to an evaporator to absorb heat from the water and produce the refrigerating effect i.e. form ice by cooling. The vapor was then subjected to a repeat cycle of compression.

In the United States of America, Dr. Gorrie in the state of Florida obtained the first patent for a refrigeration machine. A steam engine was used as a prime mover to drive the compressor. In the early twentieth century, the steam engine started becoming redundant as a power source to drive the compressor due to its low speeds. Modern compressors driven by electric motors were introduced. The vapor compression refrigeration system was invented and quickly started becoming popular. Developments in the field of refrigerants and invention of efficient designs for compressors, condensers, evaporators and expansion valves made the vapor compression refrigeration system as the most widely used system of refrigeration. In 1915 a two-stage compressor was developed and the era of multi-staging and cascade systems began.

Simultaneously, vapor absorption systems were being developed using refrigerants such as ammonia to provide economical refrigeration for areas without electricity. Further in the later half of the twentieth century, a host of non-conventional refrigeration techniques were developed. Thermo-electric refrigeration, steam jet refrigeration, vortex tube refrigeration, refrigeration by adiabatic demagnetization of salts were some techniques developed. Finally advanced research led to the development of cryogenics, a science of low temperature refrigeration.

In this present twenty-first century, energy crisis has made the utilization of energy for refrigeration systems a subject of critical scrutiny and ways and means are being devised to make refrigeration systems more energy efficient. Thus harnessing of alternative energy from sources such as solar energy, waste heat, energy from biomass and wind energy to drive refrigeration machines has become the need of the times. Research work is being done in this direction and in future more energy efficient refrigeration systems using non-conventional energy sources would become commonplace.

Units of Refrigeration

The standard unit of refrigeration is 1 ton. It is defined as "One ton of refrigeration is equivalent to the rate of heat transfer needed to produce 1 ton (2000 lbs) of ice at 0°C from water at 0°C in 24 hours. The standard symbolic representation of the unit is *TR*. Note that 1 *TR* = 12,000 Btu/Hr = 210 kJ/min = 3.5 kW.

Refrigerating Effect

It is defined as the amount of cooling produced by the system. In short it is the amount of heat removed from an application by a refrigerant. Units are same as that of heat *i.e.* kJ.

Coefficient of Performance (COP)

It is defined as the ratio of the refrigerating effect to the work input required to achieve that refrigerating effect. COP = Q/W. If we express the refrigerating capacity (TR) in tons of refrigeration and power (P) in kilowatt we get, COP = $3.5/(P/TR)$.

Refrigeration Efficiency (η_R)

It is defined as the ratio of COP of a cycle to the COP of Carnot cycle operating in the same temperature range.

Energy Efficiency Ratio (EER)

It is defined as the ratio of heat removal (Btu/hr) rating of a refrigeration system to energy input (Watt-Hours) of the machine. EER = $3.5 \times$ COP.

It is defined as the ratio of actual COP of a cycle to the ideal COP. The actual COP is lower than ideal COP due to losses in the compressor and the motor or any other prime mover driving the compressor.

Difference between Heat Engine, Heat Pump and Refrigerator

Refer figure 2.1 for the basic working philosophy of these three equipments.

(i) **Heat Engine**: It is a device, which produces work by converting energy which is supplied to it in the form of heat. Let Q_1 be the heat rejected, Q_2 be the heat supplied, and W be the work done by the engine = $Q_2 - Q_1$. The efficiency of the engine = work done/heat supplied.

Thus expression for efficiency of engine is $\eta = W/Q_2 = Q_2 - Q_1 / Q_2$

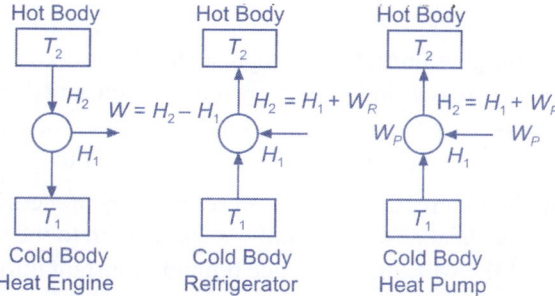

Fig 2.1 Comparison of Heat Engine, Refrigerator and Heat Pump

2.4 Cryogenics

(ii) **Refrigerator:** It is a device, which maintains the temperature of body lower than that of the surroundings. It consumes work and extracts heat from a cold body or space and rejects it to a hot body or space i.e. surrounding. Work consumed is $W = Q_2 - Q_1$, where Q_1 is the heat extracted from the cold body and Q_2 is the heat rejected to the hot body. The coefficient of performance is $(COP)_{Ref} = Q_1/W = Q_1/Q_2 - Q_1$.

(iii) **Heat Pump:** It is a device, which maintains the temperature of body higher than that of the surroundings. It consumes work and extracts heat from a cold body or space and supplies it to a hot body or space i.e. surrounding. Work consumed is $W = Q_2 - Q_1$, where Q_1 is the heat extracted from the cold body and Q_2 is the heat supplied to the hot body. The coefficient of performance is $(COP)HP = Q_2/W = Q_2/Q_2 - Q_1$.

> **Note:** Heat pump and refrigerator use same cycle but the operating temperature ranges are different. A refrigerator operates between cold body temperature T_1 and the ambient temperature T_a. The heat pump on the other hand operates between ambient temperature T_a and the hot body temperature T_2.

Reverse Carnot Cycle

This cycle is the ideal refrigeration cycle with the maximum possible COP. Figure 1.2 shows the T-S diagram for the cycle.

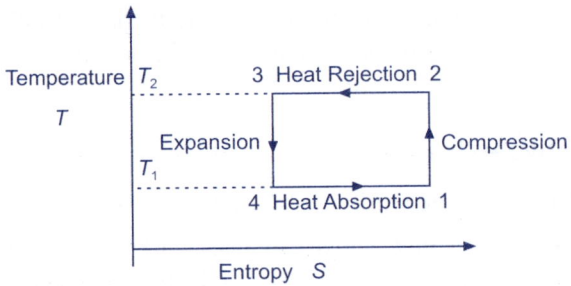

Fig. 2.2 T-S Diagram for Reverse Carnot Cycle

This cycle is hypothetical in nature and no actual refrigeration machine follows it. Hence it only serves as a benchmark of academic importance. The four processes of this cycle are as follows:

1. **Isentropic Compression:** This is represented by process 1–2. The working substance is compressed and its entropy remains constant during this reversible process. Due to compression the pressure of the working substance increases from P_1 to P_2 and the temperature increases from T_1 to T_2.

2. **Isothermal Heat Rejection:** This is represented by process 2–3. The working substance rejects heat absorbed from the application to the surroundings. The temperature of the working substance remains constant during this reversible process.

3. **Isentropic Expansion:** This is represented by process 3–4. The working substance undergoes an expansion and its entropy again remains constant during this reversible process. Due to expansion the pressure of the working substance decreases from P_2 to P_1 and the temperature decreases from T_2 to T_1.
4. **Isothermal Heat Absorption:** This is represented by process 4–1. This is the actual refrigeration process in which the working substance absorbs heat from the application. The temperature of the working substance remains constant during this reversible process of heat absorption. However the temperature of the application i.e. the space or body to be cooled reduces as desired.

Heat Rejected = $Q_2 = T_3(S_2 - S_3) = T_2(S_2 - S_3)$

Heat Absorbed = $Q_1 = T_4(S_1 - S_4) = T_1(S_2 - S_3)$.

Work Input = $(Q_2 - Q_1) = (T_2 - T_1)(S_2 - S_3)$

COP for refrigerator = $\dfrac{\text{Heat Absorbed}}{\text{Work Input}}$

COP for refrigerator = $\dfrac{T_1(S_1 - S_3)}{(T_2 - T_1)(S_2 - S_3)}$

COP for refrigerator = $\dfrac{T_1}{(T_2 - T_1)}$.

Limitations of the Reverse Carnot Cycle

This cycle is not practical since isothermal heat rejection and heat absorption are impossible to achieve in practice. This is due to the inevitability of temperature change during heat rejection and heat absorption process. Also isentropic compression and expansion absorption are impossible to achieve in practice. This is due to the irreversibility generated due to frictional losses during operation. Moreover isothermal processes need slower speeds for the compressor, whereas isentropic processes need faster speeds. This speed conflict makes implementation of this cycle impossible.

Note that COP of the reverse Carnot cycle is the ratio of temperature T_1 to the temperature difference $T_2 - T_1$. Note here that T_1 is the lower temperature and T_2 is the higher temperature. Now in order to improve the COP, either T_1 is increased or T_2 is reduced. However T_1 and T_2 cannot be varied beyond a certain operational range due to functional limitations. Lower temperature T_1 should be less than temperature of the application but above the freezing point of the working substance. Higher temperature T_2 should be more than temperature of the application but below the maximum permissible temperature of safe operation. That means T_2 should not be too high, as it will cause material failure and serious damage to the engine working on this cycle. Thus ideal COP cannot be achieved in practice and actual COP is lower. Also COP in summer is lower than that in winter.

Important Theory Questions From University Exams

Q.1. Derive the COP for a heat pump, a refrigerator and a heat engine working on the Carnot cycle.

Solution: Let Q_1 be the heat absorbed and Q_2 be the heat rejected. Thus expression for coefficient of performance of a refrigerator, heat pump and heat engine are as follows:

(i) COP for refrigerator = $\dfrac{\text{Heat Absorbed}}{\text{Work Input}}$

COP for refrigerator = $\dfrac{T_1 (S_1 - S_3)}{(T_2 - T_1)(S_2 - S_3)}$

COP for refrigerator = $\dfrac{T_1}{(T_2 - T_1)}$ ...(Ans)

(ii) COP for heat pump = $\dfrac{\text{Heat Supplied}}{\text{Work Input}}$

COP for heat pump = $\dfrac{T_2 (S_2 - S_3)}{(T_2 - T_1)(S_2 - S_3)}$

COP for heat pump = $\dfrac{T_1}{(T_2 - T_1)}$ ...(Ans)

(iii) COP for heat engine = $\dfrac{\text{Work Output}}{\text{Heat Input}}$

COP for heat engine = $\dfrac{(T_2 - T_1)(S_2 - S_3)}{T_2(S_2 - S_3)}$

COP for heat engine = $\dfrac{T_2 - T_1}{T_2}$ ...(Ans)

Q.2. State the basic principle of refrigeration. What is refrigerating effect? Define one Ton of refrigeration.

Solution: Refrigeration is the artificial removal of heat from a substance or a space that produces a temperature lower than the surrounding temperature. Refrigerating effect is the amount of cooling produced by the refrigeration system. In other words, it is the amount of heat absorbed from a space or body by the refrigeration system. The unit of refrigerating effect is the same as that of heat i.e. Joule.

One Ton of refrigeration is equivalent to the rate of heat transfer needed to produce 1 US ton of ice at 0°C from water at 0°C in 24 hours. The symbolic representation of the unit is TR. Note that 1 TR = 12,000 Btu/Hr = 210 kJ/min = 3.5 kW.

Q.3. State the various techniques of refrigeration used in practice.

Solution: Refrigeration is achieved in practice by the following methods:
- Vapor Compression Refrigeration
- Vapor Absorption Refrigeration
- Steam Jet Refrigeration
- Adiabatic Demagnetization
- Vortex Tube Refrigeration
- Thermo-electric Refrigeration

Q.4. What are the limitations of the Carnot cycle with gas as a refrigerant?

Solution: Reverse Carnot cycle is an ideal cycle, which can use any working substance. The limitations of the cycle using gas as a refrigerant are as follows: 1) Isentropic expansion and compression is not possible due to friction. 2) Isothermal heat addition or heat rejection i.e. cooling or heating at a fixed temperature is not possible. 3) Isothermal processes need slower speeds of the gas compressor, whereas isentropic processes need faster speeds of the compressor. This speed conflict makes this cycle impractical. 4) There is limitation on the extreme temperatures T_1 and T_2 of the cycle. T_1 should be lower than the temperature of the application and T_2 should be higher than the refrigerant temperature.

Q.5. What are various methods of achieving refrigeration?

Solution: The different techniques to achieve refrigeration are listed below:

1. **Dissolution of salts in water:** Certain salts like calcium chloride dissolve in water and lower its temperature to about $-50°C$. The water from the solution can be evaporated to recover the salt.

2. **Change of phase:** Refrigeration can be achieved due to latent heat transfer during change of phase of the working substance. The change of phase may be from solid to vapor, i.e. sublimation or solid to liquid i.e. melting or liquid to vapor i.e. evaporation.

3. **Throttling process:** Throttling is an isenthalpic process, where the fluid at high pressure is expanded through a throttle valve causing a drop in temperature.

4. **Expansion of a gas through a turbine:** Expansion of a gas through a turbine lowers its pressure and temperature.

5. **Magnetic effect:** Paramagnetic substances cause cooling if they are demagnetized after being subjected to a strong magnetic field.

6. **Thermo-electric effect:** If two dissimilar metals are joined together and electric current is passed through them, one junction gets heated and other gets cooled.

7. **Ranque effect:** This effect is used in the vortex tube. A high pressure gas is allowed to expand through a nozzle fitted tangentially to a pipe. This causes simultaneous discharge of cool air at the core and the hot air at the periphery.

Q.6. A reversed heat engine cycle is also a refrigeration cycle and a heat pump cycle. Explain this statement.

Solution: Refer Figure 1.1. A heat engine consumes energy in the form of heat and produces work as an output. Heat pump and refrigerator consume work and produce heating and cooling effects respectively. As far as the cycle is concerned the refrigerator and heat pump use the same cycle, but with different working temperatures. The refrigerator cools the application by working between the lower temperature T_1 and the ambient temperature T_a. The heat pump heats the application by working between the ambient temperature T_a and the higher temperature T_2.

Q.7. Derive the relationship between the COP of a heat pump and the COP of a refrigerator.

OR

Prove that the COP of heat pump is greater than one.

Solution: Refer Figure 1.1.

$$\text{COP for heat pump} = \frac{\text{Heat Supplied}}{\text{Work Input}}$$

$$\therefore \text{COP for heat pump} = \frac{Q_2}{Q_2 - Q_1}$$

$$\therefore \text{COP for heat pump} = \frac{Q_2 - Q_1 + Q_1}{(Q_2 - Q_1)}$$

$$\therefore \text{COP for heat pump} = 1 + \frac{Q_2}{Q_2 - Q_1}$$

$\therefore$ COP for heat pump = 1 + COP for refrigerator – (proved)

Thus COP of heat pump is always greater than one.

Solved Problems from University Exams

Q.1. A reverse Carnot cycle air conditioner of 4 TR capacity operates with a cooling coil of temperature 5°C. The surrounding air at 45°C is used as a cooling medium rising to temperature of 55°C. The temperature of heat rejection is 60°C. Determine the mass flow rate of air entering the heat exchanger, COP of the air conditioner and power consumption of the unit.

Solution: $T_2 = 60 + 273 = 333$ K; $T_1 = 5 + 273 = 278$ K

$$\text{COP} = \frac{T_1}{(T_2 - T_1)} = \frac{278}{(333 - 278)} = 6.05 \qquad \text{...(Ans)}$$

$$\text{COP} = \frac{Q}{W} = 6.05 \text{ ; where } W = 4 \times 3.5 = 14 \text{ kW.}$$

∴ $Q = 84.7$ kW $= m \times C_p \times (T_2 - T_1) = m \times 1 \times 10^3 \times (55 - 45)$
∴ Mass flow rate of air $= m = 0.084$ kg/s ...(Ans)

Q.2. A reverse Carnot cycle air refrigeration equipment has a compressor working with a compression ratio of 12. The temperature limits of the cycle are 300 K and 270 K. Determine the refrigerating effect and the COP of the unit.

Solution: $T_2 = 300$ K; $T_1 = 270$ K

$$\text{COP} = \frac{T_1}{(T_2 - T_1)} = \frac{270}{(300 - 270)} = 9 \quad ...(Ans)$$

$$\text{COP} = \frac{Q}{W} = 9$$

∴ $Q = 9 \times W$

Refrigerating effect $Q = 9 \times \dfrac{1}{(r)^{\gamma-1/\gamma}} = 9 \times \dfrac{1}{(12)^{1.4-1/1.4}} = 4.45$ kW ...(Ans)

Q.3. A refrigeration system produces 20 kg/hour of ice from water at 20°C. Find the refrigerating effect and tonnage of the unit. If power consumption is 1.5 kW, calculate the COP. Take enthalpy of solidification of water as 335 kJ/kg and specific heat of water as 4.19 kJ/kg°C.

Solution: Total enthalpy = enthalpy of solidification + $C_p \times \Delta T$
∴ Total enthalpy, $h = 335 + 4.19 (20 - 0) = 418.8$ kJ/kg
Refrigerating effect $= Q = m \times h = 20 \times 418.8 = 8376$ kJ/h ...(Ans)

$$\text{Capacity, } TR = \frac{Q}{3.5 \times 3600} = \frac{8376}{3.5 \times 3600} = 0.66 \text{ Tons} \quad ...(Ans)$$

$$\text{COP} = \frac{Q}{W} = \frac{8376}{1.5 \times 3600} = 1.55 = 1.55 \quad ...(Ans)$$

Q.4. A machine working on reverse Carnot cycle operates between +35°C and –20°C. Determine the COP if the machine is a refrigerator and a heat pump.

Solution: $T_2 = 35 + 273 = 308$ K; $T_1 = -20 + 273 = 253$ K

$$\text{COP for refrigerator} = \frac{T_1}{(T_2 - T_1)} = \frac{253}{(308 - 253)} = 4.6 \quad ...(Ans)$$

$$\text{COP for heat pump} = \frac{T_2}{(T_2 - T_1)} = \frac{308}{(308 - 253)} = 5.6 \quad ...(Ans)$$

Q.5. 3 HP per ton is required to maintain a temperature of –40°C in a refrigerator. If the refrigerator works on the reverse Carnot cycle, determine 1) COP 2) Temperature of sink 3) Heat rejected to the sink per ton of refrigeration.

Solution: $T_1 = -40 + 273 = 233$ K

2.10 Cryogenics

Work required = $W = 3 \times 4500/427 = 31.6$ kcal/min

∴ $W = 31.6$ kcal/min

Heat extracted = $Q = 1\ TR = 50$ kcal/min

COP = $Q/W = 50/31.6 = 1.58$ ...(Ans.)

Heat rejected to sink = $H = Q + W = 50 + 31.61 = 81.61$ kcal/min ...(Ans.)

$$COP = 1.58 = \frac{T_2}{(T_2 - T_1)} = \frac{233}{(T_2 - 233)}$$

∴ Sink temperature = $T_2 = 380.46$ K = $107.46°C$. ...(Ans.)

Q.6. Determine mass of ice produced from water per day for the following conditions: Water temperature = 22°C
Tonnage of unit = 150 Tons
Operating Temperatures = – 5°C and 28°C
Latent Heat of Ice = 330 kJ/kg
Also determine the power required to drive the unit.

Solution: $T_1 = -5 + 273 = 268$ K; $T_2 = 28 + 273 = 301$ K

$$COP = \frac{T_1}{(T_2 - T_1)} = \frac{268}{(301 - 268)} = 8.12 \quad ...(Ans.)$$

Refrigerating Effect = $Q = 150\ TR = 525\ kW$

COP = $8.12 = Q/W = 525/W$

∴ Power required to drive the unit = $W = 64.65\ kW$ ...(Ans.)

Total heat removed from water at 22°C to form ice at 0°C = H

∴ H = Latent Heat + $m \times C_p \times \Delta T = 330 + 1 \times 4.18 \times (22 - 0) = 421.96$ kJ/kg

∴ Mass of Ice produced = $M = Q/H = 525/421.96 = 1.24$ kg/s

∴ Mass of Ice produced daily = $24 \times 60 \times 60 \times 1.24$ kg/s = $1,07,498$ kg **...(Ans)**

Q.7. A cold storage plant is required to store 20 tons of fish. The fish is supplied at a temperature of 30°C. The specific heat of fish above the freezing point is 2.93 kJ/kg K. The specific heat of fish below the freezing point is 1.26 kJ/kg K. The fish is stored in cold storage that is maintained at –8°C. The freezing point of the fish is –4°C. The latent heat of solidification of the fish is 235 KJ/Kg. If the plant requires 75 kW to drive it, find 1) Capacity of the plant. 2) Time taken to achieve cooling. Assume actual COP as 0.3 times the Carnot COP for the refrigerating plant.

Solution: $T_1 = -8 + 273 = 265$ K; $T_2 = 30 + 273 = 303$ K

COP = $T_1 = 265 = 6.97$ –(Ans) $(T_2 - T_1)$ $(303 - 265)$

Actual COP = $0.3 \times 6.97 = 2.09$

Refrigerating effect = Q = Actual COP × Work Input = $2.09 \times 75 = 156.8$ kW

Tonnage of plant = $156.8/3.5 = 44.8\ TR$ ...(Ans)

Freezing point of fish = $-4 + 273 = 269°K$

H_1 = Heat removed from fish above freezing point = $m \times Cp_1 \times (T_2 - T_3)$

$\therefore H_1 = 20{,}000 \times 2.93 \times (303 - 269) = 1.992 \times 10^6$ kJ

H_2 = Heat removed from fish below freezing point = $m \times Cp_2 \times (T_3 - T_1)$

$\therefore H_2 = 20{,}000 \times 1.26 \times (269 - 265) = 0.101 \times 10^6$ kJ

H_3 = Latent heat removed from fish = $m \times L = 20{,}000 \times 235$ kJ

$\therefore H_3 = 4.7 \times 10^6$ kJ

H = Total heat removed from fish = $H_1 + H_2 + H_3 = 6.793 \times 10^6$ kJ

Total time for cooling = $H/Q = 6.793 \times 10^6 / 156.8 = 43320$ seconds

$\therefore$ Time = $43320 / 3600 = 12.03$ hours ...(Ans.)

Unsolved Theory Questions

Q.1. Define the following terms: refrigeration, one ton of refrigeration, refrigerating effect and coefficient of performance.

Q.2. Write a short note on history of refrigeration.

Q.3. Distinguish between a heat pump and a refrigerator.

Q.4. Derive a relation between the COP of a heat pump and that of a refrigerator.

Q.5. Write a short note on reverse Carnot cycle.

Q.6. Reverse Carnot cycle is not practical. Is this statement true or false? Why?

Q.7. Discuss limitations of Carnot cycle.

Q.8. Name a few refrigeration techniques.

Q.9. What do you mean by EER ?

Q.10. Prove that $COP_{HP} = COP_{REF} + 1$.

Unsolved Problems

Q.1. A refrigeration system produces 15 kg/hour of ice from water at 22°C. Find the refrigerating effect and tonnage of the unit. If power consumption is 1.5 kW, calculate the COP. Take enthalpy of solidification of water as 335 KJ/Kg and specific heat of water as 4.19 kJ/kg°C.

Q.2. A machine working on reverse Carnot cycle operates between following set of temperatures. Determine the COP if the machine is a refrigerator and a heat pump.

  A. −10°C and 35°C
  B. −8°C and 27°C
  C. −5°C and 22°C

Q.3. 2 HP per ton is required to maintain a temperature of –35°C in a refrigerator. If the refrigerator works on the reverse Carnot cycle, determine 1) COP 2) Temperature of sink 3) Heat rejected to the sink per ton of refrigeration.

Q.4. Determine mass of ice produced from water per day for the following conditions: Water temperature = 23°C.
Tonnage of unit = 175 Tons
Operating Temperatures = –8°C and 27°C
Latent Heat of Ice = 335 kJ/kg
Also determine the power required to drive the unit.

Q.5. A cold storage plant is required to store 15 tons of mangoes. The mangoes are supplied at a temperature of 28°C. The specific heat of mangoes above the freezing point is 2.98 kJ/kg K. The specific heat of mangoes below the freezing point is 1.36 kJ/kg K. The mangoes are stored in cold storage that is maintained at –6°C. The freezing point of the mangoes is –2°C. The latent heat of solidification of the mangoes is 205 kJ/kg. If the plant requires 55 kW to drive it, find.
1. Capacity of the plant.
2. Time taken to achieve cooling. Assume actual COP as 0.4 times the Carnot COP for the refrigerating plant.

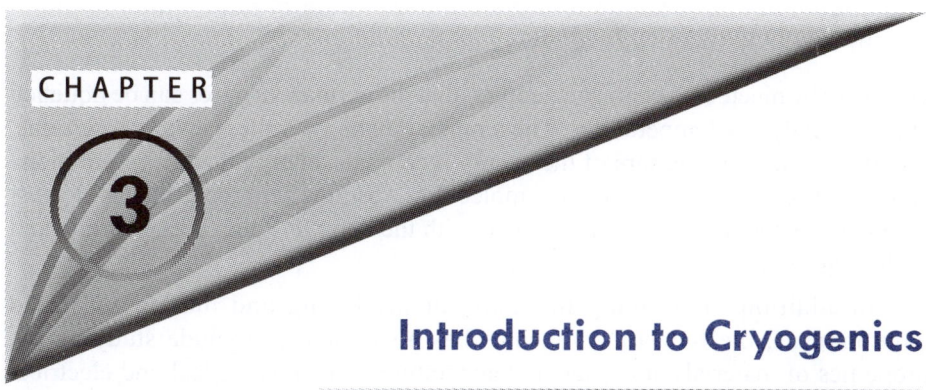

CHAPTER 3

Introduction to Cryogenics

INTRODUCTION

The term "Cryogenics" is derived from the greek word "Kryos" which means cold or frost. It is frequently applied to very low temperature refrigeration applications such as in liquefaction of gases and preservation of biological specimens. Cryogenics is defined as the scientific production of temperatures below –150°Celsius or –238° Fahrenheit or 123° Kelvin, and the study of the resulting effects on materials such as metals and plastics.

Cryogenics has thus developed as a branch of physics or engineering that studies the production of very low temperatures and the behavior of materials at those temperatures. To achieve extremely low temperatures, liquefied gases such as liquid nitrogen or liquid helium may be used. These materials are called cryogens. Some cryogens, such as liquid nitrogen, are used for the quick-freezing of foods and vaccines, or for chemical reactions that need to be carried out at low temperatures. Cryogenic oxygen and hydrogen are useful as fuels for powering rockets for space flight. Controlled cryogenic treatment of metals and other materials has increased their strength and resistance to wear. A person who studies elements under extremely cold temperature is called a cryogenicist.

The word *cryogenics* literally means "the production of icy cold." The term, however, is mainly used as a synonym for the low-temperature state. It is not well-defined at what point on the temperature scale refrigeration ends and cryogenics begins. A number of researchers define cryogenic temperatures as those ranging from –150° C (123 K or –238°F) down to –273.15°C (0 K or –460°F). Researchers at the U.S. National Institute of Standards and Technology in Boulder, Colorado have chosen to consider the field of cryogenics as that involving temperatures below –180°C (93.15 K). This dividing line was chosen because the normal boiling points of the so-called "permanent" gases (such as helium, hydrogen and normal air) lie below –180°C, while the freon refrigerants, hydrogen sulfide, and other common refrigerants have boiling points above –180°C.

More specifically, a low-temperature environment is termed a cryogenic environment when the temperature range is below the point at which permanent gases begin to liquefy. Among others, they include oxygen, nitrogen, hydrogen, and helium. The origin of cryogenics as a scientific discipline coincided with the discovery by nineteenth century scientists, that the permanent gases can be liquefied at exceedingly low temperatures. The temperature of a sample, whether it be a gas, liquid, or solid, is a measure of the energy it contains, energy that is present in the form of vibrating atoms and moving molecules. Absolute zero represents the lowest attainable temperature and is associated with the complete absence of atomic and molecular motion.

In addition to studying methods for producing and maintaining cold environments, the field of cryogenics has also come to include studying the properties of materials at cryogenic temperatures. The mechanical and electrical properties of many materials change very dramatically when cooled to 100 K or lower. For example, rubber, most plastics, and some metals become exceedingly brittle, and nearly all materials contract. In addition, many metals and ceramics lose all resistance to the flow of electricity, a phenomenon called superconductivity. Very near absolute zero (2.2 K) liquid helium undergoes a transition to a state of superfluidity, in which it can flow through exceedingly narrow passages with no friction.

Need for Cryogenics

Single stage vapor compression systems have serious limitations in the low temperature area. Some of the drawbacks are:

1. The use of vapor compression system in low temperature range is limited by the freezing points of the refrigerants. Thus refrigerant must have freezing temperature lower than the application temperature.
2. The equipment such as compressors do not perform well at lower temperatures.
3. COP is low due to high pressure ratios.
4. Pressure in the condenser is high when a refrigerant with low boiling temperature is used.
5. Evaporator pressures are low, so suction volumes are large.

Thus single stage vapor compression systems are not suitable for low temperatures. Hence cryogenic systems such as cascade systems are suitable.

History of Cryogenics

The existence of absolute zero was first pointed out in 1848 by William Thompson (later to become Lord Kelvin), and is now known to be –459°F (–273°C). It is the basis of an absolute temperature scale, called the Kelvin scale, whose unit, called a Kelvin rather than a degree, is the same size as the Celsius degree. Thus, –459°F corresponds to –273°C corresponds to 0K. Cryogenics, then, deals with producing and maintaining environments at temperatures below about 173K (–100°C).

The development of cryogenics as a low temperature science is a direct result of attempts by nineteenth century scientists to liquefy the permanent gases. One of these scientists, Michael Faraday, had succeeded, by 1845, in liquefying most of the gases then known to exist. His procedure consisted of cooling the gas by immersion in a bath of ether and dry ice and then pressurizing the gas until it liquefied. Six gases, however, resisted every attempt at liquefaction and were thus known at the time as permanent gases. They were oxygen, hydrogen, nitrogen, carbon monoxide, methane, and nitric oxide. The noble gases, helium, neon, argon, krypton, and xenon, not yet discovered.

The science of low temperature refrigeration or cryogenics started with the development of oxygen liquefaction equipment in 1877 by Louis Cailletet, a French physicist. Cailletet's work was closely paralleled by that of the Swiss chemist, Raoul Pictet. Cailletet and Pictet as well as all early cryogenic researchers, used the Joule-Thomson effect to attain low temperatures. The first step in using the Joule-Thomson effect was to cool and compress a gas as much as possible. The gas was then allowed to expand quickly as it escapes through a needle valve. Rapid expansion cools the gas even more, lowering its temperature below its boiling point. Pictet used a variation of this technique to liquefy oxygen. He used the Joule-Thomson effect with a series of gases, each with a lower boiling point, until he was able to reduce the temperature of oxygen below its boiling point. The German chemist, Carl von Linde, adopted a similar approach in constructing the first commercial refrigeration system in 1876.

Of the known permanent gases, oxygen and nitrogen, the primary constituents of air, received the most attention. For many years investigators labored to liquefy air. Finally, in 1883 the first measurable quantity of liquid oxygen was produced by S. F. von Wroblewski at the University of Cracow. Oxygen was found to liquefy at 90K and nitrogen at 77 K.

An important advance in early cryogenics research was the development of an efficient system for storing liquefied gases. That system was invented by the Scottish chemist James Dewar in 1891. The Dewar flask is well known today as the vacuum flask used to keep liquids hot or cold. The flask consists of a double-walled bottle containing a vacuum between the two silvered walls. The vacuum prevents loss or entry of heat by conduction or convection and the silver coating reduces heat transfer by reflection. With this technology, every gas had been successfully liquefied by the early 1900s except for hydrogen and helium.

Following the liquefaction of air, a race to liquefy hydrogen ensued. James Dewar, succeeded in 1898 in freezing hydrogen, thus reaching the lowest temperature achieved to that time, 14 K. Along the way, argon was discovered (1894) as an impurity in liquid nitrogen, and krypton and xenon were discovered (1898) during the fractional distillation of liquid argon. Fractional distillation is accomplished by liquefying a mixture of gases each of which has a different boiling point. When the mixture is evaporated, the gas with the highest boiling point evaporates first, followed by the gas with the second highest boiling point, and so on. Each of the newly discovered gases condensed at temperatures higher than the boiling point of hydrogen, but lower than 173 K. The last element to be liquefied was helium gas. In 1907, Linde set up the first air liquefaction plant in America. In 1908, the Dutch physicist Heike Kamerlingh Onnes finally succeeded in liquefying helium at a temperature of 4.2 K (–452°F).

3.4 Cryogenics

Further cryogenic research demanded the development of new cooling techniques. In the mid-1920s, the method of adiabatic demagnetization was proposed independently by Peter Debye and W. F. Giauque. In this process, the material to be cooled is placed in contact with a paramagnetic field and with liquid helium and then subjected to a strong magnetic field. The liquid helium is then removed and the magnetic field reduced to zero. At that point, the temperature in the test material may drop as low as 0.01 K.

Figure 3.1 shows the temperature Vs timeline of cryogenic research as presented by Phillip Leburn of CERN Labs.

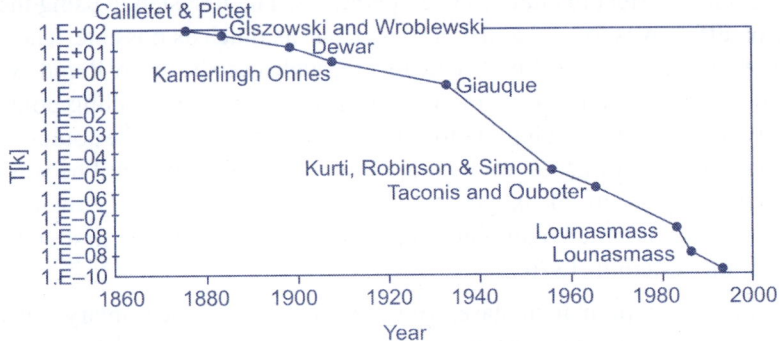

Fig: 3.1 Timeline for Cryogenic Research by Leburn

In the twentieth century a lot of researchers continued to work on improving cryogenic equipment, safety and efficiency. The most notable and persistent was Dr. Randall Barron of Louisiana Technical University. Dr. Barron has written multiple research papers about the subject, as have some of his students. These papers are widely cited in the cryogenics industry. Another early pioneer in the industry is Dr. Hugh E. Trucks. Dr. Trucks was a design specialist at General Dynamics Corporation and was later a private consultant. He wrote several articles on the subject and was also affiliated with Cryogenics International, of Scottsdale Arizona.

The field of cryogenics advanced during World War II when scientists found that metals frozen to low temperatures showed more resistance to wear. Based on this theory of cryogenic hardening, the commercial cryogenic processing industry was founded in 1966 by Ed Busch. With a background in the heat treating industry, Busch founded a company in Detroit called CryoTech in 1966. Though CryoTech later merged with 300 below to create the largest and oldest commercial cryogenics company in the world, they originally experimented with the possibility of increasing the life of metal tools to anywhere between 200% – 400% of the original life expectancy using cryogenic tempering instead of heat treating. This evolved in the late 1990s into the treatment of other parts such as amplifier valves racing engines, firearms etc. The theory was based on lowering high temperature parts to room temperature causing certain strength increases in the molecular structure. Using liquid nitrogen, CryoTech formulated the first early version of the cryogenic processor. Unfortunately for the newly-born industry, the results were unstable, as components sometimes experienced thermal shock when they were cooled too quickly. Some components in early tests even shattered because of the ultra-low temperatures. In the late twentieth century, the field improved significantly with the rise of applied research, which coupled microprocessor based industrial controls to the cryogenic processor in order to create more stable results.

Chronology of Cryogenics

Following is a chronology of some important events in cryogenics

    1845 – Liquefaction by Faraday
    1848 – Absolute Zero by Lord Kelvin
    1877 – Cailletet and Pictet liquefied Oxygen
    1883 – Wroblewski at Cracow University Poland liquefied Nitrogen
    1891 – Dewar developed a Cryogenic Vessel
    1898 – Dewar Liquefied Hydrogen
    1902 – Claude establish Air Liquid
    1907 – Linde established first liquefaction plant in America
    1908 – Onnes liquefied Helium
    1911 – Onnes discovered super conductivity
    1926 – Goddard fires the first cryogenic engine
    1933 – Adiabatic Demagnetization by Giauque
    1934 – Kapitza developed first expansion engine for helium
    1947 – Collins Cryostat developed
    1952 – Cryogenic Engineering lab established in USA
    1958 – Multilayer cryogenic insulation developed by Black
    1959 – NASA liquid hydrogen plant commissioned
    1964 – LNG tankers were commissioned
    1966 – Helium dilution refrigerator was developed by Hall
    1970 – High capacity liquid oxygen plants developed

Cryogenic Terminology

(a) **Absolute zero:** Absolute zero is the lowest temperature possible. It is associated with the absence of molecular motion and is equal to 0K [–459°F (–273°C)].

(b) **Boiling point:** The boiling point of a liquid is the temperature at which it boils, also the temperature at which its vapor condenses.

(c) **Critical Temperature:** The highest temperature at which gas can be liquefied irrespective of the pressure.

(d) **Critical Pressure:** The specific vapour pressure at which gas can be liquefied.

(e) **Bose-Einstein condensate:** A material state in which a collection of supercooled atoms fall into the same quantum state, essentially acting like a single superatom.

(f) **Cryogen:** A cryogen is a liquid that boils at temperatures below about 173K [–148°F (–100°C)].

(g) **Joule-Thomson Coefficient:** The entity denoting change in tempearture resulting from expansion of a substance from one pressure level to another.

(h) **Entropy:** The measurement of a tendency towards increased randomness and disorder.

(i) **Kelvin temperature scale:** The Kelvin temperature scale is an absolute temperature scale with the same size unit, called the Kelvin, as the Celsius scale, but shifted so that zero Kelvin (0K) corresponds to absolute zero.

(j) **Expansion valve:** An adjustable orifice through which expansion occurs.

(k) **Superconductivity:** Superconductivity is the ability of a material to conduct electricity without loss, that is, without electrical resistivity, at a temperature above absolute zero. The phenomenon occurs in certain materials when their electrical resistance drops suddenly and completely to zero at a specific cryogenic temperature, called the critical temperature.

(l) **Regenerator**: A heat exchanger for transferring heat between two fluids

Applications of Cryogenics

Some important applications of cryogenics are:
1. In preservation of semen and biological specimens
2. Production of dry ice i.e. solid CO_2
3. Liquefaction of gases
4. Cryogenic engine for space application
5. Metallurgy
6. Alternative fuels for transport
7. Food preservation
8. Cooling of supercomputers
9. Space propellants
10. In surgery

Some other applications of cryogenics are:

It is observed some metals the electrical resistance drops to zero very suddenly at temperatures above absolute zero. This effect is called superconductivity and has some very important applications in today's world. For example, superconductors are used to make magnets for particle accelerators and for magnetic resonance imaging (MRI) systems used in many hospitals. The discovery of superconductivity led other scientists to study a variety of material properties at cryogenic temperatures. Today, physicists, chemists, material scientists, and biologists study the properties of metals, as well as the properties of insulators, semiconductors, plastics, composites, and living tissue. In order to chill the samples, they are placed in an insulated container, called a dewar, and are cooled by filling the inner space with a cryogenic liquid, or by cooling it with a cryogenic refrigerator.

The recycling industry takes advantage of cryogenics by immersing recyclables in liquid nitrogen, after which they are easily pulverized and separated for reprocessing. Still another cryogenic material property that is sometimes useful is that of thermal contraction. Materials shrink when cooled. An example is the use of liquid nitrogen in the assembly of some automobile engines. In order to get extremely tight fits when installing valve seats, the seats are cooled to liquid nitrogen temperatures, whereupon they contract and are easily inserted in the engine head. When they warm up, a perfect fit results.

Cryogenic liquids are also used in the space program. For example, cryogens are used to propel rockets into space. A tank of liquid hydrogen provides the fuel to be burned and a second tank of liquid oxygen is provided for combustion. Refer Figure 3.2 for a typical rocket propelled by cryogenic propellants. A more exotic application is the use of liquid helium to cool orbiting infrared telescopes. Any object warmer than absolute zero radiates heat in the form of infrared light. The infrared sensors that make up a telescope's lens must be cooled to temperatures that are lower than the equivalent temperature of the light they are intended to sense, otherwise the telescope will be blinded by its own light. Since temperatures of interest are as low as 3K liquid helium at 1.8K is used to cool the sensors.

Fig: 3.2 Rocket using Cryogenic Propellants

Cryogenic preservation has even extended to a hotly debated topic: in vitro fertilization. In vitro fertilization is a technique that improves the chances of a woman being pregnant by implanting embryo into the uterus of a female recipient. Controversy in the medical community surrounds the value of cryogenically preserving later stage human blastocytes. Although using cryogenic preservation for maintaining embryo may lead to bioethical complications, the medical potential for preserving tissues or organs has tremendous medical benefits.

Finally, the production of liquefied gases has itself become an important cryogenic application. Cryogenic liquids and gases such as oxygen, nitrogen, hydrogen, helium, and argon all have important applications. Shipping them as gases is highly inefficient because of their low densities. This is true even at extremely high pressures. Instead, liquefying cryogenic gases greatly increases the weight of cryogen that can be transported by a single tanker.

3.8 Cryogenics

Worldwide Scenario for Cryogenic Industry

Air and Gas separation industry is now well established in the world. Today, liquid natural gas (LNG) represents one of the largest – and fast-growing – industrial domains of application of cryogenics together with the liquefaction and separation of air gases. The densification by condensation, and separation by distillation of gases remains the main driving force for the cryogenic industry, exemplified not only by liquid oxygen and nitrogen used in chemical and metallurgical processes, but also by the cryogenic liquid propellants of rocket engines and the proposed use of hydrogen as a "clean" energy vector in transportation. It is estimated that the world capacity of liquid oxygen is over 150 million tons per annum and that of LNG is over 60 million tons per annum. Gases like ethylene, helium, argon, nitrogen and hydrogen are also being consumed in large quantities throughout the world and their demand is also increasing.

The UK and France appear to be the two leading European countries in Cryogenics field. The UK also has a thriving cryogenic instrumentation industry, with a strong market share, although less strong in cryocooler technologies in the non-space market. The highly reliable, low vibration and energy efficient cryocoolers developed by UK organizations, together with associated cryogenic systems engineering skills and test facilities are the most promising to the non-space market sectors in terms of technology transfer opportunities. Table 3.1 compares some cryocoolers developed worldwide.

Table 3.1 Cryocooler Development Worldwide

Cooler	Typical temperature	Typical Heat lift	Advantages	Disadvantages
Radiator	80 K	0.5 W	Reliable low vibraton long lifetime	Complicates orbit
Cryogen	4 K	0.05 W	Stable, low vibration	Short lifetime, out-gassing, massive, complex
Stirling –1 stage	80 K	0.8 W	Efficient, heritage	Vibrations
Stirling – 2 stage	20 K	0.06 W	Intermediate temperature	Under development
Pulse tube	80 K	0.08 W	Lower vibrations	Lower efficiency than stirling
Peltier	170 K	1 W	Lightweight	High temperature, low efficiency
Joule-Thomson	4 K	0.01 W	Low vibrations	Requires hybrid design
Sorption	10 K	0.1 W	Low vibrations	Under development
Reverse Brayton	65 K	8 W	High capacity	Complex
ADR	0.05 K	0.01 mW	Only way to reach these temperatures	Large magnetic field

There are a number of companies worldwide that develop and manufacture cryogenic systems. In the main, they are not focused on space as a market, with the majority manufacturing cryogenic systems for analytical and scientific instrumentation including superconducting magnets, together with a mostly separate industry producing cryogenic systems for other purposes. There is a considerable related supply chain, with companies producing vacuum vessels and components, electronics and accessories.

Indian Cryogenics Scenario

The cryogenic activity started in India in 1935 with the advent of Indian Oxygen Company (INOX), which established air separation plants in India. The requirement of oxygen plants was felt in 1970 due to the boom in the steel sector. ISRO boosted the cryogenics activity in India by using cryogenic propellants at its Mahendragiri facility from early 1970s. Private companies like BOC also expanded the liquefied gases business in India. Currently it is estimated that there are more than 200 oxygen manufacturing plants in India with an annual capacity of around 230 million cubic meter.

The national dairy development board during its operation flood for milk also spurred the requirement of liquid nitrogen cryogenerators, which benefited the cryogenics industry in India. Cryogenic Gas separation plants have also been installed in fertilizer plants.

Cryogenics in India, at last, took a quantum jump at the beginning of 1990s in terms of the production of large volumes of liquid helium and liquid nitrogen. Huge budgets were allotted by the Govt. funding agencies. The big boost to cryogenic activity in India came from the high technology programs like the Super conducting Cyclotron development at Variable Energy Cyclotron Centre (VECC), Kolkata, the Superconducting LINAC boosters at TIFR, Mumbai and IUAC, New Delhi, and the Superconducting Tokamak at the Institute for Plasma Research (IPR), Gandhinagar.

Indian Cryogenics Council was also formed to promote cryogenics in India. It is the apex professional society of personnel of Cryogenics and Low Temperature Physics and is popularly known as ICC. The aim of ICC is to promote education, advancement and application of cryogenics in India through various activities such as organizing seminars and workshops, training courses and publication of books and journals.

The current Indian cryogenics market turnover is around INR 1100 crores and there is ample scope for further improvement.

References

1. Cryogenic Engineering – B.A. Hands.
2. Cryogenic Systems – Randall Baron.
3. Internet Site http://science.jrank.org/pages/1893/Cryogenics.html.
4. Internet Wikipedia website on Cryogenics.
5. Refrigeration and Air Conditioning – Dr.S.S.Thipse.
6. Cryogenic Engineering – Dr. P.K.Bose.
7. Nathan Hill – ESA publication on Cryogenics – 2008.
8. Paper on Cryogenics by Phillip Leburn – CERN.

Questions

Q.1. What is Cryogenics?

Q.2. Write a short note on history of Cryogenics.

Q.3. Explain the need for Cryogenics.

Q.4. Discuss the concept of superconductivity.

Q.5. Enumerate at least five applications of Cryogenics.

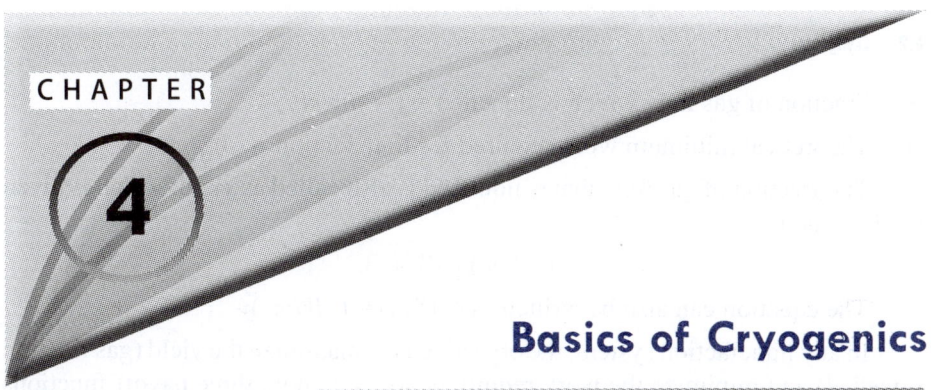

CHAPTER 4

Basics of Cryogenics

INTRODUCTION

For the science of cryogenics, the basic purpose is "To Remove Heat." In any application device or system, the load current varies over time. In general, for cryogenic systems, the cooling load also varies over time. The cryogenic system must be sized to accommodate the highest load, not just the average load. For example cryogenic systems providing 100-1000 watts of cooling at 65-80 K are required for devices utilizing high-temperature superconductors and are widely used in electric power delivery and utilization system. The efficiency of mechanical cryogenic refrigerators has been improving steadily over many years, but is typically < 20% of Carnot efficiency, defined by $\eta_c = T_c / (T_h - T_c)$ Another way to represent this is through the *Specific Power* (= $1/\eta$) which is expressed in units of watts per watt, i.e. the number of watts of input power required to remove one watt of heat at the cooling temperature.

Cryogenic refrigerators are required to be efficient, cheap, reliable, compact and acceptable to utility and industrial markets. The much more demanding task is to turn these adjectives into quantitative measures of performance. Toward that end, the major performance goals for a cryogenic equipment are: an efficiency of 30% relative to Carnot, a capital cost below $25 per watt of cooling, and an available operating time (or reliability) exceeding 99.8%. Cryogenic manufacturers have worked hard over many years to improve reliability, by eliminating moving parts at the cold end and reducing mechanical wear through use of flexure bearings and gas bearings.

In this chapter we will be focusing on performance of various cryogenic systems which is quantified by various performance parameters or payoff functions. We will be considering a liquefaction system for the analysis.

Cryogenic System Performance Parameters

For a typical cryogenic liquefaction system, the payoff functions used are as follows:
1. Work required per unit mass of gas liquefied = $-W/m_f$
2. Work required per unit mass of gas compressed = $-W/m$

4.2 Cryogenics

3. Fraction of gas flow which is liquefied $= y = m_f/m$
4. Theoretical minimum work required for liquefaction $= -W_1/m_f$

The fraction of gas flow that is liquefied is also called as yield, which is given by the equation

$$y = (-W/m)/(-W/m_f).$$

The equation can also be written as $(-W/m) = (-W/m_f)y$.

In any liquefaction system, the objective is to maximize the yield (gas fraction liquefied) and minimize the work requirements. Also note three payoff functions vary for each gas and comparison becomes difficult. The performance parameter which allows comparison of a typical cryogenic system for different fluids is "Figure of Merit (FOM)".

Figure of Merit (FOM) for a cryogenic system is defined as theoretical minimum work requirement divided by the actual work requirement of the system.

$$\text{FOM} = W_1/W = (-W_1/m_f) / (-W/m_f).$$

The FOM is a number between 0 and 1. It gives a measure of how close the performance of actual cryogenic system approaches the ideal system performance.

Apart from FOM, system performance parameters such as adiabatic and mechanical efficiencies of compressor and expander, effectiveness of heat exchanger can also be considered while estimating cryogenic system performance.

Joule-Thomson Experiment

The study of the dependence of the energy and enthalpy of real gases on volume (pressure) was done by Joule in association with Thomson who devised a different procedure. They allowed gas to expand freely through a porous plug, or frit. This experiment was carried out by Joule and J. J. Thomson to further elucidate the properties on real gases under expansion. During this porous plug experiment, a fluid is allowed to expand in a steady flow process. A sample of a gas, initially at p_1, V_1, and T_1 was forced through a porous plug at constant pressure, p_1. The gas came out of the other side of the plug at p_2, V_2, and T_2. The apparatus was insulated so that $q = 0$. Refer Figure 4.1 for the porous plug construction.

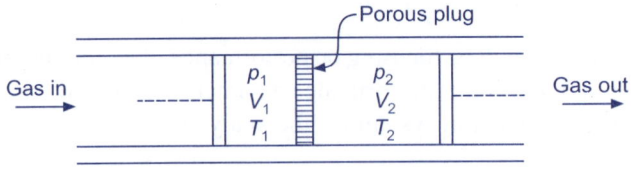

Fig. 4.1 Porous Plug Construction

The work has two terms, the work done on the system to force the gas through the plug and the work done by the system on the surroundings as it came out the other side of the plug.

The total work is
$$w = -p_1(0 - V_1) - p_1(V_2 - 0)$$
$$= p_1 V_1 - p_2 V_2$$

Since $q = 0$, the change in internal energy of the gas is,
$$\Delta U = q + w$$
$$= 0 + p_1 V_1 - p_2 V_2$$
$$\neq 0.$$

This process, unlike the Joule expansion, is not at constant internal energy. The enthalpy, however, is given by,
$$\Delta H = \Delta U + \Delta(pV)$$
$$= p_1 V_1 - p_2 V_2 + p_2 V_2 - p_1 V_1$$
$$\Delta H = 0.$$

This is therefore an isenthalpic expansion and the experiment measures directly the change in temperature of a gas with pressure at constant enthalpy which is called the Joule-Thomson coefficient.

Joule-Thomson (JT) Coefficient

We know that throttling is an isenthalpic or constant enthalpy process. Throttling involves sudden decrease in pressure. Now, what about the temperature? Does it increase or decrease? To answer this question, one has to define a quantity known as Joule-Thomson Coefficient denoted by "μ".

$$\mu = \left(\frac{\partial T}{\partial P}\right)_h = f(T, P, h).$$

Thus the Joule-Thomson coefficient is a function of temperature, pressure and enthalpy and illustrates change in temperature per unit pressure for a given value of enthalpy.

The Joule-Thomson coefficient is positive if temperature drops during throttling and cooling effect is achieved. The Joule-Thomson coefficient is negative if temperature increases during throttling and heating effect is achieved. Thus from refrigeration point of view, a positive Joule-Thomson coefficient is preferred.

Let us now derive an expression for the Joule-Thomson coefficient.

Enthalpy change for a pure substance is:

$$\therefore dh = C_p dT + \left[v - T\left(\frac{\partial v}{\partial T}\right)_p\right] dp$$

$$\therefore dT = \frac{dh}{C_p} + \frac{1}{C_p}\left[T\left(\frac{\partial v}{\partial T}\right)_p - v\right] dp \qquad \ldots(i)$$

4.4 Cryogenics

As $T = f(h, p)$

$$\therefore dT = \left(\frac{\partial T}{\partial h}\right)_P dh + \left(\frac{\partial T}{\partial P}\right)_h dp \qquad ...(ii)$$

equating equations (i) and (ii)

$$\therefore \frac{dh}{C_p} + \frac{1}{C_p}\left[T\left(\frac{\partial v}{\partial T}\right)_P - v\right]dp = \left(\frac{\partial T}{\partial h}\right)_P dh + \left(\frac{\partial T}{\partial P}\right)_h dp$$

$$\therefore \left(\frac{\partial T}{\partial P}\right)_P = \frac{1}{C_p}\left[T\left(\frac{\partial v}{\partial T}\right)_P - v\right] = \mu$$

$$\therefore \mu = \frac{T^2}{C_p}\left[\frac{\partial (v/T)}{\partial T}\right].$$

Joule-Thomson Coefficient for Real gases

For an ideal gas, the JT coefficient is equal to zero. This means the gas will not exhibit any temperature change upon expansion.

For a real gas, the enthalpy is a definite property and its value depends on the state of the system, e.g. on temperature and pressure

$$H = H(P, T)$$

$$dH = \left(\frac{\delta H}{\delta P}\right)_T dP + \left(\frac{\delta H}{\delta P}\right)_P dT$$

In the Joule-Thomson experiment H is constant, i.e. $dH = 0$

$$\mu_p = \left(\frac{\delta T}{\delta P}\right)_T = -\frac{\left(\frac{\delta H}{\delta P}\right)_T}{\left(\frac{\delta H}{\delta T}\right)_P}$$

Now $\left(\frac{\partial H}{\partial T}\right)_P$ is C_p the heat capacity at constant pressure. Also since

$$dH = TdS + VdP$$

$$\left(\frac{\delta H}{\delta P}\right)_T = T\left(\frac{\delta S}{\delta P}\right)_P + V$$

But the Maxwell's relation from

$$dG = VdP - SdT$$

is $$\left(\frac{\delta S}{\delta P}\right)_T = \left(\frac{\delta V}{\delta T}\right)_P$$

$$\therefore \quad \left(\frac{\delta H}{\delta P}\right)_T = V - T\left(\frac{\delta V}{\delta T}\right)_P$$

Thus

$$\mu_\pi = \frac{T\left(\frac{\delta V}{\delta T}\right)_P - V}{C_p}.$$

For a real gas $\left(\frac{\partial V}{\partial T}\right)_P$ may be obtained from any equation of state by differentiation as shown below for the van der Waals and Beattie-Bridgeman equations of state. A. JT coefficient for the van der Waals equation of state

$$PV = RT - \frac{a}{V} + bP + \frac{ab}{V^2}$$

cab be written in the form given below if the very small term $\frac{ab}{V^2}$ is neglected.

and the term $\frac{a}{PV}$ is replaced by $\frac{a}{RT}$

$$V = \frac{RT}{P} - \frac{a}{RT} + b$$

Differentiation w.r.t. T at constant P given

$$\left(\frac{\delta V}{\delta T}\right)_P = \frac{R}{P} + \frac{a}{RT^2}$$

Rearranging terms in equation for V gives,

$$\frac{R}{P} = \frac{V-b}{T} + \frac{a}{RT^2}$$

Substituting this in equation for differentiation of T gives

$$T\left(\frac{\delta V}{\delta T}\right)_P - V = \frac{2a}{RT} - b$$

Thus the Joule-Thomson coefficient becomes

$$\mu_\pi = \frac{\left(\frac{2a}{RT}\right) - b}{C_p}.$$

B. Beattie-Bridgeman equation of state

$$PV = RT + \frac{\beta}{V} + \frac{\gamma}{V^2} + \frac{\delta}{V^3}$$

$$\beta = RTB_0 - A_0 - \frac{RC}{T^2}$$

4.6 Cryogenics

$$\delta = RB_0 \, bc/T^2$$

has five adjustable constants A_0, B_0, a, b, c

Using similar procedure, we can derive the JT coefficient as

$$\mu_\pi = \frac{1}{C_P}\left\{-B_0 + \frac{2A_0}{RT} + \frac{4c}{T^3} + \left[\frac{2B_0 b}{RT} - \frac{3A_0 a}{(RT)^2} + \frac{5B_0 c}{RT^4}\right]JP\right\}.$$

Inversion Curve

Throttling is an isenthalpic process and the JT coefficient allows prediction of the outcome of the throttling process. Throttling a real gas through an adiabatic valve, results in the reduction of its pressure with a concurrent change in its temperature. The temperature may increase or decrease depending on the substance and its initial temperature. The coefficient of the Joule-Thomson effect is important in the liquefaction of gases because it tells whether a gas cools or heats on expansion. The resultant temperature and pressure are noted for each expansion along with enthalpy values. If constant enthalpy curves are plotted on T-P diagram as shown in Figure 4.2, we will get zero slope at their maximum point. That point is known as inversion temperature point. The JT inversion curve, a locus of points in a pressure-temperature plot at which the drop in pressure has no effect on the temperature indicates whether the temperature will increase or decrease during expansion. Figure 4.2 shows the inversion curves for different gases.

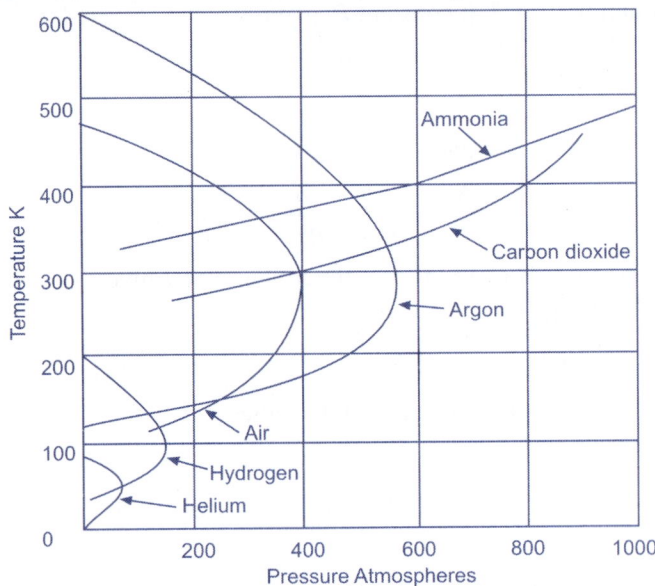

Fig. 4.2 Inversion Curves for Various Gases

For the maxima $\left(\dfrac{\partial T}{\partial P}\right)_h = 0$, these maxima are known as Inversion points. This is because the slope changes from positive to negative across the point. The slope is nothing but the Joule-Thomson Coefficient. Now if all maxima are joined together, we get a curve known as "Inversion Curve". The significance of this curve is that it separates the positive and negative values of the Joule-Thomson Coefficient.

Thermodynamically Ideal Cryogenic System

The thermodynamically ideal cryogenic liquefaction system is based on the Joule-Thomson effect. Joule-Thomson expansion process utilizes the expansion process to generate the required refrigerant for liquefaction using the inlet gas. Gas is compressed at ambient pressure and cooled in a heat exchanger. It is then passed through a throttle valve where it undergoes isenthalpic Joule-Thomson expansion producing some liquid. The liquid is removed and the cool gas is returned to the compressor via the heat exchanger. This simple system has very few components operating at cryogenic temperatures. Refer Figure 4.3 for the details of the system. Heat transfer is achieved using heat exchangers, and isenthalpic expansion follows compression in cryogenic turbines or J-T throttle valves taking advantage of the Joule-Thomson (JT) effect.

Initially, the gas is compressed at ambient pressure and pre-cooled to 80 K in a counter-flow heat exchanger using liquid nitrogen. Heat exchangers are used to lower the temperature even further, below its inversion temperature by transferring heat from the gas stream to the returning cooled gas. The cooled and compressed gas is forced to pass through a throttle valve or a mechanical expander where it undergoes an isenthalpic expansion to ambient pressure, producing some liquid. The liquid is removed and the cooled gas is returned to the compressor via the heat exchangers as shown in Figure 4.3. The process comprises an isothermal compression followed by reversible isentropic expansion to cool the gas and transform it into liquid. The work needed in this theoretical process is called the ideal work of liquefaction. The ideal work of liquefaction takes into account the energy required to reduce the temperature of the gas to its boiling point, the energy of the phase transformation from gas to liquid.

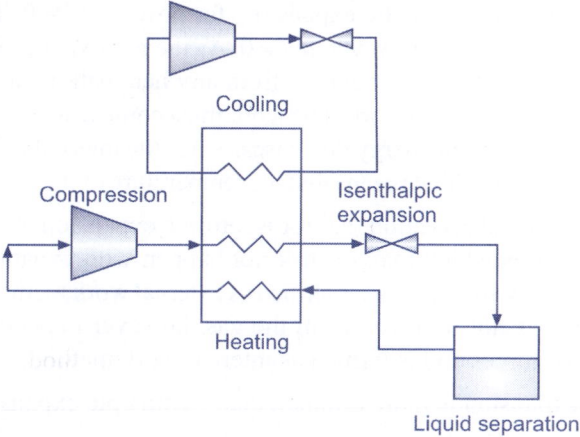

Fig. 4.3 Ideal Cryogenic Liquefaction System

4.8 Cryogenics

We can apply the first law of thermodynamics to the ideal system as follows:

$$Q_t - W_t = m(h_f - h_1)$$

Where Q_t = Total heat transfer in the system; W_t = refrigeration work done on the system; m = mass of the gas to be liquefied; h_f = enthalpy of the saturated liquid after expansion and h_1 is the enthalpy of the gas at ambient conditions.

We can also apply the second law of thermodynamics to the ideal system as follows:

$$Qr = mT_1(s_f - s_1)$$

Where Q_r = Reversible heat transfer in the system; T_1 = ambient temperature, m = mass of the gas to be liquefied; s_f = entropy of the saturated liquid after expansion and s_1 is the entropy of the gas at ambient conditions.

Thus combining expressions for first law and second law, we can write the work requirement for the ideal cryogenic system as

$$W = m[T_1(s_f - s_1) - (h_f - h_1)]$$

Simplifying we get, $W/m = [T_1(s_f - s_1) - (h_f - h_1)]$

However as liquefaction is an energy absorbing system, work done is negative, hence it is usually expressed in negative form.

$$-W/m = [T_1(s_1 - s_f) - (h_1 - h_f)]$$

Adiabatic Expansion of an Ideal Gas

Adiabatic expansion in an expansion engine is a method to produce low temperature. The expansion is also reversible and hence is termed as isentropic expansion. The isentropic expansion coefficient can be defined as temperature change due to pressure change at constant entropy.

$$\mu = \left(\frac{\delta T}{\delta P}\right)_S$$

The definition of an adiabatic expansion, for now, is $dq = 0$. That is, no heat goes in or out of the system. However, $dw \neq 0$. As the gas expands it does work on the surroundings. Since the gas is cut off from any heat bath it can not draw heat from any source to convert into work. The work must come from the internal energy of the gas so that the internal energy decreases. Since the internal energy of an ideal gas in only dependent on T that means that the temperature of the gas must decrease.

We can make the observation that for isentropic expansion through expander, temperature decreases which may or may not happen with expansion valve. The removal of heat energy from the gas is termed as external work method. In expansion valve, heat energy is not removed from the gas, however molecules move apart under influence of forces and is termed as internal work method.

Isentropic expansion is more efficient than isenthalpic expansion.

From the first law with only pV work we have

$$dU = dq - pdV$$
$$= -pdV$$

As $dq = 0$ for an adiabatic process.

Regarding U as a function of T and V, we can write $U = U(T, V)$.

$$dU = \left(\frac{\partial U}{\partial T}\right)_V dT + \left(\frac{\partial U}{\partial V}\right)_T dV$$

because of the definition of C_V and because for ideal gas second derivative vanishes.

The dU's in Equations must be equal so that

$$-pdV = C_V dT$$
$$-\frac{nRT}{V}dV = C_V dT.$$

Rearranging Equation we get

$$\frac{dT}{T} = -\frac{nR}{C_V}\frac{dV}{V}$$

By the same token, using enthalpy, we find

$$dH = dq + Vdp$$
$$= Vdp,$$

and

$$dH = \left(\frac{\partial H}{\partial T}\right)_p dT + \left(\frac{\partial H}{\partial p}\right)_T dp$$
$$= C_p dT.$$

From which we deduce that

$$\frac{dT}{T} = \frac{nR}{C_p}\frac{dp}{p}.$$

Comparing Equations we see that

$$-\frac{nR}{C_V}\frac{dV}{V} = \frac{nR}{C_p}\frac{dp}{p},$$

$$-\frac{C_p}{C_V}\frac{dV}{V} = \frac{dp}{p}$$

$$-\gamma\frac{dV}{V} = \frac{dp}{p},$$

Where we have written $C_p/C_V = \gamma$.

If we regard C_p and C_V as constant then Equation can be integrated to give,

$$-\gamma \ln \frac{V_2}{V_1} = \ln \frac{p_1}{p_2},$$

$$\gamma \ln \frac{V_1}{V_2} = \ln \frac{p_2}{p_1},$$

$$\ln \left(\frac{V_1^\gamma}{V_2^\gamma} \right) = \ln \frac{p_2}{p_1};$$

$$\frac{V_1^\gamma}{V_2^\gamma} = \frac{p_2}{p_1},$$

$$p_1 V_1^\gamma = p_2 V_2^\gamma.$$

This is the equation for the adiabatic expansion of an ideal gas.

Reverse Carnot Cycle

Reversed Carnot cycle is an ideal refrigeration cycle for constant temperature external heat source and heat sinks. Figure 4.4 shows the schematic of a reversed Carnot refrigeration system using a gas as the working fluid and Figure 4.5 shows the cycle diagram on P-V coordinates. As shown, the cycle consists of the following four processes:

Process 1-2: Reversible, adiabatic compression in a compressor

Process 2-3: Reversible, isothermal heat rejection in a compressor

Process 3-4: Reversible, adiabatic expansion in a turbine

Process 4-1: Reversible, isothermal heat absorption in a turbine.

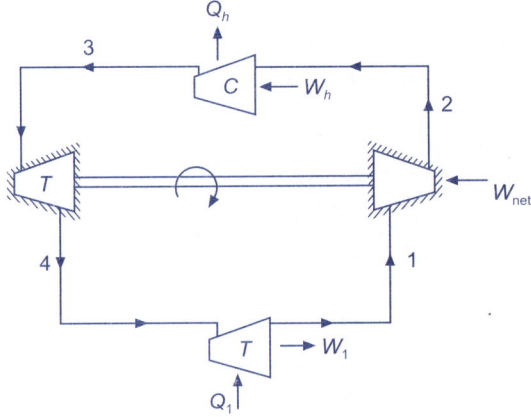

Fig. 4.4 Reverse Carnot Cycle

The heat transferred during isothermal processes 2-3 and 4-1 are given by:

$$q_{2-3} = \int_2^3 T.ds = T_h(s_3 - s_2)$$

$$q_{4-1} = \int_4^1 T.ds = T_1(s_1 - s_4)$$

$s_1 = s_2$ and $s_3 = s_4$, hence $s_2 - s_3 = s_1 - s_4$

Applying first law of thermodynamics to the closed cycle,

$$\oint \delta q = (q_{4-1} + q_{2-3}) = \oint \delta w = (w_{2-3} - w_{4-1}) = -w_{net}$$

Note that the work of isentropic expansion, w_{3-4} exactly matches the work of isentropic compression w_{1-2}. The COP of the reverse Carnot system is as follows:

$$COP_{Carnot} = \left|\frac{q_{4-1}}{W_{net}}\right| = \left(\frac{T_1}{T_h - T_1}\right)$$

Thus the COP of the reverse Carnot cycle depends only on the refrigeration (T_1) and heat rejection (T_h) temperatures only.

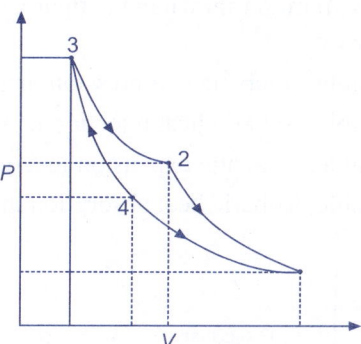

Fig. 4.5 Reverse Carnot Cycle on PV Diagram

Reverse Carnot cycle is an idealization and it suffers from several practical limitations. One of the main difficulties with reverse Carnot cycle employing a gas is the difficulty of achieving isothermal heat transfer during processes 2–3 and 4–1. For a gas to have heat transfer isothermally, it is essential to carry out work transfer from or to the system when heat is transferred to the system (process 4–1) or from the system (process 2–3). This is difficult to achieve in practice. In addition, the volumetric refrigeration capacity of the reverse Carnot system is very small leading to large compressor displacement, which gives rise to large frictional effects. All actual processes are irreversible, hence reverse Carnot cycle is an idealization yet studied as a benchmark.

Reverse Brayton Cycle or Bell Coleman Cycle

A reverse Brayton cycle that is driven in reverse, via net work input, and when air is the working fluid, is the air refrigeration cycle or Bell Coleman cycle. Its purpose

4.12 Cryogenics

is to move heat, rather than produce work. This air cooling technique is used widely in jet aircraft. Refer Figure 4.6 for the schematic of the reverse Brayton Cycle.

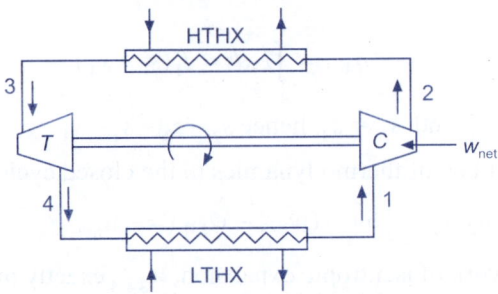

Fig. 4.6 Reverse Brayton Cycle

This is an important cycle frequently employed in gas cycle refrigeration systems. This may be thought of as a modification of reversed Carnot cycle, as the two isothermal processes of Carnot cycle are replaced by two isobaric heat transfer processes. This cycle is also called as Joule or Bell-Coleman cycle. Figure 4.7 shows the cycle on T-S diagram. As shown in the figure, the ideal cycle consists of the following four processes:

Process 1-2: Reversible, adiabatic compression in a compressor
Process 2-3: Reversible, isobaric heat rejection in a heat exchanger
Process 3-4: Reversible, adiabatic expansion in a turbine
Process 4-1: Reversible, isobaric heat absorption in a heat exchanger.

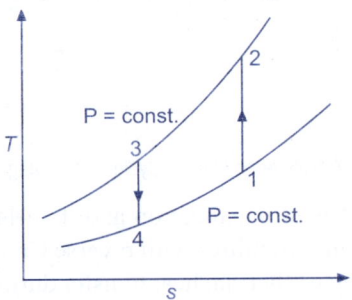

Fig. 4.7 Reverse Brayton Cycle on T-S Diagram

Process 1-2: Gas at low pressure is compressed isentropically from state 1 to state 2. Applying steady flow energy equation and neglecting changes in kinetic and potential energy, we can write:

$$W_{1-2} = m(h_2 - h_1) = mc_p(T_2 - T_1)$$

$$s_2 = s_1$$

and $$T_2 = T_1 \left(\frac{P_2}{P_1}\right)^{\frac{\gamma-1}{\gamma}} = T_1 r_p^{\frac{\gamma-1}{\gamma}}$$

where $r_p = (P_2/P_1)$ = pressure ratio.

Process 2-3: Heated high pressure gas flows through a heat exchanger and rejects heat sensibly and isobarically to a heat sink. The enthalpy and temperature of the gas drop during the process due to heat exchange, no work transfer takes place and the entropy of the gas decreases. Again applying steady flow energy equation:

$$Q_{2-3} = m(h_2 - h_3) = mc_p(T_2 - T_3)$$

$$s_2 - s_3 = c_p \ln \frac{T_2}{T_3}$$

$$p_2 = p_2.$$

Process 3-4: High pressure gas from the heat exchanger flows through a turbine, undergoes isentropic expansion and delivers network output. The temperature of the gas drops during the process from T_3 to T_4. From steady flow energy equation:

$$W_{3-4} = m(h_3 - h_4) = mc_p(T_3 - T_4)$$

$$s_3 = s_4$$

and

$$T_3 = T_4 \left(\frac{P_3}{P_4}\right)^{\frac{\gamma-1}{\gamma}} = T_4 \, r_p^{\frac{\gamma-1}{\gamma}}$$

where $r_p = (P_3/P_4)$ = pressure ratio.

Process 4-1: Cold and low pressure gas from turbine flows through the low temperature heat exchanger and extracts heat sensibly and isobarically from a heat source, providing a useful refrigeration effect. The enthalpy and temperature of the gas rise during the process due to heat exchange, no work transfer takes place and the entropy of the gas increases. Again applying steady flow energy equation:

$$Q_{4-1} = m(h_1 - h_4) = mc_p(T_1 - T_4)$$

$$s_4 - s_1 = c_p \ln \frac{T_4}{T_1}$$

$$p_4 = p_1$$

From the above equations, it can be easily shown that:

$$\left(\frac{T_2}{T_1}\right) = \left(\frac{T_3}{T_4}\right)$$

Applying 1st law of thermodynamics to the entire cycle:

$$\oint \delta q = (q_{4-1} - q_{2-3}) = \oint \delta W = (w_{3-4} - w_{1-2}) = W_{net}$$

The COP of the reverse Brayton cycle is given by:

$$\text{COP} = \left|\frac{q_{4+1}}{w_{net}}\right| = \left(\frac{(T_1 - T_4)}{(T_2 - T_1)(T_3 - T_4)}\right).$$

Using the relation between temperatures and pressures, the COP can also be written as:

4.14 Cryogenics

$$\text{COP} = \left(\frac{(T_1-T_4)}{(T_2-T_1)(T_3-T_4)}\right) = \frac{(T_4)}{(T_3-T_4)} = \left(\frac{(T_1-T_4)}{(T_1-T_4)(r_p^{\frac{\gamma-1}{\gamma}}-1)}\right) = (r_p^{\frac{\gamma-1}{\gamma}}-1)^{-1}$$

The expression for COP of Reverse Brayton cycle can also be re-written as

$$\text{COP} = (T_1-T_4)/(T_2-T_1)-(T_3-T_4) = Tr-(\Delta t_r+\Delta_t)/(T_o-T_r)+(\Delta t_r+\Delta t_c+\Delta_t)$$

Where T_o is the ambient temperature, T_r is the refrigeration temperature, Δt_r is the temperature difference in refrigerator, Δt_c is the temperature difference in cooler, Δ_t is the rise of temperature during isobaric refrigeration.

Methods of Producing Low Temperature

Refrigeration is the process of removing heat from an enclosed space, or from a substance, and moving it to a place where it is unobjectionable. The primary purpose of refrigeration is lowering the temperature of the enclosed space or substance and then maintaining that lower temperature. The term cooling refers generally to any natural or artificial process by which heat is dissipated. The process of artificially producing extreme cold temperatures is referred to as cryogenics. Cold is the absence of heat, hence in order to decrease a temperature, one "removes heat", rather than "adding cold." In order to satisfy the Second Law of Thermodynamics, some form of work must be performed to accomplish this. This work is traditionally done by mechanical work but can also be done by other means. In this section we will study some methods of producing low temperatures. These are 1) Enthalpy of mixing salt and water, 2) Expansion in turbine 3) Throttling or Isenthalpic expansion, 4) Thermoelectric cooling and 5) Adiabatic demagnetisation.

Using Enthalpy of Mixing (Mixing of Salt with Water)

This technology provides a method of cooling by using enthalpy of mixing of salt and water. The technology involves producing salt water-mixed ice by cooling salt water fed to an ice-maker. The ice is then scraped by scrapers to form small particles which can be used for refrigeration of perishables. Salt-water mixture or Saline may be considered an ideal mixture. This means that enthalpy of the substance is simply the sum of the enthalpies of the salt and the water. For each state the salinity of the water in parts per million (ppm) is known. To convert from ppm to a mass fraction of salt the equation is

mfs = mass fraction of salt = ppm / 1,000,000

The enthalpy is then $h = mfs \times hs + mfw \times hw$.

The enthalpy of water, hw is known and the enthalpy of salt, hs, can be found from:

$$hs = hs_o + cp_s(T - T_o)$$

Where hs_0 = 29.288 kJ/kg, cp_s = 0.8368 kJ/kg K, and T_o = 308 K.

Brine is water saturated or nearly saturated with a salt (usually sodium chloride). It is used to preserve vegetables, fish, and meat, in a process known as brining. Brine is a common fluid used in large refrigeration installations for the transport of heat. It is used because the addition of salt to water lowers the freezing temperature of the solution and the heat transport efficiency can be greatly enhanced for the comparatively low cost. At a concentration of 23.3%, the freezing point of NaCl brine is lowered to −21°C. The coldest temperature that Fahrenheit could reliably reproduce with brine freezing was 0 F.

Expansion in a Turbine

For many years, the traditional method to reduce pressure in a liquid stream was using a Joule-Thomson (J-T) valve. An improvement in the overall efficiency of this expansion process can be achieved by replacing the J-T valve with a turbine expander. Turbine expanders are used in process plants for the liquefaction of gases like air, nitrogen, methane, natural gas and other gases, to reduce the enthalpy of the condensed gas and to recover power. For gas liquefaction processes, thermodynamic expansion is optimized by use of a turbine expansion process for liquefied gases which can be adjusted to different flow rates and differential pressures by varying the rotational speed and/or guide vane position. The turbine expander is located in-line between an upstream system for gas liquefaction and a downstream system for liquefied gas handling including a terminal vessel for storage or phase separation. The turbine controller sets the rotational speed and/or the guide vane position of the expander depending on changes in the terminal pressure. A process plant for turbine gas expansion consists of various elements such as compressors, gas expanders, heat exchangers, valves, orifices and pipes. Refer Figure 4.8 for construction of the turbine expander. The upstream system is in general designed to cool down and to condense the gas under higher pressure, and the downstream system is designed to handle the liquefied gas. The downstream system is connected to the terminal vessel, which could be a storage or a phase separator. The terminal vessel is operated with almost constant pressure, independent of the flow rate. The pressure drop in the downstream and upstream system, due to fluid friction, depends significantly on the squared value of the flow rate and on such parameters as density, viscosity, temperature, mixture and inlet conditions for the upstream system, and it is not possible to predict the pressure drop without a certain margin of error. It is a characteristic of all rotating fluid machines, that the ratio between output and input power, the efficiency, depends on the value of the potential and kinetic fluid energy, and reaches a maximum value for a certain differential pressure and flow rate. The maximum value is called best efficiency point.

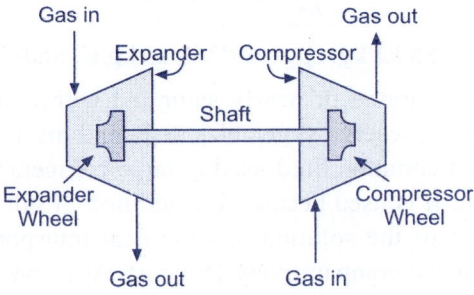

Fig. 4.8 Turbine Expander

Because of the variation and uncertainty range of the pressure drop in the system, it is necessary to install a control valve preferably between the turbine expander and the downstream system, to meet the conditions for differential pressure and flow rate in order to operate the turbine expander at the best efficiency point. The disadvantage of the turbine expansion process is that the control valve reduces the pressure through a Joule-Thomson expansion without power recovery. To optimize the thermodynamic expansion in gas liquefaction processes the overall Joule-Thomson expansion has to be minimized and replaced by expansions which reduce the enthalpy of the gas. With the installation of a turbine expander, an increase in overall process efficiency of approximately 5% can be obtained. In a typical liquefaction plant additional output is achieved through the enthalpy reduction during the expansion across the turbine. The enthalpy reduction also reduces the boil-off losses and increases the mass output. The expander is installed inside a pressure vessel designed and built to applicable pressure vessel codes. The system is suitable for installation in hazardous area locations as required and defined by applicable codes and standards. By introducing variable speed technology to the application, the efficiency of the turbine can be optimized over a wide range of flows and desired pressure reductions. This method enables automatic optimization of work transfer and performance to match changes in process conditions.

Throttling (Isenthalpic Expansion)

A throttling process is defined as an expansion process in which there is no change in enthalpy from one state to another. Furthermore no heat transfer or work takes place during the process thus the process is adiabatic. Classic example of throttling process is ideal gas flowing through a valve in mid position. Refer Figure 4.9 for the throttling process.

Fig. 4.9 Throttling Process

The throttling process is an irreversible steady flow expansion process in which a perfect gas is expanded through an orifice of minute dimensions such as a narrow throat or a slightly opened valve. Due to fall in pressure during expansion, the gas should come out with a large velocity, but due to high frictional resistance between the gas and the walls of the aperture, there is no change in velocity. The kinetic energy of gas is converted into heat, which is utilized in warming the gas to its initial temperature. As no heat is supplied or rejected during throttling process and no work is done, therefore $Q_{1-2} = 0$ and $W_{1-2} = 0$.

Thus the steady flow energy equation can be written as

$$h_1 + V_1^2/2 + gZ_1 + Q_{1-2} = h_2 + V_2^2/2 + gZ_2 + W_{1-2}.$$

Neglecting velocity changes and potential level changes $V_1 = V_2$ and $Z_1 = Z_2$, the above equation reduces to

$$h_1 = h_2$$

Thus the throttling process is a constant enthalpy process. The throttling process was first investigated by Joule and Thomson in their porous plug experiment.

Thermoelectric Cooling

The thermoelectric effect is the direct conversion of temperature differences to electric voltage and vice versa. A thermoelectric device creates a voltage when there is a different temperature on each side. Conversely when a voltage is applied to it, it creates a temperature difference (known as the Peltier effect). At atomic scale an applied temperature gradient causes charged carriers in the material (electrons) to diffuse from the hot side to the cold side causing the thermally-induced current. This effect can be used to generate electricity, to measure temperature, to cool objects, or to heat them. Because the direction of heating and cooling is determined by the sign of the applied voltage, thermoelectric devices make very convenient temperature controllers.

Traditionally, the term thermoelectric effect or thermoelectricity encompasses three separately identified effects, the Seebeck effect, the Peltier effect, and the Thomson effect all of which are thermodynamically reversible.

Thermoelectric cooling uses the Peltier effect to create a heat flux between the junction of two different types of materials. A Peltier cooler, heater, or thermoelectric heat pump is a solid-state active heat pump which transfers heat from one side of the device to the other side against the temperature gradient (from cold to hot), with consumption of electrical energy. Such an instrument is also called a Peltier device, Peltier diode, cooling diode, Peltier heat pump, solid state refrigerator, or thermoelectric cooler (TEC). Because heating can be achieved more easily and economically by many other methods, Peltier devices are mostly used for cooling. Simply connecting it to a DC voltage will cause one side to cool, while the other side warms. The effectiveness of the pump at moving the heat away from the cold side is dependent upon the amount of current provided and how well the heat can be removed from the hot side. Peltier devices can also be used to generate electricity (thermogenerator) if a temperature difference is maintained between the two sides. Figure 4.10 shows construction of a thermoelectric refrigerator.

4.18 Cryogenics

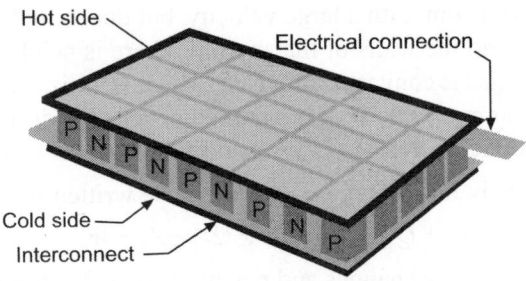

Fig 4.10 Construction of a Thermoelectric Refrigerator

Thermoelectric junctions are generally only around 5–10% as efficient as the ideal refrigerator (Carnot cycle), compared with 40–60% achieved by conventional vapor compression cycle systems. Due to the relatively low efficiency, thermoelectric cooling is generally only used in environments where the absence of moving parts outweighs pure efficiency. Peltier (thermoelectric) cooler performance is a function of ambient temperature, hot and cold side heat exchanger performance, thermal load, module (thermopile) geometry, and electrical parameters.

However recent developments suggest that series Peltier effect modules could soon surpass internal combustion engines both in efficiency and power density for fuel based power generation. Peltier devices are commonly used in camping and portable coolers and for cooling electronic components and small instruments. Some electronic equipment intended for military use in the field is thermoelectrically cooled. Thermoelectric coolers can be used to cool computer components to keep temperatures within design limits without the noise of a fan, or to maintain stable functioning. A Peltier cooler with a heat sink can cool a chip to well below ambient temperature.

Adiabatic Demagnetization Refrigeration

An Adiabatic Demagnetization Refrigerator (ADR) works by using the properties of heat and the magnetic properties of certain molecules. Refer Figure 4.11 for the construction of ADR and Figure 4.12 for an actual photograph of an ADR. Some molecules have large internal magnetic fields, or "moments". Just like a tiny bar magnet, these molecules will align themselves with an external magnetic field. The random thermal motions of the molecules, on the other hand, tend to de-align them. The higher the temperature, the more they de-align. ADRs generally use certain types of salts for the molecules, because they have particularly large magnetic moments. The salt is contained in a cylinder, usually called a "salt pill". This salt pill is thermally connected to the object to be cooled.

The salt pill is first placed in a strong magnetic field. The molecules align with the external magnetic field, and the magnetic energy of each molecule is minimal. If the strength of that field is decreased, then the thermal motion of the molecules starts to twist them out of alignment with the field. This requires energy, which comes from the thermal motion of the molecules. The thermal energy is thus transformed into magnetic energy, cooling the salt pill down. As heat flows into the salt pill from the outside, the magnetic field is slowly reduced. This allows the molecules to twist further out of alignment, absorbing more heat. The rate at which the field is reduced can be regulated so as to keep the salt pill at a constant temperature as it absorbs heat. Conversely, increasing the magnetic field will convert magnetic energy back to thermal energy, raising the temperature of the salt pill.

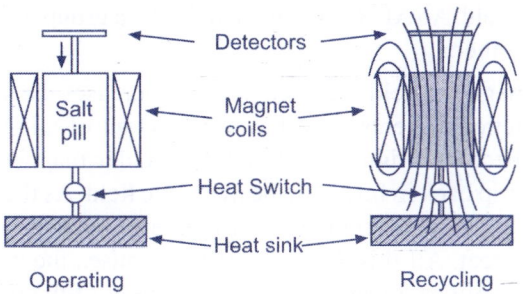

Fig. 4.11 Construction of an Adiabatic Demagnetization Refrigerator

Eventually the magnetic field is reduced to zero, and no more heat can be extracted by the salt pill. At this point, all that can be done is to increase the field, heating the salt pill. Thus far all the thermal energy is converted to magnetic energy. Now it has to be turned back into heat and disposed somewhere. If the magnetic field is increased to a much higher value than it originally was, the temperature of the salt pill will increase. This will also raise the temperature of the detectors, rendering them temporarily inoperable. However, the salt pill temperature can actually be higher than the temperature of the surrounding Dewar. At that point it can be thermally connected to the Dewar until it cools to the temperature of the Dewar. Thus the heat originally absorbed by the salt pill is dumped to the Dewar. The salt pill is then disconnected from the Dewar and the magnetic field is slowly reduced, beginning the cycle again. The temperature of the salt pill quickly reaches a level at which the detectors can operate, and the temperature is again regulated by adjusting the rate at which the magnetic field is reduced.

Fig. 4.12 Photograph of an Adiabatic Demagnetization Refrigerator

4.20 Cryogenics

The ADR does not run continuously. It stores the heat that it absorbs, both heat from cooling warm objects and heat that leaks in. In a paramagnetic substance, each molecule acts like a tiny electromagnet, with the electrons playing the part of tiny electric currents. In non-paramagnetic substances, the fields of the various electrons all cancel each other out, leaving the molecule with no overall field. In paramagnetic molecules, however, the fields don't quite cancel, so the molecule produces a small field. An ADR salt pill, then, is like a group of microscopic magnets all packed in together.

When the magnetic field is turned up, that increases the amount of energy the molecular magnetic moments must have to stay out of alignment with the field. When the field becomes high enough, the molecular magnetic moments give up their energy and flip back in line with the magnetic field. As the energy gets dumped by the molecular magnetic moments, it converts back into random molecular motion, i.e. into heat. All this dumping of heat causes the salt pill's temperature to rise. As the salt pill's temperature rises above that of the liquid helium coolant, the operator turns on the heat switch. When enough heat has flowed to the coolant bath, the operator turns off the heat switch, then reduces the magnetic field. Once again, the amount of energy needed to knock a molecular magnetic moment out of alignment is small enough that random thermal vibrations have enough energy. Thus, the molecular moments begin absorbing heat, and the salt pill cools, starting another cycle. It is possible to achieve 0.01 K with this method.

Solved Problem

Problem 1. In a reversed Brayton cycle, $T_a = 300$ K, $T_r = 200$ K, $p_a = 1$ atm, $\Delta t_r = \Delta t_c = 5$ K and overall $\Delta t = 50$ K. Assume cycle efficiency as 20%. The isentropic efficiency of compressor is 90%. Find the highest temperature, pressure, and efficiency of expander and COP for the system.

Solution:
$$\text{COP} = T_r - (\Delta t_r + \Delta t) / (T_o - T_r) + (\Delta t_r + \Delta t_c + \Delta t)$$
$$= 200 - (5 + 50) / (300 - 200) + (5 + 5 + 50)$$
$$= (200 - 55) / (100 + 60)$$
$$= 145 / 160 = 0.90$$

Thus COP of the refrigerator is 0.90 ...(Ans)

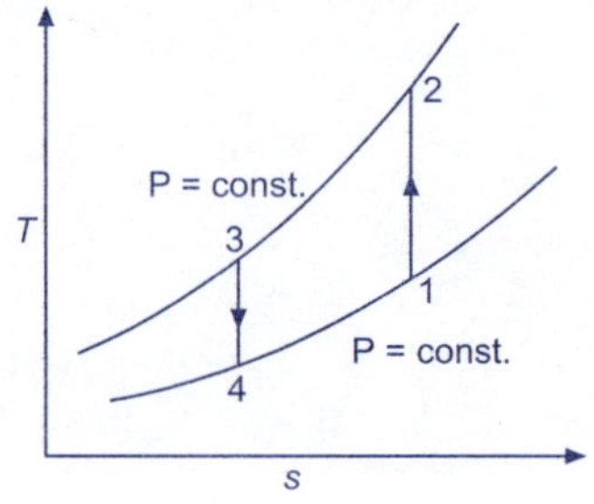

Fig. 4.13 Reverse Brayton Cycle

$$T_2/T_1 = (P_2/P_1)^{\gamma-1/\gamma}$$

Here $P_1 = 1$ bar (given) $T_1 = 300$ K (given) and $T_2 = 300 + 50 = 350$ K, $\gamma = 1.4$ for air.

Thus $350/300 = (P_2/1)^{1.4 - 1/1.4}$

and $(P_2)^{0.285} = 1.16$

Which gives $P_2 = 1.68$ bar $= P_3 =$ maximum pressure ....(Ans)

Maximum temperature $= T_3 = T_2 (P_2/P_1)^{1.4 - 1/1.4} = 350(1.68)^{0.285} = 405$ K .(Ans)

Total efficiency $= \eta = 0.2 = \eta c \times \eta e = 0.9 \times \eta e$

Thus $\eta e = 0.18 \times 100\% = 18\% =$ efficiency of the expander ...(Ans)

References

1. Cryogenic Engineering – B.A. Hands.
2. Internet Website http://www.chemistry.mcmaster.ca/~ayers/chem2PA3.
3. Cryogenic Systems – Randall Baron.
4. Internet Website http://www.chem.arizona.edu/~salzmanr/480a/jadjte.html.
5. EU Report on Hydrogen Storage, Tzimas et.al. (2003).
6. Internet Website http://en.wikipedia.org/wiki/Thermoelectric_cooling.
7. Internet website http://imagine.gsfc.nasa.gov/docs/teachers/lessons.
8. Lecture Notes on Refrigeration of IIT Kharagpur.
9. Internet Website http://www.faqs.org/patents/app/20080276629.

Questions

Q.1. What is figure of merit? How is it defined?

Q.2. Discuss the operation of Adiabatic demagnetization process.

Q.3. Enumerate some methods to produce low temperatures.

Q.4. Write short note on isenthalpic and isentropic expansion process.

Q.5. Define the Joule-Thomson coefficient and derive an expression for the same.

Q.6. What is an inversion curve? How is it important?

Q.7. In a Reverse Brayton cycle, $T_a = 300$K, $T_r = 250$K, $p_a = 1$ atm, $\Delta t_r = \Delta t_c = 10$ K and inlet temperature of compressor is 240 K. Relative efficiency is 5 % for refrigerating effect of 20 kJ/kg. The isentropic efficiency of compressor is 80%. Find the highest temperature, highest pressure, efficiency of expander.

Q.8. For Reverse Brayton cycle prove that

$$COP = T_r - (\Delta t_r + \Delta t) / (T_o - T_r) + (\Delta t_r + \Delta t_c + \Delta t).$$

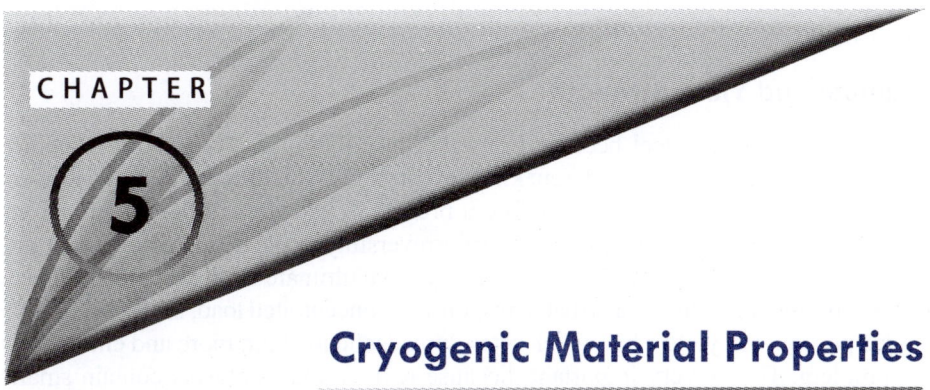

CHAPTER 5

Cryogenic Material Properties

INTRODUCTION

The explosive growth of cryogenics in the early 50's led to much interest in material properties at low temperatures. Important fundamental theory and measurements of low temperature material properties were performed in the 50's, 60's, and 70's. The results of this large amount of work has become fragmented and dispersed in many different publications, most of which are out of print and difficult to find. Old time engineers often have a file filled with old graphs; young engineers often don't know how to find this information. Since most of the work was performed before the desktop computer became available, when data can be found, it is published in simple tables or graphically, making the information difficult to accurately determine and use. NIST has begun a program to gather cryogenic material property data and make it available in a form that is useful to engineers. Initially they tried to use models based upon fundamental physics but it soon became apparent that the models could not accurately predict properties over a large temperature range and over different materials. The current approach is to choose a few simple types of equations such as polynomial or logarithmic polynomials and determine the coefficients of different materials and properties. This will allow engineers to use the equations to predict material properties in a variety of ways including commercial software packages. Integrated and average values can easily be determined from the equations. These equations are not meant to provide any physical insight into the property or to provide 'standard' values but are for working engineers that require accurate values. While there is a great deal of published data on cryogenic material properties, it is often difficult to find and not in a form that is convenient to use. At low temperatures, several effects occur in engineering materials such as vanishing of specific heats, super conductivity and ductile brittle transitions in carbon steel. None of these phenomena can be inferred from property measurements made at near ambient conditions. In this section we shall study the mechanical properties including ultimate and yield strength, fatigue strength, impact strength, hardness and ductility and elastic moduli and thermal properties like thermal conductivity, specific heat of solids, liquids and gases and coefficient of thermal expansion. Finally we shall focus on electrical properties such as electrical conductivity and superconductivity. The idea is to become familiar with low temperature behavior of materials.

5.2 Cryogenics

Ultimate and Yield Strength

Materials like carbon steel become brittle because they get stronger at cryogenic temperatures. There are two different kinds of "strength". "Ultimate strength" is how much of a load the stuff will take before it breaks. "Yield strength" is how much it will take before it starts to yield, i.e. deform irreversibly or the deformation won't go away if the load is removed. Metals generally have ultimate strengths much higher than yield strengths. This means that in response to concentrated load, the most highly-loaded portion will yield a bit, which often spreads the load out more and eliminates the problem. This is very important, because real structures always contain small flaws which concentrate loads. As temperature drops, both kinds of strength increase. However, they typically don't increase at the same rate. If yield strength grows more quickly than ultimate strength, then eventually the two curves cross over, and below the crossover temperature, the metal fails before it yields i.e. it has become brittle.

For many materials there is a definite value of stress where the strain starts increasing rapidly in a tensile test. This value of stress is called the yield strength of the material. This can also be defined as stress required to permanently deform the material in a tensile stress by 0.2%. The ultimate stress is defined as the maximum nominal stress attained during simple tensile test. Most of the engineering materials are alloys, where atoms from one material are added to another. If the size of atoms added are smaller than base material atoms, they disperse at random. These atoms make the process of dislocation or yielding of the alloy material difficult thus increasing the yield strength. When temperature is lowered, the vibrations of the atoms is damped and larger stress is required for yielding. Thus for most of the engineering materials at cryogenic temperatures yield strength and ultimate strength increases. This fact is shown in Figure 5.1, for typical alloy steel.

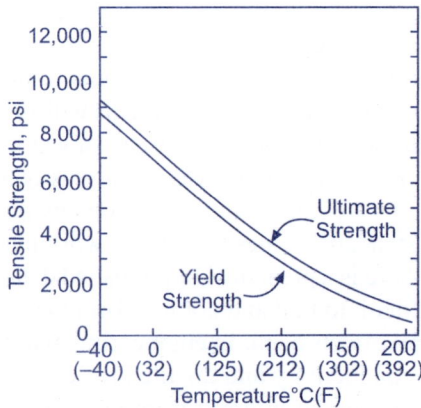

Fig. 5.1 Yield Strength and Ultimate Strength for Alloy Steel

Fatigue Strength

The stress required for failure in a simple bending test is called fatigue strength of a material. Some materials exhibit property that fatigue failure will not occur stress is maintained below the endurance limit. At cryogenic temperatures, the fatigue strength of the materials increases with reducing temperatures. This fact is shown in Figure 5.2. for two different alloy steels. Fatigue failure generally occurs in three stages after 1000 cycles in steps of microcrack formation, crack growth and rapid ductile failure. Microcracks are initiated at the surface of the material due to inhomogenous shear deformation. As temperature of the material is decreased, larger stress is required to extend the crack. For some materials like aluminum, the ratio of fatigue strength to ultimate strength remains constant and this fact can be used to generate fatigue data at cryogenic temperatures.

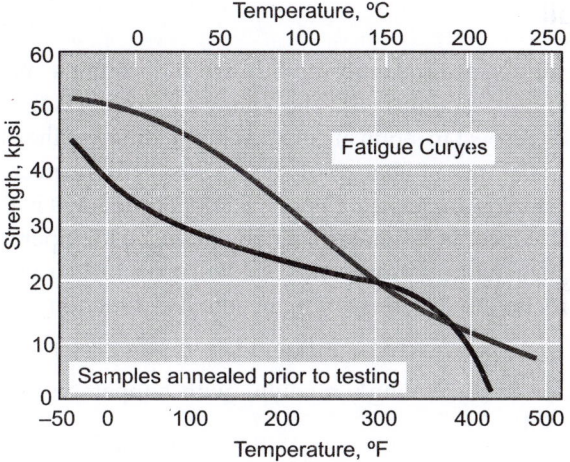

Fig. 5.2 Fatigue Strength Curves for Alloy Steels

Impact Strength

The Charpy and Izod impact tests give the measure of resistance of material to impact loading. These tests indicate the energy absorbed by the material when it is fractured by a suddenly applied force. Sometimes a ductile brittle transition occurs in some materials like carbon steel which reduces impact strength at cryogenic temperatures. Refer Figure 5.3, for Izod impact strength curves of two types of alloy steels.

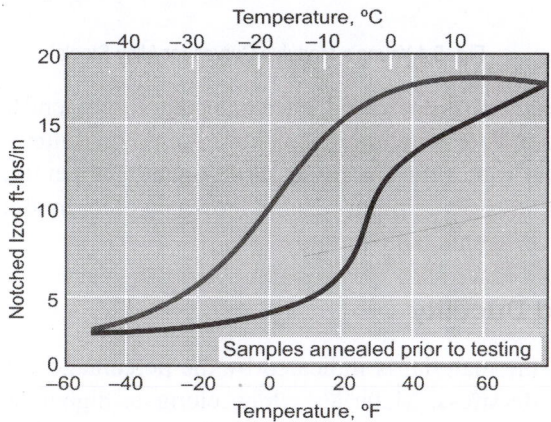

Fig. 5.3 Izod Impact Strength Curves for Alloy Steels

5.4 Cryogenics

The impact behavior of a metal depends upon its lattice structure. The face centered cubic (FCC) structure has more slip planes than the body centered cubic (BCC) structure. The materials with FCC structure tend to fail by plastic deformation in the impact test by absorbing energy at cryogenic temperatures and retain their resistance to impact. The materials with BCC structure however tend to fail by cleaving in the impact test without absorbing energy at cryogenic temperatures and become brittle.

Elastic Moduli

The three commonly used elastic moduli are the young's modulus (E), shear modulus (G) and bulk modulus (B). The Young's modulus is the rate of change of tensile stress with respect to strain at constant temperature in the elastic region. The shear modulus is the rate of change of shear stress with respect to shear strain at constant temperature in the elastic region. The bulk modulus is the rate of change of pressure with respect to volumetric strain at constant temperature in the elastic region. We can also define a property called Poisson's ratio μ which is the ratio of strain in one direction to the stress in perpendicular direction.

$$B = E/3\,(1-2\mu) \text{ and } G = E/2\,(1+\mu)$$

The variation in the Young's modulus with temperature is shown for alloy steels in Figure 5.4.

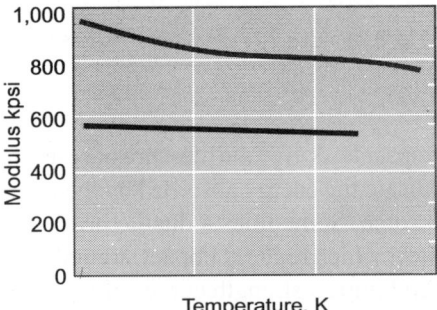

Fig. 5.4 Young's Modulus Curves for Alloy Steels

As the temperature is decreased, intermolecular forces tend to increase because of reductions in molecular vibrations. The Young's modulus which depends on the intermolecular forces thus increases at cryogenic temperatures and reduces at ambient temperatures. The similar trend is shown by other moduli as poison's ratio is unaffected at cryogenic temperatures.

Hardness and Ductility

The ductility of the material is indicated by the percentage elongation to failure in a simple tensile stress. Materials which elongate higher than 5% are called ductile where as those which elongate lower than 5% are called brittle materials.

The variation in the elongation with temperature is shown for some materials in Figure 5.5.

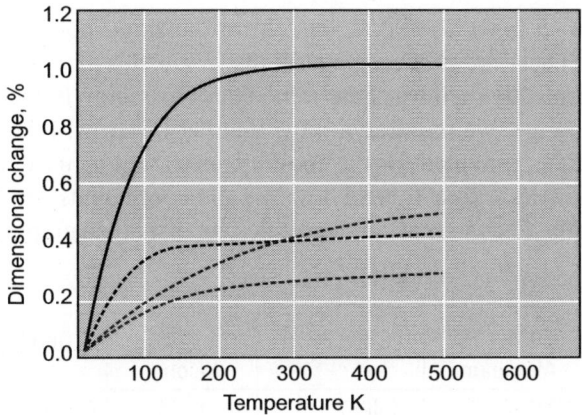

Fig. 5.5 Elongation Curves for Different Materials

The ductility of materials is lowered at low temperatures for materials which undergo ductile to brittle transition. However for other materials which do not undergo this transition, the ductility is improved. Hardness of materials is measured by the indentation made on a surface of the material with a standard indenter. The hardness of materials can be measured by Brinell, Rockwell or Vickers tests and is found to increase as temperatures are lowered.

Thermal Conductivity

Thermal conductivity of a material is defined as the heat transfer rate per unit area divided by the temperature gradient causing the heat transfer. The variation in thermal conductivity in solid materials is caused due to heat conduction by electron motion, lattice vibrational energy transport and molecular motion. Using kinetic theory of gases the expression for thermal conductivity (k) is

$$k = 1/f\,(9\gamma - 5)\,\rho\,C_v\,\upsilon\,\lambda$$

where γ is the specific heat ratio, ρ is the material density, C_v = specific heat at constant volume, υ = average particle velocity, λ = mean free path of particle, f = factor.

Widely divergent values of thermal conductivity for the same material are often reported in the literature. For comparatively pure materials (like copper), the differences are due mainly to slight material differences that have large effects on transport properties, such as thermal conductivity, at cryogenic temperatures. At 10 K, the thermal conductivity of commercial oxygen free copper for two samples can be different by more then a factor of 20 while the same samples at room temperature would be within 4%. It is also not uncommon for some experimental results to have uncertainties as high as 50%. The general form of the equation for thermal conductivity, k, is

5.6 Cryogenics

$$\log(k) = a + b\log T + c(\log T)^2 + d(\log T)^3 + e(\log T)^4 + f(\log T)^5 + g(\log T)^6 + h(\log T)^7 + i(\log T)^8$$

where a, b, c, d, e, f, g, h, and i are the fitted coefficients, and T is the temperature. These are common logarithms. While this may seem like an excessive number of terms to use, it was determined that in order to fit the data over the large temperature range, we required a large number of terms. It should also be noted that all the digits provided for the coefficients should be used, any truncation can lead to significant errors. Table 5.1 shows the coefficients for a variety of metals and non-metals. The thermal conductivities are displayed graphically in Figure 5.6.

Table 5.1 Coefficients for Conductivity

Coeff.	6061 – T6 Aluminum	304 SS	718 Inconel	Beryllium copper	Ti–6 A1–4V
a	0.07918	−1.4087	−8.28921	0.50015	−5107.8774
b	1.09570	1.3982	39.4470	1.93190	19240.422
c	−0.07277	0.2543	−83.4353	−1.69540	−30789.064
d	0.08084	−0.6260	98.1690	0.71218	27134.756
e	0.02803	0.2334	−67.2088	1.27880	−14226.379
f	−0.09464	0.4256	26.7082	−1.61450	4438.2154
g	0.04179	−0.4658	−5.72050	0.68722	−763.07767
h	−0.00571	0.1650	0.51115	−0.10501	55.796592
i	0	0	−0.0199	0	0
Data range	4–300 K	4.300 K	4–300 K	4–300 K	20–300 K

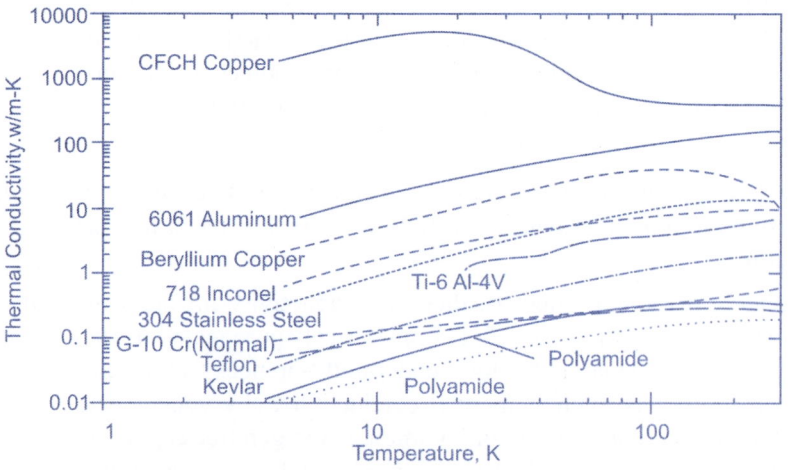

Fig. 5.6 Thermal Conductivity for Cryogenic Materials

In cases of gases, thermal conductivity decreases as temperature is lowered. This is because for a gas, specific heat is not a strong function of temperature and product of density and mean free path is constant. Thus reduced temperature lowers the mean molecular speed which in turn lowers thermal conductivity. Cryogenic liquids on the other hand show an increase in thermal conductivity with lowering of temperatures with the exception of liquid Hydrogen and liquid Helium. As far as alloys are concerned, the thermal conductivity decreases as temperature increases due to scattering of energy carriers and dislocations in the material.

Specific Heat of Solids

The specific heat is the amount of heat energy per unit mass required to cause a unit increase in the temperature of a material, the ratio of the change in energy to the change in temperature. Specific heats are strong functions of temperature, especially below 200 K. Models for specific heat began in the 1871 with Boltzmann and were further refined by Einstein and Debye in the early part of the 20th century. While there are many variations of these first models, they generally only provide accurate results for materials with perfect crystal lattice structures. The specific heat of many of the engineering materials of interest here is not described well by these simple models. Figure 5.7 graphically shows the specific heat curves for cryogenic materials.

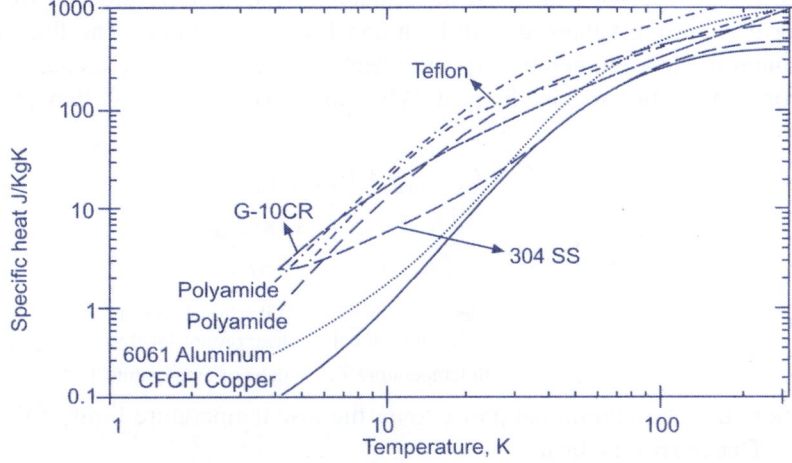

Fig. 5.7 Specific Heat Curves for Cryogenic Materials

The specific heat of the substance is measured as the energy required to change the temperature of substance by one degree at constant pressure (C_p) or constant volume (C_v). For solids the difference between C_p and C_v is small. Specific heat is represented by Debye model which assumes the solid is continuous medium and vibrational waves are limited to number of vibrational degrees of freedom. Einstein's oscillator treatment of specific heat gave qualitative agreement with experiment and gave the correct high temperature limit (the Law of Dulong and Petit). The quantitative fit to experiment was improved by Debye's recognition that there was a maximum number of modes of vibration in a solid. He pictured the vibrations as standing wave modes in the crystal, similar to the electromagnetic modes in a

5.8 Cryogenics

cavity which successfully explained blackbody radiation. The density of states for these modes, which are called "phonons", is of the same form as the photon density of states in a cavity. To impose a finite limit on the number of modes in the solid, Debye used a maximum allowed phonon frequency now called the Debye frequency v_D. In the treatment of specific heat, we define a Debye temperature by

$$T_D = \frac{hv_D}{k}$$

For low temperatures, Debye's treatment led to a specific heat expression

$$C_V = \frac{12\pi^4}{5} N_A k \left(\frac{T}{T_D}\right)^3$$

Where N_A is the number of atoms and K is the thermal conductivity. The temperature expression T/T_D is known as the Debye function. The dependence upon the cube of the temperature agreed with experimental results for nonmetals, and for metals when the electron specific heat was taken into account. The measurement of the low temperature specific heat variation with temperature has led to tabulation of the Debye temperatures for a number of solid materials. The full expression for the Debye specific heat must be evaluated by numerical procedures. It has the correct limiting values at both high and low temperatures. The final step in explaining the low temperature specific heats of metals was the inclusion of the electron contribution to specific heat. When these were combined, they produced the expression

$$C = C_{electronic} + C_{vibrational}$$

$$C_{metal} = \frac{\pi^2 N_A k^2}{2E_F} T + \frac{12\pi^4 N_A k}{5T_D^3} T^3$$

Electronic specific heat proportional to temperature T

Vibrational specific heat proportional to cabe of temperature T

Note that the vibrational part is only the low temperature limit of the more general Debye specific heat.

Specific Heat of Liquids and Gases

For cryogenic liquids, specific heat at constant volume and constant pressure decrease as the temperature is lowered. Refer Figure 5.8 for the specific heat curve for Gases and liquids. Specific heat for a cryogenic fluid is given by $C_p - C_v = T$ $(\partial v/\partial t)(\partial p/\partial t)$. The coefficient of thermal expansion for liquids and gases is large, so the specific heat increases near critical point of liquids. The coefficient of thermal expansion is given by $\beta = 1/v\,(\partial v/\partial t)$

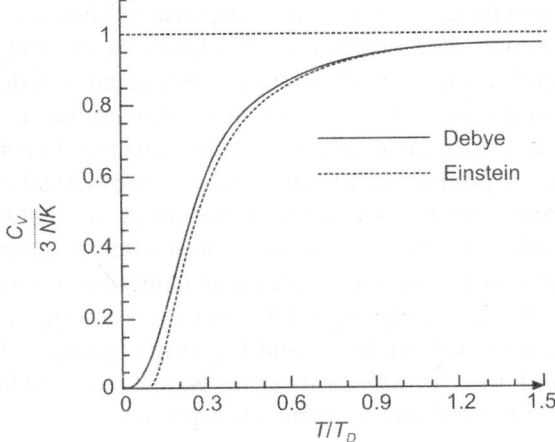

Fig. 5.8 Specific Heat Curves for Liquids and Gases

As far as gases is concerned, the molecule could be entirely described using classical mechanics, then the theorem of equipartition of energy could be used to predict that each degree of freedom would have an average energy in the amount of $(1/2)kT$ where k is Boltzmann's constant and T is the temperature. The calculation of the constant-volume heat content would be straightforward. Each molecule would be holding, on average, an energy of $(f/2)kT$ where f is the total number of degrees of freedom in the molecule. Note that $Nk = R$ if N is Avogadro's number, which is the case in considering the heat capacity of a mole of molecules. Thus, the total internal energy of the gas would be $(f/2)NkT$ where N is the total number of molecules. The heat capacity (at constant volume) would then be a constant $(f/2)Nk$. In summary, the molar heat capacity (mole-specific heat capacity) of an ideal gas with f degrees of freedom is given by

$$C_{V,m} = \frac{f}{2}R$$

This equation applies to all polyatomic gases, if the degrees of freedom are known.

Coefficient of Thermal Expansion

The volumetric coefficient of thermal expansion β is defined as fractional change in volume per unit change in temperature, while pressure remains constant. The linear coefficient of thermal expansion α is defined as fractional change in length per unit change in temperature, while pressure remains constant. For isotropic materials $\beta = 3\alpha$. From an atomic perspective, thermal expansion is caused by an increase in the average distance between the atoms. This results from the asymmetric curvature of the potential energy versus inter atomic distance. The anisotropy results from the differences in the coulomb attraction and the interatomic repulsive forces. Different metals and alloys with different heat treatments, grain sizes, or rolling directions

5.10 Cryogenics

introduce only small differences in thermal expansion. Thus, a generalization can be made that literature values for thermal expansion are probably good for a like material to within 5%. This is because the thermal expansion depends explicitly on the nature of the atomic bond, and only those changes that alter a large number of the bonds can affect its value. In general, large changes in composition (10 to 20%) are necessary to produce significant changes in the thermal expansion (~5%), and different heat treatments or conditions do not produce significant changes unless phase changes are involved. Most of the literature reports the integrated linear thermal expansion as a per cent change in length from some original length generally measured at 293 K and is given by the expression $(L_T - L_{293})/L_{293}$. Where L_T is the length at some temperature T and L_{293} is the length at 293 K. While this is a practical way of measuring thermal expansion, the more fundamental property is the coefficient of linear thermal expansion, α given by

$$\alpha(T) = \frac{1}{L}\frac{dL(T)}{dT}$$

The coefficient of linear thermal expansion is reported in the literature. Refer Figure 5.9 for the linear thermal expansion curves. One can simply take the derivative of the integrated linear thermal expansion that results in the coefficient of linear thermal expansion. The integrated linear thermal expansion can be reported as a change in length and the coefficient of linear thermal expansion can be provided.

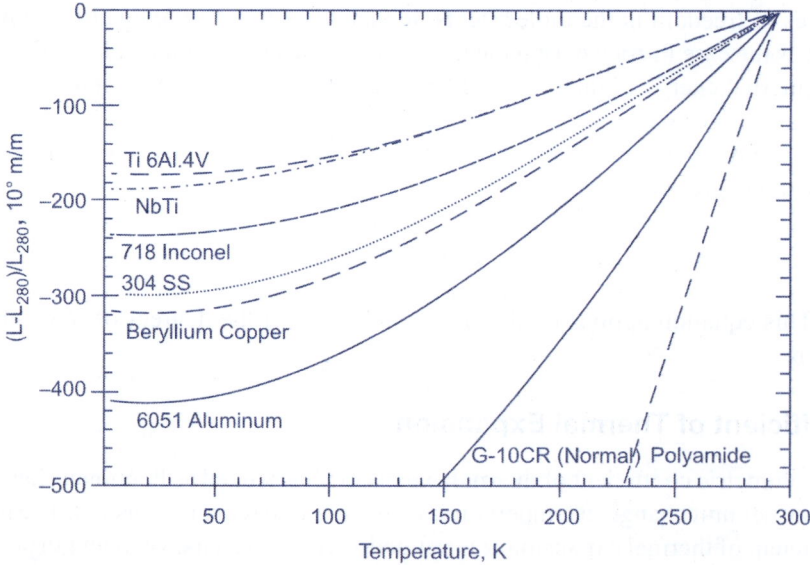

Fig. 5.9 Liner Coefficient of Thermal Expansion Curves

The temperature variation of the coefficient of thermal expansion is explained through intra molecular forces. As molecule acquires more energy by heating, its mean position relative to its neighboring molecules becomes larger resulting in expansion of material. The rate at which this spacing increases depends on the rate of temperature rise and hence the coefficient of linear expansion increases as temperatures increase and vice versa.

Electrical Conductivity

The electrical conductivity ke of a material is defined as electric current per unit cross sectional area divided by voltage gradient I the direction of current flow. The electrical resistivity is the reciprocal of electrical conductivity. Refer Figure 5.10 for the electrical resistivity curves.

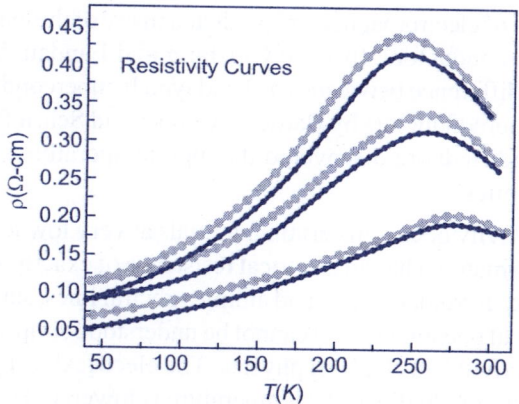

Fig. 5.10 Electrical Resistivity Curves

When an external field is applied to an electrical conductor, free electrons move in the direction of the field and this motion is opposed by the positive ions. Decreasing the temperature, lowers the resistance of positive ions thereby increasing electron motion and the electrical conductivity. The expression for electrical conductivity (ke) by Kittel was

$$ke = (N/V)\, e^2\, \lambda\, /\, mv$$

where N/V = number of electrons per unit volume, e = charge of electron, λ = electron mean free path, m = mass of electron, v = speed of the electron. The electrical conductivity is inversely proportional to the absolute temperature. The electron mean free path is given by

$$\lambda = 3/2\, \Pi\, (N/v)\, [h/2\, \Pi\, mvrX]$$

where N/V = number of electrons per unit volume, m = mass of electron, v = speed of the electron, r = ionic radius and X = amplitude of ionic vibrations. For semiconductors as the temperature is lowered in cryogenic range, effects of impurities become prominent and variation in temperature becomes complex and conductivity decreases.

Super Conductivity

Superconductivity was discovered in 1911 in a laboratory of Leiden University (Netherlands) by H. Kammerlingh Onnes. He started his carrier by building liquefiers and was the first to produce liquid helium. He later used liquid-helium cooling to investigate the electrical properties of metals at low temperatures. In 1911 it was observed that the resistance of a mercury wire, prepared by freezing a glass capillary tube filled up with liquid mercury purified by distillation, disappeared when the sample was cooled to a temperature slightly below 4.2 K. This discovery of superconductivity was followed by the observation of other metals which exhibit zero resistivity below a certain critical temperature. Superconductivity is a common phenomenon for pure metals and metallic alloys. In 1935 Fritz and Heinz London proposed an electromagnetic theory that, in conjunction with the classical Maxwell equations of electromagnetism, predicted many of the electric and magnetic properties of superconductors. In 1950 Ginzburg and Landau developed a theory that explained the difference between type I and type II superconductors. In 1957 the BCS theory for superconductors by Bardeen, Cooper and Schrieffer was developed. In 1986 Müller and Bednorz discovered the high temperature superconductors in copper oxide ceramics.

Superconductivity occurs in certain materials at very low temperatures. When superconductive, a material has an electrical resistance of exactly zero and no interior magnetic field. Like ferromagnetism and atomic spectral lines, superconductivity is a quantum mechanical phenomenon. It cannot be understood simply as the idealization of "perfect conductivity" in classical physics. The electrical resistivity of a metallic conductor decreases gradually as the temperature is lowered. However, in ordinary conductors such as copper and silver, this decrease is limited by impurities and other defects. Even near absolute zero, a real sample of copper shows some resistance. In a superconductor however, despite these imperfections, the resistance drops abruptly to zero when the material is cooled below its critical temperature. An electric current flowing in a loop of superconducting wire can persist indefinitely with no power source. Superconductivity occurs in many materials: simple elements like tin and aluminium, various metallic alloys and some heavily-doped semiconductors. Superconductivity does not occur in noble metals like gold and silver, nor in pure samples of ferromagnetic metals. In 1986, it was discovered that some ceramic materials have critical temperatures of more than 90 kelvin. These high-temperature superconductors renewed interest in the topic because the current theory could not explain them. From a practical perspective, 90 kelvin is easy to reach with the readily available liquid nitrogen (boiling point 77 kelvin). This means more experimentation and more commercial applications are feasible, especially if materials with even higher critical temperatures could be discovered.

Superconductors can be classified as 1) By their physical properties they can be Type I (if their phase transition is of first order) or Type II (if their phase transition is of second order), 2) By the theory of explanation: they can be conventional (if they are explained by the BCS theory) or unconventional, 3) By their critical temperature: they can be high temperature ($T_c > 77$ K), or low temperature (cooled under their critical temperature) and 4) By material: they can be chemical elements (as mercury), alloys (as niobium-titanium), ceramics (as magnesium diboride), or organic superconductors (as fullerenes).

In superconducting materials, the characteristics of superconductivity appear when the temperature T is lowered below a critical temperature T_c. The value of this critical temperature varies from material to material. Conventional superconductors usually have critical temperatures ranging from around 20 K to less than 1 K. Electron pairing due to phonon exchanges explains superconductivity in conventional superconductors, but it does not explain superconductivity in the newer superconductors that have a very high critical temperature. Similarly, at a fixed temperature below the critical temperature, superconducting materials cease to superconduct when an external magnetic field is applied which is greater than the critical magnetic field. This is because the Gibbs free energy of the superconducting phase increases quadratically with the magnetic field while the free energy of the normal phase is roughly independent of the magnetic field. If the material superconducts in the absence of a field, then the superconducting phase free energy is lower than that of the normal phase. More generally, a higher temperature and a stronger magnetic field lead to a smaller fraction of the electrons in the superconducting band and consequently a longer London penetration depth of external magnetic fields and currents. The penetration depth becomes infinite at the phase transition. Refer Figure 5.11 for the superconductor characteristics.

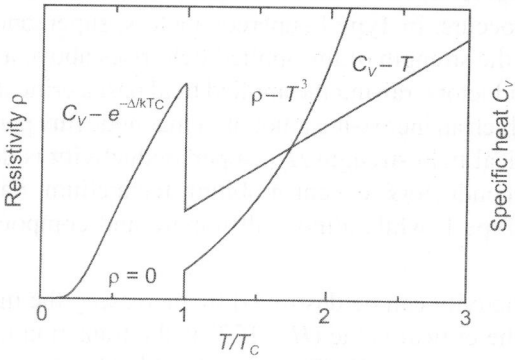

Fig. 5.11 Superconductor Characteristics

The onset of superconductivity is accompanied by abrupt changes in various physical properties, which is the hallmark of a phase transition. For example, the electronic heat capacity is proportional to the temperature in the normal (non-superconducting) regime. At the superconducting transition, it suffers a discontinuous jump and thereafter ceases to be linear. At low temperatures, it varies instead as $e^{-\alpha/T}$ for some constant α. This exponential behavior is one of the pieces of evidence for the existence of the energy gap. The order of the superconducting phase transition was long a matter of debate. Experiments indicate that the transition is second-order, meaning there is no latent heat. However in the presence of an external magnetic field there is latent heat, as a result of the fact that the superconducting phase has a lower entropy below the critical temperature than the normal phase. It has been experimentally demonstrated that, as a consequence, when the magnetic field is increased beyond the critical field, the resulting phase transition leads to a decrease in the temperature of the superconducting material.

When a superconductor is placed in a weak external magnetic field H, the field penetrates the superconductor only a small distance λ, called the London penetration depth, decaying exponentially to zero within the bulk of the material. This is called the Meissner effect, and is a defining characteristic of superconductivity. For most superconductors, the London penetration depth is on the order of 100 nm. The Meissner effect is distinct because a superconductor expels all magnetic fields, not just those that are changing. Suppose we have a material in its normal state, containing a constant internal magnetic field. When the material is cooled below the critical temperature, we would observe the abrupt expulsion of the internal magnetic field. The Meissner effect was explained by the brothers Fritz and Heinz London, who showed that the electromagnetic free energy in a superconductor is minimized provided

$$\Delta^2 H = \lambda^{-2} H$$

where H is the magnetic field and λ is the London penetration depth. This equation, which is known as the London equation, predicts that the magnetic field in a superconductor decays exponentially from whatever value it possesses at the surface. The Meissner effect breaks down when the applied magnetic field is too large. Superconductors can be divided into two classes according to how this breakdown occurs. In Type I superconductors, superconductivity is abruptly destroyed when the strength of the applied field rises above a critical value H_c. In Type II superconductors, raising the applied field past a critical value H_{c1} leads to a mixed state in which an increasing amount of magnetic flux penetrates the material. At a second critical field strength H_{c2}, superconductivity is destroyed. Most pure elemental superconductors, except niobium, technetium, vanadium and carbon nanotubes, are Type I, while almost all impure and compound superconductors are Type II.

Superconductivity can be destroyed by increasing the magnetic field around the material to the critical value (H_c). If T_o is the transition temperature, then the magnetic field has the value H_o. The critical field value can be defined as

$$H_c = H_o [1 - (T/T_o)^2]$$

For type II conductors the lower critical field is the one for which transition begins (H_{c1}) and upper critical field is one where the transition ends (H_{c2}). These can be defined as

$$H_{c1} = H_c(0.081 + \ln \acute{k}) / \acute{k} \sqrt{2} \text{ and } H_{c2} = \sqrt{2} \acute{k} H_c$$

Where $\acute{k}$ is the surface energy parameter.

Theories of Superconductivity

Since discovery of superconductivity, a great efforts have been devoted to finding out how and why it works. During the 1950s, theoretical condensed matter physicists arrived at a solid understanding of "conventional" superconductivity, through a pair of remarkable and important theories: the phenomenological Ginzburg-Landau theory (1950) and the microscopic BCS theory (1957). Generalizations of these theories form the basis for understanding the closely related phenomenon of superfluidity. The three prominent theories of superconductivity are:

1. **Bardeen-Cooper-Schrieffer (BCS) Theory:** This theory derived several important theoretical predictions that are independent of the details of the interaction, since the quantitative predictions hold for many low temperature superconductors. These have been confirmed in numerous experiments. The electrons are bound into Cooper pairs, and these pairs are correlated due to the Pauli exclusion principle for the electrons, from which they are constructed. Therefore, in order to break a pair, one has to change energies of all other pairs. This means there is an "energy gap" for "single-particle excitation", unlike in the normal metal (where the state of an electron can be changed by adding an arbitrarily small amount of energy). This energy gap is highest at low temperatures but vanishes at the transition temperature when superconductivity ceases to exist. The BCS theory gives an expression that shows how the gap grows with the strength of the attractive interaction and the (normal phase) single particle density of states at the Fermi energy. Furthermore, it describes how the density of states is changed on entering the superconducting state, where there are no electronic states any more at the Fermi energy. The energy gap is most directly observed in tunneling experiments and in reflection of microwaves from the superconductor. BCS theory predicts the dependence of the value of the energy gap E at temperature T on the critical temperature T_c. The ratio between the value of the energy gap at zero temperature and the value of the superconducting transition temperature (expressed in energy units) takes the universal value of 3.5, independent of material. Near the critical temperature the relation becomes

$$E = 352\, K_B T_c \sqrt{1-(T/T_C)}\,.$$

Due to the energy gap, the specific heat of the superconductor is suppressed strongly (exponentially) at low temperatures, there being no thermal excitations left. However, before reaching the transition temperature, the specific heat of the superconductor becomes even higher than that of the normal conductor (measured immediately above the transition) and the ratio of these two values is found to be universally given by 2.5. BCS theory correctly predicts the Meissner effect, i.e. the expulsion of a magnetic field from the superconductor and the variation of the penetration depth (the extent of the screening currents flowing below the metal's surface) with temperature. It also describes the variation of the critical magnetic field (above which the superconductor can no longer expel the field but becomes normal conducting) with temperature. BCS theory relates the value of the critical field at zero temperature to the value of the transition temperature and the density of states at the Fermi energy. In its simplest form, BCS gives the superconducting transition temperature in terms of the electron-phonon coupling potential and the Debye cutoff energy:

$$k_B T_c = 1.14\, E_D\, e^{-1/N(0)V}.$$

The BCS theory reproduces the isotope effect, which is the experimental observation that for a given superconducting material, the critical temperature is inversely proportional to the mass of the isotope used in the material.

2. **Ginzburg Landau Theory**: The Ginzburg-Landau theory, named after Vitaly Lazarevich Ginzburg and Lev Landau, is a mathematical theory used to model superconductivity. It does not purport to explain the microscopic mechanisms giving rise to superconductivity. Instead, it examines the macroscopic properties of a superconductor with the aid of general thermodynamic arguments. This theory is sometimes called phenomenological as it describes some of the phenomena of superconductivity without explaining the underlying microscopic mechanism.

Based on Landau's previously-established theory of second-order phase transitions, Landau and Ginzburg argued that the free energy F of a superconductor near the superconducting transition can be expressed in terms of a complex order parameter ψ, which describes how deep into the superconducting phase the system is. The free energy has the form

$$F = F_n + \alpha |\psi|^2 + \frac{\beta}{2}|\psi|^4 + \frac{1}{2m}|(-i\hbar\nabla - 2eA)\psi|^2 + \frac{|B|^2}{2\mu_0}$$

where F_n is the free energy in the normal phase, α and β are phenomenological parameters, m is an effective mass, A is the electromagnetic vector potential, and B is the magnetic induction.

$$B = \nabla \times A$$

By minimizing the free energy with respect to fluctuations in the order parameter and the vector potential, one arrives at the Ginzburg-Landau equations

$$\alpha\psi + \beta|\psi|^2\psi + \frac{1}{2m}(-i\hbar\nabla - 2eA)^2\psi = 0$$

$$j = \frac{2e}{m}Re\{\psi^*(-i\hbar\nabla - 2eA)\psi\}$$

where j denotes the electrical current density and Re the real part. The first equation, which bears interesting similarities to the time-independent Schrödinger equation, determines the order parameter ψ based on the applied magnetic field. The second equation then provides the superconducting current. The Ginzburg-Landau equations produce many interesting and valid results. Perhaps the most important of these is its prediction of the existence of two characteristic lengths in a superconductor. The first is a coherence length ξ, given by

$$\xi = \sqrt{\frac{\hbar^2}{2m|\alpha|}}$$

which describes the size of thermodynamic fluctuations in the superconducting phase. The second is the penetration depth λ, given by

$$\lambda = \sqrt{\frac{m}{4m_0 e^2 \psi_0^2}}$$

where ψ_0 is the equilibrium value of the order parameter in the absence of an electromagnetic field. The penetration depth describes the depth to which an external magnetic field can penetrate the superconductor. The ratio $\kappa = \lambda/\xi$ is known as the Ginzburg-Landau parameter. It has been shown that Type I superconductors are those with $0 < \kappa < 1/\sqrt{2}$, and Type II superconductors those with $\kappa > 1/\sqrt{2}$.

3. **Heinz and Fritz London theory:** Is a phenomenological theory of superconductivity. Heinz and Fritz added two new equations to the Maxwell equations:

$$\overline{E} = d(A\overline{J})/dt \text{ and } \overline{B} = -rot(A\overline{J}) \text{ with } A = m/(n_S e^2)$$

where m the electron mass, e the electron charge and n_S the super electrons density. The most important consequence of this theory is the prediction of the superficial currents. The London equation explains the Meissner effect, showing that the magnetic field decays exponentially inside the superconductor over a distance of 20–40 nm. It is described in terms of a parameter called the London penetration depth λ.

$$\lambda = \sqrt{A/\mu_0}$$

where μ_0, is the magnetic permeability of the air. However, this theory cannot explain the type II superconductors.

Applications of Superconductivity

Superconducting magnets are some of the most powerful electromagnets known. They are used in MRI and NMR machines, mass spectrometers, and the beam-steering magnets used in particle accelerators. They can also be used for magnetic separation, where weakly magnetic particles are extracted from a background of less or non-magnetic particles, as in the pigment industries. In the 1950s and 1960s, superconductors were used to build experimental digital computers using cryotron switches. More recently, superconductors have been used to make digital circuits based on rapid single flux quantum technology and RF and microwave filters for mobile phone base stations. Superconductors are used to build Josephson junctions which are the building blocks of SQUIDs (superconducting quantum interference devices), the most sensitive magnetometers known. SQUIDs are used in scanning SQUID microscopes. Series of Josephson devices are used to define the SI volt. Depending on the particular mode of operation, a Josephson junction can be used as a photon detector or as a mixer. The large resistance change at the transition from the normal to the superconducting state is used to build thermometers in cryogenic micro-calorimeter photon detectors. Other early markets are arising where the relative efficiency, size and weight advantages of devices based on high-temperature superconductivity outweigh the additional costs involved. Promising future applications include high-performance smart grid, electric power transmission, transformers, power storage devices, electric motors (e.g. for vehicle propulsion, as in maglev trains), fault current limiters, nanoscopic materials such as buckyballs, nanotubes, composite materials and superconducting magnetic refrigeration.

References

1. Cryogenic Systems – Randall Baron.
2. Internet Website www.wikipedia.com on superconductivity.
3. Cryogenic Engineering – B.A. Hands.
4. Solvay Advance Polymers – Amodal Guide for Materials.
5. INL report on tensile strength of SS 316 (2007).
6. Dupont Properties Handbook for Tefzel.
7. Internet website http://hyperphysics.phy-astr.gsu.edu.
8. Das et.al. Solid State Communications v 134 (2005) pp 837–842.

Questions

Q.1. Enlist at least five cryogenic material properties.

Q.2. Describe any two mechanical properties of cryogenic materials.

Q.3. Discuss specific heat properties of solids, liquids and gases at cryogenic temperatures.

Q.4. What is superconductivity? What are its applications?

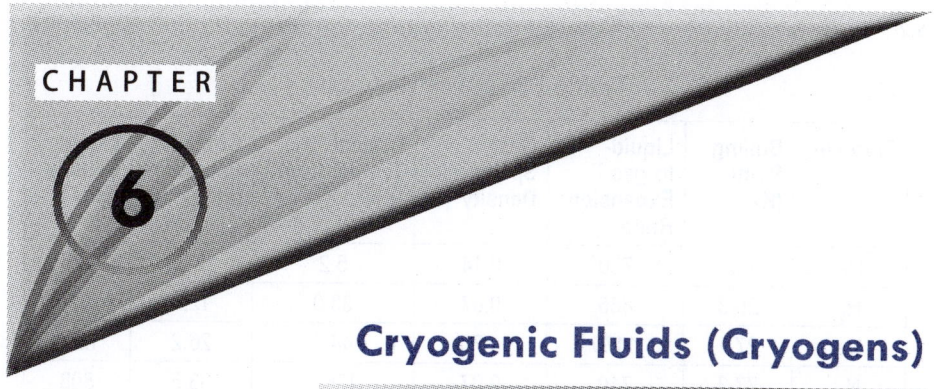

CHAPTER 6

Cryogenic Fluids (Cryogens)

INTRODUCTION

Cryogenic fluid is defined as a fluid having a boiling point lower than –89.9°C at an absolute pressure of 101.3 kPa or a chemical that induces freezing. These fluids are also known as cryogens and are stored in cryogenic containers. The cryogenic vessels are low-pressure tanks, which are designed to contain a cryogenic fluid and maintain the operating pressure and contents within liquid phase by venting, insulation or refrigeration. Today, liquefied natural gas (LNG) is the largest used cryogen in the world. Cryogens like liquid oxygen, liquid nitrogen, helium 3, helium 4, and liquid hydrogen are also being consumed in large quantities and their demand is increasing. Safety is critical when handling cryogens. It is important that cryogenic fluids which could give rise to oxygen deficient atmospheres are stored in adequately ventilated rooms. In determining whether ventilation is adequate, consideration should be given to the normal evaporation of the cryogenic fluids in the room and filling losses from filling vessels in warm conditions as well as spillage due loss of vacuum insulation or rupture of a bursting disc. Cryogens expand into large volumes of gas that can displace air. For example, 1 L of liquid nitrogen forms nearly 700 L of nitrogen gas at room temperature.

Properties of Cryogens

The important properties of some leading cryogens are given in Table 6.1. Cryogens have several specific and often common properties. These include:

1. They are extremely cold-most cryogens boil below 200 K.
2. They have large liquid-to-gas expansion ratios (> 700 for most cryogens) – a small liquid spill produces a very large volume of gas that can displace air in a confined space.
3. Some cryogens are chemically reactive (e.g. F_2 and O_2).
4. Some cryogens are flammable (e.g. H_2 and CH_4).

The hazards associated with handling cryogenic liquids can be minimized by understanding their properties in detail.

6.2 Cryogenics

Table 6.1 Properties of Cryogens

Cryogen	Boiling Point (K)	Liquid-to-gas Expansion Ratio	Gas Specific Density	Critical Temperature (K)	Critical Pressure (atm)	Liquid Density (g/l)
He	4.2	780	0.14	5.2	2.2	125
H_2	20.3	865	0.07	33.0	12.8	71
Ne	27.1	1470	0.70	44.4	26.2	1206
N_2	77.3	710	0.97	126.3	33.5	808
Ar	87.3	860	1.39	150.9	48.3	1402
O_2	90.2	875	1.11	154.8	50.1	1410
CO_2	194.7	790	1.70	304.2	72.8	1560

CRYOGEN 1: LIQUID NITROGEN (LN_2)

Liquid nitrogen is inert, colorless, odorless, non-corrosive, nonflammable, and extremely cold. Nitrogen makes up the major portion of the atmosphere (78.03% by volume). Nitrogen is inert and will not support combustion; however, it is not life supporting. Nitrogen is inert except when heated to very high temperatures where it combines with some of the more active metals, such as lithium and magnesium, to form nitrides. It will also combine with oxygen to form oxides of nitrogen and, when combined with hydrogen in the presence of catalysts, will form ammonia. Liquid Nitrogen has specific gravity of about 0.8. Its temperature at ambient pressure is about 77 K. The latent heat of vaporization is about 160 kJ/L, so a heat input of 1 watt will boil off about 23 cc of liquid per hour. Although nitrogen is nontoxic and inert, it can act as a simple asphyxiant by displacing the oxygen in air to levels below that required to support life. Inhalation of nitrogen in excessive amounts can cause dizziness, nausea, vomiting, loss of consciousness, and death. Death may result from errors in judgment, confusion, or loss of consciousness that prevents self-rescue.

Properties of liquid nitrogen are:
- Molecular Weight: 28.01
- Boiling Point @ 1 atm: –320.5°F (–195.8°C, 77°K)
- Freezing Point @ 1 atm: –346.0°F (–210.0°C, 63°K)
- Critical Temperature: –232.5°F (–146.9°C)
- Critical Pressure: 492.3 psia (33.5 atm)
- Density, Liquid @ BP, 1 atm: 50.45 lb/scf
- Density, Gas @ 68°F (20°C), 1 atm: 0.0725 lb/scf
- Specific Gravity, Gas (air = 1) @ 68°F (20°C), 1 atm: 0.967
- Specific Gravity, Liquid (water = 1) @ 68°F (20°C), 1 atm: 0.808

- Specific Volume @ 68°F (20°C), 1 atm: 13.80 scf/lb
- Latent Heat of Vaporization: 2399 BTU/lb mole
- Expansion Ratio, Liquid to Gas, BP to 68°F (20°C): 1 to 694.

For efficient storage and handling, LN_2 is kept in vacuum-insulated dewars of various sizes. Fig. 6.1 shows a typical cryogenic liquid nitrogen cylinder. Cryogenic liquid cylinders are insulated, vacuum-jacketed pressure vessels. They come equipped with safety relief valves and rupture discs to protect the cylinders from pressure build-up. These containers operate at pressures up to 350 psig and have capacities between 80 and 450 liters of liquid. Product may be withdrawn as a gas by passing liquid through an internal vaporizer or as a liquid under its own vapor pressure. Cryogenic containers are equipped with pressure relief devices to control internal pressure. Under normal conditions, these containers will periodically vent product. Containers should be handled and stored in an upright position.

The LN_2 dewar being filled must be raised so the tube extends well into it so the liquid does not simply blow back out the top. Once this is done, the transfer assembly is attached to the top of the small dewar. Evaporation of LN_2 will slowly build up sufficient pressure (3-6 psi) to force the liquid through the flexible line when the valve is opened. Alternatively, a source of pressurized nitrogen gas can be attached to the small vent valve on the transfer assembly to force the liquid out of the dewar. This is illustrated in Figure 6.1.

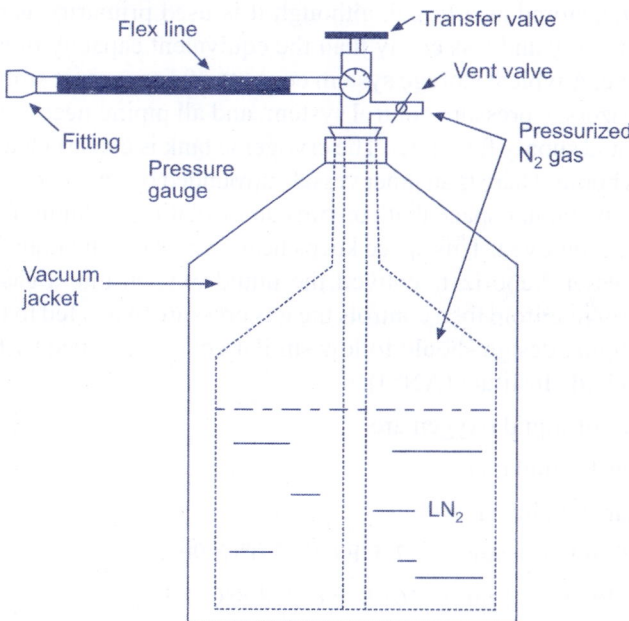

Fig. 6.1 Liquid Nitrogen Dewar Vessel

Because of its temperature and latent heat, LN_2 can cause severe frostbite when it contacts the skin for more than a second or so. In addition, cold transfer lines can cause instantaneous frostbite, so gloves must be worn when handling any cold metallic lines. In general, one should exercise care in working with LN_2, especially when filling storage or instrument dewars, where it is pressurized.

CRYOGEN 2: LIQUID OXYGEN (LO_2)

Oxygen is the second largest component of the atmosphere, comprising 20.8% by volume. Liquid oxygen is pale blue and extremely cold. Refer Figure 6.2 for a photograph of liquid oxygen. Although nonflammable, oxygen is a strong oxidizer. Oxygen is necessary to support life. Oxygen will react with nearly all organic materials and metals usually forming an oxide. Materials that burn in air will burn more vigorously in oxygen. Equipment used in oxygen service must meet stringent cleaning requirements and systems must be constructed of materials that have high ignition temperatures and that are non-reactive with oxygen under the service conditions. Vessels should be manufactured to American Society of Mechanical Engineers (ASME) codes and designed to withstand the process temperatures and pressures. Liquid oxygen is a cryogenic liquid. Cryogenic liquids are liquefied gases that have a normal boiling point below –238°F (–150°C). Liquid oxygen has a boiling point of –297.3°F (–183.0°C). Because the temperature difference between the product and the surrounding environment is substantial–even in the winter–keeping liquid oxygen insulated from the surrounding heat is essential. Oxygen is often stored as a liquid, although it is used primarily as a gas. Liquid storage is less bulky and less costly than the equivalent capacity of high-pressure gaseous storage. A typical storage system consists of a cryogenic storage tank, one or more vaporizers, a pressure control system, and all piping necessary for the fill, vaporization, and supply functions. The cryogenic tank is constructed, in principle, like a thermos bottle. There is an inner vessel surrounded by an outer vessel. Between the vessels is an annular space that contains an insulating medium, from which all the air has been removed. This space keeps heat away from the liquid oxygen held in the inner vessel. Vaporizers convert the liquid oxygen into a gaseous state. A pressure control manifold then controls the gas pressure that is fed to the process or application. Piping design should follow similar codes, as issued by the American National Standards Institute (ANSI).

Properties of liquid oxygen are:
- Molecular Formula: O_2
- Molecular Weight: 31.999
- Boiling Point @ 1 atm: –297.4°F (–183.0°C, 90°K)
- Freezing Point @ 1 atm: –361.9°F (–218.8°C, 54°K)
- Critical Temperature: –181.8°F (–118.4°C)
- Critical Pressure: 729.1 psia (49.6 atm)
- Density, Liquid @ BP, 1 atm: 71.23 lb/scf
- Density, Gas @ 68°F (20°C), 1 atm: 0.0831 lb/scf

- Specific Gravity, Gas (air = 1) @ 68°F (20°C), 1 atm: 1.11
- Specific Gravity, Liquid (water = 1) @ 68°F (20°C), 1 atm: 1.14
- Specific Volume @ 68°F (20°C), 1 atm: 12.08 scf/lb
- Latent Heat of Vaporization: 2934 BTU/lb mole
- Expansion Ratio, Liquid to Gas, BP to 68°F (20°C): 1 to 860
    Solubility in Water @ 77°F (25°C), 1 atm: 3.16% by volume.

The hazards associated with liquid oxygen are exposure to cold temperatures that can cause severe burns; over-pressurization due to expansion of small amounts of liquid into large volumes of gas in inadequately vented equipment; oxygen enrichment of the surrounding atmosphere; and the possibility of a combustion reaction if the oxygen is permitted to contact a non-compatible material. The low temperature of liquid oxygen and the vapors it releases not only pose a serious burn hazard to human tissue, but can also cause many materials of construction to lose their strength and become brittle enough to shatter. The large expansion ratio of liquid-to-gas can rapidly build pressure in systems where liquid can be trapped. This necessitates that these areas be identified and protected with pressure relief. This expansion ratio also allows atmospheres of oxygen-enriched air to form in the area surrounding a release. It is important to note that fire chemistry starts to change when the concentration of oxygen increases to as little as 23%. Materials easily ignited in air not only become more susceptible to ignition, but also burn with added violence in the presence of oxygen. These materials include clothing and hair, which have air spaces that readily trap the oxygen. Oxygen levels of 23% can be reached very quickly and all personnel must be aware of the hazard. Any clothing that has been splashed or soaked with liquid oxygen or exposed to high oxygen concentrations should be removed immediately and aired for at least an hour. Clothing saturated with oxygen is readily ignitable and will burn vigorously.

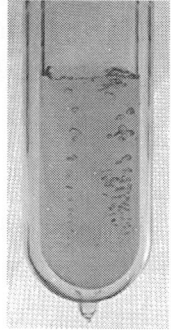

Fig. 6.2 Liquid Oxygen

Liquid oxygen has a pale blue color and is strongly paramagnetic and can be suspended between the poles of a powerful horse shoe magnet. In commerce, liquid oxygen is classified as an industrial gas and is widely used for industrial and medical purposes. Liquid oxygen is obtained from the oxygen found naturally in air by fractional distillation. Liquid oxygen has an expansion ratio of 861:1 at 68 °F

(20°C); and because of this, it is used in some commercial and military aircraft as a source of breathing oxygen. LOX used in space rockets (and probably in aerospace) is a mixture of liquid oxygen with up to 25% liquid ozone and several additives to stabilize this liquid oxidator.

CRYOGEN 3: LIQUID HYDROGEN (LH$_2$)

Liquid hydrogen is the liquid state of the element hydrogen. To exist as a liquid, H$_2$ must be pressurized and cooled to a very low temperature, −252.87°C. Liquid hydrogen is typically used as a concentrated form of hydrogen storage. Once liquefied it can be maintained as a liquid in pressurized and thermally insulated containers. Liquid hydrogen consists of 99.79% parahydrogen, 0.21% orthohydrogen. Hydrogen is colorless as a liquid. Its vapors are colorless, odorless, tasteless, and highly flammable. Liquid hydrogen is non-corrosive. Special materials of construction are not required to prevent corrosion. However, because of its extremely cold temperature, equipment must be designed and manufactured of material, which is suitable for extremely low temperature operation. Vessels and piping should be designed to the American Society of Mechanical Engineers (ASME) Code and the American National Standards Institute (ANSI) Pressure Piping Code or Department of Transportation (DOT) Codes for the pressure and temperatures involved. Hydrogen gas is odorless and nontoxic but may produce suffocation by diluting the concentration of oxygen in air below levels necessary to support life. The amount of hydrogen gas necessary to produce an oxygen-deficient atmosphere is well within the flammable range, making fire and explosion the primary hazards associated with hydrogen and air atmospheres. The wide flammability range, 4% to 74% in air, and the small amount of energy required for ignition necessitate special handling to prevent the inadvertent mixing of hydrogen with air. Care should be taken to eliminate sources of ignition, such as sparks from electrical equipment, static electricity sparks, open flames or any extremely hot objects. Hydrogen and air mixtures within the flammable range can explode and may burn with a pale blue, almost invisible flame.

Properties of liquid hydrogen are:
- Molecular Weight: 2.016
- Boiling Point @ 1 atm: −423.0°F (−252.8°C, 20°K)
- Freezing Point @ 1 atm: −434.5°F (−259.2°C, 14°K)
- Critical Temperature: −399.8°F (−239.9°C)
- Critical Pressure: 188 psia (12.9 atm)
- Density, Liquid @ B.P., 1 atm: 4.23 lb./cu.ft.
- Density, Gas @ 68°F (20°C), 1 atm: 0.005229 lb./cu.ft.
- Specific Gravity, Gas (Air = 1) @ 68°F (20°C), 1 atm: 0.0696
- Specific Gravity, Liquid @ B.P., 1 atm: 0.0710

- Specific Volume @ 68°F (20°C), 1 atm: 192 cu. ft./lb.
- Latent Heat of Vaporization: 389 Btu/lb. mole
- Flammable Limits @ 1 atm in air 4.00%: –74.2% (by Volume)
- Detonable Limits @ 1 atm in air 18.2%: –58.9% (by Volume)
- Autoignition Temperature @ 1 atm: 1060°F (571°C)
- Expansion Ratio, Liquid to Gas, B.P. to 68°F (20°C): 1 to 848.

Liquid hydrogen is normally vaporized and used as a gas. Withdrawal of liquid from a tanker, tank, or liquid cylinder requires the use of a closed system, with proper safety relief devices, which can be evacuated and/or purged to eliminate the possibility of creating a flammable atmosphere or explosive mixture of liquid air and liquid hydrogen. Purging should be done with helium since liquid hydrogen can solidify other gases, such as nitrogen, and cause plugging and possible rupture of the transfer line or storage vessel. Liquid transfer lines must be vacuum insulated to minimize product loss through vaporization or the formation of liquid air on the lines with subsequent oxygen enrichment. All equipment must be electrically grounded and bonded before transferring liquid. The hazards associated with handling liquid hydrogen are fire, explosion, asphyxiation, and exposure to extremely low temperatures. The potential for forming and igniting flammable mixtures containing hydrogen may be higher than for other flammable gases because hydrogen migrates quickly through small openings.

The minimum ignition energy for flammable mixtures containing hydrogen is extremely low. Burns may result from unknowingly walking into a hydrogen fire. The fire and explosion hazards can be controlled by appropriate design and operating procedures. Preventing the formation of combustible fuel-oxidant mixtures and removing or otherwise inerting potential sources of ignition (electric spark, static electricity, open flames, etc.) in areas where the hydrogen will be used is essential. Careful evacuation and purge operations should be used to prevent the formation of flammable or explosive mixtures. Adequate ventilation will help reduce the possible formation of flammable mixtures in the event of a hydrogen leak or spill and will also eliminate the potential hazard of asphyxiation. Protective clothing should be worn to prevent exposure to extremely cold liquid and cold hydrogen vapors. Cold burns may occur from short contact with frosted lines, liquid air that may be dripping from cold lines or vent stacks, vaporizer fins, and vapor leaks. Gaseous and liquid hydrogen systems must be purged of air, oxygen, or other oxidizers prior to admitting hydrogen to the systems, and purged of hydrogen before opening the system to the atmosphere. Purging should be done to prevent the formation of flammable mixtures and can be accomplished in several ways. Piping systems and vessels intended for gaseous hydrogen service should be inerted by suitable purging or evacuation procedures.

Heat, open flames, electrical sparks, and static electricity easily ignite hydrogen. It will burn with a pale blue, almost invisible flame. Most hydrogen fires will have the flame characteristic of a torch or jet and will originate at the point where the hydrogen is discharging. If a leak is suspected in any part of a system, a hydrogen flame can be detected by cautiously approaching with an outstretched broom, lifting it up and down. The most effective way to fight a hydrogen fire is to shutoff the flow of gas. If it is necessary to extinguish the flame in order to get to a place where the flow of hydrogen can be shut off, a dry powder extinguisher is recommended. However, if the fire is extinguished without stopping the flow of gas, an explosive mixture may form, creating a more serious hazard than the fire itself should re-ignition occur from the hot surfaces or other sources. The usual fire fighting practice is to prevent the fire from spreading and let it burn until the hydrogen is consumed. Dry powder fire extinguishers should be available in the area.

Hydrogen molecules exist in two forms, Para and Ortho, depending on the electron configurations. At hydrogen's boiling point of 20 K, the equilibrium concentration is almost all Para-hydrogen, however at room temperature or higher the equilibrium concentration is 25% Para-hydrogen and 75% Ortho-hydrogen. Ortho to Para-hydrogen conversion releases a significant amount of heat (527 kJ/kg). Room temperature hydrogen consists mostly of the orthohydrogen form. After production, liquid hydrogen is in a metastable state and must be converted into the parahydrogen isomer form to avoid the exothermic reaction that occurs when it changes at low temperatures, this is usually performed using a catalyst like ferric oxide, activated corbon, platinized asbestos, rare earth metals, uranium compounds, chromic oxide, or some nickel compounds. An intrinsic property of the hydrogen molecule is the distinction between ortho and para species. The difference arises from the requirement for the total molecular wave function to be anti-symmetric under exchange of the protons. Ortho-H_2 molecules have nuclear spin $I = 1$ and possess only anti-symmetric rotational states with odd rotational quantum number J, while J is even for para-molecules, whose nuclear spin I is zero. The proportion of ortho-H_2 in the gas at thermodynamic equilibrium at any temperature can be easily calculated from the rotational energy eigenvalues. It amounts to about 75% at room temperature, and decreases rapidly when lowering the temperature below 140 K, to become zero at $T = 0$. After a temperature variation of the sample, ortho-para conversion starts, until thermodynamic equilibrium is reached again.

The process of the intrinsic conversion of an ortho-H_2 molecule into a para one is induced by the presence of another ortho molecule among its nearest neighbors, which can provide the necessary inhomogeneous magnetic field. The speed of conversion is then proportional to the square of the average ortho concentration $x(t)$, and the equation describing its kinetic, in the event that the temperature is not low enough to neglect the equilibrium ortho fraction x_e,

$$\frac{dx}{dt} = -K\frac{x(x-x_e)}{1-x_e}.$$

The solution of this differential equation is

$$x(t) = x_e\left[1 - \frac{x(0)-x_e}{x(0)}\exp\left[-\frac{x_e}{1-x_e}Kt\right]\right]^{-1}$$

which reduces to the simpler linear relation of $1/x$ vs. time in the limit $x_e \to 0$, K being the usual conversion rate constant.

CRYOGEN 4: LIQUID ARGON (LAr)

Liquid argon is tasteless, colorless, odorless, non-corrosive, nonflammable, and extremely cold. Belonging to the family of rare inert gases, argon is the most plentiful of the rare gases, making up approximately 1% of the earth's atmosphere. It is monatomic and extremely inert, forming no known chemical compounds. Materials of construction must be selected to withstand the low temperature of liquid argon. Vessels and piping should be designed to American Society of Mechanical Engineers (ASME) specifications or the Department of Transportation (DOT) codes for the pressures and temperatures involved. Although used more commonly in the gaseous state, argon is commonly stored and transported as a liquid, affording a more cost effective way of providing product supply. When argon is converted to liquid form it becomes a cryogenic liquid. Liquid argon has a boiling point of –302.6°F (–185.9°C). The temperature difference between the product and the surrounding environment, even in winter, is substantial. Keeping this surrounding heat from the product requires special equipment to store and handle liquid Argon. A typical system consists of the following components: a cryogenic storage tank, one or more vaporizers, a pressure control system, and all of the piping required for fill, vaporization, and supply. The cryogenic tank is constructed like a vacuum bottle. It is designed to keep heat away from the liquid that is contained in the inner vessel. Vaporizers convert the liquid argon to its gaseous state. A pressure control manifold controls the pressure at which the gas is fed to the process.

Being odorless, colorless, tasteless, and nonirritating, argon has no warning properties. Humans possess no senses that can detect the presence of argon. Argon is nontoxic and largely inert. It can act as a simple asphyxiant by displacing the oxygen in air to levels below that required to support life. Inhalation of argon in excessive amounts can cause dizziness, nausea, vomiting, loss of consciousness, and death. Death may result from errors in judgment, confusion, or loss of consciousness that prevents self-rescue. Extensive tissue damage or burns can result from exposure to liquid argon or cold argon vapors.

Properties of liquid argon are:
- Molecular Weight: 39.95
- Boiling Point @ 1 atm: –302.6°F (–185.9°C, 87°K)
- Freezing Point @ 1 atm: –308.8°F (–189.4°C, 85°K)
- Critical Temperature: –188.4°F (–122.4°C)
- Critical Pressure: 705.8 psia (48.0 atm)
- Density, Liquid @ B.P., 1 atm: 87.40 lb/scf
- Density, Gas @ 68°F (20°C), 1 atm: 0.1034 lb/scf
- Specific Gravity, Gas (air = 1) @ 68°F (20°C), 1 atm: 1.38

6.10 Cryogenics

- Specific Gravity, Liquid (water = 1) @ 68°F (20°C), 1 atm: 1.40
- Specific Volume @ 68°F (20°C), 1 atm: 9.67 scf/lb
- Latent Heat of Vaporization: 2804 BTU/lb mole
- Expansion Ratio, Liquid to Gas, B.P. to 68°F (20°C): 1 to 840.

Liquid argon is stored, shipped, and handled in several types of containers, depending upon the quantity required by the user. The types of containers in use are the Dewar, cryogenic liquid cylinder, and cryogenic storage tank. Storage quantities vary from a few liters to many thousands of gallons. Since heat leak is always present, vaporization takes place continuously. Rates of vaporization vary depending on the design of the container and the volume of stored product. Containers are designed and manufactured according to the applicable codes and specifications for the temperatures and pressures involved. Liquid product is typically removed through insulated withdrawal lines to minimize the loss of liquid product to gas. Insulated flexible or rigid lines are used to withdraw product from storage tanks. Connections on the lines and tanks vary by manufacturer. Liquid cylinders designed to dispense gaseous argon have valves equipped with standard Compressed Gas Association (CGA) outlets. Valves provided for the withdrawal of liquid product are also equipped with standard CGA outlets, but are different than the connections used for gaseous discs to protect the cylinders from pressure build-up. These containers operate at pressures up to 350 psig and have capacities between 80 and 450 liters of liquid. Argon may be withdrawn as a gas by passing liquid through an internal vaporizer or as a liquid under its own vapor pressure. Figure 6.3 shows a typical cryogenic liquid argon cylinder. Cryogenic liquid cylinders are insulated, vacuum-jacketed pressure vessels. They come equipped with safety relief valves and rupture disks.

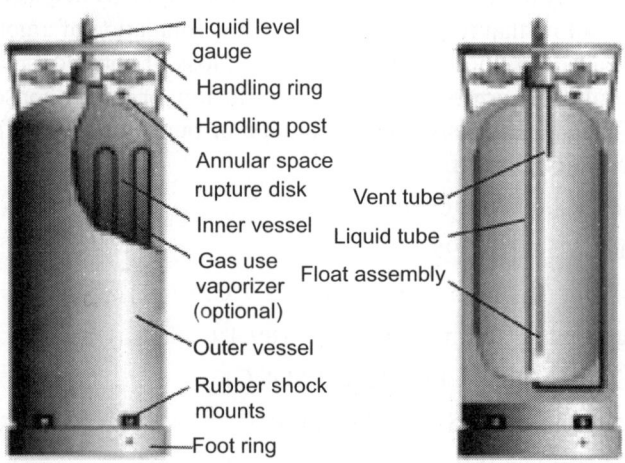

Fig. 6.3 Liquid Argon Cylinder

CRYOGEN 5: LIQUID HELIUM (LHE)

Liquid helium is inert, colorless, odorless, non-corrosive, extremely cold, and nonflammable. Helium will not react with other elements or compounds under ordinary conditions. Since helium is non-corrosive, special materials of construction are not required. However, materials must be suitable for use at the extremely low temperatures of liquid helium. Being odorless, colorless, tasteless, and nonirritating, helium has no warning properties. Although helium is non-toxic and inert, it can act as a simple asphyxiant by displacing the oxygen in air to levels below that required to support life. Inhalation of helium in excessive amounts can cause dizziness, nausea, vomiting, loss of consciousness and death. At low oxygen concentrations, unconsciousness and death may occur in seconds and without warning. Personnel, including rescue workers, should not enter areas where the oxygen concentration is below 19.5%, unless provided with a self-contained breathing apparatus or air-line respirator.

Properties of liquid helium are:

- Molecular Symbol: He
- Molecular Weight: 4.003
- Boiling Point @ 1 atm: −452.1°F (−268.9°C, 4°K)
- Freezing Point @ 367 psia: −459.7°F (−272.2°C, 0°K)
- Critical Temperature: −450.3°F (−268.0°C)
- Critical Pressure 33.0 psia: (2.26 atm)
- Density, Liquid @ B.P., 1 atm: 7.798 lb./cu.ft.
- Density, Gas @ 32°F (0°C), 1 atm: 0.0103 lb./cu.ft.
- Specific Gravity, Gas (Air = 1) @ 32°F (0°C), 1 atm: 0.138
- Specific Gravity, Liquid @ B.P., 1 atm: 0.125
- Specific Volume @ 32°F (0°C), 1 atm: 89.77 cu.ft./lb.

Liquid helium is most frequently provided to customers in liquid containers, which consist of two cylindrical vessels, one within the other. The annular space is evacuated and contains multiplayer insulation. Helium is typically withdrawn as a liquid but may be withdrawn as a gas at low flow rates and pressure. Multiple pressure relief devices are installed on these liquid containers to protect against overpressure. Liquid helium container valves should never be left open to atmosphere for extended periods. The fill/withdrawal vent outlets should be closed to prevent contamination. The system needs to be checked regularly for frost accumulation. Blockage or restriction of openings or vents may lead to excessive vessel pressure and subsequent rupture.

Helium comes in two forms Helium-3 and Helium-4. Helium-3, as an isotope, was postulated to be radioactive, until helions from it were accidentally identified as a trace "contaminant" in a sample of natural helium (which is mostly Helium-4) from a gas well, by Luis W. Alvarez of the Lawrence Berkeley National Laboratory, in 1939. A Helium-3 refrigerator uses Helium-3 to achieve temperatures of 0.2 to 0.3 kelvin. A dilution refrigerator uses a mixture of Helium-3 and Helium-4 to reach cryogenic temperatures as low as a few thousandths of a kelvin. An important property of Helium-3, which distinguishes it from the more common Helium-4, is that its nucleus is a fermion since it contains an odd number of spin 1/2 particles. Helium-4 nuclei are bosons, containing an even number of spin 1/2 particles. Properties of liquid helium has long since been investigated and many interesting phenomena are known to us. Recently it has become possible to obtain the isotope He_3 in pure form and a lot of interesting experiments have been done for various concentrations of He_3 in He_4. At first, ratio of the two components in the vapor and liquid phases was investigated and it was proposed by Taconis, that below the λ-point He_3 mixes only with the normal part of He_4. On the other hand, change of λ-point with the content of He_3 was investigated both experimentally and theoretically. He_4 particles obey Bose-Einstein statistics and He_3 particles which obey Fermi-Dirac statistics. He_4 is the more common isotope of helium. The Figure 6.4 shows the phase diagram of He_4 at low temperatures. He_4 remains liquid at zero temperature if the pressure is below 2.5 Mpa. The liquid has a phase transition to a superfluid phase, also known as He_2, at the temperature of 2.17 K (at vapor pressure). The solid phase has either hexagonal close packed (hcp) or body centered cubic (bcc) symmetry.

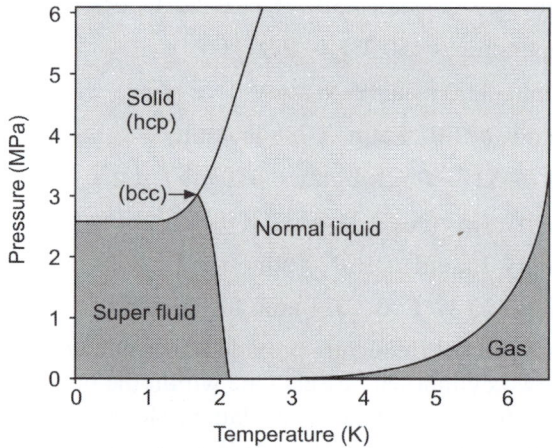

Fig. 6.4 Phase Diagram of Helium

CRYOGEN 6: LIQUID NEON (LNe)

Neon is a rare atmospheric gas and as such is non-toxic and chemically inert. Neon poses no threat to the environment, and can have no impact at all because it's chemically unreactive and forms no compounds. Liquefied neon is commercially used as a cryogenic refrigerant in applications not requiring the lower temperature range attainable with more extreme liquid helium refrigeration. Liquid neon is actually quite expensive, and nearly impossible to obtain in small quantities for laboratory tests. For small quantities, liquid neon can be more than 55 times more expensive than liquid helium. The driver for expense is actually rarity of the gas, not the liquefaction process. Liquid neon has properties unique among other low-temperature cryogenic fluids in that its latent heat of vaporization is over three times that of hydrogen (actually 3.27) and 41.03 times that of helium on a BTU/gallon ratio basis. Liquid neon also has the greatest gas-to-liquid ratio of the atmospheric gases, 1445:1, meaning one liquid litre will evolve 1,445 gas litres (at room temperature). The space program uses liquid neon as a coolant for ultra-sensitive infrared imaging and detection equipment and for creating deep-space environmental temperatures during satellite testing.

Neon gas is inert and is classified as a simple asphyxiant. Inhalation in excessive concentrations can result in dizziness, nausea, vomiting, loss of consciousness, and death. Death may result from errors in judgment, confusion, or loss of consciousness, which prevent self-rescue. At low oxygen concentrations, unconsciousness and death may occur in seconds without warning.

Properties of liquid neon are:

- Atomic number: 10
- Atomic mass: 20.179 $g.mol^{-1}$
- Density: 0.9×10^{-3} $g.cm^{-3}$ at 20°C
- Melting point : –249°C
- Boiling point : –246°C
- Van Der Waals' radius : 0.16 nm
- Isotopes: 3
- Energy of first ionization: 2080 $kJ.mol^{-1}$.

CRYOGEN 7: LIQUID FLUORINE (LF)

Fluorine is a Group 17 element. Fluorine is the most electronegative and reactive of all elements. It is a pale yellow, corrosive gas, which reacts with practically all organic and inorganic substances. Finely divided metals, glass, ceramics, carbon, and even water burn in fluorine with a bright flame. Reasonably safe handling techniques for fluorine are now available and one can transport liquid fluorine by the ton. Compounds of fluorine with noble gases such as xenon, radon, and krypton are known. Elemental fluorine and the fluoride ion (in quantity) are highly toxic. Fluorine is available commercially in cylinders but is very difficult to handle. Fluorine may be recovered with difficulty as a highly reactive and corrosive pale yellow gas by electrolysis of hot molten mixtures (1:2) of potassium fluoride (KF) and hydrogen fluoride (HF). The electrolyte is corrosive, so is the product. Liquid fluorine is difficult to store as it reacts with most materials but steel and Monel metal containers may be used as the metal surfaces deactivate through the formation of unreactive surface fluorides.

6.14 Cryogenics

Properties of liquid fluorine are:
- Atomic weight : 18.998403
- Boiling point, °C : –188.13
- Freezing point, °C: –219.61
- Critical temperature, °C: –129.2
- Critical pressure, atm : 55
- Density of Liquid g/ml : 1.505
- Dissociation energy, kcal/mol : 36.8
- Heat of vaporization, cal/mol : 1510.

Fluorine reacts with considerable violence with most hydrogen-containing compounds, such as water, ammonia, and all organic chemical substances whether liquids, solids, or gases. The reaction of fluorine with water is very complex, yielding mainly hydrogen fluoride and oxygen with less amounts of hydrogen peroxide, oxygen difluoride, and ozone. Fluorine displaces other nonmetallic elements from their compounds, even those nearest fluorine in chemical activity. It displaces chlorine from sodium chloride, and oxygen from silica, glass, and some ceramic materials. In the absence of hydrofluoric acid, however, fluorine does not significantly etch quartz or glass even after several hours at temperatures as high as 200°C. Fluorine is a very toxic and reactive element. Many of chlorine compounds are toxic and can cause severe burns. Care must be taken to prevent liquid fluorine from coming in contact with the skin or eyes.

CRYOGEN 8: LIQUID CARBON-DIOXIDE (LCO$_2$)

Gaseous CO_2 is purified in several stages, compressed and cooled to produce liquid CO_2 at a pressure of about 20 bar gauge and a temperature of about –20°C. It is collected in a large tank which is isolated, analysed and released for distribution. Liquid carbon dioxide is stored and transported at constant temperature and pressure, i.e. about 20 bar and –20°C. Static tanks and road tankers are insulated to keep the product cold, and most static tanks have a small refrigeration plant to offset heat ingress. Generally, this means CO_2 tanks are closed unlike most other bulk gas tanks. Liquid CO_2 cannot exist at atmospheric pressure and flashes to a mixture of cold gas and solid CO_2 (dry ice) when de-pressurised. It has solvent properties similar to hexane and becomes "supercritical" when pumped and heated above the critical point. Prolonged exposure to dry ice can cause severe skin damage through cold burns, and the fog produced may also hinder attempts to withdraw from contact in a safe manner. Because it sublimates into large quantities of carbon dioxide gas, which could displace oxygen-containing air and pose a danger of asphyxiation, dry ice should only be exposed to open air in a well-ventilated environment.

Liquid carbon dioxide (CO_2) has excellent cleaning properties and dissolves stains as well as traditional dry cleaning solvents. However, CO_2 is safe to handle.

It is non-toxic, non-flammable and odorless. It does not produce hazardous waste or emissions that require special disposal. Liquid CO_2 based dry cleaning technology has been in development since 1998 and is now the best method available for cleaning delicate clothing.

Properties of liquid carbondioxide are:

- Boiling point : 217 K
- Thermal conductivity W/mk : 1950
- Specific heat J/kgK : 0.017
- Dynamic viscosity 10^6 Pas : 0.18
- Density of Liquid kg/m³ : 1180
- Surface tension N/m : 0.88
- Enthalpy of vaporization, KJ/kg : 350

Liquid carbon dioxide is stored in tanks designed as per the recommendations in ASME Sec VIII codes and the material of construction is SS-304. The vessel can be horizontally and vertically mounted with thick insulation and aluminum cladding. The surfaces are wire brushed and sand blasted before painting with 2 coats of zinc chromate primer followed by 2 coats of enamel paint. Figure 6.5 shows a typical liquid carbon dioxide tank.

Fig. 6.5 Liquid Carbon Dioxide Tank

Superfluidity

Superfluids are, like superconductors, related to the behavior of materials at very low temperatures. Superfluids can only observed at much lower temperatures than superconductors, Helium-4 doesn't display superfluid behavior until nearly below 2K and these temperatures are not easy to reach. When a material does become a superfluid, it displays some very strange behavior such as:

1. If it is placed in an open container it will rise up the sides and flow over the top.

2. If the fluid's container is rotated from stationary, the fluid inside will never move as the viscosity of the liquid is zero.
3. If a light is shone into a beaker of superfluid and there is an exit at the top the fluid will form a fountain and shoot out of the top exit.

Superfluidity is a phase of matter or description of heat capacity in which unusual effects are observed when liquids, typically of Helium-4 or helium-3, overcome friction by surface interaction when at a stage (known as the "lambda point" for Helium-4) at which the liquid's viscosity becomes zero. Figure 6.6 shows a superfluid Helium-4.

Fig. 6.6. Superfluid Helium-4

Superfluidity was discovered by Pyotr Kapitsa, John F. Allen and Don Misener in 1937 and has been described through phenomenological and microscopic theories. There are other interesting facts about superfluids, the point at which a liquid becomes a superfluid is named the lambda point. This is because at around this area the graph of specific heat capacity against temperature is shaped like the greek letter Lambda.

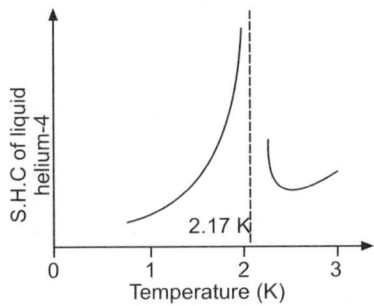

Fig. 6.7 Lambda Curve

The lambda curve specifies the relationship at the superfluidity boundary of Helium-4, between its specific heat capacity and its temperature. The graph has an asymptote at its lambda point; this is when the atoms are moving slowly enough to form a single energy state, and are then unaffected by other atoms. Above the lambda point the atoms all move about in a disordered way, but below this temperature they all move in an organized way. We can see these effects in Helium-4 because it is still a liquid at these low temperatures; most materials solidify above these temperatures.

L. D. Landau's phenomenological and semi-microscopic theory of superfluidity of ^{4}He earned him the Nobel Prize in Physics in 1962. Assuming that sound waves are the most important excitations in ^{4}He at low temperatures, he showed that ^{4}He flowing past a wall would not spontaneously create excitations if the flow velocity was less than the sound velocity. In this model, the sound velocity is the "critical velocity" above which superfluidity is destroyed. From the momentum and flow velocity of the excitations he could then define a "normal fluid" density, which is zero at zero temperature and increases with temperature. At the so-called Lambda temperature, where the normal fluid density equals the total density, the ^{4}He is no longer superfluid. Although the phenomenologies of the superfluid states of Helium-4 and Helium-3 are very similar, the microscopic details of the transitions are very different. Helium-4 atoms are bosons, and their superfluidity can be understood in terms of the Bose statistics that they obey. On the other hand, helium-3 atoms are fermions, and the superfluid transition in this system is described by a generalization of the theory of superconductivity. A unified description of superconductivity and superfluidity is possible in terms of gauge symmetry breaking.

Superfluids, such as supercooled Helium-4, exhibit many unusual properties. Superfluid acts as if it were a mixture of a normal component, with all the properties associated with normal fluid, and a superfluid component. The superfluid component has zero viscosity, zero entropy and infinite thermal conductivity. One of the most spectacular results of these properties is known as the thermomechanical or "fountain effect". If a capillary tube is placed into a bath of superfluid helium and then heated, even by shining a light on it, the superfluid helium will flow up through the tube and out the top as a result of the Clausius-Clapeyron relation. A second unusual effect is that superfluid helium can form a layer, 30 nm thick, up the sides of any container in which it is placed. A more fundamental property than the disappearance of viscosity becomes visible if superfluid is placed in a rotating container. Instead of rotating uniformly with the container, the rotating state consists of quantized vortices. That is, when the container is rotated at speed below the first critical velocity the liquid remains perfectly stationary. Once the first critical velocity (the speed of sound in the superfluid) is reached, the superfluid will very quickly begin spinning at the critical speed. The speed is quantized, that is a superfluid can only spin at certain "allowed" or critical speed values. In simplified terms, if the container is rotated to a certain allowed speed, the superfluid will rotate very quickly along with the container, otherwise, if the speed is too slow, then the superfluid will not move at all, unlike how a normal fluid like water will rotate along with its container from the start.

The normal phase of non-zero entropy and superfluid phase of zero entropy coexist. This gives a strange phenomenon of two-fluid model. A consequence of this is that, when the two-fluid passes through a capillary, effectively only the superfluid can flow through due to its zero viscosity along the way. This gives a strange situation where we have mass transfer without heat transfer (since there is no entropy transfer). Recently in the field of chemistry, superfluid Helium-4 has been successfully used in spectroscopic techniques as a quantum solvent. Superfluids are also used in high-precision devices such as gyroscopes, which allow the measurement of some theoretically predicted gravitational effects. The Infrared Astronomical Satellite IRAS, launched in January 1983 to gather infrared data was cooled by 720 litres of superfluid helium, maintaining a temperature of 1.6 K (–271.4°C).

References

1. Cryogenic Engineering – B.A. Hands.
2. Internet Website www.wikipedia.com.
3. Cryogenic Systems – Randall Baron.
4. Internet Website www.airproducts.com..
5. Furr, A. K. 2000. CRC Handbook of Laboratory Safety, 5th Edition. CRC Press.
6. Internet Website www.noao.edu/kpno/manuals/irim/cryo.html.
7. Internet Website - http://www.egglescliffe.org.uk/physics/superfluids.html.
8. Internet Website www-safety.deas.harvard.edu/services/oxygen.html.
9. Internet website www.lenntech,com/periodic/elements.
10. Paper by Grazzi et.al, "Measurement of the ortho-to-para hydrogen conversion in the high-pressure compound Ar(H_2)" Physics letters (2000).
11. Internet Website www.mellcon.com/liquid-co_2-storage-tank.asp.

Questions

Q.1. What are cryogens? Give examples.

Q.2. Write a short notes on properties of liquid nitrogen, liquid helium 3 and liquid hydrogen.

Q.3. Explain the concept of super-fluidity for cryogenic fluids.

CHAPTER

7

Gas Liquefaction Systems

INTRODUCTION

Liquefaction of Gases is an important area in cryogenic engineering. The production process of liquefied gases start with gas being transported to the plant site as feedstock. After filtration and metering in the feedstock reception facility, the feedstock gas enters the liquefaction plant and is distributed among the identical liquefaction systems. Each liquefaction process plant consists of reception, acid gas removal, dehydration/Mercaptan i.e. odourant removal, mercury removal, gas chilling and liquefaction, refrigeration, fractionation, nitrogen rejection and sulfur recovery units. In an effort to design, engineer, and manufacture the most cost effective, space and weight efficient liquefaction facility possible, many factors must be considered. The first thing that must be determined is what detrimental contaminants exist in the entering gas stream. These contaminants can include, but are not limited to, oxygen, nitrogen, water, carbon dioxide (CO_2), hydrogen sulfide (H_2S), mercury, arsenic and/or heavy hydrocarbons (C^{3+}). Each of these components can create significant problems for the operation of an liquefaction plant. For example, the CO_2 content in a gas stream entering an LNG Plant must be reduced to less than 50 ppm to avoid the formation of dry ice within the system which can plug off equipment and shutdown the plant. Similarly, mercury in the gas stream can attack the aluminum components often used in liquefaction plant heat exchangers and other equipment. Depending on the amount of H_2S contained in the inlet gas, an H_2S scavenger system may be used to remove the sulfur before entering any other part of the plant system. Depending on the amount of CO_2 contained in the inlet gas stream and the volume of gas entering the plant, it may be beneficial to remove the bulk amount of CO_2 using a membrane treating system in order to minimize the size of the downstream plant and reduce the overall energy consumption of the plant.

The next step is to dry the gas to the point that it will contain less than 1 ppm of H_2O. A Molecular Sieve (mol sieve) dry desiccant is the industry standard for performing this function. The number of beds is generally determined by the volume of gas being dehydrated and the water content in the inlet gas stream. One or more of the dehydration beds operate in the adsorption phase, where water vapor is adsorbed onto the desiccant, while one bed is heat regenerated to strip water from the mol sieve. Mol sieves can also be designed to remove trace amounts of CO_2,

7.2 Cryogenics

H$_2$S and mercaptan, if it is known before hand that these contaminants are present and need to be removed. Properly designed mol sieve systems will remove water to less than 1 ppm. If mercury is present, a separate vessel, filled with activated carbon, is typically used to remove mercury gas stream. These beds are typically located downstream of the mol sieve system to keep water from deactivating the bed. Mercury removal systems are generally designed to reduce the mercury content in a gas stream to less than 10 nano-grams per cubic meter. Arsenic removal systems are a virtual duplicate of mercury removal systems in appearance, but utilize a different bed material to remove arsenic and the various arsines that may be present in the gas stream.

Finally, depending on the quality of the inlet gas and how "clean" of a liquefied gas product is desired, the gas may be "conditioned" to remove the heavy-end hydrocarbons from the gas stream before it enters the actual liquefaction plant. Thus the process through which liquefied gases is produced consists of three main steps, namely: 1. **Transportation of Gas**: The best place to install the plant is near the gas source. The gas is basically transported through pipelines or by truck and barge. 2. **Pretreatment of Gas:** The liquefaction process requires that all components that solidify at liquefaction temperatures must be removed prior to liquefaction. This step refers to the treatment the gas requires to make it liquefiable and includes compression, filtering of solids, removal of liquids and gases that would solidify under liquefaction, and purification which is removal of non-methane gases and finally. 3. **Liquefaction of Gas**: The liquefaction of the gas follows the liquefaction curve. There exist a vast number of gas liquefaction plants designs, but, all are based on the combination of heat exchange and refrigeration. The gas being liquefied, however, takes the same liquefaction path. The dry, clean gas enters a heat exchanger and exits as liquefied gas. The capital invested in a plant and the operating cost of any liquefaction plant is based on the refrigeration techniques. Currently the following gas liquefaction process based plants are available

- Simple Linde-Hampson System
- Precooled Linde-Hampson System
- Linde Dual Pressure System
- Cascade System
- Claude System
- Kapitza System
- Heylandt System
- Claude Dual Pressure System

The selection of a particular gas liquefaction process is driven by customer requirements, plant capacity, and the availability of resources. In this chapter all the liquefaction systems are elaborated in detail and comparison of performance is also carried out.

Liquefaction System Performance Parameters

There are three main performance parameters for a liquefaction system
1. Work required per unit mass of gas compressed $= -W/m$
2. Work required per unit mass of gas liquefied $= -W/m_f$
3. Fraction of the total flow of gas liquefied = yield $= y = m_f/m$

Thus yield for a liquefaction system is given by $y = (-W/m)/(-W/m_f)$

The liquefaction system should require minimum work and should maximise the yield or the gas to be liquefied.

The figure of merit for a liquefaction system can be defined as the theoretical minimum work requirement (W_1) divided by the actual work requirement (W) for the system.

$$\text{FOM} = W_1/W = -W_1/m_f / W/m_f.$$

The figure of merit is having a value between 0 and 1 and gives measure of how closely the actual system approaches the ideal system performance. The factors affecting figure of merit are compressor and expander thermal and mechanical efficiency, effectiveness of heat exchanger, pressure drop through piping, heat transfer from the ambient.

Thermodynamically Ideal Liquefaction System

In order to have a means of comparison of liquefaction systems through the figure of merit, it is important to analyze the thermodynamically ideal liquefaction system. This system is ideal in the thermodynamic sense only and not for practical systems. The perfect cycle in thermodynamics is the Carnot cycle. Liquefaction is essentially an open system process consisting of a reversible isothermal compression followed by a reversible isentropic expansion. The gas to be liquefied is compressed reversibly and isothermally from ambient conditions (point 1) to some high pressure (point 2). This high pressure is selected so that gas will become saturated liquid upon reversible isentropic expansion through the expander (point f). The pressure attained at the end of isothermal compression is extremely high in the order of 70 bar, which is the reason it is not an ideal process for a practical system. Refer Figure 7.1 for the thermodynamically ideal liquefaction system.

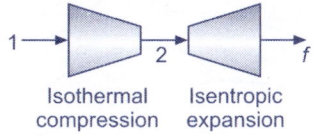

Fig. 7.1 Thermodynamically Ideal Liquefaction System

Applying the first law of thermodynamics to the ideal system provides the equation

$$Q_r - W = -m(h_1 - h_f)$$

7.4 Cryogenics

Applying the second law of thermodynamics to the ideal system provides the equation

$$Q_r = -m T_1 (s_1 - s_f)$$

The work requirement for the ideal system becomes

$$-W/m = -W_1/m_f = T_1 (s_1 - s_f) - (h_1 - h_f)$$

Thus for an ideal system if 100% gas compressed is liquefied then $m = m_f$ and yield $y = 1$. The liquefaction systems abosrbs work, thus work is negative and the term $-W/m$ is positive.

Joule-Thomson (J-T) Effect

The Joule-Thomson effect describes the temperature change of a gas or liquid when it is forced through a valve or porous plug while kept insulated so that no heat is exchanged with the environment. This procedure is called a throttling process or Joule-Thomson expansion process. At room temperature, all gases except hydrogen, helium and neon cool upon expansion by the Joule-Thomson process. The J-T effect is named for James Prescott Joule and William Thomson (Lord Kelvin) who discovered it in 1852 following earlier work by Joule in which a gas undergoes free expansion in a vacuum. Temperature change of either sign can occur during the Joule-Thomson process. Each real gas has a Joule-Thomson (Kelvin) inversion temperature above which expansion at constant enthalpy causes the temperature to rise, and below which such expansion causes cooling. This inversion temperature depends on pressure; for most gases at atmospheric pressure, the inversion temperature is above room temperature, so most gases can be cooled from room temperature by isenthalpic expansion. For a real gas we can assume isenthalpic expansion through valves. In one region temperature increases during expansion and in another region temperature reduces after expansion. The curve separating the two regions is called the inversion curve. Figure 7.2 shows the inversion curve.

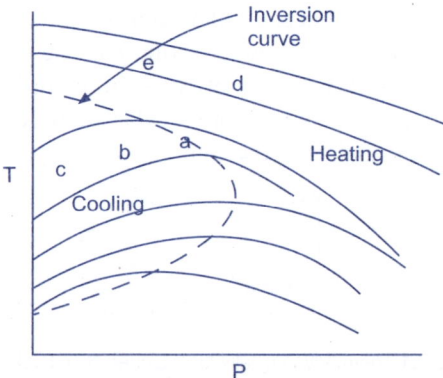

Fig. 7.2 Inversion Curve

As a gas expands, the average distance between molecules grows. Because of intermolecular attractive forces, expansion causes an increase in the potential energy of the gas. If no external work is extracted in the process and no heat is transferred, the total energy of the gas remains the same because of the conservation of energy. The increase in potential energy thus implies a decrease in kinetic energy and therefore in temperature. A second mechanism has the opposite effect. During gas molecule collisions, kinetic energy is temporarily converted into potential energy. As the average intermolecular distance increases, there is a drop in the number of collisions per time unit, which causes a decrease in average potential energy. Again, total energy is conserved, so this leads to an increase in kinetic energy (temperature). Below the Joule-Thomson inversion temperature, the work done internally against intermolecular attractive forces dominates, and free expansion causes a decrease in temperature. Above the inversion temperature, gas molecules move faster and so collide more often, and the reduced collisions causing a decrease in the average potential energy dominates: Joule-Thomson expansion causes a temperature increase.

The rate of change of temperature T with respect to pressure P in a Joule-Thomson process (that is, at constant enthalpy H) is the Joule-Thomson (Kelvin) coefficient μ_{JT}. This coefficient can be expressed in terms of the gas's volume V, its heat capacity at constant pressure C_P, and its coefficient of thermal expansion α as

$$\mu_{JT} \equiv \left(\frac{\partial T}{\partial P}\right)_H = \frac{V}{C_P}(\alpha T - 1).$$

The Joule-Thomson coeffcient is a derivative interpreted as change in temperature due to change in pressure at constant enthalpy. The value of μ_{JT} is typically expressed in K/Pa and depends on the type of gas and on the temperature and pressure of the gas before expansion. Its pressure dependence is usually only a few percent for pressures up to 100 bar. All real gases have an inversion point at which the value of μ_{JT} changes sign. The temperature of this point, the Joule-Thomson inversion temperature, depends on the pressure of the gas before expansion. In practice, the Joule-Thomson effect is achieved by allowing the gas to expand through a valve which must be very well insulated to prevent any heat transfer to or from the gas. No external work is extracted from the gas during the expansion.

For an Ideal gas,

$$\left(\frac{\delta H}{\delta P}\right)_T = -\left(\frac{\delta H}{\delta T}\right)_P \left(\frac{\delta T}{\delta P}\right)_H = -C_P \mu_\pi$$

From first law of thermodynamics

$$\left(\frac{\delta H}{\delta P}\right)_T = \left(\frac{(\delta U + PV)}{\delta P}\right)_T = \left(\frac{\delta U}{\delta P}\right)_T + \left(\frac{\delta(\delta V)}{\delta P}\right)_T = \left(\frac{\delta U}{\delta P}\right)_T$$

Thus the Joule-Thomson coeffcient $\mu_{JT} = 0$ for Ideal gas. Thus ideal gas will not exhibit any change in temperature on expansion.

The JT effect is applied in the liquefaction industry as part of the Linde standard process, where the cooling effect is used to liquefy gases, and also in many cryogenic applications (e.g. for the production of liquid oxygen, nitrogen, and argon). Only when the Joule-Thomson coefficient for the given gas at the given temperature is greater than zero can the gas be liquefied at that temperature by the Linde cycle. In other words, a gas must be below its inversion temperature to be liquefied by the Linde cycle. For this reason, simple Linde cycle liquefiers cannot normally be used to liquefy helium, hydrogen, or neon.

Adiabatic Expansion in Turbine

The second method of producing low temperatures is the adiabatic expansion of gas through work producing device such as a turbine or expansion engine. The expansion is reversible and adiabatic and hence is termed as isentropic expansion. The governing equation of such an expansion is PV^γ = constant, where γ is the specific heat ratio. Figure 7.3 shows the P-V diagram for reversible adiabatic expansion.

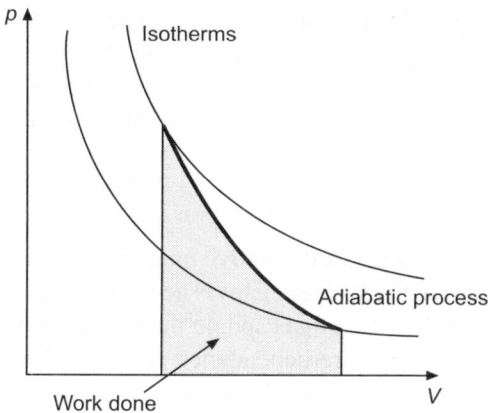

Fig. 7.3 P-V Diagram for Reversible Adiabatic Expansion

The isentropic expansion coefficient which expresses temperature change due to pressure change at constant entropy is given by the expression

$$\mu s = (\partial T/\partial P)s.$$

The isentropic expansion coeffcient can also be written as the product of volumetric coeffcient of thermal expansion and specific volume by the expression

$$\mu s = T/C_p \, (\partial v/\partial T)p.$$

The isentropic expansion coeffcient can also be written for ideal gas by the expression

$$\mu s = v/C_p.$$

For any gas, isentropic expansion through an expander always results in a temperature decrease unlike the isenthalpic JT process where temperature may increase or decrease. During the reversible adiabatic process, energy is removed from the gas as external work and so this method of production of low temperatures is termed as external work method. The isentropic expansion is always more efficient than isenthalpic expansion between two thermal reservoirs. However considering expansion problems associated with two phase flows of refrigerants in refrigeration systems, isenthalpic valves are preferred over isentropic expanders.

Simple Linde-Hampson (L-H) System

Linde cycle is the simplest cryogenic cycle used for liquefaction of gases. The schematic of the cycle is shown in Figure 7.4 and its T-S diagram is shown in Figure 7.5. A basic differentiation between the various refrigeration cycles lies in the expansion device. This may be either an expansion engine like expansion turbine or reciprocating expansion engine or a throttling valve. The expansion engine approaches an isentropic process and the valve an isenthalpic process. In the Linde's system, the basic principle of isenthalpic expansion is incorporated.

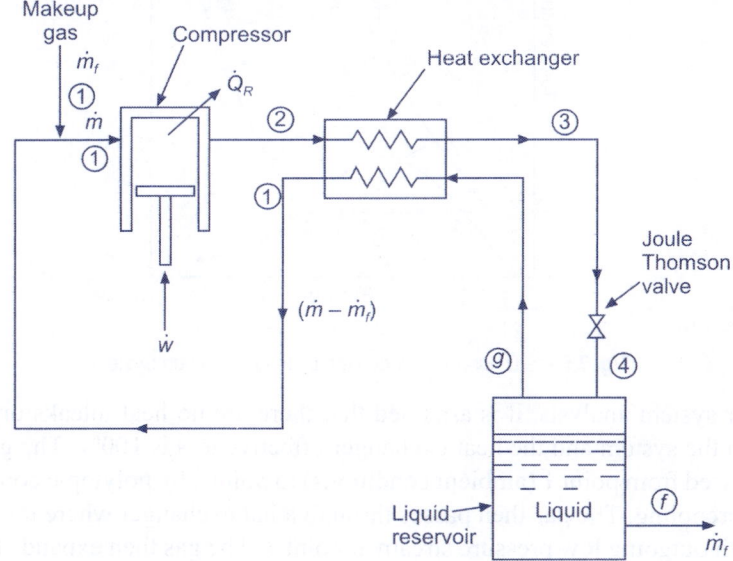

Fig. 7.4 Simple Linde-Hampson Liquefaction System and P-H Diagram

The components of the simple L-H system are:

1. **Compressor:** It is a device used to reduce the volume of gases and increase their pressure. Generally for cryogenic application compressors with high compression ratios are used. To achieve high stage compression ratio, a number of compressor stages are used in series rather than using a single compressor which reduces work consumption.

2. **Heat Exchanger:** Heat exchangers are devices which transfer heat from hot fluid stream to cold fluid stream. The hot fluid is cooled for throttling process and cold fluid is heated up for compression process.

3. **Valve:** A throttling valve is used to reduce the pressure of the compressed gas so that liquid gas can be produced and stored. The process is assumed to be isenthalpic expansion.

7.8 Cryogenics

4. **Separator:** In this chamber liquid is separated and removed and the gaseous part is again recirculated in the system.
5. **Mixer:** It is a device helps to maintain a constant flow rate of gas into the compressor. The extra amount of gas also called make up gas is added into incoming stream from separator. The process is assumed to be isobaric.

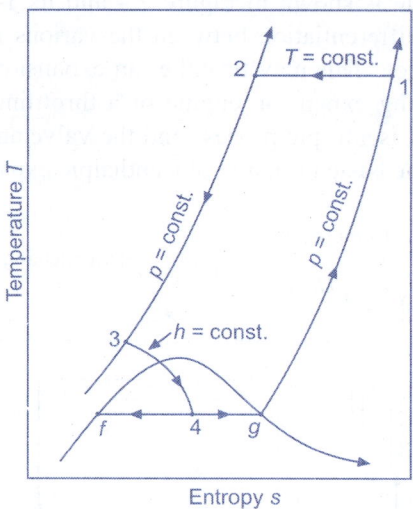

Fig. 7.5 T-S Diagram of Linde-Hampson Liquefaction System

For system analysis, it is assumed that there are no heat inleaks or pressure drops in the system and the heat exchanger effectiveness is 100%. The gas is first compressed from point 1 (ambient conditions) to point 2 by polytopic compression and aftercooling. The gas then passes through a hat exchanger where it exchanges heat with outgoing low pressure stream at point 3. The gas then expands through a J-T or expansion valve to point 4 which is having same pressure as ambient. The expansion causes cooling and some of the gas is liquefied and is withdrawn from the receiver at saturated condition "f". the rest of the gas is recycled in the system from the receiver. This cold gas is heated by absorbing energy from incoming high pressure stream. Applying the first law for the system we get

$$m_f/m = y = (h_1 - h_2)/(h_1 - h_f)$$

The term y which denotes fraction of gas liquefied is called yield of the system. This depends on ambient conditions and gas conditions after compression. For maximum yield. The following condition needs to be satisfied

$$(\partial h/\partial P)_T = \mu_{JT} C_p = 0$$

Where μ_{JT} = Joule-Thomson Coefficient.

The simple Linde Hampson cycle is unsuitable for gases like neon, hydrogen and helium as the maximum inversion temperature of these gases is below room temperature and expansion through the J-T Valve does not produce any cooling and subsequent liquefaction. The first law of thermodynamics as applied to this system gives the expression

$$Q_r - W = m(h_2 - h_1)$$

Work requirement for the system is given by

$$-W/m = T_1(s_1 - s_2) - (h_1 - h_2)$$

Work requirement per unit mass of gas liquefied is given by

$$-W/m_f = -W/m_y = (h_1 - h_f / h_1 - h_2)[T_1(s_1 - s_2) - (h_1 - h_2)].$$

Pre-cooled Linde-Hampson System

This method is patented by Linde Corporation. This is a generic method for liquefying a gas stream such as hydrogen to produce liquid gases. The liquefaction of the gas stream takes place countercurrent to a refrigerant mixture cycle cascade consisting of two refrigerant mixture cycles. The first refrigerant mixture cycle is used for pre-cooling and a second refrigerant mixture cycle is used for liquefaction and super cooling of the gas stream to be liquefied. No additional process steps are involved in a heat exchange between the gas stream which is to be precooled and the refrigerant mixture of the first refrigerant mixture cycle. The term "precooling" means the cooling of the gas stream down to a temperature at which the separation of heavy impurities takes place. The subsequent further cooling of the gas stream to be liquefied hereinafter comes under the term "liquefaction". The basic Linde-Hampson cycle uses a Joule-Thomson orifice. A gas could be cooled by letting it expand freely against the atmosphere. This could be explained by that the gas does work against the atmosphere by lifting and/or heating it, and thereby loosing energy in form of heat. The precooled Linde-Hampson system is quite desirable for small-scale liquefaction plants. This type of precooled system was used in several of the earlier helium liquefiers. A schematic diagram of the pre-cooled Linde-Hampson system is represented in Figure 7.6. This system consists of two independent circuits, the upper one is the pre-cooled system and the lower one is the basic circuit of the gas to be liquefied.

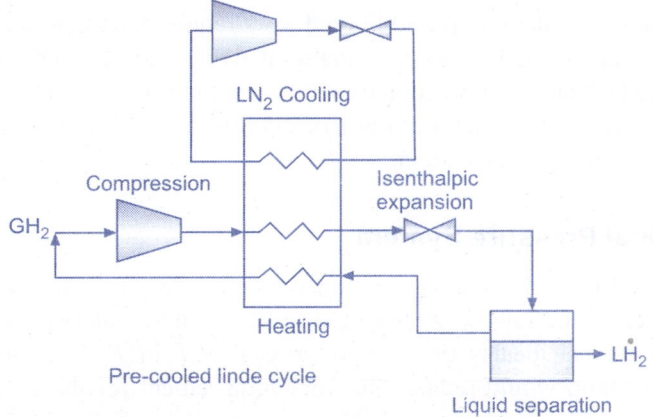

Fig. 7.6 Precooled Linde Hampson Liquefaction System

7.10 Cryogenics

The precooled system is filled with a precoolant fluid that has a triple-point temperature below that of the maximum inversion temperature of gas to be liquefied. Liquid nitrogen is a suitable fluid to fill the precooler system for liquefying industrial gases. The basic circuit consists of a compressor employed to raise the enthalpy (so the temperature and the pressure are also raised) of the gas to be cooled, then the gas is cooled through three heat exchangers, two of them involving the precooling system. The gas then passes through a valve to a tank, where the liquefied gas is extracted within a channel. The remaining gas is sent to the compressor after being heated through two heat exchangers. The performance of a simple Linde-Hampson cycle is improved if gas enters the heat exchanger at a temperature lower than ambient temperature which is achieved by precooling. The precooling is achieved by a separate refrigeration circuit of liquid nitrogen or ammonia or carbondioxide. The critical temperature of the auxiliary refrigerant should be above ambient temperature to achieve proper condensation of the refrigerant. A three channel heat exchanger is used for the pre-cooling.

The refrigerant mass flow rate is defined as the ratio of mass flow of auxiliary refrigerant to the mass flow rate of gas.

$$r = m_r / m$$

The yield is given by $y = m_f/m$

Applying first law to the system the expression for yield becomes

$$y = (h_1 - h_2 / h_1 - hf) + r(h_d - h_1 / h_1 - h_f)$$

Where h_d is the enthalpy at outlet of heat exchanger. The first term is the yield of simple L-H system and the second term indicates the additional yield due to pre-cooling. The maximum yield is given by

$$y_{max} = (h_b - h_3) / (h_b - h_f)$$

Where h_d is the enthalpy of boiling refrigerant and h_f is enthalpy of liquefied gas.

The work requirement for the cycle per unit mass of gas compressed is given by

$$-W/m = T_1 (s_1 - s_2) - (h_1 - h_2) + r(h_b - h_d)$$

The last term denotes the additional work required in the auxiliary compressor for the pre-cooled L-H system. Thus total work required per unit mass of gas compressed is higher for precooled L-H system than simple L-H system. However the total work required per unit mass of gas liquefied is lower for pre-cooled L-H system than simple L-H system.

Linde Dual Pressure System

The simple Linde-Hampson process can be modified such that not all of the gas is expanded to the lowest pressure but to some intermediate pressure. The work requirement for an ideal isothermal compressor is $RT_1 \ln (P_2/P_1)$. Thus any reduction in pressure ratio would reduce the work requirement for the L-H process. This concept is used in dual pressure L-H process as shown in Figure 7.7.

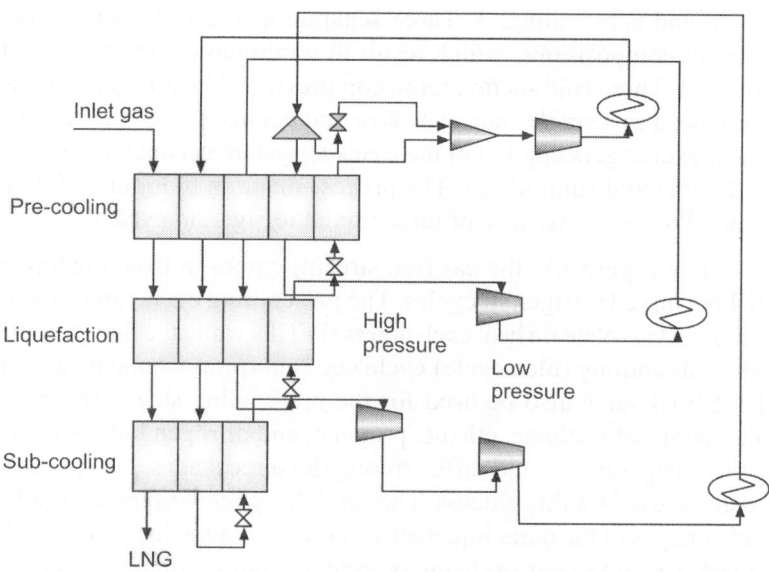

Fig. 7.7 Dual Pressure Linde-Hampson Liquefaction System

The gas is first compressed to an intermediate pressure and then to the high pressure of the cycle after a return system is added. The high pressure gas is passed through a three channel heat exchanger and expanded to intermediate pressure, where some gas is liquefied. The saturated liquid and vapor are separated in a liquid receiver and vapor is returned to the second compressor through the three channel heat exchanger, while liquid is expanded further to the low pressure of the cycle.

Applying first law of thermodynamic we get the yield as

$$y = (h_1 - h_3 / h_1 - h_f) - i(h_1 - h_2 / h_1 - h_f)$$

Where i = intermediate pressure stream flow rate = m_i / m

The second term denotes reduction in yield of LH cycle due to splitting of the flow at intermediate pressure. The work requirement per unit mass of gas compressed is

$$-W/m = [T_1(s_1 - s_3) - (h_1 - h_3)] - i [T_1(s_1 - s_2) - (h_1 - h_2)]$$

Thus work requirements for dual pressure Linde cycle are lower than simple L-H cycle. For practical systems the quantity i is of the order of 0.7 to 0.8.

Cascade System

The cascade process uses mixed component refrigerant cycles thereby improving the thermodynamic efficiency and operational flexibility. The cascade process is highly efficient due to the low shaft power consumption of the three mixed refrigerant cycle compressors. The process equipment is comprised of 1. Plate-fin heat exchangers for gas pre-cooling, 2. Coil-wound heat exchangers for the gas liquefaction and sub-cooling, 3. Three separate mixed refrigerant cycles, each with different compositions, which result in minimum compressor shaft power requirement, 4. Three cold suction turbo compressors. Up to 12 mtpa liquefied gas can be produced in a single train. The size and complexity of the separate spiral wound heat exchangers applied in the cascade system is considerably less when compared with single unit plants. The process diagram in Figure 7.8 is showing the Cascade Process consisting of three mixed refrigerant cycles.

As seen in Figure 7.8, the gas (red stream) comes in from the top and goes through three mixed refrigerant cycles. The pre-cooling cycle (green cycle) cools the gas through two plate fin heat exchangers (PFHE) while the liquefaction (purple cycle) and sub-cooling (blue cycle) cycle cool via spiral wound heat exchangers (SWHE). SWHE may also be used for the pre-cooling stage. The refrigerants are made mainly of methane, ethane, propane, and nitrogen but the composition ratio of the refrigerants would differ among the three stages. No specific kind of compressors is used for this process. The cascade system has the same advantages and disadvantages of the other liquefaction processes but it has an extra advantage that the spiral wound heat exchangers used are more efficient as compared to the single unit heat exchangers. That allows larger single compressors to handle refrigerant over a larger temperature range.

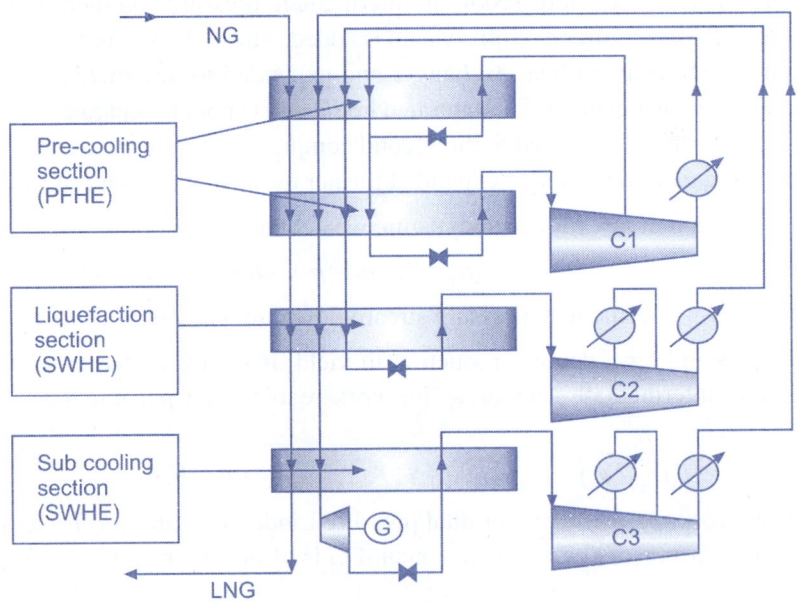

Fig. 7.8 Cascade Liquefaction Plant

The cascade system can be thought of as an extension of the pre-cooled system in which the pre-cooled system is cooled by other refrigeration systems. This system was first used to produce liquid air. This system uses three refrigerants namely ammonia, methane and nitrogen to cool natural gas. The cascade system is desirable from thermodynamic point of view as it approaches closely the ideal reversible system. The lower pressure requirements and smaller pressure drops are advantages of this system. The disadvantages include high cost due to additional stages and increased possibility of leaks leading to safety hazards. Each loop of the system must be leak proof and maintenance of such a system is difficult.

Claude System

This is the most adopted cycle for large gas liquefaction plants in the world. Gas is used not only in the feed line to be liquefied but also in the recycling line which generates necessary cold for the liquefaction. The feed gas is compressed to approximately 5 MPa, and then cooled down to 80 K. Finally the feed gas is liquefied at 0.1 MPa, 20.4 K by expansion at the J-T valve. In order to improve the process efficiency, a process with a super critical expansion turbine installed in the feed line has been developed. Refer Figure 7.9 for the Claude process and its T-S Diagram. In the Claude process some of the high pressure gas is passed through an expansion engine and is then sent to a heat exchanger for cooling. This process referred to as ideal liquefaction uses a reversible expansion process to reduce the energy required for liquefaction. It consists of an isothermal compressor followed by an isenthalpic expansion to cool the gas and produce a liquid. In practice, an expansion engine can be used only to cool the gas stream, not to condense it because excessive liquid formation in the expansion engine would damage the turbine blades.

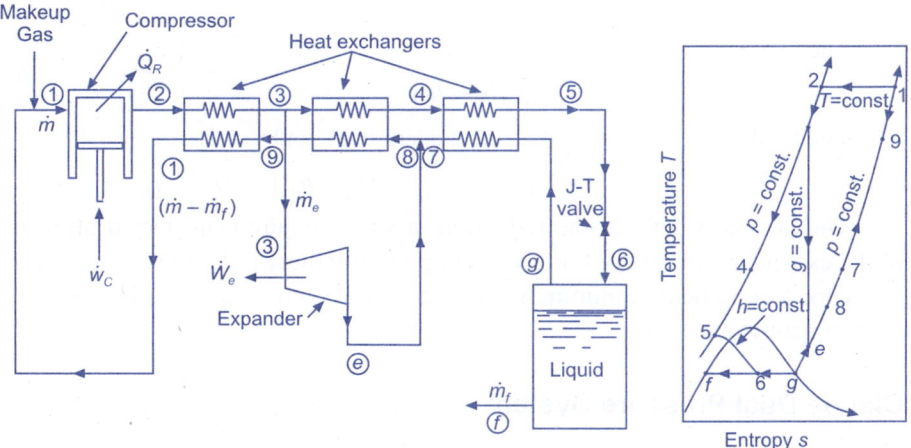

Fig. 7.9 Claude Liquefaction System with T-S Diagram

The expansion through an expansion valve is an irreversible process. Thus if we wish to approach closer to the ideal performance, we must seek a better process to produce lower temperatures. In the Claude system energy is removed from the gas stream by allowing it to do some work in an expansion engine or an expander. If the expansion engine is reversible and adiabatic, the expansion process

7.14 Cryogenics

is isentropic and a much lower temperature is attained than that for an isenthalpic expansion. The Claude liquefaction system differs from the simple Linde's system by the addition of an expander and a second heat exchanger. In this system, the gas is compressed isothermally to approximately 4 Mpa pressure between points 1 and 2. The high pressure nitrogen is partially cooled by passing through the first heat exchanger between points 2 and 3. Around 60 to 80% of gas is diverted from the mainstream, expanded through an expander and reunited with the return stream below the second heat exchanger. The remaining gas to be liquefied passes through the second and third heat exchanger and is expanded by the expansion valve and collected in the liquid receiver. The cold vapor from the liquid receiver is returned through the heat exchangers to cool the incoming gas.

An expansion valve is necessary in the Claude system as larger liquid quantities cannot be tolerated in the expander. This is due to the fact that liquids are incompressible and may cause stresses in the moving parts of the expanders. In order to reduce work requirement of the system, the work produced by the expanders can be used for gas compression. The expander flow fraction can be defined as the fraction of mass flow of gas that passes through the expander.

$$x = m_e/m$$

Applying first law, we can get the yield as

$$y = m_f/m = (h_1 - h_2 / h_1 - h_f) + x(h_3 - h_e / h_1 - h_f)$$

The second term denotes improvement in the yield due to expansion in the expanders.

The expansion work is given by

$$W_e = m_e(h_3 - h_e)$$

The total work requirement of the system is given by

$$-W/m = [T_1 (s_1 - s_2) - (h_1 - h_2)] - x(h_3 - h_e)$$

The last term denotes the reduction in work requirement due to utilization of the expander work output. For a given pressure, there is a value of x for which the work requirement is minimum. As pressure is increased, the minimum work requirement reduces.

Claude Dual Pressure System

The dual pressure Claude system is shown in Figure 7.10. It is similar in construction to the dual pressure Linde system.

In this system, only the gas that is passed through the expansion valve is compressed to high pressure. The gas that is circulated through the expander is compressed to some intermediate pressure. This reduces the work requirement per unit mass of gas liquefied. Optimum performance is achieved for this system for a typical gas like nitrogen which is compressed from 1 atm to 35 atm and around 75% of the flow is diverted through the expander.

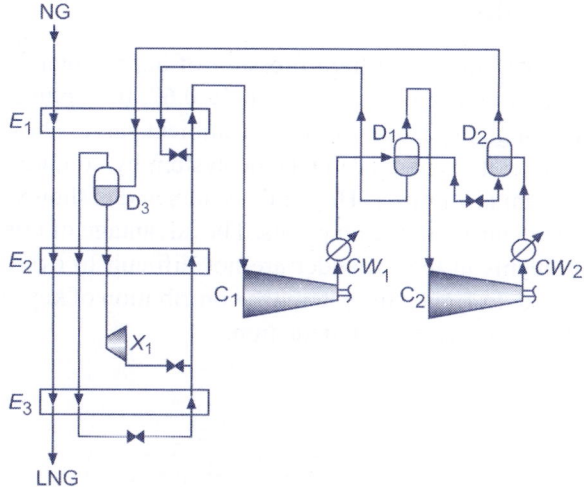

Fig. 7.10 Dual Pressure Claude Liquefaction System

Kapitza Liquefaction System

Kapitza in 1939 modified the basic Claude system by eliminating the third or low temperature heat exchanger. Refer Figure 7.11 for the schematic of Kapitza Cycle. Several practical modifications were also introduced in the system. A rotary expansion engine was used instead of a reciprocating expander. The high temperature heat exchanger in the Kapitza system was actually a set of valved regenerators, which combined the cooling process with purification process. The incoming warm gas was cooled in one unit and impurities were deposited there. The outgoing stream warmed up in the other unit and flushed out the frozen impurities. After some time a valve is opened for switching of high and low pressure streams between the two units. The Kapitza system operates at lower pressure of the order of 700 kPa.

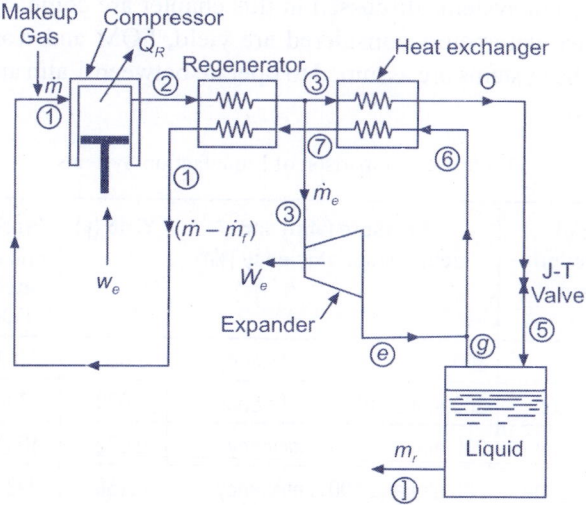

Fig. 7.11 Kapitza Liquefaction System

Heylandt Liquefaction System

Heylandt in 1949 noted that for a high pressure of approximately 200 atm, and an expansion engine flow ratio of approximately 0.60, the optimum temperature before expansion through expander was near to ambient value. Thus it was possible to eliminate the first heat exchanger in Claude system by compressing the gas to 20 MPa. Such a system is known as Heylandt system and is shown in Figure 7.12. This system is used in air liquefaction plants. The advantage of this system is that the lubrication problems in the expander are not difficult to overcome and light lubricants can be used. In Heylandt system the contribution of key components i.e. expander and JT valve is equal for liquefaction.

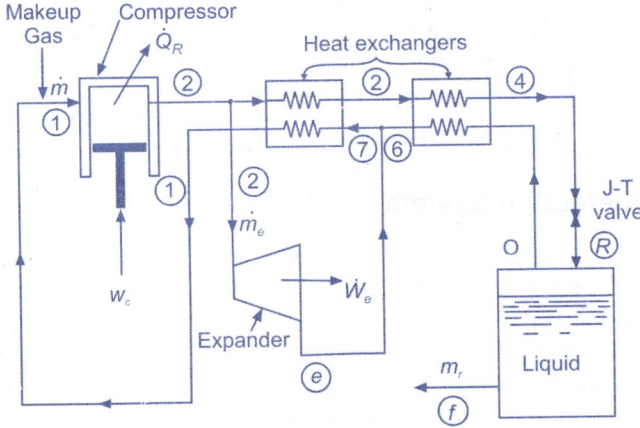

Fig. 7.12 Heylandt Liquefaction System

Comparison of Liquefaction Systems

All the liquefaction systems discussed in this chapter are compared in Table 7.1. The performance parameters considered are yield, FOM and work required per unit mass. All the systems are assumed to operate between 1 atm and 20°C and the conditions stated.

Table 7.1 Comparison of Liquefaction Systems

Sr. No	Name of Liquefaction System	Pressure (atm) and compressor efficiency (%)	Yield (y)	Work per unit mass liquefied (kJ/kg)	Figure of Merit (FOM)
1.	Ideal System	200 atm and 100% efficiency	1.000	738.9	1.000
2.	Linde-Hampson	200 atm and 100% efficiency	0.079	5739	0.129
3.	Linde Hampson	200 atm and 70% efficiency	0.062	10573	0.070
4.	Pre-cooled Linde Hampson	200 atm and 100% efficiency	0.158	2928	0.251

Contd...

5.	Pre-cooled Linde Hampson	200 atm and 70% efficiency	0.143	4691	0.158
6.	Linde Dual Pressure	60, 200 atm and 100% efficiency	0.057	3056	0.242
7.	Linde Dual Pressure	60, 200 atm and 70% efficiency	0.039	6535	0.113
8.	Claude system	40 atm and 100% efficiency	0.262	815	0.907
9.	Claude system	40 atm and 70% efficiency	0.198	1906	0.388
10.	Heylandt system	200 atm and 100% efficiency	0.377	873	0.846
11.	Heylandt system	200 atm and 70% efficiency	0.305	1839	0.402
12.	Cascade system	200 atm and 100% efficiency	0.045	3256	0.221

Problems

Problem: 1 Determine the liquid yield, work per unit mass compressed and work per unit mass liquefied for simple Linde Hampson and pre-cooled Linde-Hampson system. The working fluid is Nitrogen and Freon-12 is the refrigerant. The operating conditions for nitrogen are 101.3 kPa and 300 K and 20.3 Mpa at point 2. The standard enthalpies are

h_a = 207.94 kJ/kg at 101.3 kPa and 300 K

h_b = 250.20 kJ/kg at 681.7 kPa and 373 K

h_c = 61.23 kJ/kg at 300 K and saturated liquid

Point d is at 101.3 kPa and 243 K in two phase region, refrigerant flow rate ratio, $r = 0.10$.

Solution:

A. For simple Linde-Hampson cycle

From T-s diagram for Nitrogen,

h_1 = 462 J/g at 1 atm and 300 K

h_2 = 432 J/g at 200 atm and 300 K

h_f = 29 J/g at 1 atm and saturated liquid

s_1 = 4.42 J/gK at 1atm and 300 K.

s_2 = 2.74 J/gK at 200 atm and 300 K

s_f = 0.42 J/gK at 1 atm and saturated liquid

The liquid yield is given by $m_f/m = y = (h_1 - h_2)/(h_1 - h_f)$

$y = (462 - 432)/(462 - 29) = 0.0693$

The work requirement per unit mass compressed is given by

$- W/m = T_1(s_1 - s_2) - (h_1 - h_2) = 300 [(4.42 - 2.74) - (462 - 432)] = 474$ J/g

7.18 Cryogenics

The work requirement per unit mass liquefied is given by
$W/m_f = -(W/m)/y = 474/0.0693 = 6840$ J/g

The ideal work requirement per unit mass liquefied is given by
$W_1/m_f = T_1(s_1 - s_f) - (h_1 - h_f) = 300(4.42 - 0.42) - (462 - 29) = 767$ J/g

Figure of Merit (FOM) is given by
$FOM = (W_1/m_f)/(W/m_f) = 767/6840 = 0.1121$ ...(Ans.)

B. For pre-cooled Linde-Hampson cycle

The liquid yield is given by $m_f/m = y = (h_1 - h_2/h_1 - h_f) + r(h_a - h_1/h_1 - h_f)$
$y = (462 - 432)/(462 - 29) + (0.10)(207.94 - 61.23)/(462 - 29) = 0.1032$

The work requirement per unit mass compressed is given by
$-W/m = T_1(s_1 - s_2) - (h_1 - h_2) + r(h_b - h_d)$
$-W/m = 300[(4.42 - 2.74) - (462 - 432)] + 0.10(250.2 - 207.94) = 470$ J/g

The work requirement per unit mass liquefied is given by
$-W/m_f = -(W/m)/y = 470/0.1032 = 4554$ J/g

The ideal work requirement per unit mass liquefied is given by
$W_1/m_f = T_1(s_1 - s_f) - (h_1 - h_f) = 300(4.42 - 0.42) - (462 - 29) = 767$ J/g

Figure of Merit (FOM) is given by
$FOM = (W_1/m_f)/(W/m_f) = 767/4554 = 0.1684$ ...(Ans.)

Thus FOM of pre-cooled Linde-Hampson system is better than that of simple Linde-Hampson system.

Problem: 2 Determine the liquid yield, work per unit mass compressed and work per unit mass liquefied for Linde dual pressure system. The working fluid is Nitrogen and the operating conditions are 101.3 kPa and 300 K and 20.3 Mpa. The intermediate pressure is 5.07 Mpa and the intermediate pressure flow rate ratio is 0.80.

Solution:

From T-s diagram for Nitrogen,

$h_1 = 462$ J/g at 1 atm and 300 K
$h_2 = 452$ J/g at 50 atm and 300 K
$h_3 = 432$ J/g at 200 atm and 300 K
$h_f = 29$ J/g at 1 atm and saturated liquid
$s_1 = 4.42$ J/gK at 1atm and 300 K.
$s_2 = 3.23$ J/gK at 50 atm and 300 K.
$s_3 = 2.74$ J/gK at 200 atm and 300 K

The liquid yield is given by $m_f/m = y = (h_1 - h_3/h_1 - h_f) - i(h_1 - h_2/h_1 - h_f)$
$y = (462 - 432)/(462 - 29) - (0.80)(462 - 452)/(462 - 29) = 0.0508$

The work requirement per unit mass compressed is given by

$-W/m = [T_1 (s_1 - s_3) - (h_1 - h_3)] - i[T_1 (s_1 - s_2) - (h_1 - h_2)]$
$-W/m = 300 [(4.42 - 2.74) - (462 - 432)] - (0.80) (300)$
$\quad\quad [(4.42 - 3.23) - (462 - 452)]$
$\quad\quad = 196.4 \text{ J/g}.$

The work requirement per unit mass liquefied is given by
$W/m_f = -(W/m)/y = 196.4 / 0.0508 = 3866 \text{ J/g}$

The ideal work requirement per unit mass liquefied is given by
$W_1/m_f = T_1(s_1 - s_f) - (h_1 - h_f) = 300 (4.42 - 0.42) - (462 - 29) = 767 \text{ J/g}$

Figure of Merit (FOM) is given by
$\text{FOM} = (W_1/m_f) / (W/m_f) = 767 / 3866 = 0.1984$ ...(Ans.)

Problem : 3 Determine the liquid yield, work per unit mass compressed and work per unit mass liquefied for Claude system. The working fluid is Nitrogen and the operating conditions are 101.3 kPa and 300 K and 5.06 Mpa (50 atm). The expander flow rate ratio is 0.60 and expander work is utilised to compress the incoming gas. The condition of gas at inlet of expander is 270 K and 50 atm.

Solution:

From T-S diagram for Nitrogen,

$h_1 = 462$ J/g at 1 atm and 300 K

$h_2 = 452$ J/g at 50 atm and 300 K

$h_3 = 418$ J/g at 50 atm and 270 K

$h_e = 238$ J/g at 1 atm and 86 K

$h_f = 29$ J/g at 1 atm and saturated liquid

$s_1 = 4.42$ J/gK at 1atm and 300 K.

$s_2 = 3.23$ J/gK at 50 atm and 300 K.

$s_3 = se = 3.11$ J/gK at 50 atm and 270 K

The liquid yield is given by $y = m_f/m = (h_1 - h_2 /h_1 - h_f) + x(h_3 - h_e / h_1 - h_f)$
$y = (462 - 452) / (462 - 29) + (0.60) (418 - 238) / (462 - 29) = 0.2725$

The work requirement per unit mass compressed is given by
$-W/m = [T_1 (s_1 - s_2) - (h_1 - h_2)] - x(h_3 - h_e)$
$-W/m = 300 [(4.42 - 3.23) - (462 - 452)] - (0.60) (418 - 238) = 239 \text{ J/g}$

The work requirement per unit mass liquefied is given by
$W/m_f = -(W/m)/y = 239 / 0.2725 = 877 \text{ J/g}$

The ideal work requirement per unit mass liquefied is given by
$W_1/m_f = T_1(s_1 - s_f) - (h_1 - h_f) = 300 (4.42 - 0.42) - (462 - 29) = 767 \text{ J/g}$

Figure of Merit (FOM) is given by
$\text{FOM} = (W_1/m_f) / (W/m_f) = 767 / 877 = 0.897$ ...(Ans.)

References

1. Cryogenic Engineering – B.A. Hands.
2. Internet website www.wikipedia.com.
3. Linde Brochure on LNG plants.
4. Cryogenic Systems – Randall Baron.
5. Cryogenic Engineering – P.K. Bose.
6. Internet website www.Kryopak.com.
7. Shimko et.al. USDOE Report on Innovative Hydrogen Liquefaction.
8. Internet website www.chemistry.mcmaster.ca.
9. S.M.Dash – Report on cryogenic cycles – NIT, Rourkela.
10. Evaluation of LNG Technologies – Report by Valerie Reviera (2008).
11. Thesis by Balaji Kumar Chowdhary, NIT Rourkela.

Questions

Q.1. Write a short note on liquefaction of gases.

Q.2. Discuss the operation of a simple Linde-Hampson liquefaction plant.

Q.3. Write a short note on Joule-Thomson effect.

Q.4. Explain the differences between Cascade plant and Linde dual pressure plant.

Q.5. Discuss the Kapitza and Heylandt liquefaction plants with a neat sketches.

Q.6. Write a short note on thermodynamically ideal liquefaction system.

Q.7. Discuss the operation of a pre-cooled Linde-Hampson plant.

Q.8. What are the system performance parameters for liquefaction plants?

Q.9. Determine the liquid yield, work per unit mass compressed and work per unit mass liquefied for Claude system. The working fluid is Nitrogen and the operating conditions are 101.3 kPa and 300 K and 10.12 MPa (100 atm). The expander flow rate ratio is 0.50 and expander work is utilised to compress the incoming gas. The condition of gas at inlet of expander is 270 K and 100 atm.

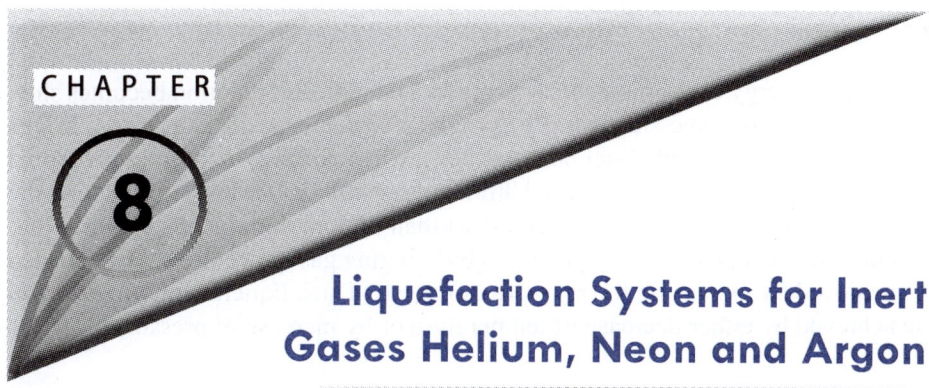

CHAPTER 8

Liquefaction Systems for Inert Gases Helium, Neon and Argon

INTRODUCTION

During the late 1700s, scientists learned that when a gas is cooled, its volume is reduced by a predictable amount. Cooling slows down the motion of the gas molecules, so they take up less space. Similarly, pressurizing a gas, or forcibly squeezing its molecules closer together, reduces its volume. Eventually, through cooling and compression, the volume of a gas can be reduced by so much that its molecules collapse upon each other and come into contact, changing into a liquid. Compression and cooling soon became the twin tools of scientists attempting to liquefy gases. Near the end of the century, a new family of elements called the *inert*, or nonreactive, gases was discovered by British chemist Sir William Ramsay and his co-workers. These gases–argon, helium and neon presented challenge to scientists interested in liquefaction. Dewar came very close to liquefying helium, but his sample of the gas also contained neon, which froze and blocked the valves on his equipment. A few years later, in 1908, Dutch physicist Heike Kamerlingh Onnes succeeded in producing liquid helium using techniques similar to Dewar's. The compressed helium gas was pre-cooled by liquid hydrogen before undergoing expansion cooling, as in the Linde process. Onnes' equipment was rather elaborate and later other scientists developed simpler helium liquefiers that could produce greater amounts of the liquid gas.

Liquefaction of inert gases includes a number of phases used to convert a gas into a liquid state. Liquid inert gases are used for scientific, industrial and commercial purposes. Many gases can be put into a liquid state at normal atmospheric pressure by simple cooling; a few, require pressurization as well. Liquefaction is used for analyzing the fundamental properties of gas molecules (intermolecular forces), for storage of gases, The liquefaction of gases is a complicated process that uses various compressions and expansions to achieve high pressures and very low temperatures; using for example turboexpanders. For inert gases, there are large empty spaces (voids) separating the tiny molecules of gases from one another. Each molecule enjoys an almost independent existence. Molecules are in a state of continuous rapid motion and have negligible attractive forces between them due to wide separation. This is particularly so, when temperature is high and pressure is low. When the temperature of the gas is lowered, both the volume of the gas and

8.2 Cryogenics

the kinetic energy of the molecules decrease. The molecular motion becomes slow and molecules become sluggish. The progressive decrease of temperature brings the molecules closer and closer because they are unable to resist the attractive force that starts operating between them. Ultimately, at sufficiently low temperature, the voids between the molecules become less than 10^{-5}cm and the gas changes into liquid state. This process of liquefaction by bringing gas molecules closer can also be achieved by increasing the pressure of the gas. Thus, liquefaction of gases can be achieved by either decrease of temperature or by increase of pressure.

LIQUEFACTION OF ARGON

Argon (Ar) is a monatomic, chemically inert gas composing slightly less than 1% of the air. Its gaseous specific gravity is 1.38 and its boiling point is $-185.9°C$. Argon is colorless, odorless, tasteless, non corrosive, nonflammable, and nontoxic. Commercial argon is the product of cryogenic air separation, where liquefaction and distillation processes are used to produce a low-purity "crude" argon product which is then purified to the commercial product. Argon is used primarily for its properties as an inert gas in applications such as arc welding, steel making, heat treating, and electronics manufacturing. The key to the liquid argon manufacture is the air separation process within the Air Separation Unit (A.S.U.). Here, the Air is separated to Argon, Oxygen, and Nitrogen by virtue of the difference in their liquefaction temperatures. The design and efficiency of the A.S.U. determines the quality of the final product i.e. Argon in terms of quantity and quality. Current designs of A.S.U. features highly efficient fin type aluminum heat exchangers and column, wherein the entire upper column is a packed column thus reducing the operational pressure and increasing Argon recovery. Argon is recovered by a revolutionary technique employing full rectification without using the Hydrogen and De-Oxo unit thus saving further on power costs, operational costs and investment. Refer to Chapter 12 for detailed discussions on A.S.U. In this section we will focus on another technique for generating Argon i.e. from the purge gas recovery of ammonia plants.

ARGON MANUFACTURE FROM PURGE GAS RECOVERY OF AMMONIA PLANTS

Argon's normal boiling point is a very cold i.e. $-185.9°C$. The gas is approximately 1.4 times as heavy as air and is slightly soluble in water. Argon's freezing point is only a few degrees lower, $-199.3°C$. Plants for Argon recovery from purge gas of Ammonia plants is a well established technique giving Argon of 99% purity levels. Cryogenic technology has a number of advantages compared to membrane technology. The cryogenic technology has in particular a higher economic efficiency, an unlimited service life within the expected life of a plant as well as a high degree of separation for Argon. Figure 8.1 shows a typical argon production unit. Argon is introduced into ammonia plants as part of the air stream, that along with natural gas, are the raw materials used to produce a mix of nitrogen and hydrogen, which is then reacted at high pressure to produce ammonia. The recoverable quantity of Argon from the purge gas for typical operation mode of the plant is governed by the laws of physics and the quantity is mainly determined by the pressure of the fuel gas and hydrogen. This purge recovery process allows separation and production of pure liquid argon. Due to the development in the field of air separation plants, argon recovery from the purge gas has lost some of its significance.

One unit of Argon recovery requires between 5 000 and 60 000 Nm^3/h of purge gas. Combination of purge gas quantities from several NH_3 plants/units could result in substantial savings. The process stages and the equipment are calculated from data bases and engineering design is performed via CAD-systems. Plant process design and calculation ensure operability also in case of considerably reduced purge gas quantities. The main item of the plant is the cold box which is equipped with special low temperature plate fin heat exchangers ensuring a high product yield and large processing quantities for all usual pressure stages up to 100 bar. Temperatures as low as –195°C can be generated for the process. For special cases of operation in higher pressure ranges, coil heat exchangers are provided. A continuous quality control and quality assurance system is ensured to produce high quality argon. Also equipment for removal of moisture and NH_3 traces by adsorption from Argon is provided.

Fig. 8.1 Linde Argon Production Unit from Purge Gas of Ammonia Plant

8.4 Cryogenics

LIQUEFACTION OF NEON

Neon is the chemical element that has the symbol Ne and atomic number 10. Although a very common element in the universe, it is rare on Earth. A colorless, inert noble gas under standard conditions, neon gives a distinct reddish-orange glow when used in neon lamps. It is commercially extracted from air, in which it is found in trace amounts. Neon (Greek meaning "new one") was discovered in 1898 by Sir William Ramsay (1852–1916) and Morris W. Travers (1872–1961) in London. Neon was discovered when Ramsay chilled a sample of the atmosphere until it became a liquid, then warmed the liquid and captured the gases as they boiled off. In December 1910, French engineer Georges Claude made a lamp from an electrified tube of neon gas.

Liquefied neon is commercially used as a cryogenic refrigerant in applications not requiring the lower temperature range attainable with liquid helium refrigeration. Liquid neon is actually quite expensive, and nearly impossible to obtain in small quantities for laboratory tests. The driver for expense is actually rarity of the gas, not the liquefaction process. Liquefaction of air is a large-scale commercial process today. This process provides the primary source of liquid oxygen at –118°C, and liquid nitrogen at –150°C. At –150°C, neon remains in the gas phase (boiling point of neon is –229°C). In fact, at this temperature, the gas phase is comprised of roughly 75% neon, 24% helium, and a trace of hydrogen. By passing the cool gas through activated charcoal helium and hydrogen can be removed. Commercial liquid neon is thus produced as a secondary product at liquid N_2/liquid O_2 plants. In this section we will study the pre-cooled Linde Hampson system and Claude turbo-expander system for Neon liquefaction.

PRE-COOLED LINDE HAMPSON (LH) CYCLE FOR NEON LIQUEFACTION

The Linde-Hampson method is a thermodynamic process, where isothermal compression and subsequent isobaric cooling is done in a heat exchanger. Joule-Thomson expansion connected with an irreversible change in entropy is used as the refrigeration process. Despite its simplicity and reliability, this method has become less attractive compared to modern ones, where cooling is carried out in reversible processes (expander) at reduced energy consumption. Pre-cooled Linde-Hampson method is patented by Linde Corporation and involves precooling with the standard LH cycle. This is a generic method for liquefying a neon stream to produce liquid neon. The liquefaction of the neon stream takes place countercurrent to a refrigerant mixture cycle cascade consisting of two refrigerant mixture cycles. The first refrigerant mixture cycle is used for precooling and a second refrigerant mixture cycle is used for liquefaction and super cooling of the neon stream to be liquefied. Refer Figure 8.2 for the representation of the LH process on the P-H (Pressure–Enthalpy) diagram. No additional process steps are involved in a heat exchange between the neon stream which is to be pre-cooled and the refrigerant mixture of the first refrigerant mixture cycle. The term "precooling" means the cooling of the neon stream down to a temperature at which the separation of heavy impurities takes place. The subsequent further cooling of the neon stream to be liquefied hereinafter comes under the term "liquefaction".

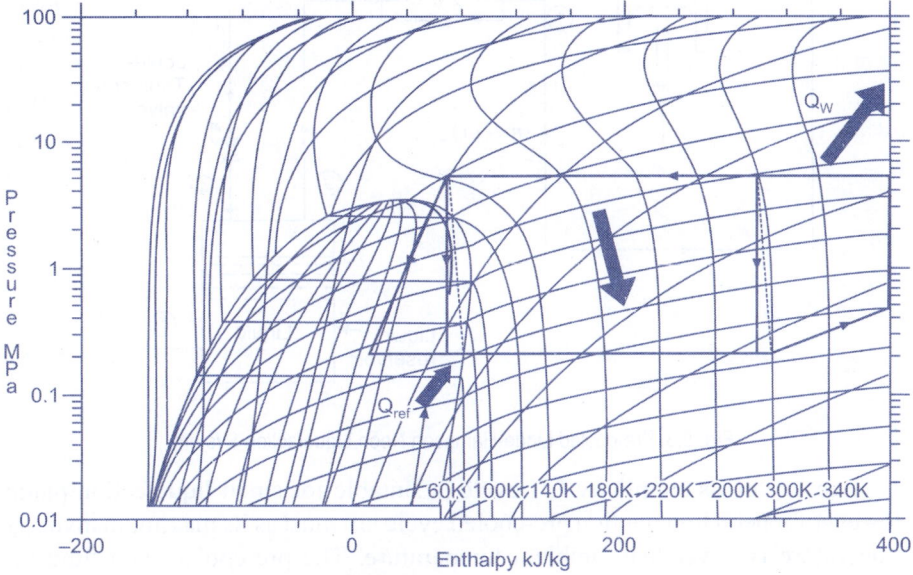

Fig. 8.2 Pre-Cooled Linde Hampson Cycle on the P-H Diagram

The basic Linde-Hampson cycle uses a Joule-Thomson orifice. A gas could be cooled by letting it expand freely against the atmosphere. This could be explained by that the gas was doing a work against the atmosphere by lifting and/or heating it, and thereby loosing energy in form of heat. The Linde-Hampson system is one of the simplest cryocoolers so it is quite desirable for small-scale liquefaction plants. It can be used to liquefy neon but also it is used to liquefy hydrogen, nitrogen and helium. This type of pre-cooled system was used in several of the earlier helium liquefiers. A scheme of the Linde-Hampson system is represented in Figure 8.3. This system consists of two independent circuits, the upper one is the pre-cooled system and the lower one is the basic circuit of the gas that we want to liquefy. The first one is employed to lower the temperature of the neon (basic circuit) below the ambient temperature. The pre-cooled system is filled with a pre-coolant fluid that has a triple-point temperature below that of the maximum inversion temperature of neon. Liquid nitrogen is a suitable fluid to fill the pre-cooler system for liquefying neon. The basic circuit consists of a compressor employed to raise the enthalpy (so the temperature and the pressure are also raised) of the neon, then it is cooled through three heat exchangers, two of them involving the pre-cooling system. Then neon passes through a valve to a tank, were liquid helium is extracted within a channel. In this tank, neon gas is sent to the compressor after being heated through two heat exchangers.

8.6 Cryogenics

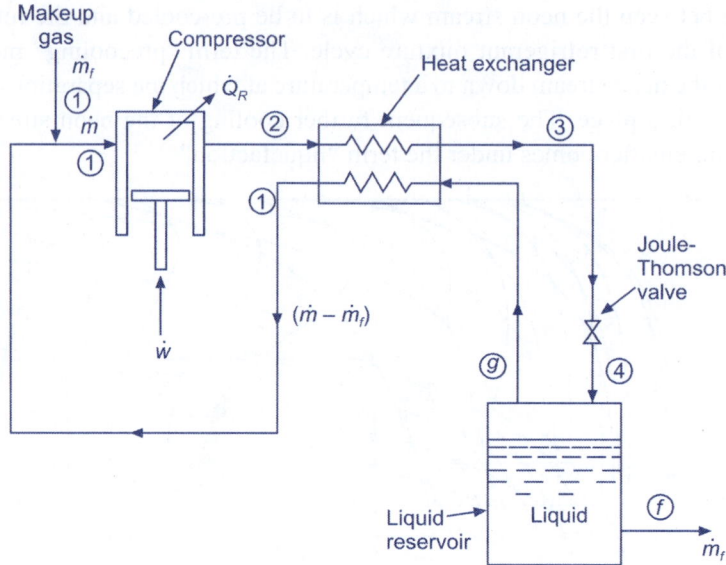

Fig. 8.3. Pre-cooled Linde-Hampson Neon Liquefaction System

Because of its simplicity LH cycle is suitable for small liquefaction plants. However for neon liquefaction pre-cooled cycle is a must as its maximum inversion temperature is lower than ambient temperature. The pre-coolant is a fluid like nitrogen which has a triple point lower than maximum inversion temperature of neon. The yield of the cycle is given by the equation $y = (h_1 - h_4)/(h_1 - h_f)$. The cycle yield could be improved by lowering the temperature at the entrance of the heat exchanger.

Claude Cycle for Neon Liquefaction

A commonly applied method in large-scale liquefaction neon plants is the Claude process, where the necessary refrigeration is provided in four main steps: 1. Compression of neon gas, removal of compression heat; 2. Pre-cooling with liquid nitrogen (80 K); 3. Cooling a fraction of the neon in an turbo expander (30 K). 4. Expanding the residual neon in a Joule-Thomson valve (20 K). Joule-Thomson expansion is applied for the final step to avoid two-phase flow in the expander. Further improvement in efficiency is expected with the development of new materials and new compression/expansion technology. Refer Figure 8.4 for the Claude refrigeration cycle and its T-s diagram and Figure 8.5 for a typical turbo-expander used in the cycle.

This is the most adopted cycle for large neon liquefaction plants in the world. Neon gas is used not only in the feed line to be liquefied but also in the recycling line which generates necessary cold for the liquefaction. The feed gas is compressed to approximately 5 MPa, and then cooled down to 80 K. Finally the feed neon gas is liquefied at 0.1MPa, 20.4K by expansion at the J-T valve. In order to improve the process efficiency, a process with a super critical expansion turbine installed in the feed line has been developed. In the Claude process some of the high-pressure gas is passed through an expansion engine and is then sent to a heat exchanger for cooling.

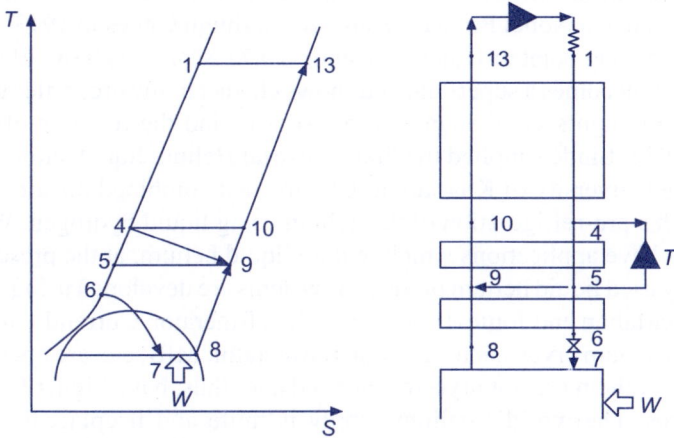

Fig. 8.4 Claude Neon Liquefaction System

This process referred to as ideal liquefaction uses a reversible expansion process to reduce the energy required for liquefaction. It consists of an isothermal compressor followed by an isenthalpic expansion to cool the gas and produce a liquid. In practice, an expansion engine can be used only to cool the gas stream, not to condense it because excessive liquid formation in the expansion engine would damage the turbine blades. Claude system has figure of merit 50% higher than pre-cooled Linde-Hampson System.

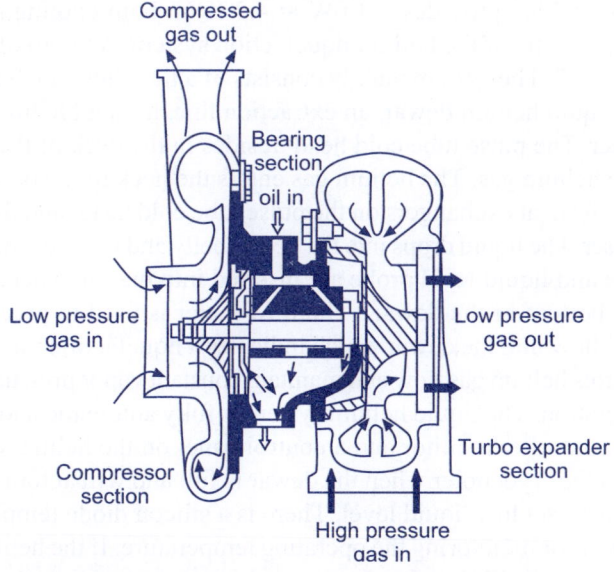

Fig. 8.5 Turbo Expander used in Claude Neon Liquefaction System

LIQUEFACTION OF HELIUM

Helium first liquefies at a temperature of 272.2 degrees below zero Celsius, in other words close to the absolute zero point. Liquefaction of Helium with the Hampson-Linde cycle led to a Nobel Prize for Heike Kamerlingh Onnes in 1913. At ambient pressure the boiling point of liquefied helium is 4.22 K or −268.93°C. Below 2.17 K liquid Helium becomes a superfluid and shows characteristic properties such as heat conduction through second sound, zero viscosity and the fountain effect among others. In 1932, Linde supplied the first industrial Helium liquefaction plant in the world to the University of Kharkov in Ukraine that combined the air liquefaction process with a pre-refrigeration of the helium using liquid hydrogen. While many laboratories have applications which require liquid helium, at the present time the most widely used liquid helium producing systems are developed using a combined Gifford- McMahon and Joule-Thomson cycle refrigerator. Currently many helium Dewars and helium cryostats for dilution refrigerator, MRI, etc. are operating in the field. Some of them are not cryo-refrigerated and, thus, have higher liquid helium boil-off rates. The world's helium supply is finite and irreplaceable. Growing demand for helium worldwide continuously increases pressure on costs and supply in recent years and in the near future. One of the promising solutions is recovery and recycling of helium by using small helium liquefaction and recondensing systems. In this section we shall study some helium liquefaction systems.

PULSE TUBE (PT) HELIUM LIQUEFACTION SYSTEM

Pulse Tube helium liquefaction system have been developed to provide the liquefaction rate of approx 18 L/day. The PT system employs a two-stage 4 K pulse tube cryocooler which provides ≥ 1.0W at 4.2 K on both cooling stages. Figure 8.6 shows a schematic of the helium liquefaction system. A photo of the system is given in Figure 8.7. The system mainly consists of a liquefier which is a pulse tube cryocooler, a liquid helium dewar, an extraction line, a liquid helium level sensor and a controller. The pulse tube cold head resides in the neck of the dewar where it liquefies the helium gas. The helium gas enters the neck of the dewar and is first pre-cooled by the heat exchangers on the pulse tube cold head, and then condensed on the condenser. The liquid drops into the dewar belly and is stored there. The liquid extraction line and liquid level probe are inserted into the same neck of the dewar and reach the bottom of the dewar. The flow meter is used to precisely measure the helium gas flow into the dewar which indicates a liquefaction rate. The pressure regulator controls helium gas flow and maintain constant vapor pressure in the dewar during the operation. The liquid helium system is fully automatic and controlled by a liquid level controller/monitor and a control panel on the helium compressor. It will shut down the cryocooler when the dewar is full and will automatically restart the system at a preset low liquid level. There is a silicon diode temperature sensor on the condenser for monitoring the operating temperature. If the helium gas supply stops, the cryocooler will shut down at preset low temperature on the control panel due to the low temperature reading on the condenser.

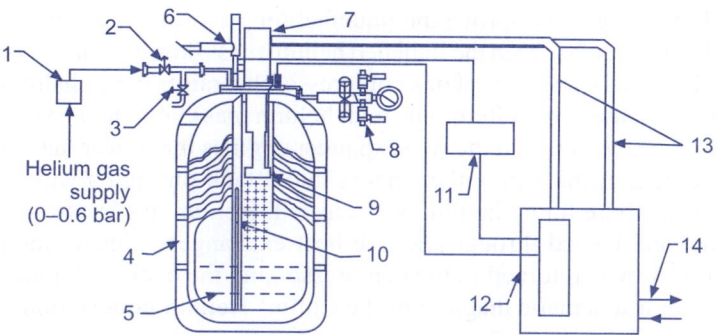

Schematic of helium liquefaction system: 1. Mass flow meter; 2. Pressure regulator; 3. Vacuum valve for dewar pumping; 4. Liquid helium dewar; 5. Liquid helium; 6.Liquid withdrawal line; 7.Cold head of pulse tube cryocooler; 8.Safety unit; 9. Temperature sensor; 10. Liquid helium level sensor; 11. Liquid helium level indicator/controller; 12. Helium compressor of cryocooler; 13. High & low SS flexible lines of cryocooler; 14. Cooling water.

Fig. 8.6 Schematic of Pulse Tube Helium Liquefaction System

Natural convection causes flow of helium gas in the neck of the dewar. The helium gas is pre-cooled and liquefied in the neck of the dewar by the cold head of the pulse tube cryocooler. The surfaces of the pulse tubes cool the helium gas and generate a strong downwards flowing boundary layer. A portion of the downward flow enters into the condenser and is liquefied. The helium gas before entering into the condenser is pre-cooled down to the temperatures of 5-6 K. Other portions of the downward flow enhance the thermo-siphon mass flow in the neck. This flow enhancement will increase the vapor cooling of the dewar neck and reduce the heat leak and boil-off of the liquid bath.

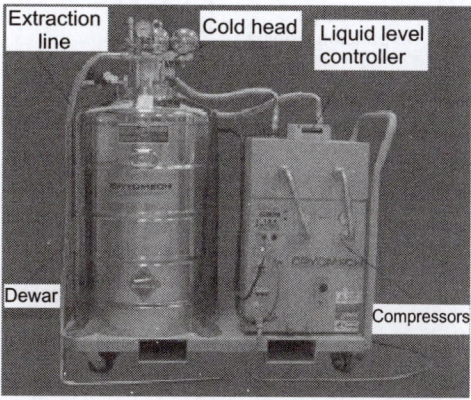

Fig. 8.7 Photograph of Pulse Tube Helium Liquefaction System

COLLINS HELIUM LIQUEFACTION SYSTEM

The development of this liquefier was done at the Massachusetts Institute of Technology. This liquefier is an extension of the Claude system. Depending upon the helium inlet pressure, generally five heat exchangers and two expansion engines are used in this system. In some cases a pre-cooling bath is employed in order to reduce

8.10 Cryogenics

the cool down time and improve the liquefaction performance of the system. The liquid yield (the ratio between the liquefied helium mass flow and the total mass flow) can be tripled. This cycle consist of a compressor that raises the enthalpy along with the pressure and the temperature. The total helium mass flow (m) is passed through the five heat exchangers and the two expander engines that lower the temperature. The cooled helium mass flow then enters into a tank through a valve (to reduce the pressure), where liquid helium is evacuated. Helium gas is extracted through the channel and heated through the five heat exchangers. Finally the remaining helium mass flow is returned to the compressor and the cycle is completed. Refer Figure 8.8 for a schematic diagram of the Collins Helium liquefaction system.

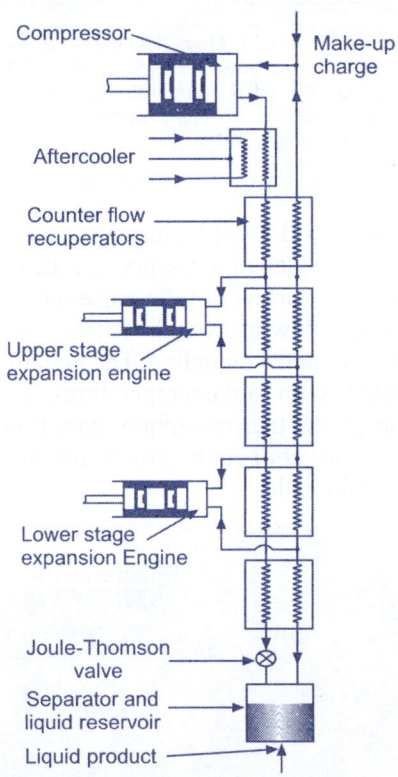

Fig. 8.8 Schematic Diagram of Collins Helium Liquefaction System

The Collins Helium Cryostat designed in 1946 was a major innovation. First, the flexible rod expander was refined by moving the warm mechanism above the expansion cylinder rather than below the cylinder, Secondly, the cold components of the liquefier were suspended in the low-pressure helium in the long neck of a wide mouth Dewar vessel. Third, the heat exchanger was also suspended in the helium in the Dewar neck and surrounded the expanders. Fourth, the main heat exchanger was spiral wound with tubing with helically wound fins over the outside diameter of the tubes. The final innovation was the arrangement of the flow to the two expanders at different temperatures, so that two stages of pre-cooling were provided for the Joule-Thomson stage rather than a single stage of pre-cooling. Refer Figure 8.9 for the photograph of Collins helium liquefier.

Fig. 8.9 Photograph of Collins Helium Liquefaction System

SIMON HELIUM LIQUEFACTION SYSTEM

In 1932 Simon designed a new method for liquefying helium in a single shot isentropic expansion, shown schematically in Figure 8.10. In a container B helium gas is compressed to a pressure of 100 atm and cooled to a temperature of about 12 K of liquid hydrogen under reduced pressure, the heat being transferred through a helium exchange gas in space Z. It is then thermally isolated by means of evacuating Z. The container B is connected with the outer space by means of a capillary that is closed with a valve. If this valve is opened the helium in the container performs work (external and against the Van Der Waals forces) during the expansion to a pressure of 1 atm, and after the expansion a substantial fraction of the space B is filled with liquid helium. In this temperature range this method is particular suited because of the small heat capacity of the metal vessel B. An experimental fact unknown during Onnes's first liquefaction.

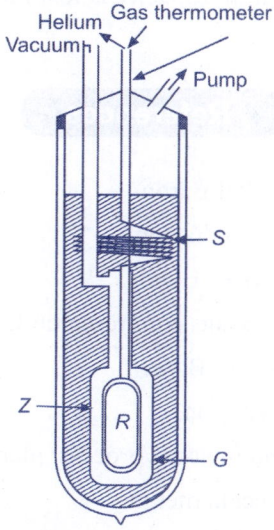

Fig. 8.10 Schematic Diagram of Simon's Helium Liquefaction System

In the same way, in this period the suggestion was made of adiabatic isentropic demagnetization of a paramagnetic salt in 1926 independently by Debye and Giauque in 1933. A paramagnetic salt is magnetized in a strong magnetic field at a temperature of about 1 K, the heat of magnetization is transferred through helium exchange gas to the helium bath. After the exchange gas is pumped away, the magnetic field is slowly reduced to zero and the temperature of the salt drops very strongly down. Simon and Kurti even coupled the isentropic demagnetization to his single isentropic expansion of the helium.

This system is employed to liquefy small quantities of helium. It does not operate as a steady-flow system, but produces helium in a batch process. The Simon Liquefaction system consists of the following processes: Process 1-2: Helium gas is introduced into a heavy-wall container at a pressure of 150 atm, Process 2-3: Liquid nitrogen is introduced into the enclosing bath. The vacuum space is filled with helium gas at atmospheric pressure, to act as a heat transfer medium. Liquid nitrogen cools the system to 77 K. Then the vacuum pressure provides thermal isolation. Process 3-4: The upper part of the inner container is filled with liquid hydrogen in order to cool helium to 20 K. Process 4-5: The pressure of the liquid hydrogen is reduced to 10 K, so the liquid hydrogen boils as the pressure is lowered until the hydrogen solidifies. The heavy-walled container is isolated. Process 5-6: The pressure of the helium contained in the heavy-walled container is reduced to atmospheric pressure (1 atm) by opening the helium exhaust valve. The released helium is vented, and the remaining helium liquefies. This system is only recommended to liquefy helium, for two reasons. The first one is that the ratio v_g/v_f is smaller for helium than any other gas at the normal boiling point. ($v_g/v_f = 7.5$ for helium, 53.3 for hydrogen and 175 for nitrogen), which means that for a given liquid yield, the relation between liquid mass flow and total mass flow is larger for helium than for other gases. The second one is that specific heats of metals is extremely small at helium temperatures, so only a small amount of cooling capacity is lost into the walls of the vessel for helium liquefaction.

References

1. Cryogenic Systems – Randall Baron.
2. Internet Website www.wikipedia.com.
3. Cryogenic Engineering – B.A. Hands.
4. Internet Website www.bookrags.com/research/liquefaction-of-gases.
5. Cryogenic Engineering – P.K. Bose.
6. Air Products Brochure on Argon.
7. Internet Website www.linde-le.com/process_plants/.
8. Internet Website www.veccal.ernet.in.
9. MIT Report by Smith on Helium Liquefaction.

10. The 25th International Conference on Low Temperature Physics (LT25) IOP Publishing Journal of Physics: Conference Series 150 (2009) Number 012053.
11. Ouboter et.al. J. Phys.: Condensed. Matter 21 (2009) 164221.

Questions

Q.1. Write a short note on liquefaction of inert gases.

Q.2. Discuss the operation of a Collins Helium liquefaction plant.

Q.3. Write a short note on Pre-cooled helium liquefaction plant for Neon.

Q.4. Discuss the Claude liquefaction plant for Neon with a neat sketch.

Q.5. Write a short note on liquefaction of Argon.

Q.6. Discuss the operation of a Pulse tube cryocooler Liquid Helium plant.

CHAPTER 9

Liquefaction Systems for Natural Gas and Hydrogen

INTRODUCTION

Natural gas in its liquid state is called liquefied natural gas (LNG). LNG comprises of liquid hydrocarbons that are recovered from natural gases in gas processing plants. These hydrocarbons involves propane, pentanes, ethane, butane and some other heavy elements. LNG accounts for about 4% of natural gas consumption worldwide, and is produced in dozens of large-scale liquefaction plants. It is produced by cooling natural gas to a temperature of minus 160 Celsius. At this temperature, natural gas becomes liquid and its volume reduces by 615 times. LNG occupies 1/600th the volume of natural gas at atmospheric temperature and pressure. LNG has high energy density, which makes it useful for energy storage in double-walled, vacuum-insulated tanks. The production process of LNG starts with natural gas, being transported to the LNG plant site as feedstock. After filtration and metering in the feedstock reception facility, the feedstock gas enters the LNG plant and is distributed among the identical liquefaction systems. Each LNG process plant consists of reception, acid gas removal, dehydration/mercaptan removal, mercury removal, gas chilling and liquefaction, refrigeration, fractionation, nitrogen rejection and sulfur recovery units.

In an effort to design engineer, and manufacture the most cost effective, space and weight efficient LNG facility possible, many factors must be considered. The first thing that must be determined is what detrimental contaminants exist in the entering gas stream. These contaminants can include, but are not limited to, oxygen, nitrogen, water, carbon dioxide (CO_2), hydrogen sulfide (H_2S), mercury, arsenic and/or heavy hydrocarbons (C_3+). Each of these components can create significant problems for the operation of an LNG plant. For example, the CO_2 content in a gas stream entering an LNG plant must be reduced to less than 50 ppm to avoid the formation of dry ice within the system which can plug off equipment and shutdown the plant. Similarly, mercury in the gas stream can attack the aluminum components often used in LNG plant heat exchangers and other equipment. Depending on the amount of H_2S contained in the inlet gas, an H_2S scavenger system may be used to remove the sulfur before entering any other part of the plant system. Oxygen is typically not found in the gas stream feeding an LNG plant, but this must be verified before proceeding further. If oxygen is present, it must be removed before entering the downstream amine unit where it would form undesirable byproducts.

9.2 Cryogenics

Depending on the amount of CO_2 contained in the inlet gas stream and the volume of gas entering the plant, it may be beneficial to remove the bulk amount of CO_2 using a membrane treating system in order to minimize the size of the downstream amine plant and reduce the overall energy consumption of the plant. An amine plant is used to remove essentially all of the CO_2 and H_2S from the inlet gas stream. In order for the LNG plant to operate properly and reliably, the CO2 should be removed to a level of less than 50 ppm. Upon leaving the amine plant, the oxygen, CO_2 and H_2S have all been removed to acceptable levels to enter the LNG plant.

The next step is to dry the gas to the point that it will contain less than 1 ppm of H_2O. A Molecular Sieve (mol sieve) dry desiccant is the industry standard for performing this function. The number of beds is generally determined by the volume of gas being dehydrated and the water content in the inlet gas stream. One or more of the dehydration beds operate in the adsorption phase, where water vapor is adsorbed onto the desiccant, while one bed is heat regenerated to strip water from the mol sieve. Mol sieves can also be designed to remove trace amounts of CO_2, H_2S and mercaptan, if it is known before hand that these contaminants are present and need to be removed. Properly designed mol sieve systems will remove water to less than 1 ppm. If mercury is present, a separate vessel filled with activated carbon, is typically used to remove mercury gas stream. These beds are typically located downstream of the mol sieve system to keep water from deactivating the bed. Mercury removal systems are generally designed to reduce the mercury content in a gas stream to less than 10 nano-grams per cubic meter. Arsenic removal systems are a virtual duplicate of mercury removal systems in appearance, but utilize a different bed material to remove arsenic and the various arsines that may be present in the gas stream. Figure 9.1 shows a typical LNG plant.

Fig. 9.1 LNG Plant

Finally, depending on the quality of the inlet gas and how "clean" of an LNG product is desired, the gas may be "conditioned" to remove the heavy-end hydrocarbons from the gas stream before it enters the actual LNG liquefaction plant. The recovery of these heavy-end hydrocarbons can be accomplished using something as simple as a propane refrigeration plant to a full-scale cryogenic gas plant, complete with turbo-expander. Depending on the extent that the ethane and heavier components (C_2+) are removed, the feed to the LNG liquefaction plant may consist of only methane and nitrogen. The nitrogen will be separated from the methane in the LNG liquefaction plant.

Thus the process through which Liquefied Natural Gas is produced consists of three main steps, namely

1. **Transportation of Gas:** The best place to install the plant is near the gas source. The gas is basically transported through pipelines or by truck and barge.
2. **Pretreatment of Gas:** The liquefaction process requires that all components that solidify at liquefaction temperatures must be removed prior to liquefaction. This step refers to the treatment the gas requires to make it liquefiable and includes compression, filtering of solids, removal of liquids and gases that would solidify under liquefaction, and purification which is removal of non-methane gases and finally
3. **Liquefaction of Gas:** The liquefaction of the gas follows the liquefaction curve as shown in Figure 9.2. The graph proves that the area above the LNG cooling curve represents work. So the bigger the area between the starting temperature (T_{high}) and the curve, the more work the system has to perform.

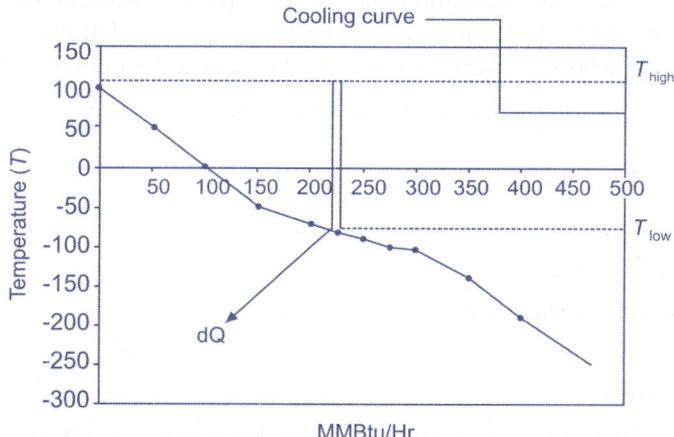

Fig. 9.2 LNG Liquefaction Curve

The work done for liquefaction of natural gas (W) is given by the following equations:

$$W = dQ\left(1 - \frac{T_{Low}}{T_{High}}\right)$$

$$W = dQ\left(\frac{T_{High} - T_{Low}}{T_{High}}\right)$$

$$W = \frac{dA}{T_{High}}$$

Where T high and T Low are the extreme temperatures, dA is the area under the curve a dQ is heat transfer during the liquefaction process.

LIQUEFIED NATURAL GAS PLANTS

There exist a vast number of natural gas liquefaction plants designs, but, all are based on the combination of heat exchange and refrigeration. The gas being liquefied, however, takes the same liquefaction path. The dry, clean gas enters a heat exchanger and exits as LNG. The capital invested in a plant and the operating cost of any liquefaction plant is based on the refrigeration techniques. Though, Liquefied Natural Gas can also be extracted from cryogenic hydrocarbon extraction and petro-chemical processes, but it requires careful consideration at these facilities to assure the process gas is liquefiable.

A typical LNG plant is comprised of the following units: 1) Feed gas compression unit to boost the feed pressures in case the natural gas pressure is low, 2) A Carbondioxide and moisture removal unit which operates mostly by a wash process and drying (CO_2 and H_2O would otherwise freeze and cause clogging in the downstream liquefaction equipment), 3) Natural gas liquefaction unit, 4) LNG storage facility, 5) LNG loading and metering stations, 6) Utilities such as gas turbine with waste heat recovery for hot oil heating and boil off gas handling system.

Currently the following natural gas liquefaction process based plants are available:
1. Joule-Thomson expansion cycle
2. Linde multistage mixed refrigerant cycle
3. Conoco Phillips cascade process
4. Mixed fluid cascade process
5. APCI
6. Shell DMR.

The selection of a particular gas liquefaction process is driven by customer requirements, plant capacity, and the availability of resources. It is expected that by the end of 2012, there would be 100 liquefaction plants with total capacity of 297.2 MMTPA. The majority of these plants use either APCI or mixed fluid cascade technology for the liquefaction process. The other processes, used in a small minority of some liquefaction plants include Shell's DMR technology. We shall now study each natural gas liquefaction process in greater detail.

Joule-Thomson Expansion Cycle LNG Plant

The Joule-Thomson expansion process utilizes the expansion process to generate the required refrigerant for liquefaction using the inlet gas. Gas is compressed at ambient pressure and cooled in a heat exchanger. It is then passed through a throttle valve where it undergoes isenthalpic Joule-Thomson expansion producing some liquid. The liquid is removed and the cool gas is returned to the compressor via the heat exchanger. This simple system has very few components operating at cryogenic temperatures, when compared to alternative methods for creating LNG, resulting in a facility that is less complicated complex to own and operate. Figure 9.3 shows the schematic of a Joule-Thomson expansion cycle LNG plant [6]. The process is less efficient when compared to a mixed refrigerant cycle, using 15% more power to create the final LNG product.

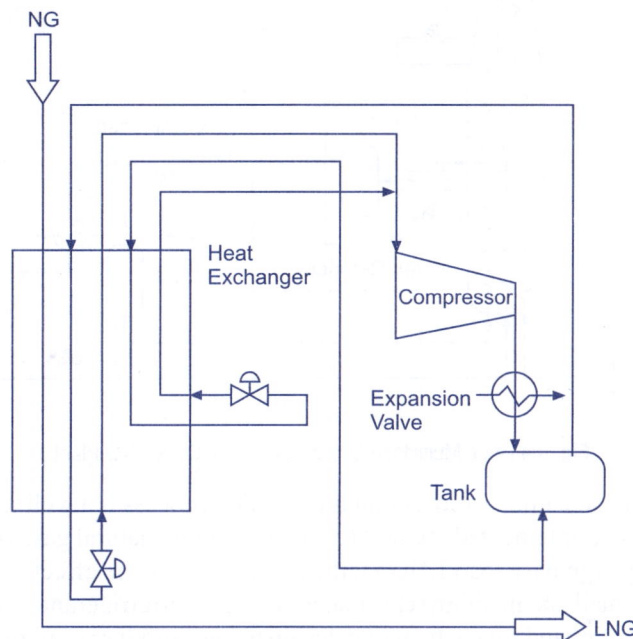

Fig. 9.3 Joule-Thomson Expansion Cycle LNG Plant

Linde Multistage Mixed Refrigerant Cycle LNG Plant

The Linde multistage mixed refrigerant cycle process is comprised of the following:

1. A coil-wound heat exchanger where the natural gas is pre-cooled, liquefied and sub-cooled against various fractions of a single mixed refrigerant cycle.
2. A medium pressure refrigerant separator, from which the liquid is used to provide the pre-cooling after J-T (Joule-Thomson) expansion to the lower section of the heat exchanger.
3. A high pressure refrigerant separator, from which the gas is cooled and partially condensed in the lower section of the heat exchanger.
4. A low temperature refrigerant separator, from which the liquid is used to provide the natural gas liquefaction duty after J-T expansion. The gaseous refrigerant stream from this separator is used to provide the sub-cooling after condensation and J-T expansion in the upper section of the heat exchanger.
5. A two-stage compressor with intercooling for compressing the combined refrigerant cycle stream from the bottom of the heat exchanger and after cooling

with water. Refer Figure 9.4 for the schematic of a Linde multistage mixed refrigerant cycle based LNG plant [6].

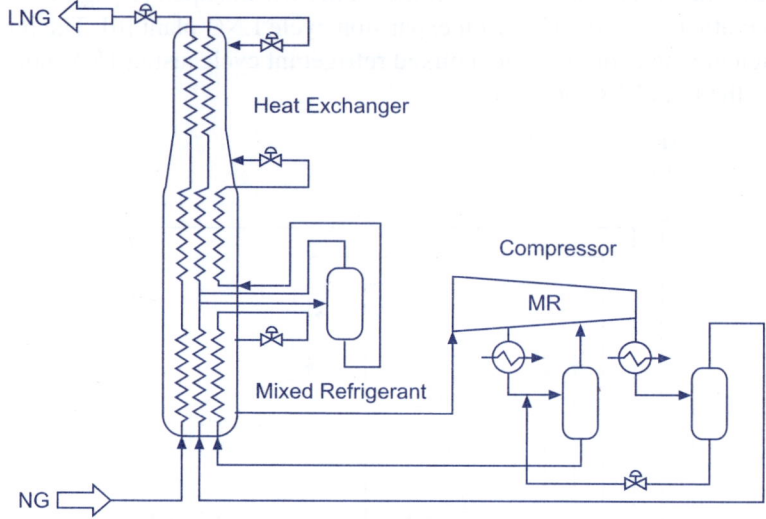

Fig. 9.4 Linde Multistage Mixed Refrigerant Cycle LNG Plant

The multistage mixed-refrigerant liquefaction process is a well tested, proven, and extensively implemented method for creating liquid natural gas. This process is well suited to large base load LNG facilities, but can also be effectively utilized in the design of small and medium scale plants. The mixed-refrigerant (MR) solution is blended and optimized for each unique design using a mixture of pure hydrocarbon components. The MR "cocktail" is compressed in a multi-stage compressor unit to a high operating pressure, which is then let down to a lower pressure across a control valve to produce the required refrigeration to liquefy the natural gas stream. The MR mixture is comprised of Methane, Ethane, i-Butane, n-Butane, and Nitrogen mixed from 99.95% pure components at varying concentrations. The availability of the MR components listed above can influence the decision to use a mixed-refrigerant design, as cost and availability vary with location.

Conoco Phillips Cascade Cycle LNG Plant

Conoco Phillips Cascade Cycle is a reliable and popular LNG plant system. As shown in Figure 9.5, the process uses a three-stage pure component refrigerant cascade of propane (purple), ethylene (green), and methane (red). The pretreated natural gas (black stream) enters the first cycle or cooling stage which uses propane as a refrigerant. This stage cools the natural gas to about −35°C and it also cools the other two refrigerants to the same temperature. Propane is chosen as the first stage refrigerant because it is available in large quantities worldwide and it is one of the cheapest refrigerants. The natural gas then enters the second cooling stage which uses ethylene as the refrigerant and this stage cools the natural gas to about

−95°C. At this stage the natural gas is converted to a liquid phase (LNG) but the natural gas needs to be further sub cooled so fuel gas would not exceed 5% when the LNG stream is flashed.

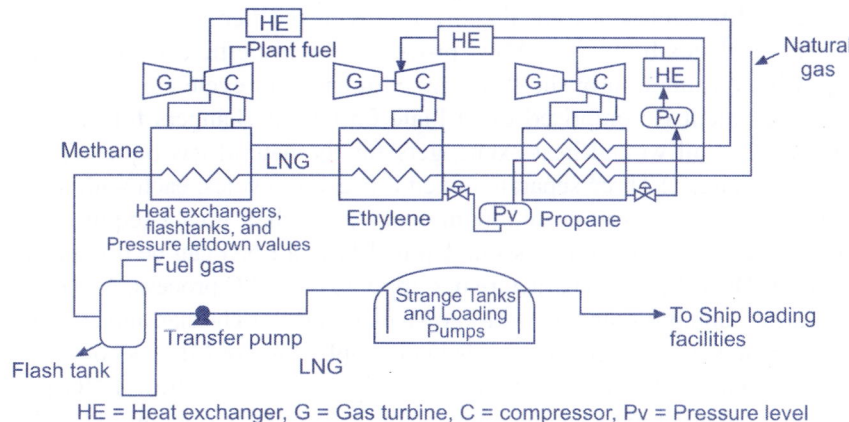

HE = Heat exchanger, G = Gas turbine, C = compressor, Pv = Pressure level

Fig. 9.5 Conoco Phillips Cascade Cycle LNG Plant

Ethylene is used as the second stage refrigerant because it condenses methane at a pressure above atmospheric and it could be also condensed by propane. After methane has been condensed by ethylene, it is sent to the third stage where it sub-cools the natural gas to about −155°C then it is expanded through a valve which drops down the LNG temperature to about −160°C. Methane is sent back to the first cooling stage and the LNG stream is flashed into about 95% LNG (which is sent to storage tanks) and 5% fuel gas used as the liquefaction process fuel. Methane is used as the sub-cooling stage refrigerant because it could sub-cools up to −155°C and it is available in the natural gas stream so it is available at all times and at lower costs. Simulation can be done to obtain how much heat duty every refrigeration cycle needs before designing the heat exchangers. The heat exchangers between the natural gas stream and the refrigeration cycles are designed to cool the natural gas to a temperature 10°C lower than the refrigerant boiling point. The refrigerants boiling points are another reason for picking these three refrigerants since no other pure refrigerant would meet these design temperatures.

Conoco Phillips cascade LNG system normally uses gas turbines rated at 30 MW which are single or double shaft gas turbines. LNG heat exchangers used in this LNG plant are brazed aluminum plate fin type. This specifically allows for low pressure drops inside the heat exchanger and a high thermal efficiency. The LNG heat exchangers are then fitted with headers to force transfer of liquid through all inlet holes. The LNG heat exchanger is then installed into a cold box to provide high levels of thermal efficiency and safety measurements. The cold box is made of perlite with circulating nitrogen gas moving through the open space. Any trace of natural gas or refrigerant that has leaked out of the LNG heat exchanger is carefully caught alerting the operator.

9.8 Cryogenics

Mixed Fluid Cascade (MFC) Cycle LNG Plant

The Mixed Fluid Cascade (MFC) process is a classic cascade process with the important difference that mixed component refrigerant cycles replaces single component refrigerant cycles, and thereby improving the thermodynamic efficiency and operational flexibility. The MFC process is highly efficient due to the low shaft power consumption of the three mixed refrigerant cycle compressors. The process equipment is comprised of 1) Plate-fin heat exchangers for natural gas precooling, 2) Coil-wound heat exchangers for the natural gas liquefaction and LNG sub-cooling, 3) Three separate mixed refrigerant cycles, each with different compositions, which result in minimum compressor shaft power requirement, 4) Three cold suction turbo compressors. Up to 12 mtpa LNG can be produced in a single train. The following characteristics apply to the MFC process: A) The MFC process is new and promoted by well-known elements. B) The size and complexity of the separate spiral wound heat exchangers applied in the MFC is considerably less when compared with today's single unit used in dual flow LNG plants. C) This technology allows single compressors to handle refrigerant over a larger temperature scale. The process diagram in Figure 9.6 is showing the Mixed Fluid Cascade Process consisting of three mixed refrigerant cycles.

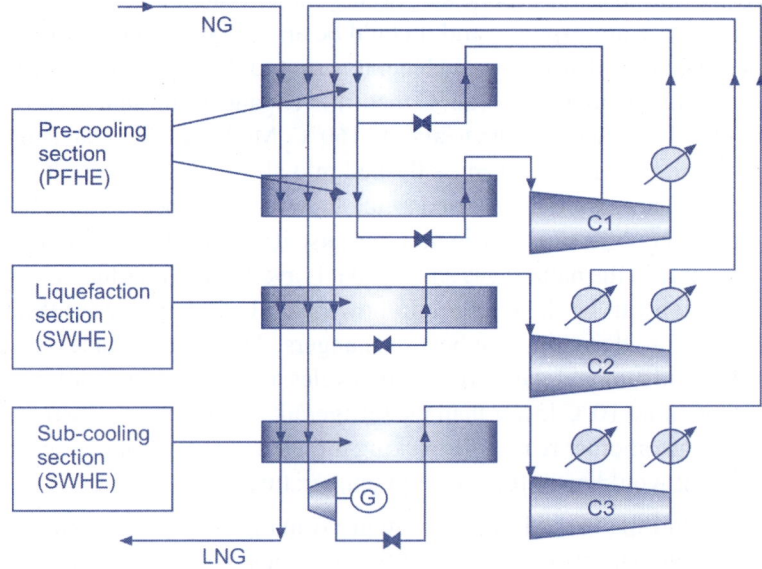

Fig. 9.6 Mixed Fluid Cascade Cycle LNG Plant

As seen in Figure 9.6, the natural gas (red stream) comes in from the top and goes through three mixed refrigerant cycles. The pre-cooling cycle (green cycle) cools natural gas through two plate fin heat exchangers (PFHE) while the liquefaction (purple cycle) and sub-cooling cycle (blue cycle) cool via spiral wound heat exchangers (SWHE). SWHE is made and patented by Linde and may also be used for the pre-cooling stage. The only possible reason PFHE is used is because it is less expensive than (SWHE) sub-cooling. The refrigerants are made mainly of methane, ethane, propane, and nitrogen but the composition ratio of the refrigerants would differ among the three stages. No specific kind of compressors is used for this process. The MFC has the same advantages and disadvantages of the other LNG processes but it has an extra advantage that the spiral wound heat exchangers used are more efficient as compared to the single unit heat exchangers used in the APCI or DMR. That allows larger single compressors to handle refrigerant over a larger temperature range.

Air Products and Chemicals Inc (APCI) Cycle LNG Plant

APCI (Air Products and Chemicals Inc.) LNG cycle is also called as Propane Pre-cooled Mixed Refrigerant Process (PPMR). It is used in around 88% of the liquefaction plants in the world. APCI plants currently produce 107.5 mtpa of LNG with 53 trains in operation. APCI technology uses a three stage refrigerant cooling powered by two 85 MW compressors as shown in Figure 9.7.

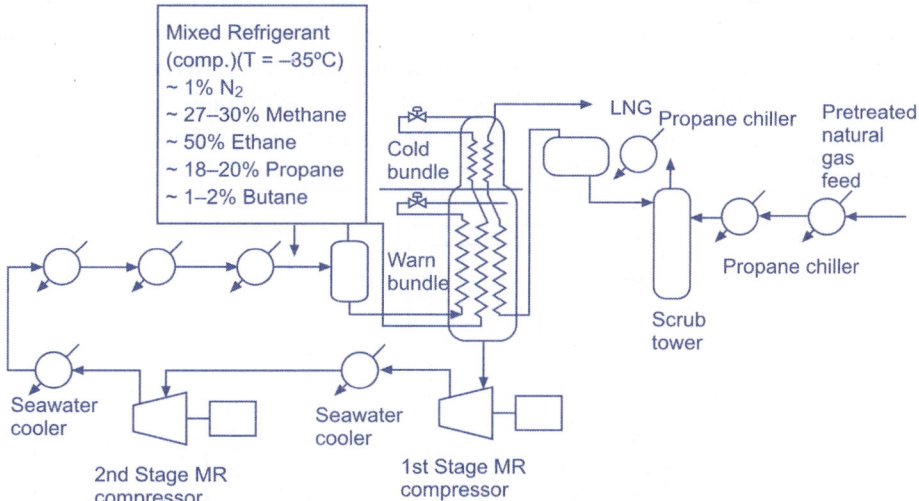

Figure 9.7 Air Products and Chemicals Inc (APCI) Cycle LNG Plant

The first stage is a pre-propane cooling stage that cools the mixed refrigerant and inlet natural gas to around −35 F. The next two cooling stages, based in a heat exchanger tower, use mixed refrigerants (MR) of about 27% methane, 50% ethane, 20% propane, 2% butane, and 1% nitrogen to cool and condense the natural gas. Propane chillers chills the gas during pre-treatment. As seen in Figure 9.7 a flash tank is used to separate the mixed refrigerant to a heavy coolant (bottom/red stream) and a lighter coolant (top/green stream). The heavy coolant (propane, butane, and some ethane) takes care of the cooling in the warm bundle i.e. the bottom part of the heat exchanger tower which cools down the natural gas stream (blue stream) to about −50°C. The light coolant is sprayed on the streams of the warm bundle via valves to ensure that the natural gas stream is cooled to the maximum point possible for

9.10 Cryogenics

this mix. The light coolant (methane, ethane and nitrogen) cools down the natural gas stream to –160°C in the cold bundle and this temperature is the point where natural gas is converted to LNG. Similar to the warm bundle, the light coolant is also sprayed on the streams in the cold bundle of the heat exchanger tower and that is the end of the cooling cycle. The liquid coming out of the top of the heat exchanger tower is then separated via flash tank to LNG (bottom stream) and light fuel (top stream) which is later sent for fractionation. The overall process requires around 200,000 hp out of the three compressors and the heat exchangers in the tower are Spiral Wound Heat Exchangers (SWHE). APCI controls the majority of liquefaction plants in operation today.

Shell Dual Mixed Refrigerant (DMR) Cycle LNG Plant

DMR (Dual Mixed Refrigerant) is very similar to the APCI liquefaction process. The process capacity is about 4.5 MMmtpa and there is only one DMR train in operation in Russia. As seen in Figure 9.8, the natural gas stream (red stream) is cooled via two stages. The first stage, which is the left column, cools the natural gas to –50°C while the second column cools the natural gas to LNG at –160°C. The composition of the pre-coolant cycle is 50/50 of Ethane/Propane on molar basis and that is why it is called the Dual Mixed Refrigerant process due to having two different refrigerants.

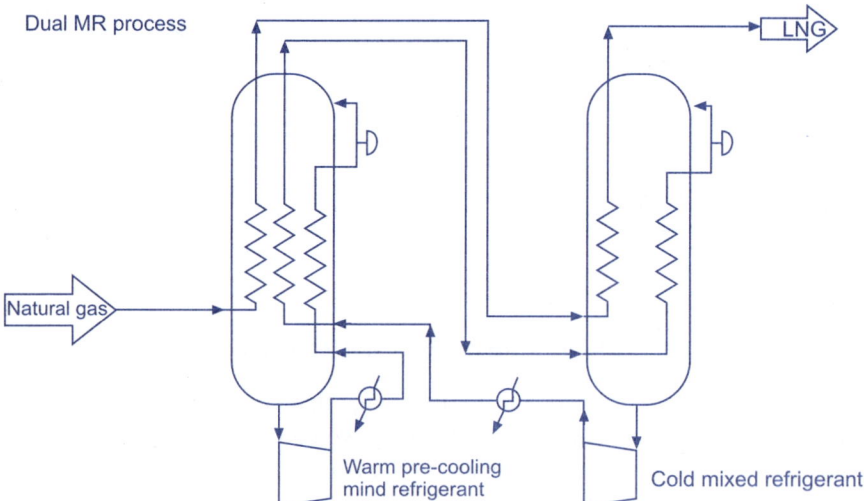

Fig. 9.8 Shell Dual Mixed Refrigerant (DMR) Cycle LNG Plant

It is seen also from Figure 9.8 that the process is similar to the APCI but in this process the heat exchanger tower is divided into two sections and this concept allows the designer to choose the load on each refrigeration cycle through controlling the two compressors. The heat exchangers used in this process are also spiral wound type as used in the APCI. Shell has also developed the LNG plant equipment with double casing to optimize the production to 5 mtpa. The advantages and disadvantages of this process is similar to the APCI, however as it has two separate compressors and heat exchangers it is more reliable than the APCI. If one compressor or heat exchanger is affected and malfunctions then the other would do the liquefaction but with less production rate.

This completes our discussion of LNG technologies.

HISTORY OF LIQUEFACTION OF HYDROGEN

Hydrogen Gas liquefaction techniques up to 1895 involved three basic steps: 1) compressing the gas to a high pressure, usually 50 atmospheres or more, 2) chilling the compressed gas to as low a temperature as possible using various cooling methods, and 3) expanding the chilled, compressed gas slowly from a high to a lower pressure by means of a needle valve. The cooling methods included, for example, evaporating an either solid carbon dioxide mixture or evaporating liquid ethylene, which Wroblewski used in liquefying oxygen. The third step made use of the Joule-Thomson effect for gases, based on experiments by Joule in 1845 and later refined by Thomson. They found that a gas, in slowly expanding from a high to a lower pressure, undergoes a change in temperature. The gas may be either heated or cooled by the expansion, depending upon the initial temperature and the particular gas. The temperature below which an expansion produces cooling is called the inversion temperature; it is high for most gases, but for hydrogen, it is about 193 K. Hydrogen, therefore, must be cooled below this temperature before it is expanded. In 1895, a breakthrough occurred in gas liquefaction techniques. The technique was to employ regenerative cooling that uses a fluid as the coolant in a process in which the fluid is itself involved. In the liquefaction of gases, it means that the gas that is cooled by the Joule-Thomson expansion process is later used to cool the incoming compressed gas before expansion. The regenerative cooling concept was an old idea, first introduced by Siemens in 1857 and used by Kirk, Coleman, Solvay, Linde, and others in refrigeration apparatus. In 1895, and within two weeks of each other, William Hampson in England and Carl von Linde in Germany obtained patents for equipment to liquefy air using tile Joule-Thomson expansion process and regenerative cooling. Linde described his apparatus to physicists and chemists at Munich in 1895.

Regenerative cooling proved to be the technological link needed to liquefy hydrogen. On May 10, 1898, James Dewar used it to become the first to statically liquefy hydrogen. Using liquid nitrogen he pre-cooled gaseous hydrogen, under 180 atmospheres, then expanded it through a valve in an insulated vessel, also cooled by liquid nitrogen. The expanding hydrogen produced about 20 cubic centimeters of liquid hydrogen, about 1 percent of the hydrogen used. Dewar measured the density of liquid hydrogen at 0.07 kilogram per liter, the modern value, which is 1/14 the density of water and about 1/12 the density of kerosene or gasoline. The insulated vessel Dewar used was the vacuum container flask he developed earlier which became known as "Dewar flasks" or Dewars. Dewars are double-walled vessels with a vacuum in the annular space to minimize heat transfer by conduction and convection. The walls are silvered to reflect radiant heat. Dewar vessels, with engineering refinements, are used today to transport liquid hydrogen with very low loss rates. By 1900, then, many of the major properties of liquid hydrogen were known.

Introduction to Liquefaction of Hydrogen

Hydrogen has been a key feedstock in the chemical industry for over 100 years. Mixtures of hydrogen and air are not only nontoxic, tasteless and odorless, they are also combustible – and the only combustion product is water vapor. Unlike dwindling fossil fuels, hydrogen is available in virtually unlimited quantities. So there are good reasons why it is increasingly regarded as vital to the future of the energy economy. Hydrogen has the most varied sources: It is a byproduct of chemical processes and is obtained by electrolysis, but most of all it comes from the steam reforming of natural gas. It almost always needs one or more stages of purification before further processing. For highest density and economy space storage, hydrogen must be cooled and liquefied. Industrial hydrogen liquefaction uses a variety of processes with helium, hydrogen or gas mixtures as coolant. The hydrogen feed to the process is first cooled by liquid nitrogen, then further cooled in multi-stage heat exchangers, where the cooling power is provided by turbo expanders. Liquefaction is finally accomplished by throttling in a Joule-Thomson valve. The liquid hydrogen is stored in an insulated tank for further distribution. Capacities of hydrogen liquefiers range from 150 l/h more than 20,000 l/h.

There are two dominant methods for the efficient storage of hydrogen: in tanks under high pressure at ambient temperature, and in insulated vessels at low pressure and extremely low temperature (–253°C). Cryogenic liquefied hydrogen (LH_2) has the advantages of a much greater energy content per unit volume and a smaller volume requirement for storage. The challenge for long-term storage of cryogenic hydrogen is to reduce the evaporation rate, which means insulating the storage vessel. The special cryotanks developed have advantages based on the design – double walled metal vessels with a vacuum between the two containers – and the use of a special insulation. Storage of liquid hydrogen can also be done in mobile tanks mounted in hydrogen-driven vehicles as well as large, stationary insulated tanks (as much as 300 m_3 in capacity). Currently the following hydrogen liquefaction process based plants are available:

1. Pre-cooled Linde Hampson cycle
2. Claude cycle
3. Helium Hydrogen Condensing cycle
4. Ortho-Para Hydrogen cycle
5. Magnetic Refrigeration cycle

Pre-cooled Linde Hampson Cycle for Hydrogen Liquefaction

This method is patented by Linde Corporation. This is a generic method for liquefying a hydrogen stream to produce liquid hydrogen. The liquefaction of the hydrogen stream takes place countercurrent to a refrigerant mixture cycle cascade consisting of two refrigerant mixture cycles. The first refrigerant mixture cycle is used for pre-cooling and a second refrigerant mixture cycle is used for liquefaction and super cooling of the hydrogen stream to be liquefied. No additional process steps are involved in a heat exchange between the hydrogen stream which is to be pre-cooled and the refrigerant mixture of the first refrigerant mixture cycle. The term "pre-cooling" means the cooling of the hydrogen stream down to a temperature at which the separation of heavy impurities takes place. The subsequent further cooling of the hydrogen stream to be liquefied hereinafter comes under the term "liquefaction".

The basic Linde Hampson cycle uses a Joule-Thomson orifice. A gas could be cooled by letting it expand freely against the atmosphere. This could be explained by that the gas was doing a work against the atmosphere by lifting and/or heating it, and thereby loosing energy in form of heat. The Linde-Hampson system is one of the simplest cryocoolers so it is quite desirable for small-scale liquefaction plants. It can be used to liquefy hydrogen but also it is used to liquefy neon, nitrogen and helium. This type of pre-cooled system was used in several of the earlier helium liquefiers. A scheme of the Linde-Hampson system is represented in Figure 9.9. This system consists of two independent circuits, the upper one is the pre-cooled system and the lower one is the basic circuit of the gas that we want to liquefy. The first one is employed to lower the temperature of the hydrogen (basic circuit) below the ambient temperature.

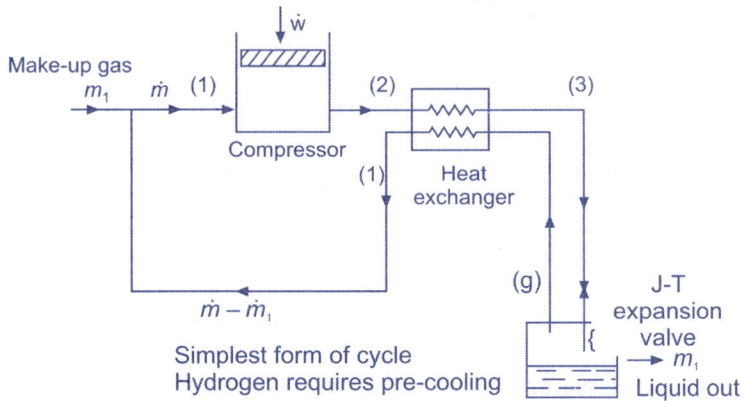

Fig. 9.9 Pre-cooled Linde Hampson Hydrogen Liquefaction System

9.14 Cryogenics

The pre-cooled system is filled with a pre-coolant fluid that has a triple-point temperature below that of the maximum inversion temperature of hydrogen. Liquid nitrogen is a suitable fluid to fill the precooler system for liquefying hydrogen. The basic circuit consists of a compressor employed to raise the enthalpy (so the temperature and the pressure are also raised) of the hydrogen, then it is cooled through three heat exchangers, two of them involving the precooling system. Then hydrogen passes through a valve to a tank, where liquid helium is extracted within a channel. In this tank, hydrogen gas is sent to the compressor after being heated through two heat exchangers.

Claude Cycle for Hydrogen Liquefaction

This is the most adopted cycle for large hydrogen liquefaction plants in the world. Hydrogen gas is used not only in the feed line to be liquefied but also in the recycling line which generates necessary cold for the liquefaction. The feed gas is compressed to approximately 5 MPa, and then cooled down to 80 K and converted to 47% para hydrogen by an ortho-para hydrogen converter (O-P converter) at the same time. Finally the feed hydrogen gas is liquefied at 0.1MPa, 20.4K by expansion at the J-T valve. In order to improve the process efficiency, a process with a super critical expansion turbine installed in the feed line has been developed. Refer Figure 9.10 for the Claude process and its T-S Diagram.

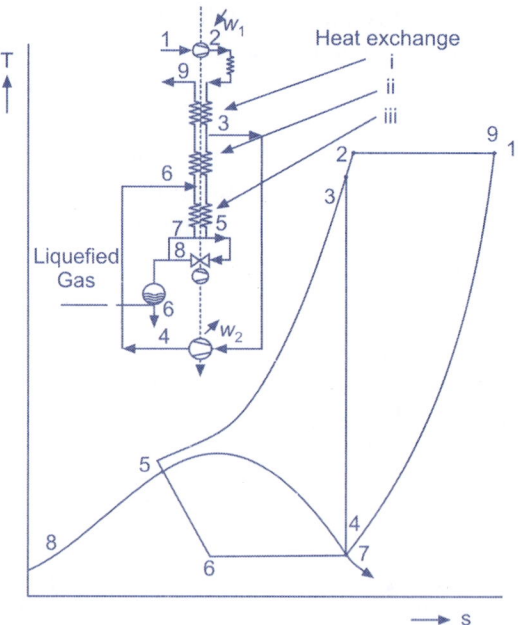

Fig. 9.10 Claude Hydrogen Liquefaction System

In the Claude process some of the high-pressure gas is passed through an expansion engine and is then sent to a heat exchanger for cooling. This process referred to as *ideal liquefaction* uses a reversible expansion process to reduce the energy required for liquefaction. It consists of an isothermal compressor followed by an isenthalpic expansion to cool the gas and produce a liquid. In practice, an expansion engine can be used only to cool the gas stream, not to condense it because excessive liquid formation in the expansion engine would damage the turbine blades.

Helium- Hydrogen Condensing Cycle for Hydrogen Liquefaction

Helium-Hydrogen condensing system is used in state of the art hydrogen liquefiers. Turbo-expanders or expansion engines are now used at most industrial gas plants to provide the necessary refrigeration for liquefaction. The expansion across a turbo-expander is ideally isentropic, or in other words, some useful work is done in expansion. But turbo-expanders cannot tolerate any liquid condensing at the outlet as the turbine wheels often rotate at speeds exceeding 100,000 rpm. Therefore, a clever combination of isentropic and isenthalpic expansion is required to generate a practical efficient process when the expansion process is applied directly on the hydrogen gas stream. In the lower capacity range, between 150 and 1000 l/h, a closed helium circuit supplies the refrigerating capacity necessary to cool the hydrogen. Hydrogen in the form of a pure gas at approx. 20 bar is fed into the vacuum-insulated cold box and, after a certain degree of super cooling at the end of the refrigeration process, it expands through a Joule-Thomson valve into the tank where liquefaction subsequently takes place. The helium refrigeration cycle and hydrogen liquefaction are therefore completely separate. The main components of the system are a helium compressor with oil purification, a cold box with heat exchangers and a helium gas-bearing expansion turbine, the liquid hydrogen and an attachment for filling transport containers. In the over 1000 l/h range, the refrigeration capacity is attained through cooling and ortho-para conversion of hydrogen by means of a hydrogen Claude process, i.e. at the cold end of the process the H_2 feed gas can be combined with the H_2 of the refrigeration process. The following main components are used for this process: piston compressors, oil-bearing expansion turbines, liquefier cold box, liquid nitrogen precooling. For larger-scale systems, it is an advantage to plan separate cold boxes for cooling the ambient temperature first to 80 K and then from 80 down to 20 K. Further system components such as hydrogen purification, storage tanks and filling devices are also supplied and installed. Refer Figure 9.11 for the schematic of the process.

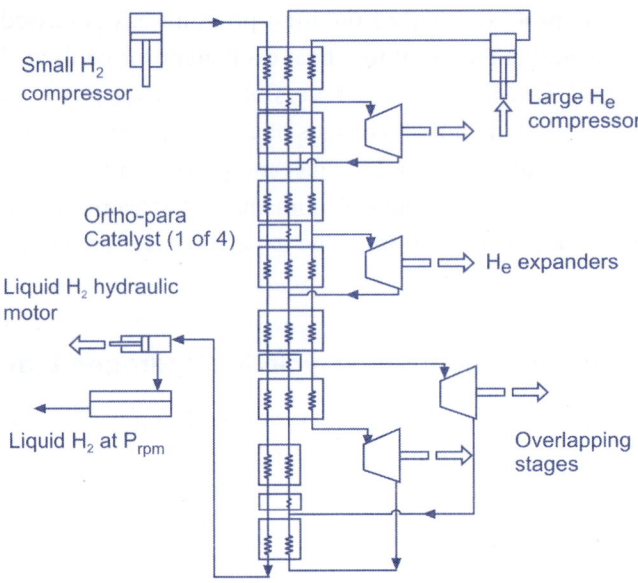

Fig. 9.11 Helium-Hydrogen Condensing Cycle

Ortho-Para Conversion for Liquid Hydrogen

Natural hydrogen gas is a mixture of molecules in two distinct forms—ortho and para-hydrogen—corresponding to the different total spin of the nuclei. Refer Figure 9.12 for the ortho and para forms of hydrogen. Physical and chemical properties of ortho- and para-hydrogen are different and the development of efficient techniques for separation or inter conversion of the hydrogen spin-isomers is important for hydrogen gas production and storage technologies with particular relevance on the eve of the hydrogen fuel era. Furthermore the study of Ortho to Para conversion is also essential for understanding the chemistry of the Universe and critical for fundamental studies of low-temperature physics. The lower energy para-form can be obtained by cryogenic distillation through ortho-para conversion, whereas a separation technique is used to enrich the gas to obtain the more energetic ortho-form.

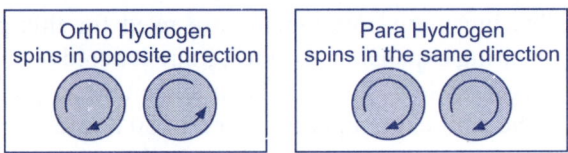

Fig. 9.12 Ortho and Para Forms of Hydrogen

These two hydrogen states have different energy levels (higher energy for the ortho form) so that the content of each species at equilibrium is temperature dependent. "Normal" hydrogen, which corresponds to the equilibrium mixture at room temperature, is 25% para and 75% ortho hydrogen. At 21 K normal boiling point of hydrogen the para content is almost 100%. The spontaneous conversion is a very slow reaction in liquid phase, inexistent in the gaseous phase. However, in 24 h, 18% of the liquid will evaporate if the liquefied hydrogen is normal hydrogen. Thus it is important to convert this hydrogen into pure para to avoid boil off losses during the storage time. A rapid cooling and liquefaction of normal hydrogen will still consist of approximately the same amount of ortho hydrogen as before the operation started, due to the slow ortho to para conversion. Catalysts and magnetic fields can increase the conversion rate considerably. Example of catalysts are hydroxides of Fe(III), Co(III), Ni(II), Cr(III), Mn(IV) or active charcoal /1/. However, due to price, performance and other criteria either iron oxide or nickel-silica catalysts are used in most liquefaction systems.

Practically, there are three main types of converters:

1. **Adiabatic converter:** Reactor with a simple bed of catalyst. As no heat is exchanged, the stream temperature will increase. Low cost for each bed but many beds required.

2. **Isothermal converter:** Reactor with bed of catalyst in a boiling refrigerant (usually liquid nitrogen or hydrogen) that keeps the temperature constant. This is more efficient but the capital cost is higher.

3. **Heat exchanger/converter:** Heat exchanger fitted with catalyst (inside tubes). Both sensible and conversion heat are removed. Capital costs are higher but this solution minimizes the power cost for removing the heat of conversion. It is difficult to change of the catalyst in this solution so it has to be reactivated in place.

The ideal solution is a continuous conversion. At every temperature, the hydrogen mixture is at equilibrium and the conversion releases little heat as possible. Continuous conversion is very difficult to realize, but can be approximated. The first solution would be to use heat exchangers with an integrated catalyst and the second one to use many converters. Apparently a choice must be done, between efficiency and investment cost.

Magnetic Refrigeration Cycle for Hydrogen Liquefaction

Magnetic refrigeration which is based on the magnetocaloric effect of solids has the potential to achieve high thermal efficiency for hydrogen liquefaction in comparison with conventional systems using a Joule-Thomson expansion. A magnetic refrigerator for hydrogen liquefaction has been developed in Japan, which cools down hydrogen gas from liquid natural gas temperature and liquefies at 20K. The magnetic liquefaction system consists of two magnetic refrigerators: Carnot magnetic refrigerator (CMR) and active magnetic regenerator (AMR) device. CMR with Carnot cycle has succeeded in liquefying hydrogen at 20K. Above liquefaction temperature, a regenerative refrigeration cycle is necessary to precool hydrogen gas, because adiabatic temperature change of magnetic material is reduced due to large lattice specific heat. AMR device has been tested as the precooling stage and it was confirmed for the first time that AMR cycle worked around 20 K.

9.18 Cryogenics

In order to realize a hydrogen liquefaction system using magnetic refrigeration, cascading several magnetic refrigerators is necessary to cover wide temperature ranges from heat sink to liquid hydrogen temperature (20.3 K). At this stage, the heat sink used is LNG with boiling temperature at 112 K. There are two kinds of magnetic refrigeration cycles; CMR and AMR. Hydrogen gas is pre-cooled to near the boiling point with an AMR refrigerator and then liquefied with a CMR. The apparatus shown in Figure 9.13 has been built and tested to investigate both liquefaction process of the CMR and AMR cycle by changing the experimental conditions such as magnetic material, heat transfer gas and so on. This apparatus consists of a magnetic refrigerant, superconducting magnet with a maximum field of 6 T cooled by mechanical cooler, an electric motor that gives field change to magnetic refrigerant, and a G-M cryocooler which absorbs heat exhausted by magnetic refrigerator.

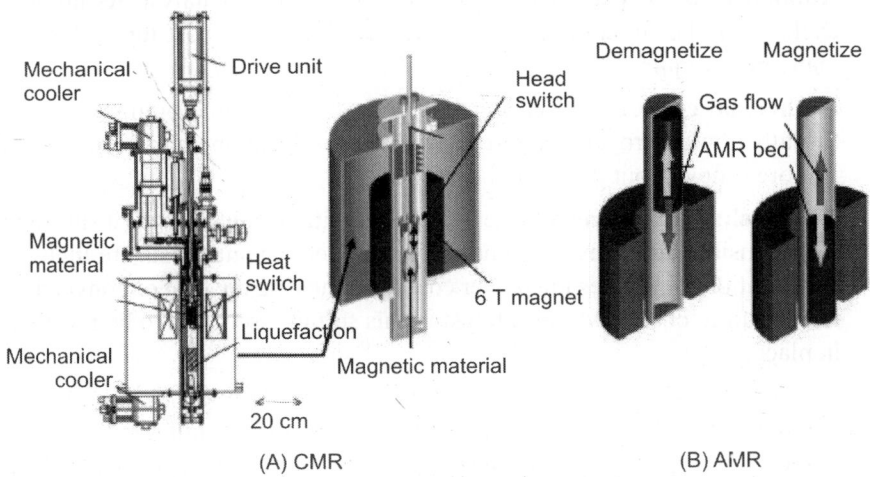

Fig. 9.13 Magnetic Refrigeration Cycle LH$_2$ Apparatus

Figure 9.13 shows the schematic drawing of the CMR operation. The liquefaction principal of the Carnot magnetic refrigerator is based on the thermosiphon method, in which liquid hydrogen is condensed directly on the surface of magnetic refrigerants and drops downward. Plates made of gadolinium were used as magnetic refrigerant. Hydrogen gas that was pre-cooled to a temperature slightly above the boiling point was successfully liquefied by CMR. Thermal efficiency, which was determined by a ratio of cooled heat to that in an ideal Carnot demagnetization process was 50 to 60 % including condensation losses. In high temperatures, Carnot cycle is not useful because large specific heat of magnetic material limits the adiabatic temperature change in most of the rare earth compounds. For this reason, a regenerative cycle is required to achieve a wide temperature operation. An AMR separates the functions of heat pumping and temperature spanning and is considered as a combination of many cascaded magnetic refrigerators and regenerative heat exchangers. AMR refrigerators have been intensively studied in room temperature and also thought to be a useful magnetic refrigerator for precooling hydrogen gas down to liquefaction temperature.

References

1. Cryogenic Engineering – B.A. Hands.
2. Internet website www.wikipedia.com.
3. Linde Brochure on LNG plants.
4. Cryogenic Systems – Randall Baron.
5. Cryogenic Engineering – P.K.Bose.
6. Internet website www.Kryopak.com.
7. Linde Brochure on LH_2 plants.
8. LNG Report from Rabeh et.al, Bolivia.
9. Matsumoto et.al. "Magnetic Refrigerator for Hydrogen Liquefaction".
10. Internet website http://www.faqs.org/patents/app/20090019888.
11. WHEC June 2006 No.16 Paper on Hydrogen Liquefaction by Jacob Stang.
12. Shimko et.al. USDOE Report on Innovative Hydrogen Liquefaction.

Questions

Q.1. Write a short note on liquefaction of natural gas.

Q.2. Discuss the operation of a Linde Multistage mixed refrigeration LNG plant.

Q.3. Write a short note on Joule-Thomson Expansion LNG plant. What are its advantages?

Q.4. Explain the differences between Conoco Phillips Cascade plant and Shell DMR plant.

Q.5. Discuss the APCI natural gas liquefaction plant with a neat sketch.

Q.6. Write a short note on liquefaction of Hydrogen.

Q.7. Discuss the operation of a Linde Hampson LH_2 plant.

Q.8. Distinguish between Ortho and Para hydrogen with a neat sketch.

Q.9. How is magnetic refrigeration used to produce liquid hydrogen?

Q.10. Explain the helium hydrogen condensing cycle for liquid H_2 plant.

CHAPTER 10

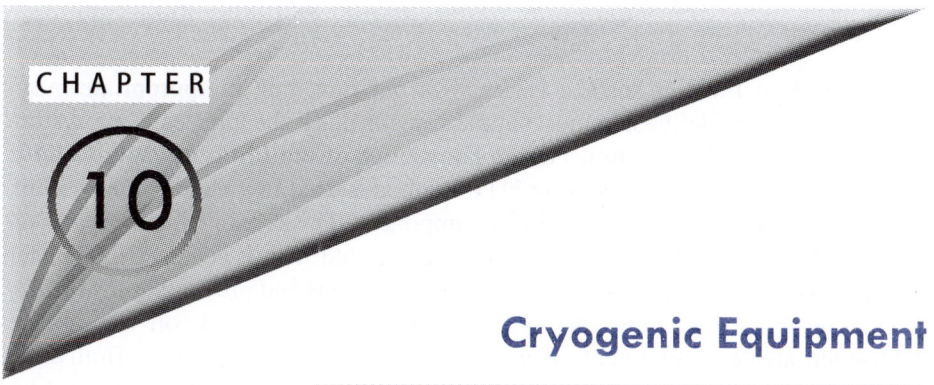

Cryogenic Equipment

INTRODUCTION

The cryogenic industry designs and manufactures a very wide range of equipment and systems such as storage vessels, columns, towers, compressors, alternators, heat exchangers, multi-tubular reactors, evaporators, expanders, pumps and valves. Any cryogenic plant contains all these components. Cryogenic process equipment has been manufactured since the early days, when Carl von Linde liquefied air on an industrial scale for the first time in Munich, Germany in May 1885. Improvements in aluminum welding technology in the late 1950s made it possible to change from rather expensive and heavy copper materials for fabrication to the cheaper and lighter all-aluminum-designs. Plate-fin and coil-wound heat exchangers are key components of cryogenic process plants worldwide. Several thousand coil-wound heat exchangers for various application and in diverse materials such as stainless steel, special alloys, copper and aluminum, with heating surfaces of up to 20,000 m^2 and unit weights of up to 170 metric tons have been fabricated since for the cryogenic process industry. In this chapter we shall study cryogenic equipment in detail.

CRYOGENIC HEAT EXCHANGERS

A heat exchanger is a device built for efficient heat transfer from one medium to another. The medium may be separated by a solid wall, so that they never mix, or they may be in direct contact. They are widely used in cryogenic applications. Heat exchangers may be classified according to their flow arrangement. In parallel-flow heat exchangers, the two fluids enter the exchanger at the same end, and travel in parallel to one another to the other side. In counter-flow heat exchangers the fluids enter the exchanger from opposite ends. The counter current design is most efficient, in that it can transfer the most heat from the medium. In a cross-flow heat exchanger, the fluids travel roughly perpendicular to one another through the exchanger. For efficiency, heat exchangers are designed to maximize the surface area of the wall between the two fluids, while minimizing resistance to fluid flow through the exchanger. The exchanger's performance can also be affected by the addition of fins or corrugations in one or both directions, which increase surface area and may channel fluid flow or induce turbulence. The driving temperature

10.2 Cryogenics

across the heat transfer surface varies with position, but an appropriate mean temperature can be defined. In most simple systems this is the log mean temperature difference (LMTD). Sometimes direct knowledge of the LMTD is not available and the Number of Transfer units or NTU method is used. In a liquefaction system, the heat exchanger is the most critical component. The primary function of heat exchanger in cryogenic applications is to conserve cold. This is achieved by transfer of heat from compressed gas to low pressure return gas and preventing heat leaks. Several designs of heat exchangers are available. Plate-fin and coil-wound heat exchangers are key components of cryogenic process plants worldwide. Both these heat exchangers are discussed in greater detail along with Giaque Hampson heat exchanger and Collins Heat exchanger.

PLATE AND FIN HEAT EXCHANGER

A plate-fin heat exchanger is a stack of alternating flat and corrugated plates. The corrugations (fins) form the flow channels for the diverse process fluids. Each process stream occupies a certain number of passages within the stack. These are collected by half-pipe headers and nozzles to single point connections on the inlet and the outlet of the respective process stream. In this way process fluids can exchange heat in only one heat exchanger block. Figure 10.1 is a sketch of the various types of fins.

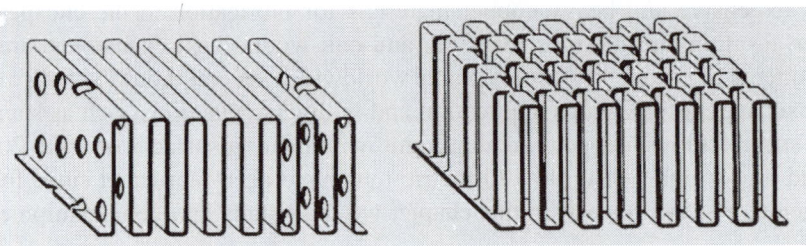

Fig. 10.1 Heat Exchanger Fins–Perforated and Serrated

Usually this type of heat exchanger is made of aluminum alloys. It is important to know that with these standard materials the upper design temperature is limited to +65°C due to code requirements. In order to meet the required performance about 50 different fin designs are available. In general a distinction is made between perforated fins and serrated fins. Serrated fins have higher heat transfer coefficients than perforated fins. However, serrated fins are more prone to fouling and result in higher pressure drop. In order to bond the loose stack of plates and fins to a rigid block, vacuum brazing is used. The manufacturing process includes spanning the loose stack together and heating it in a vacuum furnace up to a temperature of about 600°C. The filler material is clad by rolling on both sides of each parting sheet. The fins are pure aluminum alloy without any cladding. After the vacuum brazing the blocks are completed by welding all the attachments such as headers, nozzles to the block. Vacuum brazed plate-fin heat exchangers made of aluminum offer a number of advantages. They are extremely compact due to the use of aluminum and highly efficient fins. Thus this type of heat exchanger is perfectly suitable for installations which require compact design. The wide selection of heat transfer fins combines high heat transfer rates with low pressure drops (i.e. low energy consumption) in tailor made heat exchangers. The ability to combine up to 10 process streams in only one heat exchanger system can eliminate the need for multiple heat exchanger arrangements and the interconnecting piping. The use of high strength aluminum alloy results in light weight units thus reducing drastically the foundation and support requirements. Due to their arrangement as a large and rigid aluminum block, and considering the small gaps inside, this type of heat exchanger cannot be recommended for cases of operation such as high temperature gradients and process streams which are corrosive and containing particles. Refer Figure 10.2 for the photograph of a plate and fin heat exchanger from Linde Corporation.

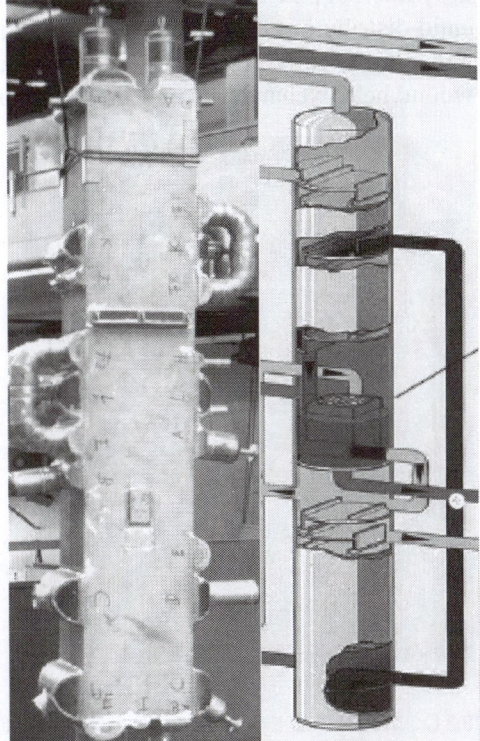

Fig. 10.2 Plate and Fin Heat Exchanger from Linde Corporation

COIL WOUND HEAT EXCHANGER

Coil-wound heat exchangers have been manufactured since the early days. Improvements in aluminum welding technology in the late 1950s made it possible to change from rather expensive and heavy copper to the cheaper and lighter all-aluminum-designs. Coil-wound heat exchangers for various application and in diverse materials such as stainless steel, special alloys, copper and aluminum, with heating surfaces of up to 20,000 m² and unit weights of up to 170 metric tons have been fabricated. A coil-wound heat exchanger is, in general, a tubular heat exchanger; however, the bundle does not consist of not using a straight tubes. Tubes of relatively long length and small diameter are wound in alternating directions around a centre pipe. In parallel a pressure vessel shell is prepared and the complete tube bundle is inserted. All single tubes start and terminate in tube sheets which are integral parts of the pressure vessel shell. Due to the flexible tube bundle arrangement these heat exchangers can bear temperature gradients and differences clearly exceeding the limits of other heat exchanger types (e.g. plate-fin heat exchangers). Over the specified design life no considerable bundle sagging is to be expected. Optimal liquid distribution of the shell side 2-phase stream over the whole cross section of the bundle is achieved by internal phase separation and special liquid distribution systems. The latest liquid distributor design minimizes the liquid hold-up on top of the bundles, thus reducing negative thermal effects. Refer Figure 10.3 for a photograph of a coil wound heat exchanger from Linde Corporation.

Fig. 10.3 Coil Wound Heat Exchanger from Linde Corporation

The tube bundles are designed and fabricated to be vibration-proof and self-draining. The pressure vessel shell is typically made of aluminum alloy 5083. For the tubes, special aluminum alloy and a non-standard (but approved) fabrication procedure is used for tube manufacturing. Each tube bundle is freely suspended from a special support system on top of each bundle. Thus shrinkage and expansion of the tube bundles due to rapid temperature changes during start-up or shut-down occur with minimum stresses between the tube bundle and the shell. The support system is designed to carry the weight of the tube bundles, the fluids and the pressure drops. Each tube bundle is wrapped into a "shroud" which is seal welded on the upper side of the shell to avoid any refrigerant passing between the tube bundle and the

Giaque Hampson Heat Exchanger

The type of Counter flow heat exchanger that is used in the unit to cool sensors is called a Giaque-Hampson coil tube. A length of fine-bore circular tube with copper fins soldered to the outside is wound in the form of a closely-coiled helix around a cylindrical mandrel. It is important to maintain the spacing of the small tubes in a uniform manner or the low pressure stream will be unevenly distributed. Punched brass spacer strips can be used to maintain uniform spacing between tubes. The compressed gas flows through the coiled tube and expands through the Joule-Thomson nozzle at the cold end of the heat exchanger. The expanded low-pressure, cold gas flows back transversely over the coiled tube thus establishing a good heat transfer to the warm, high-pressure gas. Equal length for all parallel paths were obtained by varying the pitch of the helix. The JT-nozzle can be realized in a number of ways such as by flattening the end of the high-pressure tube, or by inserting a small rod or wire into the end of the tube. Refer Figure 10.4 for the construction of the Giaque Hampson heat exchanger. The flow of gas into the heat exchanger is manually or automatically controlled. In the heat exchanger the compressed fluid at high pressure is cooled by heat transfer to the low-pressure return fluid. For a perfect heat exchanger

$$QL = m(h_A - h_B)$$

It is clear that for cooling $h_A > h_B$ is required, as the warm-end temperature of the heat exchanger is below the inversion temperature of the working fluid.

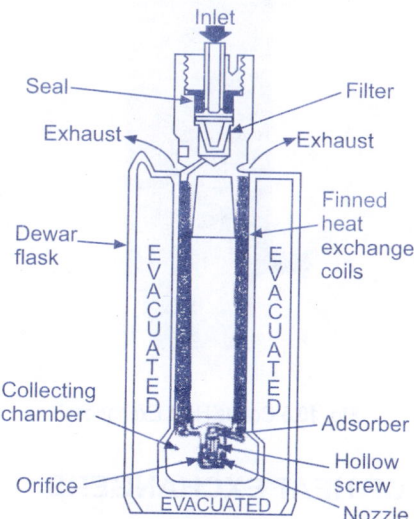

Fig. 10.4 Giaque Hampson Heat Exchanger

10.6 Cryogenics

COLLINS HEAT EXCHANGER

The Collins or gravity flow is a simple heat exchanger design for heat recovery that was developed under a grant from the DOE Inventions Program. This straightforward design is a vertical, counter flow heat exchanger that extracts heat out of warm fluid and applies it to preheat the cold fluid. The design consists of a 3- or 4-inch central copper pipe (that carries the warm fluid) with 1/2 inch. copper coils wound around the central pipe. Heat is transferred from the warm fluid passing through the large, central pipe to cold incoming fluid simultaneously moving upward through the coils on the outside of the pipe. The coils are flattened a little where they touch the pipe to increase the contact area and improve heat transfer. The key to this patented device was the inventor's observation that warm fluid clings in a film-like fashion to the inside wall of the pipe as it undergoes gravity flow in the open drain, and this warm, falling film transfers heat through the pipe wall to the incoming cold fluid that passes through the copper coil wound around the pipe. The Collins heat exchanger has a number of advantages such as rugged construction with no moving parts, Compactness as it can be installed vertically and is available with multiple parallel coils outside the central pipe to reduce pressure drop. Refer Figure 10.5 for a schematic of the Collins heat exchanger.

Fig. 10.5 Collins Heat Exchanger

CLASSIFICATION OF HEAT EXCHANGERS

The two types of heat exchangers most used are the re-heat systems, the Recuperative and Regenerative heat exchange type systems. Recuperative systems are also known as shell and tube type heat exchangers. In these heat exchangers, a stream of cold process gas passes through a series of tubes and is heated by another stream of gas which passes over the tubes on the shell side. These types of systems are generally used for thermal oxidizers with low to medium process flow rate, and generally can provide up to 80% thermal efficiency. Regenerative type heat exchangers use a media to absorb heat given off by one hotter fluid and transfer it to another colder fluid. Typical media used are packed towers of ceramic material with required gaps for the gases to pass through them. The operation of regenerative heat exchangers is cyclic. In the first cycle hot gases/fluids passing through the media heat up the media. In the following cycle, the cold gases pass through the media and they are heated by the already hot media. Regenerative systems can operate with process flow rates in the low or high range, and can yield a thermal efficiency between 80% and 95%. Both the recuperative and regenerative heat exchangers are explained in greater detail in the subsequent section.

Recuperative Heat Exchangers

Recuperative heat exchanger systems are most commonly available. The technology for this kind of system has been around for more than 15 years. These systems can yield up to 80% thermal energy recovery (effectiveness). These systems rely on a shell and tube type heat exchange, where the hotter gas passes over the shell, heating up the cool gas passing through the tubes, using convection. This is an example of a recuperative heat exchanger used in conjunction with a Thermal Oxidizer. The two segments for the heat exchanger represent the shell, and then the tube sections. The transfer of heat allows the burner less requirement for thermal input, and so uses less fuel. Refer Figure 10.6 for a photograph of the recuperative heat exchanger.

Fig. 10.6 Recuperative Heat Exchanger

Shell and tube type of heat exchangers are further divided on the basis of their operation. In parallel type heat exchangers, cold gas which is required to be heated passes through tubes, which are arranged in several passes. Hot gases flow on the shell side of the heat exchanger in a straight line. This heat exchanger is counter type of heat exchanger, where the hot gases incoming the heat exchanger come in contact with cold gases leaving the heat exchanger. Conversely, hot gases entering the system heat up the cold gases. Counter type of heat exchangers give the highest efficiency for heat exchangers. In series type heat exchangers, hot gases coming out of the thermal oxidizer pass through tubes and cold gases pass on the shell side. A number of passes are arranged on the shell side for the cold gases to pass over much larger area of tubes. The hot gases pass through the tubes, which are arranged longitudinally.

Regenerative Heat Exchanger

Regenerative technology is a newer technology than recuperative, however these systems are more efficient than recuperative systems. Regenerative systems are not steady state, like recuperative, rather they rely on a cycling gas which flows between at least two fixed packed beds. In this case, at least one bed would be an inlet, while the other would be an outlet bed. A common combustion/retention chamber connects the two. These vessels are connected by inlet and outlet control valves to direct air flow through the different beds. The hot air from the combustion chamber would flow through one of the beds, heating it up, while the other discharges, then cold process air would flow through the heated ceramic, reaching near the temperature needed for combustion. Then after being combusted, the air would flow through the cold ceramic again to heat it back up. This would be a continuous cycle of retention and purging. The most important aspect of this cycle is the way which each chamber is purged of the air and then re-heated. Continued preheating of the process gas causes the ceramic media canister to cool down, at the same time the ceramic media in the next chamber is heated. The inherent nature of the regenerative systems involves discontinuous or cyclic operation. Refer Figure 10.5 for a photograph of the regenerative heat exchanger.

 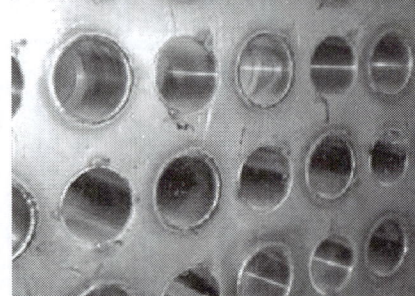

Fig. 10.7 Regenerative Heat Exchanger

The general process for a regenerative heat exchanger is as follows. The process gas is brought through the pre-heated ceramic bed and is combusted. It then flows out over another ceramic bed, which is subsequently heated. After an allotted time the valves are closed and the process gas flows through the ceramic bed which had just been heated, thus pre-heating the gas and flows over another ceramic bed. The three ceramic beds alternate the duty of pre-heating and being heated, as their valves are opened and closed. The entire process, although more complicated that recuperative heat exchange, is far more efficient, yielding an effectiveness of over 95%.

Heat Transfer Coefficients and Friction Factors for Heat Exchangers

Heat transfer coefficients are normally expressed in terms of dimensionless numbers or factors. In this section we will review some important numbers and friction factors.

1. The first important number is the Prandtl Number given by $P_r = \mu C_p/k$, where μ is the fluid dynamic viscosity, C_p is the specific heat at constant pressure and k is the thermal conductivity.

2. The Colburn factor is given by $J = (h/G\, C_p)\, Pr^{2/3}$, where h = convective heat transfer coefficient, C_p is the specific heat at constant pressure, Pr is the Prandtl number, G = mass flow rate per unit flow area.

3. The Nusselt number is given by $N = hD/k$, where h = convective heat transfer coefficient, D = equivalent diameter and k is the thermal conductivity.

4. The Reynolds number is given by $R_e = \rho VD/\mu$, where ρ = density, V = velocity, D = diameter and μ is the fluid dynamic viscosity.

5. Fanning friction factor $f = \Delta p/L\, (G^2/2g\rho D)$, where $\Delta p/L$ = pressure drop per unit length, g = gravitational acceleration, G = mass flow rate per unit area, D = diameter and ρ = density.

For circular tubes with laminar flow the Nusselt number is calculated by the Hausen equation

$$Nu = 3.658 + [0.068\, (D/L)\, R_e\, P_r]/[1 + 0.04\, (D/L)\, R_e\, P_r]^{2/3}$$

The Colburn factor for circular tubes with outer diameter D_o and Inner diameter D_i is calculated as

$$J = 0.023\, Re^{-0.2}\, [1 + 3.5(D_o/D_i)]$$

For staggered tubes in Giaque Hampson heat exchanger, the Colburn factor is

$$J = 0.33\, Re^{-0.4}$$

For coiled tubes in Collins heat exchanger, the Colburn factor is

$$J = 0.118\, Re^{-0.3}$$

The friction factor for flow in circular tube is given by

$f = 64/Re$ for laminar flow

$f = 0.316\, Re^{-0.25}$ for turbulent flow

The friction factor for circular tubes with outer diameter D_o is calculated as

$$f = \Delta p/N\, (G^2/2g\rho D_o)$$

For staggered tubes in Giaque Hampson heat exchanger, the friction factor is

$$f = [\,1 + 0.47\, (D_o - 1)^{-1.08}\,]\, Re^{-0.16}$$

For coiled tubes in Collins heat exchanger, the friction factor is

$$f = 1.904\, Re^{-0.2}$$

Variables Affecting Heat Exchanger Performance

1. **Fin effectiveness:** In heat exchangers, fins are used to improve heat transfer rate by increasing surface area. The entire fin is not effective in transferring heat due to variation in temperature of fin. Thus fin effectiveness characterizes the heat transfer efficiency of a fin and is given by

   $\eta_f = \tanh(mL)/mL$, where $M = h/k\delta$

   L = length of fin, h = convective heat transfer coefficient, k = thermal conductivity, δ is the volume of fin.

2. **Surface effectiveness:** The entire surface area of the fin is also not effective in transferring heat due to variation in temperature of the fin surface. Thus surface effectiveness of fin characterizes the heat transfer efficiency of the surface of the fin and is given by

   $\eta_o = 1 - (A_f/A_o)(1 - \eta_f)$

   A_f = total area of fin, A_o = surface area of fin and η_f is fin effectiveness.

3. **Overall Heat Transfer Coefficient:** The heat transfer between the hot stream and cold stream in the heat exchanger is calculated with the overall hat transfer coefficient which is given by

   $U = Q/A_o \Delta T$

   Q = total heat transferred, A_o = surface area of heat exchanger, ΔT is the temperature difference.

LMTD for Heat Exchanger

The log mean temperature difference (LMTD) is used to determine the temperature driving force for heat transfer in flow systems in heat exchangers. The LMTD is a logarithmic average of the temperature difference between the hot and cold streams at each end of the exchanger. The larger the LMTD, the more heat is transferred. The use of the LMTD arises straightforwardly from the analysis of a heat exchanger with constant flow rate and fluid thermal properties.

We assume that a generic heat exchanger has two sides (which we call "A" and "B") at which the hot and cold streams enter or exit; then, the LMTD is defined by the logarithmic mean as follows:

$$\text{LMTD} = \frac{\Delta T_A - \Delta T_B}{\ln\left(\dfrac{\Delta T_A}{\Delta T_B}\right)}$$

where ΔT_A is the temperature difference on side A, and ΔT_B on side B. This equation is valid boths for parallel flow, where the streams enter from the same side, and for counter-current flow, where they enter from different sides. A third type of flow is cross-flow, in which one system, usually the heat sink, has the same nominal temperature at all points on the heat transfer surface. This follows similar mathematics, in its dependence on the LMTD, except that a correction factor F often needs to be included in the heat transfer relationship. There are times when the four temperatures used to calculate the LMTD are not available, and the NTU method may then be preferable. Once calculated, the LMTD is usually applied to calculate the heat transfer in an exchanger according to the simple equation:

$$Q = U \times A \times \text{LMTD}$$

where Q is the exchanged heat duty (in watts), U is the heat transfer coefficient (in watts per kelvin per square meter) and A is the exchange area.

NTU for Heat Exchanger

The Number of Transfer Units (NTU) Method is used to calculate the rate of heat transfer in heat exchangers (especially counter current exchangers) when there is insufficient information to calculate the Log-Mean Temperature Difference (LMTD). In heat exchanger analysis the fluid inlet and outlet temperatures are specified or can be determined by simple mass balance using the LMTD method, but when this information is not available the NTU or the effectiveness method is used. To define the effectiveness of a heat exchanger we need to find the maximum possible heat transfer which can be hypothetically achieved in a counter flow heat exchanger of infinite length. Therefore one fluid will experience the maximum possible temperature difference which is the difference between the inlet temperature of the hot stream and the inlet temperature of the cold stream. The method proceeds by calculating the heat capacity rates (i.e. mass flow rate multiplied by specific heat) for the hot and cold fluids respectively, and denoting the smaller one as C_{min}. The reason for selecting smaller heat capacity rate is to include maximum feasible heat transfer among the working fluids during calculation.

$$q_{max} = C_{min}(T_{h,i} - T_{c,i})$$

Here q_{max} is the maximum heat which could be transferred between the fluids. According to the above equation in order for the maximum heat to be transferred the heat capacity should be minimized. The effectiveness (E) is the ratio between the actual heat transfer rate and the maximum possible heat transfer rate and is given by the equation

$$E = \frac{q}{q_{max}}$$

where

$$q = C_h(T_{h,i} - T_{h,o}) = C_c(T_{c,o} - T_{c,i})$$

and

$$q_{max} = C_{min}(T_{h,i} - T_{c,i})$$

Effectiveness is dimensionless quantity between 0 and 1 inclusive. If E is known for a particular heat exchanger with the inlet conditions of the flows then the amount of heat being transferred between the fluids can be calculated as

$$q = EC_{min}(T_{h,i} - T_{c,i}).$$

For any heat exchanger it can be shown that

$$E = f\left(NTU, \frac{C_{min}}{C_{max}}\right)$$

For given geometries, E can be calculated using correlations in terms of the 'heat capacity ratio'

$$C_r = \frac{C_{min}}{C_{max}}$$

and the number of transfer units,

$$NTU = \frac{UA}{C_{min}}$$

where U is the overall heat transfer coefficient and A is the heat transfer area. We can obtain a specific equation for the effectiveness of a parallel flow heat exchanger as

$$E = \frac{1 - \exp[-NTU(1 + C_r)]}{1 + C_r}.$$

Similar effectiveness relationships can derived for concentric tube heat exchangers and shell and tube heat exchangers as well. Such relationships differ from each other depending on the type of the flow i.e. counter current, concurrent, cross flow. Note that $C_r = 0$ is a special case scenario where in the heat exchanger condensation or vaporisation is occurring. Hence in this special case heat exchanger behavior is independent of the flow arrangement. Therefore, the effectiveness is given by

$$E = 1 - \exp[-NTU].$$

In a cross flow heat exchanger fluid is said to be mixed when it is not confined to flow in a definite channel such as a tube and is said to be unmixed when it is confined to flow in a definite channel. The effectiveness is given for the three cases as follows:

Both unmixed : $E = 1 - \exp\{-Ntu^{0.22}[1 - \exp(-C_r Ntu^{0.78})]/C_r\}$

C_{max} mixed and C_{min} unmixed : $E = (1/C_r)[1 - \exp[-C_r(1 - e^{-NTU})]]$

C_{max} unmixed and C_{min} mixed : $E = [1 - \exp(1/C_r)][1 - \exp(-Ntu\ C_r)]$

Effect of Effectiveness on Cryogenic System Performance

Effectiveness of heat exchanger directly affects the performance of a cryogenic system. Refer a typical cryogenic liquefaction system in Figure 10.8.

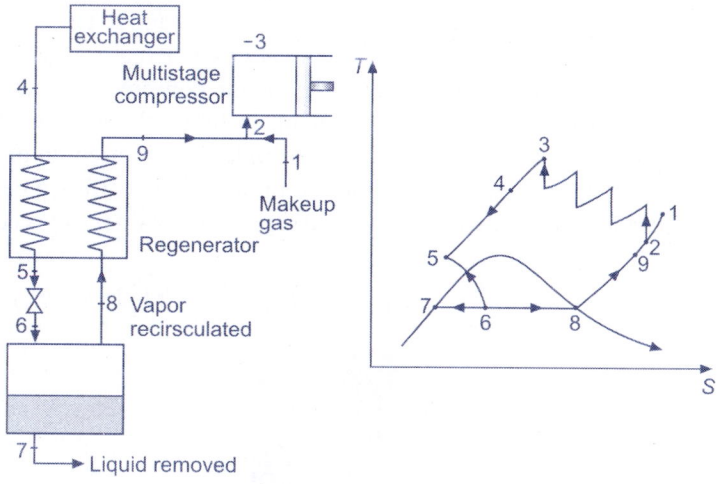

Fig. 10.8 A Cryogenic Liquefaction System

Assuming a liquefaction system we can define specific heats for the hot and cold stream as

$cph = (h_2 - h_3) / (T_2 - T_3)$ and $cpc = (h_1 - h_g)/(T_1 - T_g)$

The capacity rates are $C_h = m\, cph$ and $C_c = m\,(1-y)cpc$.

The yield is given by $y = h_1 - h_2/h_1 - h_f$

The heat exchanger effectiveness is defined as $E = 1 - (h_1 - h_2/h_1 - h_g)$

The work done is given by $\Delta W = m(1 - E)(h_1 - h_2)$

Thus increased heat exchanger effectiveness leads to increased yield and work done.

CRYOGENIC COMPRESSORS

Cryogenic applications use centrifugal, hermetic cold gas compressors. Compressors that process hydrogen, nitrogen, natural gas, and helium at temperatures approaching 5 K (−450°F) are available. Hermetic designs are extremely desirable for cryogenic applications because they eliminate the need for mechanical shaft seals and as a result, completely eliminate cryogen leakage. Additionally, hermetic compressors eliminate air infiltration and are inherently explosion proof. High speed designs and the use of variable frequency drives result in efficient operation across a wide variety of head or flow conditions. Single shaft, direct drive designs are highly reliable due to their simplicity and are extremely stable throughout their entire operating range. Low vapor pressure rolling element grease packed and proprietary dry lubricated bearings provide long term, reliable service without contaminating the process fluid. Friction free magnetic bearings eliminate wear items and allow machines to operate at extremely high speeds for many years without maintenance. User friendly designs allow the compressor to be serviced without breaking the cold box vacuum. Refer Figure 10.9 for a typical cryogenic compressor.

10.14 Cryogenics

Cryogenic compressors are of two types reciprocating and rotary. Reciprocating compressors are well suited for lower flow applications, whereas rotary compressors are more suited to large flow applications. Cryogenic compressors feature heavy-duty crosshead construction with two vertical, in-line cylinders and full water cooling. The guide cylinders are cast into the crankcase for strength and rigidity. The compressor is fabricated with heavier components and offers a center main bearing for additional horsepower and rod load capacity. Work done by a compressor is given by

$$-W/m = T_1(S_1 - S_2) - (h_1 - h_2) = RT_1 \ln(P_2/P_1)$$

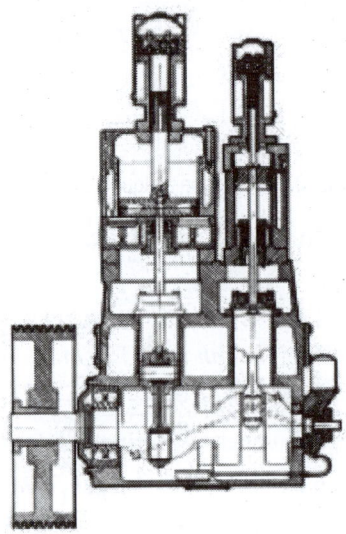

Fig. 10.9 Cryogenic Compressor

CRYOGENIC TURBO EXPANDERS

Turbo machinery is essential to the success of any cryogenic process. Radial inflow turbo expanders provide refrigeration for some of the most efficient air separation and liquefaction facilities. Reliability results from proven designs which are easy to operate and maintain. High-pressure turbo expanders, usually installed in pairs, are typically used to provide refrigeration for the production of cryogenic liquids. The expander extracts work from a high-pressure gas stream providing refrigeration to the process. The work removed from the gas stream by the expander is used to provide compression power to cycle. These applications demand high efficiency and reliability at a low installed cost. The efficiency of the turbo expander has a significant impact on the cost of the cryogenic liquid produced. The family of process compressor-loaded expanders is ideal for applications where power and capital cost are evaluated highly. Advanced design tools and capabilities, combined with streamlined project management processes, allow for the best match between efficiency and capital investment. Expander efficiencies are at the leading edge of technology, guaranteed to 91.5% in some applications. Designed for high-pressure service, our expanders will accommodate a wide range of flow, pressure ratios, and cryogenic gases, ensuring unsurpassed performance. Refer Figure 10.10 for a typical cryogenic turbo expander

Fig. 10.10 Cryogenic Turbo Expander

The benefits of turbo expanders are low capital and installed costs, ease of operation and maintenance, high-efficiency in excess of 91.5%, energy recovery through high efficiency compression, low seal gas consumption, high refrigeration load capacity, and proven equipment reliability. High-pressure turbo expanders are available in configurations ranging from stand-alone expanders to fully skidded packages, including seal gas and lubrication systems. Available in configurations up to 10 stages and up to 3.5 MW generators, the turbine expander provides increased process efficiency and improved mechanical performance over other designs. From its corrosion resistant stainless steel construction to its high efficiency induction generators, the turbo expander offers total life cycle value. The turbo expander assembly is composed of four basic sub-assemblies—the intake and discharge barrel, the turbine, the generator stand and the seal housing. The intake and discharge shell is a barrel within a barrel configuration. It is designed to direct the flow of liquid up by an annular assembly into a 360° guided inlet to the first stage turbine runner. After passing through all of the energy-absorbing stages, the flow is directed to the outlet and discharge flange. This design allows the complete turbine power and rotating assembly to be removed from the barrel while leaving the piping and insulation undisturbed. Each turbine drives a high efficiency, vertical solid shaft induction generator. These units are designed for the high ambient temperature and humidity prevalent at the most extreme climates. The tilting pad pivot shoe thrust and journal bearings are suitable for full runaway speed. They are capable of carrying the hydraulic thrust generated by the turbine over the entire operation range of the application and are lubricated by clean oil. Generators and tandem mechanical gas seal assemblies are isolated from the cryogenic portion of the turbine by an engineered warm-up chamber. Refer Figure 10.11 for the cross sectional view of a cryogenic turbo expander.

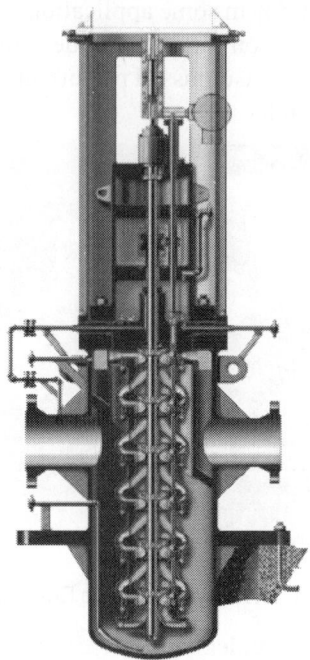

Fig. 10.11 Cross Section of a Cryogenic Turbo Expander

Expanders can be classified as axial flow or radial flow. Expanders are in use since 1902 for liquefying air. Improvements were made in early designs by lengthening the cylinder and by using flexible piston rods made of stainless steel. The work done by an expander is

$$-W/m = (h_2 - h_1) = C_p(T_2 - T_1).$$

Efficiences for Compressor and Expander

We can define the following efficiencies for the compressor and expander

1. **Isothermal efficiency for Compressor:** This can be defined as the ratio of work requirement for the reversible and isothermal compression to the actual work excluding mechanical losses.

$$\eta_i = (W/m)r / (W/m)$$

2. **Adiabatic efficiency for Expander:** This can be defined as the ratio of actual enthalpy decrease during expansion to ideal enthalpy reduction during isentropic expansion.

$$\eta_t = (h_1 - h_2) \text{ actual}/(h_1 - h_2)\text{ideal}$$

3. **Mechanical efficiency for Compressor:** This can be defined as the ratio of brake or actual work to the indicated or ideal work including mechanical losses.

$$\eta_m = (W/m) \text{ Brake}/(W/m) \text{ Indicated}$$

4. **Mechanical efficiency for Expander:** This can be defined as the ratio of brake or actual work output to the indicated or ideal work including mechanical losses.

$$\eta_m = (W/m)\text{Brake}/(h_1 - h_2)\text{ideal}$$

5. **Overall efficiency for Compressor:** This can be defined as the product of isothermal efficiency and mechanical efficiency for the compressor

$$\eta_o = \eta_i \times \eta_m$$

6. **Overall efficiency for Expander:** This can be defined as the product of adiabatic efficiency and mechanical efficiency for the expander

$$\eta_o = \eta_t \times \eta_m$$

Losses for Compressors and Expanders

For reciprocating compressors the losses are as follows:

1. **Inlet and outlet valve losses:** These losses are due to pressure drops across the inlet and exhaust valves and depend on the gas flow rate and area of cross-section of the inlet and exhaust valves. The pressure drop Δp across the inlet and outlet valve is given by the equation

$$(\Delta p/p) = K_L V^2/2gRT$$

where P = total pressure, K_L = Loss coefficient, V = Velocity, T = absolute temperature, g = gravitational acceleration and R = gas constant.

2. **Incomplete expansion losses:** During the expansion, the fluid control valve is opened at a cut-off point before the expansion stroke is completed. This causes losses due to incomplete expansion.

3. **Heat transfer losses:** During the expansion, the heat transfer to the working fluid reduces enthalpy drop of the fluid during expansion. This causes heat transfer losses.

$$Q = W = m(h_2 - h_1)$$

4. **Piston friction losses:** Piston friction losses include the energy that is dissipated as an increase of internal energy of piston and cylinder and is returned as heat to the working fluid. The piston friction losses are dependant upon the material of piston and cylinder and the clearance between them.

For rotary compressors the losses are as follows:

5. **Inlet valve losses:** These losses are due to pressure drops across the inlet valves and depend on the gas flow rate and area of cross-section of the inlet valve. The pressure drop Δp across the inlet valve is given by the equation

$$(\Delta p/p) = K_L V^2/2gRT$$

where P = total pressure, K_L = Loss coefficient, V = Velocity, T = absolute temperature, g = gravitational acceleration and R = gas constant.

6. **Disk friction losses:** Disk friction losses include the frictional energy that is dissipated between the rotor and the gas in the space between the rotor and

the housing. The Stodola equation gives the relation between disk friction and enthalpy change (Δh)

$$(W/m)/\Delta h = 0.004 \, (D/H_b) \, (U/C_o)^3$$

where, W/m = work done per unit mass, D = blade tip diameter, H_b = blade height, U = mean blade speed and C_o = ideal blade velocity after expansion.

7. **Impeller losses:** Impellor losses are dependant on blade angles at inlet and exit, velocity of fluid at inlet and exit, and type of impellor.
8. **Leaving losses:** These losses are due to energy wasted by discharge of the kinetic energy of outlet stream.

$$\Delta h = V^2/2g$$

Effect of Compressor and Expander Inefficiencies on System Performance

The compressor efficiency has no direct bearing on the liquid yield (y) of the cryogenic system. However compressor work requirements are directly related to compressor efficiency and can be given by the equation

$$-W/m = 1/y \, [T(s_1 - s_2) - (h_1 - h_2)]$$

The term in the bracket is the ideal isothermal work requirement. Actual work requirements however vary in inverse proportion to the overall compressor efficiency. For an expander the overall efficiency affects both the liquid yield and work requirements. The adiabatic efficiency of expander affects the enthalpy drop across expander and is related to the yield by the equation.

$$y = (h_1 - h_2)/(h_1 - h_f) + x\eta(h_3 - h_4)/(h_1 - h_f)$$

here y = yield, x = expander flow rate ration and η is the adiabatic efficiency. The second term of the above equation is dependent on the adiabatic efficiency. Thus the liquid yield can increase or decrease with corresponding increase or decrease in adiabatic efficiency.

CRYOGENIC PUMPS

Cryogenic pumps are designed to move coolants and cryogenic liquids. They are built to withstand and operate in extremely cold temperatures. Cryogenic pumps feature hermetically sealed designs to minimize heat leakage from the motor or contamination by process fluids into the cryogenic fluid. Long shaft cryogenic pumps are designed with the pump motor and mounting flange separated from the pump impeller by a long shaft. The pump impeller is submerged in the cryogen or freezing liquid. This minimizes the leaking of heat from the motor into the frozen or freezing cryogenic fluid. Long shaft cryogenic pumps may be welded or bolted to a variety of cryogenic equipment, including dewars and cryostats. A centrifugal pump is typically used to transfer cryogenic liquids between a storage tank or tanker car because of their ability to produce and maintain a high flow rate. Refer Figure 10.12 for the photograph, cross-sectional view and impeller construction of a typical cryogenic pump. Cryogenic pumps may also be submersible. A submersible cryogenic pump is frequently used in applications where heat leak is not the most important factor. Submersible cryogenic pumps are used as pumps in vehicles that use liquefied natural gas or in the liquid hydrogen propellant system in a rocket. Cryogenic pumps for use in extremely cold environments are usually constructed with a vacuum housing to provide a barrier between the motor and the cryogenic fluid. Cryogenic pumps are used to circulate coolant in a variety of applications, including cooling high temperature superconducting cables or magnets and in prototype slush hydrogen applications.

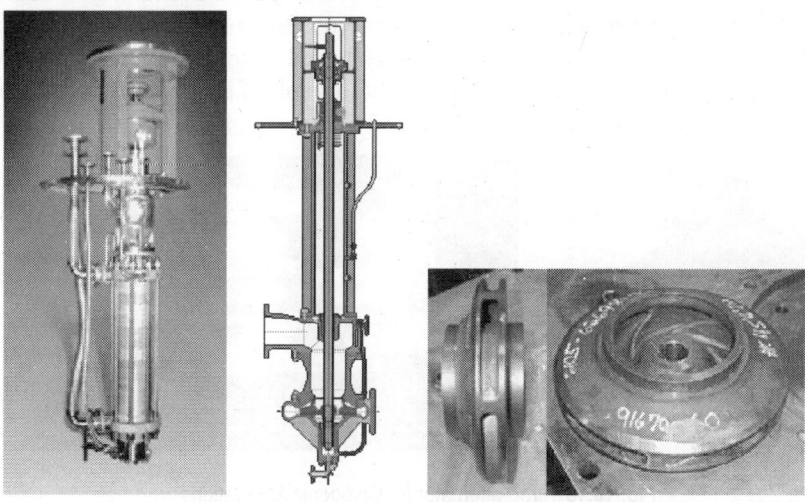

Fig. 10.12 Pumps for Cryogenic Applications

Industrial cryogenic pumps are used for handling liquids such as Oxygen and Nitrogen, at temperatures as low as –320°F. The top suction design of the pump has become a standard in the industry for handling fluids which are near or at boiling point without vapor binding. Design considerations for the cryogenic pumps include special metallurgical considerations due to low temperatures and flammability of liquefied gases such as liquid oxygen, thermal expansion and contraction and material behavior at cryogenic temperatures, and extremely low NPSH. Impellers may be either of the semi-open or enclosed design and are keyed to the shaft. Both single and multi-stage cryogenic pumps can be designed to incorporate an inducer for low NPSH applications. Inducers are individually sized for each application. Sensitive thermocouples are used to detect and monitor the liquid level in the pump column. This safety feature ensures that the labyrinth seals operate properly in the vapor phase. Icing at the support plate is thus eliminated ensuring long life of the cryogenic pump. Easily replaceable cartridge assemblies are mounted in the bearing frame above the top mounting plate. This configuration provides greater support for the pump shaft and eliminates the reliance of the drive motor's bearings. Shaft sealing is provided by a precision labyrinth seal which operates at the suction pressure of the pump and eliminates the requirement for costly and failure-prone high pressure gas seals.

CRYOGENIC TURBO ALTERNATORS

Claude cycle is used for the production of refrigeration at 4.4K. Temperatures at this level are suitable for numerous applications of superconductivity. A principle component of the equipment used in Claude cycle is the high-speed dynamic turbo-alternator, in which all rotating parts are suspended on self-acting gas bearings. Such a system avoids sliding contact and wear during operation and avoids the contamination problems associated with oil lubricated compressors used in conventional cryogenic refrigerators. Sometimes a system consisting of multiple cryogenic turbo alternators, is used to improve efficiency. A typical cryogenic turboalternator is developed and successfully tested to temperatures in the range of 80 to 100 K. Refer Figure 10.13 for the photograph of a turbo-alternator and its cross-section view.

 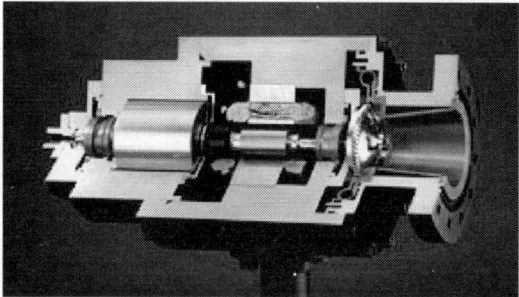

Fig. 10.13 Turbo-alternator for Cryogenic Applications

High power turbo-alternators utilizing gearbox reduction have been developed since 1980s. Recent designs have focused on high-speed, single-shaft, permanent magnet machines utilizing radial in-flow, and axial impulse turbines. These applications have been developed for waste heat recovery and Brayton cycle applications. An increasing number of applications require the efficiency, size, and hermetic characteristics of a single shaft turbo-alternator. The decreasing cost of the power electronics utilized in the rectification/inverter equipment has made this an economically viable alternative for many systems that have traditionally utilized gearboxes and low-speed generators. The hermetically sealed design eliminates maintenance intensive shaft seals. Power levels from 3 kW to 75 kW have been built and test. The turbo-alternator is usually situated between the turbine wheel and the fly wheel. When the unit is on standby the alternator acts as an electrical motor, consumes a small amount of power, and maintains the unit's rotational speed at 60,000 rpm. In the event of a power failure, the turbo-alternator, driven by the fly wheel's inertia, immediately begins generating electrical power. The turbo-alternator continues generating electricity for as long as is necessary.

CRYOGENIC SYSTEM OPTIMIZATION

With the present competitive product markets in the gas processing industry, process optimization to yield the lowest possible capital and operating costs is an absolute necessity. To achieve this level of optimization, detailed process simulation exploring all possibilities is required. In cryogenic operations, some of the major factors to be considered are the heating, cooling, and refrigeration loads. These loads are usually interconnected and are optimized by the use of a network of cross exchangers. The most successful designs are using increasing numbers of cross exchangers with progressively closer approach temperatures to achieve heating and cooling without causing demands on plant utilities. Unfortunately, in many cases the heating and cooling potential is not fully realized because the heat delivery versus temperature curve causes impossible temperature crosses in conventional cross exchangers.

Any cryogenic system optimization program has three broad goals.

1. Increase reliability of cryogenic equipment to allow availability of 99.8%.

   This is the availability of the cryogenic subsystem including controls. Recent feedback indicates that the power components require 99.8-99.9 % availability. This means the cryogenic subsystem needs 99.9% and higher reliability; this is about one 8-hour shift of unavailability per year.

2. Increase efficiency of cryogenic equipment to achieve 30% of Carnot.

   Some ways to increase efficiency may lower reliability: for example going from 1 to 2 expanders in a Brayton Cycle. The reliability can be increased through redundancy but this impacts cost.

3. Decrease cost of cryogenic equipment from $100 to $25/W at ~ 65-80 K.

   The easy way to do this is to increase the production base and use standardized components.

In some cases, the cooling system is not necessarily the optimum for the eventual long lifetime utility or industrial application but is chosen based on available equipment or minimal cost. The refrigeration load at the operating temperature is a small fraction of the electrical rating of equipment. When the application output is increased, the thermal load increases but at a slower rate than the linear increase in application rating. Losses and radiation from ambient surfaces increase less than linearly with size of cryogenic equipment. For synchronous machines an additional heating term has to be accommodated in cryogenic system capacity or by thermal inertia. A load following capability is desired in the cryogenic refrigerator system. When the application is operating at a reduced load, existing cryocooler technology that operates at a single compressor setting will tend to cause operation at lower temperatures, due to decreased losses. If the cryogenic refrigerator can be turned down to match a lower load on the application, and reduce the energy consumed by the refrigerator, an additional energy savings is achieved. With present cryocooler technology, refrigerators and compressors are sized for the peak load and run with a near constant compressor power even when the heat load is much reduced. As much as a factor of four savings in compressor power for refrigeration of the heat load is possible if the compressor operation could be adjusted to run in proportion to the load. Standardization across electric power devices is strongly recommended where feasible. Thermal and structural analysis tools can be used for the sizing of insulation and structural geometry of integrated cryogenic equipment system.

Problems

Problem 1. A Counter flow heat exchanger in a liquefaction system operates under following conditions. Warm gas at 2027 kPa and temperature 22 K. The flow rate of warm gas is 10 g/s with specific heat 6.274 kJ/kgK. Cold fluid at 101.3 kPa and inlet temperature 4.2 K and outlet temperature is 21.1 K. The flow rate of cold fluid is 9 g/s with specific heat 5.49 kJ/kgK. Overall heat transfer coefficient is 170 W/m²K. Determine heat transfer area.

Solution:

Capacity rates can be calculated as

$$C_h = m \text{ cph} = (0.010)(6274) = 62.74 \text{ W/°C} = C_{max}$$
$$C_c = m \text{ cpc} = (0.009)(5490) = 49.41 \text{ W/°C} = C_{min}$$
$$C_r = C_{min}/C_{max} = 49.41/62.74 = 0.7875$$

Effectiveness can be calculated as

$$E = (T_{c2} - T_{c1})/(T_{h1} - T_{c1}) = (21.1 - 4.2)/(22 - 4.2) = 0.94$$

NTU can be calculated as

$$\text{NTU} = 1/(1 - C_r) [\ln(1 - E\, C_r/1 - E)]$$
$$\text{NTU} = 1/(1 - 0.7875) [\ln(1 - (0.94)(0.7875)/1 - 0.94]$$
$$\text{NTU} = 7.562 = UA/C_{min}$$
$$7.562 = 170 \text{ A} / 49.41$$
$$A = 2.19 \text{ m}^2$$

This the required heat transfer area.

References

1. Cryogenic Systems – Randall Baron.
2. Internet website www.wikipedia.com on heat exchangers.
3. Cryogenic Engineering – B.A. Hands.
4. Internet website http://www.barber nichols.com/products.
5. Linde Brochure on Cryogenic Heat Exchangers.
6. Internet website http://www.lawrencepumps.com/products/cryogenic.htm.
7. Technical paper from Bryan Research and Engineering.
8. Internet website http://www.airproducts.com/Products/Equipment/.
9. FlowServe Brochure on Turbo Expanders.
10. Internet website http://www.epconlp.com/asme/heat-exchangers.
11. Internet website http://lt.tnw.utwente.nl/research/linde_hampson.doc.
12. US DOE report no. DOE/EE-0247 on Heat Exchangers.

Questions

Q.1. Enlist at least five cryogenic process equipment.

Q.2. Describe any two types of cryogenic heat exchangers.

Q.3. Discuss the losses and inefficiencies of compressors and expanders.

Q.4. Write a short note on cryogenic pumps.

Q.5. A Counter flow heat exchanger in a liquefaction system operates under following conditions. Warm gas at 2000 kPa and temperature 25 K. The flow rate of warm gas is 12 g/s with specific heat 6.5 kJ/kgK. Cold fluid at 110 kPa and temperature 3.5 K. The flow rate of cold fluid is 8 g/s with specific heat 4.5 kJ/kgK. Overall heat transfer coefficient is 150 W/m²K. Determine heat transfer area.

CHAPTER

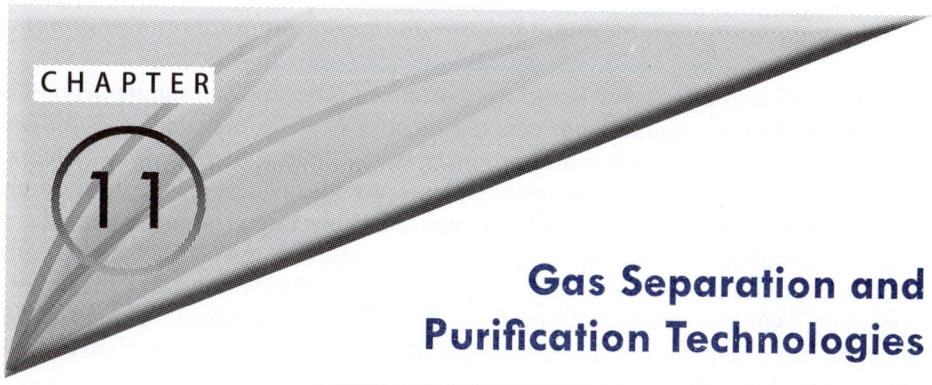

Gas Separation and Purification Technologies

INTRODUCTION

Large scale commercial production of industrial gases from the atmosphere generally involves cryogenic processing of air. In addition to desired products such as oxygen, nitrogen, argon and the rare gases, feed air used as a starting material for cryogenic processing also contains impurities or undesirable components such as water vapor, carbon dioxide and one or more hydrocarbon species. These impurities must be removed before processing of feed air can be completed because the impurities interfere with continuous and efficient operation of the cryogenic equipment, or may present hazardous conditions which imperil the safety of operators or damage equipment. A significant portion of the cost of a gas separation plant is associated with cleaning or pre-purifying the gas and cooling to cryogenic temperatures. Cryogenic gas separation plants produce nitrogen, oxygen and argon products using very low temperature distillation for separation and purification. Cryogenic gas separation plants are most commonly used to produce high purity products at medium to high production rates. Non-cryogenic plants produce gaseous nitrogen or oxygen products using near-ambient-temperature separation processes. There are two major types of non-cryogenic processes, using either selective adsorption or differential permutation through membranes to produce relatively pure gases like nitrogen. These processes use differences in properties such as molecular structure, size and mass to achieve the desired degree of separation. To enable gas mixtures to be separated into its constituents by means of rectification–the actual separation process–a large part of the air volume used must be liquefied. A gas can only be transformed into a liquid state at temperature and pressure conditions below those of its critical point. The boiling point and condensation conditions of gas mixtures such as air are not identical. There is a condensation line and a boiling point line which delineate the boiling point range.

Cryogenic gas separation plant design is an extremely critical process. System optimization does not stop with specification of the cryogenic plant but a fully-optimal facility design will reflect all aspects of anticipated operating conditions at the site. Plant supply solutions which represent the best combination of cost and performance, as determined by customer priorities and criteria are desirable. The complete plant design integrates the gas separation unit (and if included, the liquefaction system) with appropriately configured and sized auxiliary equipment

11.2 Cryogenics

such as pipeline compressors, liquid storage tanks for backup and supplementation of plant production, liquid vaporizers (ambient or heated), elevated pressure gas storage, and bulk merchant liquid products storage and trailer loading systems. Well-optimized cryogenic gas separation plants provide operational flexibility, high operating reliability, and cost-effective operation. When large volumes of liquid are required to support customer operations, the gas separation unit can be coupled with a liquefier. Liquefiers can be supplied as stand-alone units, which allow maximum operating flexibility, which can be particularly useful for "piggyback" gas and liquid plants, or they may be fully integrated with the separation units to realize capital cost savings for plants.

In this chapter some of the gas separation and purification systems will be examined in detail. The cryogenic separation is the most economical method of gas separation of mixtures. Commercial gases like liquid oxygen, argon, nitrogen are obtained by rectification of liquid air. Purification methods which are effective at cryogenic temperatures such as physical adsorption have also been analyzed. The calculation of minimum work required to separate the mixture of gases is a critical requirement for designing a cryogenic gas separation plant.

THERMODYNAMICALLY IDEAL GAS SEPARATION SYSTEM

The thermodynamically ideal separation system is examined in order to have a basis for comparison of the various practical separation systems used. All the processes in the ideal system are reversible. Mixing of gases is an irreversible process as it requires some work to separate the mixture into its original constituents. However reversible mixing is possible with semi-permeable membranes which allow passage of one gas only from a mixture and work obtained while mixing is used to unmix the gases. Refer to a typical gas separation system in Figure 11.1. The column is assumed to operate reversibly. The incoming stream of mixture M is separated into two streams A and B. It is necessary to determine the work required to unmix the two gases reversibly. Heat exchange with an external reservoir is required to maintain the temperature of the separation process constant. The energy spent to separate the gases reversibly is same as the energy required to compress each gas component which is dependent on the ratio of partial pressure of the component to the total mixture pressure.

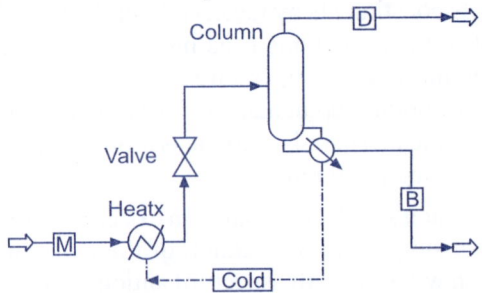

Fig. 11.1 Gas Separation System

The vertical column houses the membrane which allows passing of only one type of gas thereby achieving separation. The mixture at temperature T_m and pressure p_m is separated into two pure gases at the same temperature and pressure. The thermodynamic work of separation is the reversible isothermal work (W_i) and is given by

$$-W_i/m = T_m(s_1 - s_2) - (h_1 - h_2)$$

where subscript 1 denotes conditions before separation and subscript 2 denotes conditions after separation. The total work required to separate a unit mass of mixture m in components A and B would be

$$-W_i/m_m = T_m[(m_a/m_m)(s_{1a} - s_{2a}) + (m_b/m_m)(s_{1b} - s_{2b})]$$
$$- [(m_a/m_m)(h_{1a} - h_{2a}) + (m_b/m_m)(h_{1b} - h_{2b})].$$

Thus the ideal work of separation is decreased as the separation temperature is decreased. T_m is usually limited to ambient temperature. If the ideal gas assumption is applied then the equation for ideal work simplifies to

$$-W_i/m_m = T_m[(m_a/m_m) R_a \ln(p_m/p_{1a}) + (m_b/m_m) R_b \ln(p_m/p_{1b})]$$

The individual pressure ratios can be expressed in terms of mass ratios and mole fractions as the separation process is isothermal in nature.

$$(p_m/p_{1a}) = m_m R_m / m_a R_a = n_m / n_a = 1/y_a$$

Where $y_a = n_a/n_m$ = mole fraction of gas A.

The work requirement per unit mole of mixture is given by

$$-W_i/n_m = RT_m [y_a \ln(1/y_a) + y_b \ln(1/y_b)]$$

This equation can be extended to the complete separation of a mixture of several ideal gases. If j is the total number of components of the mixture, the equation becomes

$$-W_i/n_m = RT_m \sum y_j \ln(1/y_j)$$

The figure of merit for the separation system is given by

$$FOM = -(W_i/n_m)/-(W/n_m)$$

GENERAL CHARACTERISTICS OF MIXTURES

A mixture is when two or more different substances are mixed together but not combined chemically. The molecules of two or more different substances are mixed as solutions, suspensions, and colloids. While there are no physical changes in a mixture, the chemical properties of a mixture, such as its melting point may differ from those of its components. Mixtures can be separated into its original components by mechanical means. Mixtures are either homogenous or heterogeneous. A single substance requires only two thermodynamic properties to specify its state. For mixtures of two or more substances in a single phase, the mole or mass fractions of all but one component are required. For mixtures of two or more substances in two phases, the properties are interdependent. For a pair of phases, there exists one relationship between properties.

For more than one phase and more than one substance the Gibbs Phase rule is applied to determine the number of properties required to fix the state of the system. The Gibbs phase rule states

$$F = C - P + 2$$

11.4 Cryogenics

where F = number of properties, C = number of components and P = number of phases. Refer Figure 11.2 for the phase equilibrium curve for two component mixture.

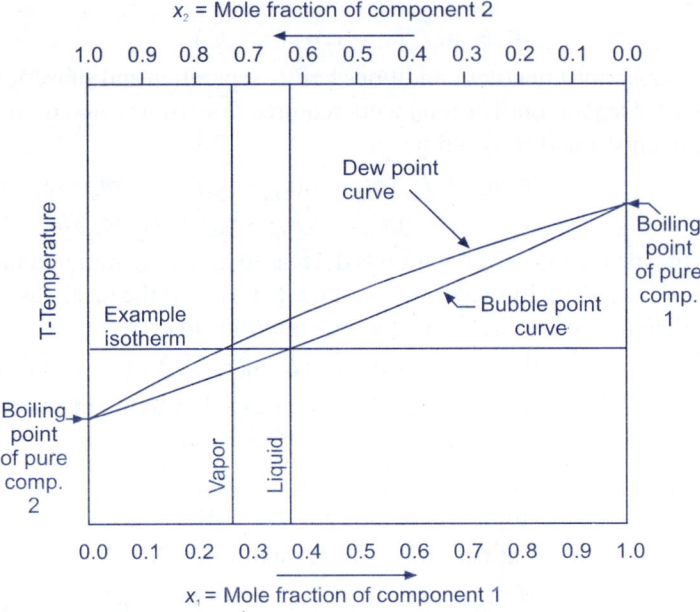

Fig. 11.2 Phase Equilibrium Curve for two Component Mixture

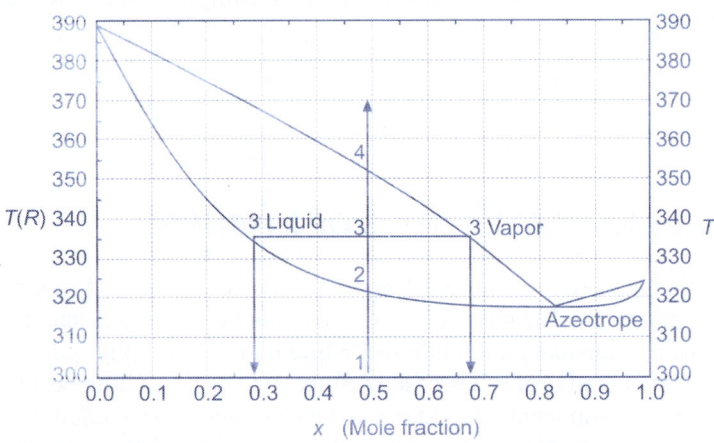

Fig. 11.3 Phase Equilibrium Curve for two Component Azeotropic Mixture

The phase equilibrium curve shown in Figure 11.2 is typical for mixtures in which the two components have critical points far above the conditions specified for the curve such as air. Figure 11.3 shows a phase equilibrium curve for a mixture that is azeotropic at certain compositions, i.e. behaves as a single pure substance at that compositions. Example of such an azeotropic mixture is chloroform – acetone. Azetropic mixtures are difficult to separate by rectification. Cryogenic fluids do not form azeotropic mixtures.

Temperature Composition Diagrams

The phases of gas mixtures can be plotted on temperature-composition diagrams. Refer Figure 11.4 for a temperature composition diagram for a mixture of two gases. The mixture is in gaseous phase at point 1. The temperature of the mixture decreases upon cooling, until the upper phase separation line is reached. This line is also called as dew line as the mixture starts to condense on cooling. If the mixture is at uniform temperature, the higher boiling point component condense first. The lower phase separation line is also called as bubble line as the mixture starts to boil upon heating.

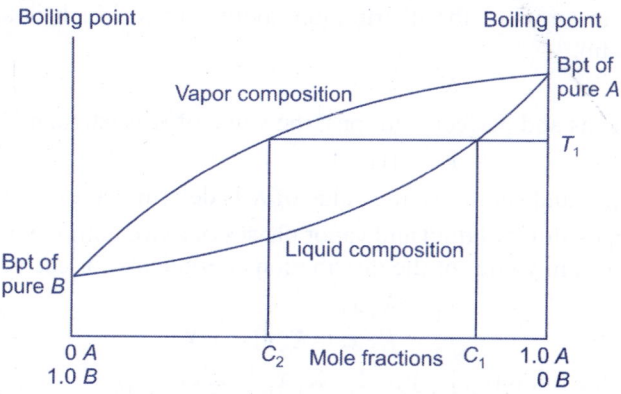

Fig. 11.4 Temperature-Composition Diagram for Mixture of two Gases

As the mixture is further cooled, the lower boiling point component begins to condense. The condensation of mixtures is not an isothermal process as the temperature decreases during cooling unlike pure substances. The mixture is separated by partial condensation as the liquid is enriched with higher boiling component and vapor is enriched with lower boiling component. Repeating the condensation process multiple times provides complete separation and this is the principle of rectification tower. The shape of the phase equilibrium curves for any substance depends upon intramolecular forces between the molecules making up the mixture. For liquid mixtures, where intramolecular forces are weak the Raoult's Law is valid. Raoult's law states the vapor pressure of an ideal solution is dependent on the vapor pressure of each chemical component and the mole fraction of the component present in the solution.

$$p_j = \Pi_j x_j$$

Where p_j = partial pressure of the jth component in the mixture, Π_j = vapor pressure of the jth component in the mixture and x_j = mole fraction of the jth component in the liquid phase.

Mixture that obey Raoult's law are called perfect solutions. If the vapor phase of the mixture is considered as an ideal gas then the partial pressure can be related to total pressure of the mixture by the Gibbs-Dalton Law.

$$p_j = p\, y_j$$

11.6 Cryogenics

Where p_j = partial pressure of the jth component in the mixture and y_j = mole fraction of the jth component in the vapor phase.

Thus
$$y_j = (\Pi_j/p)\, x_j$$

The total pressure for a mixture of ideal gases is the sum of partial pressures.
$$p = \Sigma p_j = \Sigma \Pi_j x_j$$

Thus for a two component mixture, $p = \Pi_1 x_1 + \Pi_2 (1 - x_1)$.

Distribution Coefficient

The relationship between the mole fractions in the vapor and liquid phases may be expressed in terms of the distribution coefficient K which is also known as equilibrium ratio.
$$K_j = y_j / x_j$$

For ideal gas and perfect solutions, the value of K is calculated as
$$K_j = (\Pi_j/p)$$

For real gas and solutions, the value of K is determined experimentally.

The composition of liquid and vapor phases of a two component mixture may be determined from values of the distribution coefficients for each component.
$$y_1 = K_1 x_1$$
$$y_2 = K_2 x_2 = K_2 (1 - x_1)$$

Thus for the mixture $y_1 + y_2 = 1 = K_1 x_1 + K_2 (1 - x_1)$

Solving for the mole fraction of component 1 in liquid phase
$$x_1 = (1 - K_2)/(K_1 - K_2)$$

Solving for the mole fraction of component 1 in vapor phase
$$y_1 = K_1 x_1 = K_1 (1 - K_2)/(K_1 - K_2).$$

Enthalpy Composition Diagrams

The set of thermodynamic curves important in the study of mixtures is the enthalpy composition diagram. A typical enthalpy composition (H-C) diagram for a mixture of two gases is shown in Figure 11.5 and an H-C diagram for ethanol and water system is shown in Figure 11.6. The enthalpy for saturated liquid and vapor can be listed separately in tabular form as well.

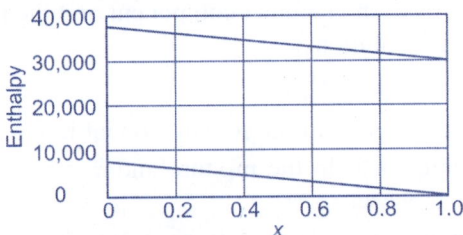

Fig. 11.5 Enthalpy-Composition Diagram for Mixture of two Gases

In Figure 11.5 two curves are shown. The top curve is the dew line for saturated vapor and the bottom line is the bubble line for saturated liquid. As a two component mixture is condensed, the temperature of the mixture decreases. Thus the lines of constant temperature (Isotherms) in the region between dew line and bubble line on the enthalpy composition diagram are having some finite slope and are not vertical. This fact can be seen in the H-C diagram for Ethanol-Water mixtures in Figure 11.6.

Simple Condensation or Evaporation

The separation of mixtures can be attained by partial condensation if the substances have varying boiling points. The effectiveness of this process depends upon the shape of the phase separation curves on the temperature-composition (T-C) diagram. A typical T-C diagram for a nitrogen-helium system is shown in Figure 11.7. The boiling point of helium is lower than nitrogen and thus the liquid that first condenses is made up of practically all nitrogen. By lowering the temperature of the mixture to 80 K, it is observed that the composition of the liquid shows increased nitrogen content. In some applications purity levels of the order of 94% are tolerated, which are achievable by condensation method. In case of air, the condensation method is not effective to separate the constituent gases like nitrogen and oxygen. Condensation method is suitable for nitrogen-helium or ammonia-hydrogen mixtures.

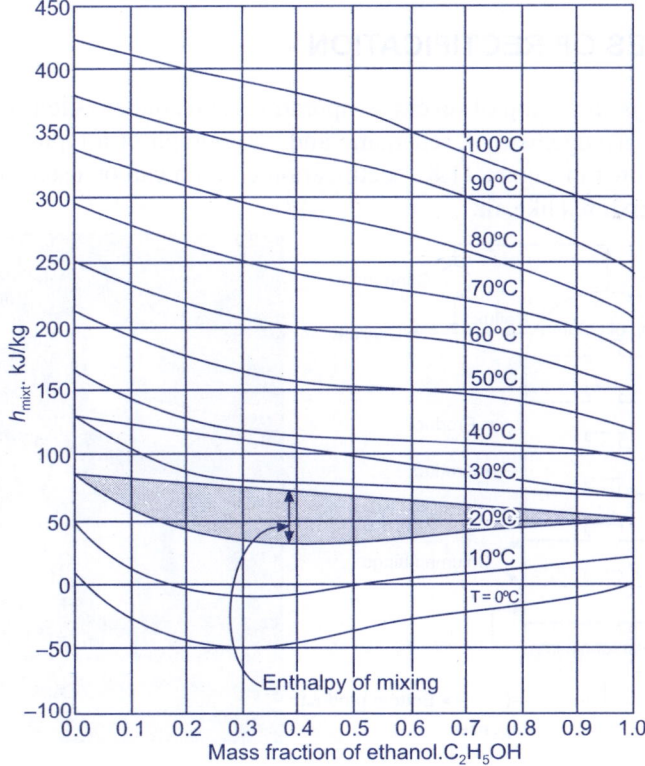

Fig. 11.6 Enthalpy-Composition Diagram for Ethanol-Water Mixture

11.8 Cryogenics

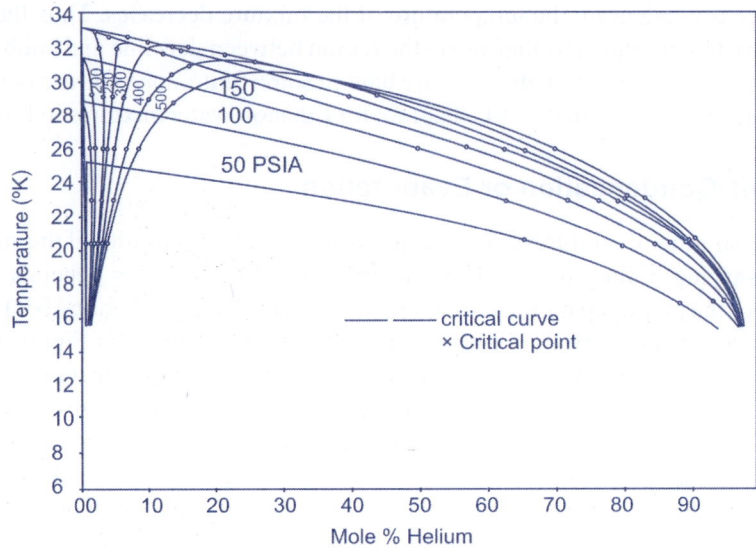

Fig. 11.7 Temperature-Composition Diagram for Nitrogen-Helium Mixture

PRINCIPLES OF RECTIFICATION

Rectification is cascading of several evaporations and condensations carried out in counter flow arrangement. A schematic and photograph of a typical rectification column is shown in Figure 11.8. Rectification column can be used for separation of gaseous mixtures like air.

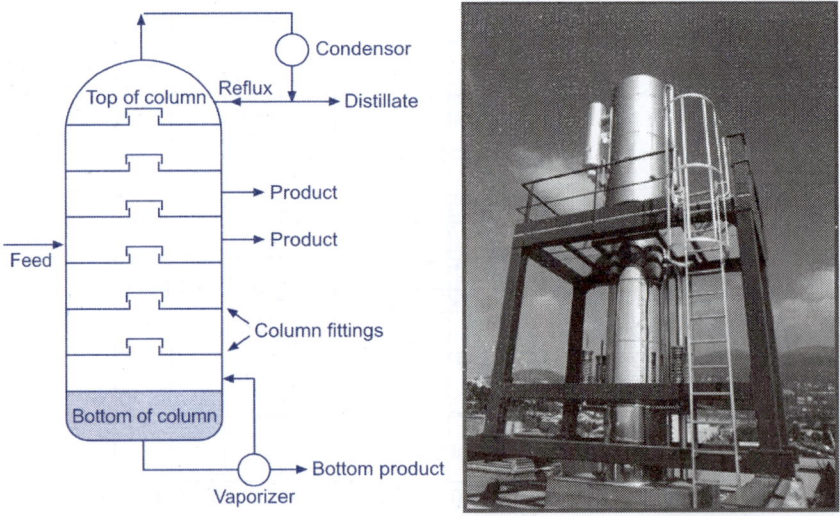

Fig. 11.8 Schematic and Photograph of Rectification Column

The temperature composition diagram is shown in Figure 11.9. The saturated mixture is introduced in the rectification column. In ideal steady state operation, the liquid layer above feed inlet has same composition as the feed. The vapor flows upward in the column through the liquid, whereas the liquid flows in the downward direction in layers (trays) through a tube. When the vapor begins to bubble through the liquid layer, the higher temperature vapor transfers heat to the liquid. This heat transfer from the bubble causes the condensation of high boiling point component and evaporation of low boiling point component from the liquid. Thus the bubble becomes richer in one component, whereas the liquid becomes richer in the other component thus achieving separation. By achieving a large number of layers (plates) high purity separation can be achieved. The bottom part of the column is heated to produce vapor and the top part of the column is cooled to produce liquid. For an ideal plate, the vapor should leave the plate at the same temperature as fluid temperature on the plate. For an actual plate the vapor does not leave the plate at the same temperature as fluid temperature on the plate. Thus actual requirement of plates is higher than theoretical plate requirement. The measure of perfection of a plate or a layer of liquid is given by the *Murphree efficiency*. The Murphree efficiency is defined as the ratio of actual change in mole fraction of the lower boiling point component in the vapor during its travel through the liquid layer to the maximum possible change in mole fraction.

$$\eta_M = (y_j - y_{j-1}) / (y^0_j - y_{j-1})$$

where y_j is the mole fraction of more volatile component in vapor phase leaving the jth plate, y_{j-1} is the mole fraction of more volatile component in vapor phase rising to the jth plate and y^0_j is the composition of vapor in equilibrium with liquid leaving the jth plate.

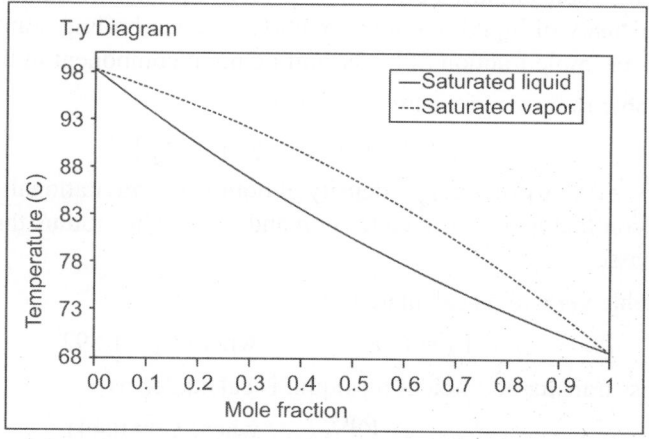

Fig. 11.9 Rectification Process on Temperature-Concentration Diagram

The heat transfer from the bubble is given by

$$Q = UA(T_g - T_f) = \rho_g V c_g (dT_g/dt)$$

Where U = overall heat transfer coefficient of bubble, A = area of cross section of bubble, T_g = temperature of vapor, T_f = liquid temperature, ρ_g = density of gas, V = volume of bubble, c_g = specific heat of vapor and t = time.

11.10 Cryogenics

The temperature distribution in the bubble is given by

$$(T_g - T_f)/(T_o - T_f) = \exp[-(27\, U\, gc\, \mu\, L_f)/(2\, R_b^3\, \rho_g\, \rho_f g\, c_g)]$$

where U = overall heat transfer coefficient of bubble, T_g = temperature of vapor, T_f = liquid temperature, T_o = initial temperature of vapor, ρ_g = density of gas, ρ_f = density of liquid, g = gravitational acceleration, g_c = conversion factor, c_g = specific heat of vapor μ = viscosity of liquid, L_f = thickness of liquid layer and R_b = radius of bubble.

For maximum Murphree efficiency for a given fluid, the radius of bubble should be small and thickness of liquid layer should be large. The bubbles are made small by passing the vapor through small holes. The depth of liquid column is usually 40 mm. Murphree efficiency can also be expressed by Geddes equation

$$\eta_M = 1 - \exp(-3\, U_M\, t^o / R_b)$$

where U_M = overall mass transfer coefficient between bubble and liquid, t^o = time required for bubble to pass through liquid layer and R_b = radius of bubble.

The overall mass transfer coefficient can be calculated as

$$1/U_M = (1/h_{Mg}) + (co_g/co_f)/h_{Mf}$$

where co_g = concentration of vapor, co_f = concentration of liquid, h_{Mg} = mass transfer coefficient of gas and h_{Mf} = mass transfer coefficient of liquid.

The ratio of concentrations can be given by

$$co_g/co_f = y\, \rho_g / x\, \rho_f$$

where co_g = concentration of vapor, co_f = concentration of liquid, ρ_g = density of gas, ρ_f = density of liquid, y = mole fraction of lower boiling point component in vapor and x = mole fraction of lower boiling point component in liquid.

The bubble radius is given by

$$R_b = 1.14\, [R_o\, \sigma\, gc/g\, (\rho_f - \rho_g)]^{1/3}$$

where ρ_g = density of gas, ρ_f = density of liquid, g = gravitational acceleration, g_c = conversion factor, σ = surface tension and R_o = orifice radius through which the bubble passes.

The bubble velocity is calculated as

$$V_b = C_1 R_b^{0.37}, \quad \text{where } C_1 = 1.892$$

The mass transfer coefficient for liquid is calculated as

$$h_{Mf} = C_2 V_b^{0.5}, \quad \text{where } C_2 = 0.441$$

The time required for bubble to pass through liquid layer is given by

$$t^o = L_f / V_b$$

The mass transfer coefficient for gas is calculated as

$$h_{Mg} = -R_b/3\, t^o\, \ln[6/\Pi^2\, \Sigma\, (1/n^2)\, \exp(-n^2\, \Pi^2\, D\, t^o/R_b^2)]$$

where R_b = radius of bubble, n = number of plates, t^o = time required for bubble to pass through liquid layer and D = self diffusion coefficient for vapor.

For higher Murphree efficiency, R_b should be small and t^o and U_M should be large.

Flash Calculations

When the feed stream is introduced in a rectification column, it is expanded through an expansion valve before entering the column. If the feed stream is a two phase fluid then it is important to determine the condition of the stream i.e. fraction of flow that is in liquid phase. The solution to this problem is called the *flash calculation* and it involves iterations as the temperature is unknown. The feed stream enters the expansion valve at molar flow rate F, molar enthalpy h_f and mole fraction x_{fj} for jth component. The feed stream may be a vapor or liquid or two-phase flow. Considering two phase flow, the liquid molar flow rate is given by L, the vapor molar flow rate is V and enthalpy of vapor is denoted by H and enthalpy of liquid is denoted by n. The mole fraction of the jth component in vapor phase is y_j and the mole fraction of the jth component in liquid phase is x_j.

The distribution coefficient = $K = y_j / x_j$

Applying mass balance to the system, $x_j F = h L + H V = h L + H (F - L)$

Solving the equation, the expression obtained is $y_j = x_j / [1 + L/F (1/K - 1)]$

Applying energy balance to the system, $x_j F = h L + H V = h L + H (F - L)$

Solving the equation, the expression obtained is $L/F = (H - h_f) / (H - h)$

The final governing equation for the flash calculation is

$$\Sigma y_j = 1 = \Sigma x_j / [1 + L/F(1/K - 1)]$$

To determine the quantity of vapor and liquid in the feed stream, an initial flash temperature is assumed in between the bubble temperature and dew temperature. Then the liquid fraction y_j is first determined. This value is substituted in the final governing equation to obtain a value. If the value obtained is unity, then the assumed temperature is correct. If not then the process is repeated and by trial and error and the flash temperature is calculated. The vapor and liquid mole fractions x_j and y_j can be calculated.

Theoretical Plate Calculations for Columns

There are two basic methods used to determine the number of theoretical plates required in a rectification column. The first method is Ponchon and Savarit method, which is an exact method requiring detailed enthalpy data and hence is not favored. The second method is the McCabe and Thiele method, which is a general method and requires only equilibrium concentration data. The McCabe-Thiele method is considered the simplest and perhaps most instructive method for analysis of binary distillation. This method uses the fact that the composition at each theoretical tray (or equilibrium stage) is completely determined by the mole fraction of one of the two components. The McCabe-Thiele method is based on the assumption of constant molar overflow which requires that: 1) the molar heats of vaporization of the feed

11.12 Cryogenics

components are equal, 2) for every mole of liquid vaporized, a mole of vapour is condensed, 3) heat effects such as heats of solution and heat transfer to and from the distillation column are negligible. Before starting the construction and use of a McCabe-Thiele diagram for the distillation of a binary feed, the vapor-liquid equilibrium (VLE) data must be obtained for the lower-boiling component of the feed. Figure 4.10 displays the rectification column used for the McCabe-Thiele method.

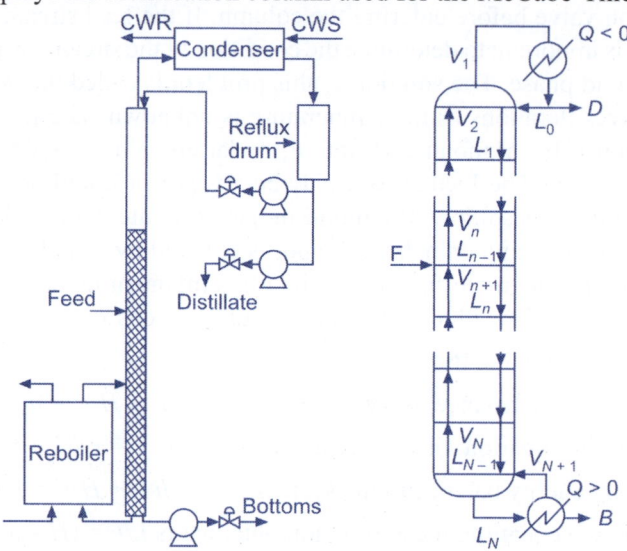

Fig. 11.10 Rectification Column for McCabe-Thiele Method

The liquid layers (plates) in the rectification column for the enriching section above the feed plate are denoted with subscript n and plates in the rectification column for the stripping section below the feed plate are denoted with subscript m. For the nth plate the liquid and vapor leaving the plate are denoted by L_n and V_n respectively. Applying the conservation of mass principle to the upper portion of the rectification column the following equation is obtained.

$$V_n = L_{n+1} + D \qquad \text{...(11.1)}$$

where D is the total mole flow rate of product out of the top column. Applying the conservation of mass principle to the upper portion of the rectification column the following equation is obtained.

$$y_n V_n = x_{n+1} L_{n+1} + x_D D \qquad \text{...(11.2)}$$

where x_D is the mole fraction of lower boiling point component in the top product. Applying first law of thermodynamics and assuning no heat inleaks the following equation is obtained.

$$H_n V_n = h_{n+1} L_{n+1} + h_D D + Q_D.$$

where H_n is the enthalpy of vapour leaving the nth plate, h_{n+1} is the enthalpy of liquid leaving the $(n+1)$th plate, h_D is the enthalpy of the top product and Q_D is the heat transfer rate in the condenser at the top of the column. Dividing the above equation by D and substituting equation $L_{n+1}/D = V_n/D - 1$.

$$(H_n - h_{n+1})(V_n/D) = (Q_D/D) + h_D - h_{n+1}.$$

Rearranging the terms the following equation is obtained

$$D/V_n = (H_n - h_{n+1})/(Q_D/D) + h_D - h_{n+1} \qquad ...(11.3)$$

It is known that $L_{n+1}/V_n = 1 - (D/V_n)$.

Dividing equation 11.2 on both sides by the vapor phase flow rate, the following equation is obtained.

$$y_n = (L_{n+1}/V_n)x_{n+1} + (D/V_n)x_D \qquad ...(11.4)$$

If the dew line and bubble line are horizontal then quantities H_n and h_{n+1} are constants. Thus it is observed that D/V_n and L_{n+1}/V_n are also constants. L_{n+1}/V_n is termed as the reflux ratio. Equation 11.3 then becomes a straight line with slope L/V_n and intercept $(D/V)x_D$. This straight line of the form $y = ax + b$ is called as the operating line for the upper or enriching section of the column. The plot of equilibrium mole fraction in the vapor phase y_n and the equilibrium mole fraction in the liquid phase x_n is termed as the equilibrium line for the column. Rearranging equation 11.4 the following equation is obtained.

$$y_n = (L_{n+1} + D)(x_D/V_n) = x_D.$$

This means the operating line passes through the point y_n on the 45 degree diagonal line drawn on x-y coordinates. A similar analysis is applied to the lower portion of the rectification column.

$$L_{m+1} = V_m + B \qquad ...(11.5)$$

where B is the total mole flow rate of product out of the bottom column. Applying the conservation of mass principle to the upper portion of the rectification column the following equation is obtained.

$$x_{m+1}L_{m+1} = y_m V_m + x_B B \qquad ...(11.6)$$

where x_B is the mole fraction of lower boiling point component in the bottom product. Applying first law of thermodynamics and assuming no heat inleaks the following equation is obtained.

$$h_{m+1}L_{m+1} + Q_B = H_m V_m + h_B B.$$

where H_m is the enthalpy of vapour leaving the mth plate, h_{m+1} is the enthalpy of liquid leaving the $(m+1)$th plate, h_B is the enthalpy of the bottom product and Q_B is the heat transfer rate in the condenser at the bottom of the column. Using similar analysis as top column, the following equations are obtained.

$$y_m = (L_{m+1}/V_m)x_{m+1} - (B/V_m)x_B \qquad ...(11.7)$$

Rearranging the terms the following equation is obtained

$$B/V_m = (H_m - h_{m+1})/(Q_B/B) - h_B + h_{m+1} \qquad ...(11.8)$$

It is known that the reflux ratio, $L_{m+1}/V_m = 1 + (B/V_m)$

This means the operating line passes through the point y_{nm} on the 45 degree diagonal line drawn on x-y coordinates. The intersection of the operating lines of the top and bottom column yields the following equation

11.14 Cryogenics

$$F = V_n - V_m + L_{m+1} - L_{n+1}$$

Where F is the mole flow rate of feed into the column. The parameter q is defined as the ratio of difference in liquid flow between the upper and lower column.

$$q = (L_{m+1} - L_{n+1})/F$$

The difference in vapour flows is given by

$$V_n - V_m = (1-q)F$$

Solving the above equation for vapor flows in top and bottom column the following equation is obtained.

$$V_n - V_m = (L_{n+1} - L_{m+1})(x/y) + (x_D D + x_B B)/y$$

For the column, $x_F F = x_D D + x_B B$

The locus of intersection of the top line and bottom line is given by

$$y = [(x_F/1-q) - (q/1-q)]x \qquad \ldots(11.9)$$

The value of parameter q can be determined from energy balance in the middle portion of the column.

$$h_f = (1-q)H + qh$$

Thus $\quad q = (H - h_f)/(H - h)$

The value of q varies depending on the condition of feed stream as shown in Figure 11.11.

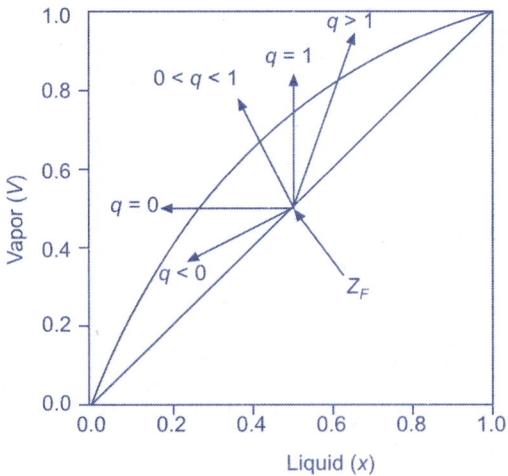

Fig. 11.11 Plot of Feed Lines

For subcooled liquid, $q > 1$
For saturated liquid, $q = 1$
For two phase mixture, $0 < q < 1$
For saturated vapor, $q = 0$
For superheated vapor, $q < 0$

The feed line (q-line) is useful in constructing the operating line for the lower section of the column, however the point of intersection of q line and operating line does not establish the location of feed inlet. A McCabe-Thiele plot is given in Figure 11.12 and the McCabe. Thiele method is explained as follows. The first step is to draw equal sized vertical and horizontal axes of a graph. The horizontal axis denotes the mole fraction (x) of the lower-boiling feed component in the liquid phase. The vertical axis denotes the mole fraction (y) of the lower-boiling feed component in the vapor phase. The next step is to draw a straight line from the origin of the graph to the point where x and y are both equal 1.0, which is the $x = y$ line in Figure 11.12. This 45 degree line is used simply as a graphical aid for drawing the remaining lines. The equilibrium line is drawn using the VLE data points of the lower boiling component, representing the equilibrium vapor phase compositions for each value of liquid phase composition. The next step is to draw vertical lines from the horizontal axis up to the $x = y$ line for the feed and for the desired compositions of the top distillate product and the corresponding bottoms product (shown in red in Figure 11.12).

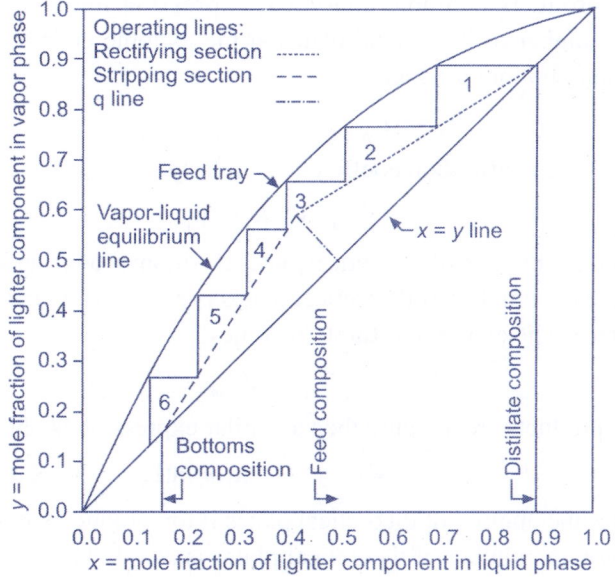

Fig. 11.12 McCabe-Thiele Diagram for Binary Feed

The next step is to draw the operating line for the rectifying section (the section above the feed inlet) of the distillation column, (shown in green in Figure 11.12). Starting at the intersection of the distillate composition line and the $x = y$ line, the rectifying operating line is drawn at a downward slope ($\Delta y/\Delta x$) of $L / (D + L)$ where L is the molar flow rate of reflux and D is the molar flow rate of the distillate product. The next step is to draw the blue q-line (seen in Figure 11.12) from the $x = y$ line so that it intersects the rectifying operating line. The parameter q is the mole fraction of liquid in the feed and the slope of the q-line is $q/(q - 1)$.

11.16 Cryogenics

For example, if the feed is a saturated liquid it has no vapor, thus $q = 1$ and the slope of the q-line is infinite which means the line is vertical. As another example, if the feed is all saturated vapor, $q = 0$ and the slope of the q-line is 0 which means that the line is horizontal. Next, as shown in Figure 11.12, the purple operating line is drawn for the stripping section of the distillation column (i.e., the section below the feed inlet). Starting at the intersection of the red bottoms composition line and the $x = y$ line, the stripping section operating line is drawn up to the point where the blue q-line intersects the green operating line of the rectifying section operating line. Finally, as exemplified in Figure 11.12, the steps between operating lines and the equilibrium line are drawn and counted. Those steps represent the theoretical plates (or equilibrium stages). The required number of theoretical plates is 6 for the binary distillation depicted in Figure 11.12. In continuous distillation with varying reflux ratio, the mole fraction of the lighter component in the top part of the distillation column will decrease as the reflux ratio decreases. Each new reflux ratio will alter the slope of the rectifying section operating line. When the assumption of constant molar overflow is not valid, the operating lines will not be straight.

For pure products a simple analytical expression must be developed to determine the number of theoretical plates at the extreme ends of the operating lines. For the pure bottom product

$$y_m = (L_{m+1}/V_m K_1) x_{m+1}$$

where K_1 is the distribution coeffcient. Similarly,

$$x_B = (L_{m+1}/V_m K_1)^M y_m$$

Where M is the number of theoretical plates if bottom product is in vapor phase. For liquid phase the number of theoretical plates is $M + 1$. The factor 1 represents the boiler surface. The expression for M becomes

$$M = \ln(x_B/y_m) / \ln(L_{m+1}/V_m K_1)$$

If the top products are are pure then a similar expression is obtained

$$N = \ln(1 - y_n/1 - x_D) / \ln(L_{n+1}/V_n K_2)$$

Where N is the number of theoretical plates if top product is in vapour phase. For liquid phase the number of theoretical plates is $N + 1$. The factor 1 represnts the condenser surface.

Minimum Number of Theoretical Plates

The number of theoretical plates for any given separation approaches a minimum as the slope of the operating line approaches unity. This is because the step size becomes smaller between top and bottom of the column. The minimum number of theoretical plates can be calculated by the equation

$$N_{min} = \ln[\{x_D(1 - x_B)/x_B(1 - x_D)\} / \ln(K_1/K_2)] - 1.$$

Where the ratio of distribution coeffcients K_1/K_2 is termed as relative volatility which is a function of temperature. It is also desired to keep reflux ratio L_{n+1}/V_n as small as possible to reduce refrigeration requirements. Figure 11.13 shows a McCabe Thiele curve of normal type which is concave downward throughout its length. The condition of minimum reflux occurs when two operating lines and the feed line the equilibrum line at point with slope = $1/s$. Although reflux is minimum for this condition, required number of theoretical plates is infinite. In practical systems there is a compromise between minimum theoretical plates (infinite refrigeration) and minimum refrigeration (infinite plates). The optimum reflux ratio is 1.1 to 1.2 times the minimum reflux ratio for practical cryogenic rectification plants.

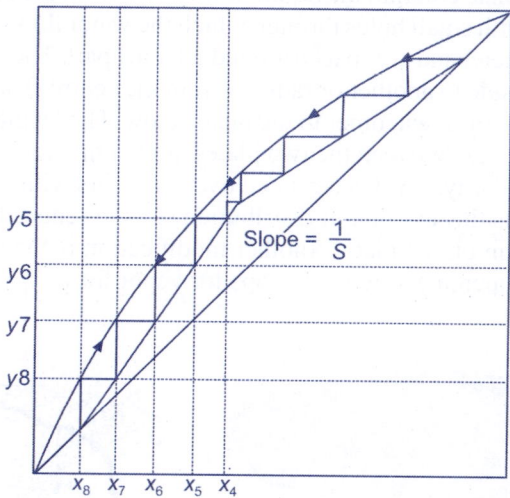

Fig. 11.13 McCabe-Thiele Diagram for Minimum Reflux

TYPES OF RECTIFICATION COLUMNS

There are two types of rectification columns, namely the packed column and plate column. Packed columns are cylinders filled with a packing material such as metallic rings or wire gauze. The packing material provides a large surface area for separation. This type of column is not frequently used in cryogenic systems. The plate column consists of several plates across which liquid flows. Some provision is made in the plates to allow the vapor to bubble through the liquid on the plate. This is accomplished by use of small holes or perforations in the plates or by use of a fine mesh screen or bubble caps. The schematic and photograph of a bubble cap tray is shown in Figure 11.14 and the schematic and photograph of a perforated tray is shown in Figure 11.15. Recent modifications in bubble cap trays include provision of risers which are large tubes of size 50 to 75 mm to channel the vapor. The risers are fitted with bubble caps with slotted edges for passage of vapor.

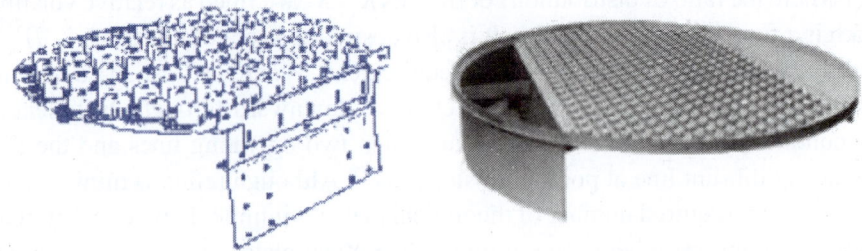

Fig. 11.14 Schematic and Photograph of Bubble Cap Tray

In perforated plate column, the liquid flows down from the upper section of the column to a plate with small holes through which the vapor flows. Typical holes are 1.5 to 4 mm in diameter and are spaced around 20 mm apart. The liquid flows across the plate from one side to another or radially from the centre. The liquid then flows down a vertical tube or downcomer to the plate below. The length of downcomer is such that it forms a seal between the two plates, preventing the entry of vapor. The design of perforated trays has been refined over the years with modifications. One such modification is the addition of distribution slots to reduce hydraulic gradient and avoid stagnation on the plate. Another modification is the provision of valve tray with variable openings fitted with caps for vapor flow.

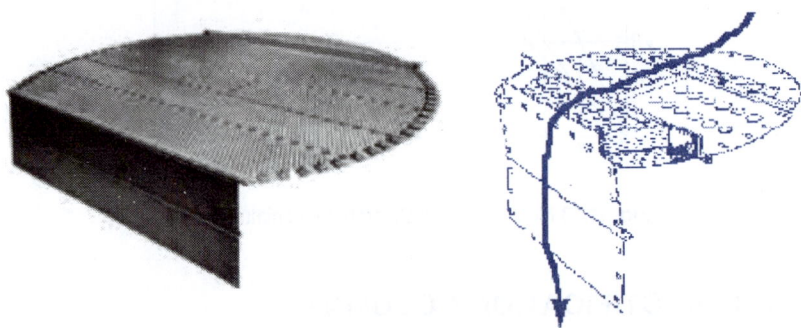

Fig. 11.15 Schematic and Photograph of Perforated Tray

Each type of plate has advantages and disadvantages. The perforated plate tray produces smaller bubbles and hence has higher Murphree efficiency. The perforated plate tray is sensitive to changes in vapor flow rate and is susceptible to liquid trickling through perforations or bubble formation. The bubble cap tray is insensitive to changes in vapor flow rate and can retain liquid on the tray in case of system shutdown.

Linde Single Column Air Separation System

Air separation by rectification can be carried out in a Linde single column. Based on his air liquefaction principle Carl von Linde constructed the first air separation plant for oxygen production in 1902, using a single column rectification system.

Refer Figures 11.16 and 11.17 for the schematic of the Linde single column and Linde single column separation system respectively.

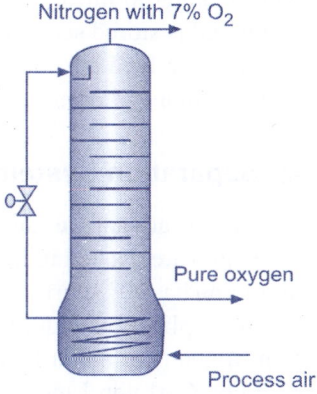

Fig. 11.16 Schematic of Linde Single Column

Single column Linde system is the simplest air separation system. The entering air is compressed, the water vapor and carbon dioxide are removed and the air is passed through a heat exchanger in which the incoming gas is cooled. The heat exchanger is a three channel type as shown in Figure 11.17.

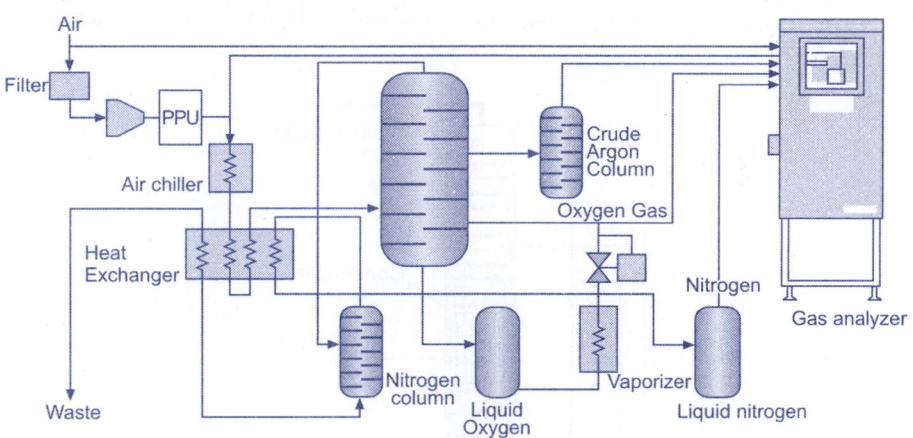

Fig. 11.17 Schematic of Linde Single Column Separation System

The cold oxygen gas is used to help cool down the incoming air. This produces oxygen as a gaseous product. If liquid oxygen is desired then, the return line is omitted and the heat exchanger is a two channel type. The liquid oxygen is withdrawn from the lower section of the column known as kettle. The cool air is partially condensed by exchanging energy with the liquid in the kettle. This partial condensation increases liquid yield and causes boiling of some liquid from the kettle without exposure to ambient. The cold air is expanded through a Joule-Thomson valve, after which some more liquid is formed. This liquid is then passed to the top of the column, where it begins to ravel down the column across the plates and gets enriched in oxygen by contact with upward flowing vapor. When the liquid reaches the bottom of the column, a portion is evaporated by the incoming air. This vapor bubbles through the liquid picking up nitrogen. This nitrogen rich vapor is passed through the heat exchanger and is stored separately. The oxygen rich liquid is also withdrawn from the column. The feed air must be compressed to 20 MPa for effective operation of the Linde single column separation unit.

Linde Double Column Air Separation System

The Linde single column system has disadvantages such as only pure oxygen can be used and large quantities of oxygen are wasted in impure nitrogen exhaust gas. The nitrogen leaving the single column has 90% purity level which is not acceptable for some applications and for some applications liquid nitrogen yield is desirable. The loss of oxygen with waste nitrogen is of the order of 26% which is a sizeable loss. These problems were solved by Carl von Linde in 1910 when he developed the Linde double column as shown in Figure 11.18. Two rectification columns are stacked on top of the other. The lower column is operated at a pressure of the order of 500 kPa to 600 kPa and the upper column s operated at a pressure of 100 kPa. At around 500 kPa, the boiling point of nitrogen is higher than that of oxygen and hence the refrigeration required by the nitrogen in the lower column is supplied by boiling oxygen in the upper column. Liquid nitrogen is produced in the lower column and some of it is withdrawn and introduced in the upper column to provide reflux liquid there.

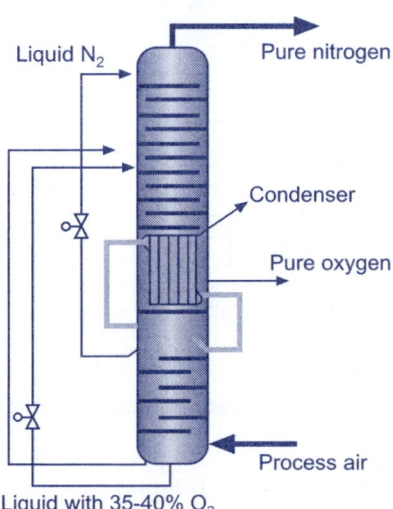

Fig. 11.18 Schematic of Linde Double Column

The operation of the double column system is similar to that of the single column system except the rectification column. Air is first compressed and carbon dioxide and water vapor are removed. The warm air is then passed through a heat exchanger for cooling the air. The heat exchanger is three channel if oxygen is desired as a gas and two channel if oxygen is desired as a liquid. The cooled air is further cooled by passing through the heat exchanger i.e. boiler located in the kettle of the lower column. Some of the kettle liquid is evaporated to provide upgoing vapor for the lower column. The cooled air is next expanded through an expansion valve and the resulting liquid vapor mixture is introduced in the middle portion of the lower column. Refer Figure 11.19 for the schematic of the Linde double column separation system. The liquid-vapor mixture is separated at this point and the liquid joins the downcoming liquid in the column and flows across the plates to the kettle. The vapor joins the upward vapor stream and reaches the condenser at the top of the lower column where it is condensed. Some of the condensed liquid is removed and passed on to the upper column and rest flows down the lower column as reflux fluid. Some nitrogen rich liquid can also be removed and stored separately. Enriched liquid air is removed from the kettle of the lower column and expanded through expansion valve and introduced in the middle of the upper column. Upon expansion in the expansion valve some kettle liquid is flashed into vapor. The vapor flows upward in the upper column and is enriched in nitrogen, whereas the liquid flows downward and is enriched in oxygen. Pure oxygen and nitrogen can be withdrawn in the process and high purity levels can be obtained by using a sufficient number of plates. For ensuring high purity, argon should be removed separately.

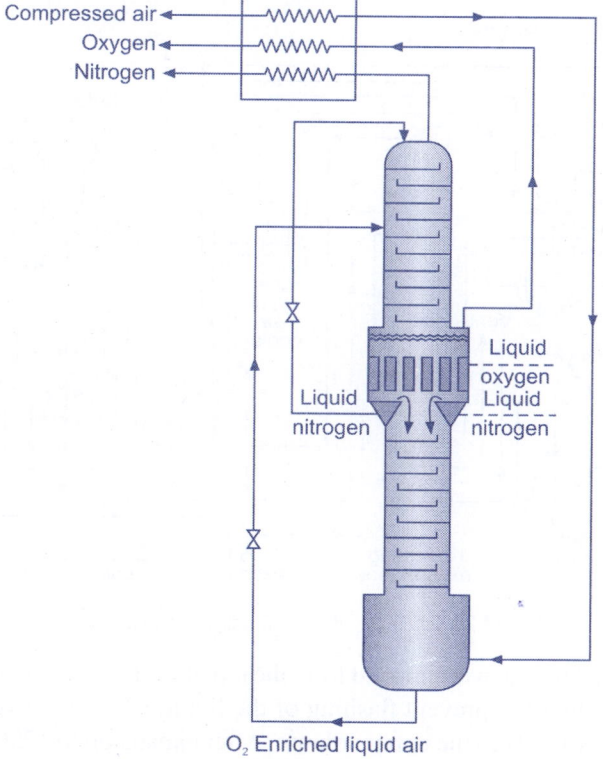

Fig. 11.19 Schematic of Linde Double Column Separation System

Linde-Frankl Air Separation System

The Linde-Frankl system was developed in 1930 to meet the demand of large quantity of oxygen (around 10 m^3/s) for the steel industry. The schematic of the Linde-Frankl system is shown in Figure 11.20. The system uses regenerators developed and patented by Frankl in 1925. Using ammonia precooling and an expansion engine, the power consumption for this system was reduced to approximately half of the power consumption of the Linde double column system. The air is first filtered and compressed to 560 kPa in a rotary compressor which can handle large flow rates and small pressure ratios. Around 96% of the total flow is diverted and passed through the two pairs of regenerators in which incoming air is cooled by outgoing pure oxygen and nitrogen gas streams. Water vapor and carbon dioxide are removed from the air in the regenerators in the first cycle by evaporation and in the second cycle air is cooked by switching with gaseous nitrogen and oxygen streams. The cold saturated air is introduced into the lower part of the lower column. The remaining 4% of the air is passed in absorbers where CO_2 is removed. The air is then compressed to 12.7 MPa and passed through a series of heat exchangers, a pre-cooler, an ammonia forecooler and two nitrogen heat exchangers. The cold air is then expanded through a Joule-Thomson valve and introduced with the regenerator stream into the lower portion of the lower column operating at 500 kPa.

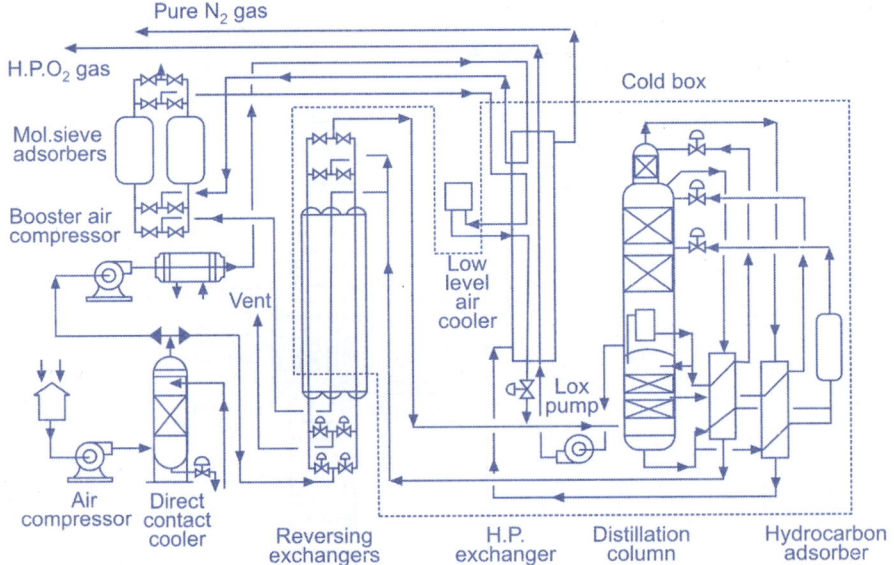

Fig. 11.20 Schematic of Linde-Frankl Separation System

Pure liquid nitrogen is removed from the top of the lower column and subcooled in a heat exchanger to prevent flashing of the liquid when it is expanded through the expansion valve into the upper column. After expansion to 1230 kPa, the liquid is introduced into the top of the upper column. Practically pure gaseous nitrogen is also removed from the top of the upper column and passed through the heat exchanger to cool the incoming air and expanded through expansion engine to 100 kPa to reduce gaseous nitrogen temperature.

Heylandt Air Separation System

Heylandt air-separation system is commonly used in USA. Figure 11.21 shows the schematic diagram of the Heylandt system. The products from this system are liquid nitrogen and liquid oxygen. The air is first filtered and compressed to 10 MPa in a four stage reciprocating compressor. Water vapor is removed from the air in the intercoolers and aftercooler of the compressor. The air is then further compressed to 13.7 MPa in a compressor driven by expansion engine. The air is then passed through a pre-cooler, an ammonia forecooler and cooled to –40°C. Before the airstream is passed through the main heat exchanger, about half of the air is bypassed through an expansion engine and expanded to conditions of 700 kPa and 110 K. The other half of the air stream passes through the main heat exchanger and is expanded through an expansion valve to 700 kPa. The two air streams are united and passed through a scrubber in which carbon dioxide is removed by contact with liquid air. The gaseous air from the scrubber is introduced near the bottom of the lower column and liquid air from the scrubber is passed through a filter to remove the hydrocarbon particles. The filtered air is united with the kettle liquid from the lower column and both streams are introduced in the upper column.

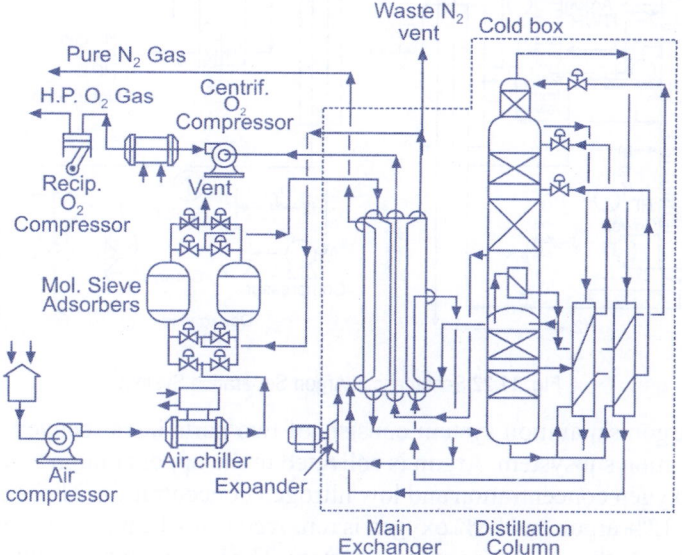

Fig. 11.21 Schematic of Heylandt Separation System

To provide reflux for the upper column, liquid nitrogen is removed from the condenser in the lower column and subcooled in a heat exchanger by the leaving gaseous nitrogen from the upper column and then introduced at the top of the upper column after expansion through an expansion valve. Liquid oxygen product is removed from the reboiler of the upper column and liquid nitrogen product is removed from the condenser of the lower column. The waste nitrogen gas is removed from top of the upper column and is exhausted to the atmosphere after passing through subcooler, main heat exchanger and precooler. For the heylandt system the lower column operates at 700 kPa and upper column operates at 200 kPa. The liquid oxygen obtained is 99.96% pure and the liquid nitrogen obtained is 93% pure. In some systems the waste nitrogen gas can be passed through a rectification column with sufficient number of plates to obtain high purity nitrogen gas which can be bottled and stored. The addition of the nitrogen rectification column adds to the cost of the system.

Argon Separation System

Argon is present as a rare gas in air in an amount of 0.93% by volume. The boiling point of argon is 87.3 K and it lies between the boiling points of oxygen and nitrogen. Argon cannot be removed from air by mechanical means as the rectification column would require large number of plates. Refer Figure 11.22 for schematic of an argon separation system.

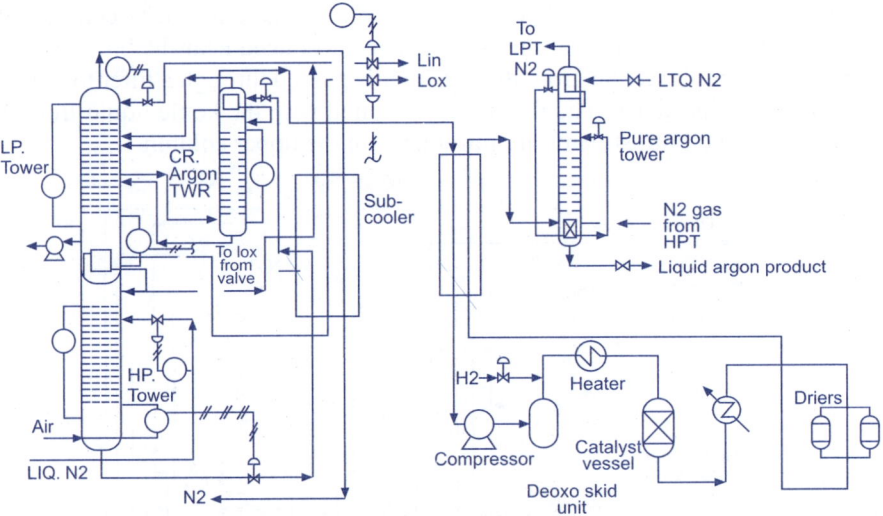

Fig. 11.22 Schematic of Argon Separation System

The argon separation system consists of two parts i.e. a recovery subsystem and purification subsystem. Argon is collected in the upper column where the fluid has high oxygen concentration and low nitrogen concentration. The argon mixture containing 12% argon and 88% oxygen is removed from the main air separation unit and passed on to the argon recovery subsystem. The argon mixture is introduced into the lower part of the argon column and is rectified there. Some of the kettle liquid from the main column is used to furnish the refrigeration necessary for supply of reflux liquid in the argon column. The argon collects in the top of the argon column as boiling point of argon is lower than that of oxygen. The collected crude argon is then passed to the argon purification subsystem.

Argon purification is achieved by catalytic combustion or by adsorption methods. In the catalytic combustion method, hydrogen is added to argon and the mixture of hydrogen, argon and oxygen is compressed to 500 kPa. The compressed mixture is passed through a catalytic combustion furnace where hydrogen and oxygen combine in the combustion reaction to form water vapor. The water vapor is removed by a water separator and an alumina dryer. Some trace impurities of hydrogen remain in the argon. The argon is then cooled by heat exchange with the crude argon and is rectified in a column to remove trace impurities of hydrogen. The extracted hydrogen is recycled and is used in the combustion process. The final argon product is obtained with a purity level of less than 40 ppm impurities by volume. In the second purification system that is adsorption method, oxygen is removed from the mixture by adsorption. The crude argon is first passed through a rectification column to remove nitrogen from the mixture. The oxygen-argon mixture is removed from the kettle of the rectification column and passed through a heat exchanger. The mixture is next passed through an adsorbent trap filled with crystalline zeolite adsorbent where the oxygen is removed. The argon gas is then passed through a filter to remove any adsorbent particles and then condensed in a condenser. Liquid air is removed from the kettle to cool the adsorbent and liquid nitrogen is removed from the main column for providing reflux liquid and for the refrigeration necessary to liquefy the argon.

Neon Separation System

The large scale production of neon began in 1960. Liquid neon is popular as a refrigerant as it is inert in nature and less expensive per unit heat of vaporization. Furthermore precise temperature control can be obtained by varying saturation pressure of liquid neon. The neon separation system consists of two parts i.e. a recovery subsystem and purification subsystem. Refer Figure 11.23 for schematic of a neon separation system.

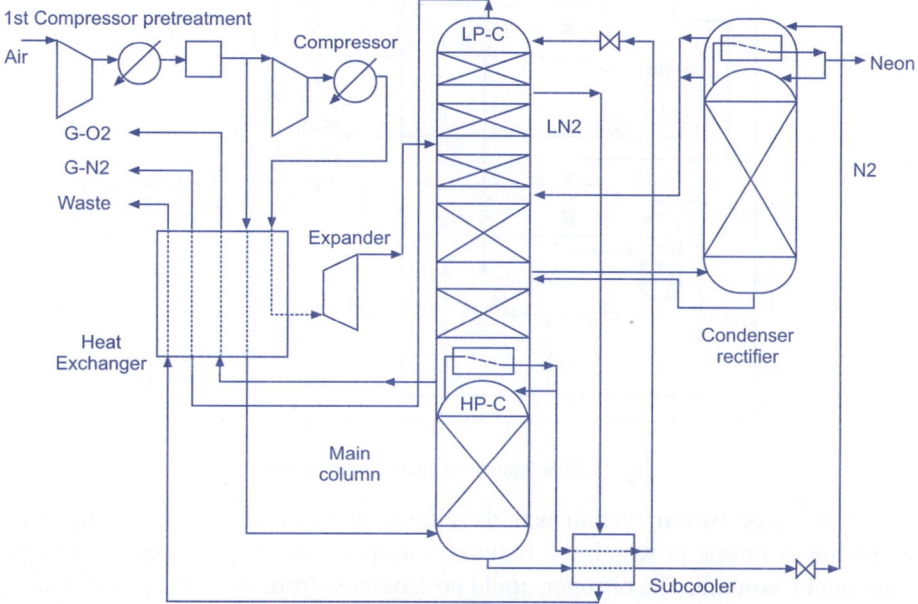

Fig. 11.23 Schematic of Neon Separation System

In the recovery subsystem, the vent gas from the condenser reboiler of the main column is passed into a small condenser-rectifier that is refrigerated by liquid nitrogen drawn from the condenser of the main column. The composition of the neon mixture leaving the condenser rectifier is 46% neon, 19% helium and 33% nitrogen and 2% hydrogen. The crude neon mixture is then transferred to the purification system. The hydrogen is removed by combustion with oxygen in a catalytic combustion chamber to produce water vapor. The water vapor is condensed in a cooler and removed from the system in a drier. Excess oxygen and nitrogen is removed from the neon mixture when the stream flows through a liquid nitrogen cooled adsorbent trap. The neon mixture at the exit of adsorbent trap contains 70% neon and 30% helium. The neon mixture is then further cooled by a liquid hydrogen refrigeration system, where it gets condensed in the condenser. Pure neon liquid is recovered from the bottom of the condenser, while a gaseous mixture of 80% helium and 20% neon is vented from the dome of the condenser.

Linde-Bronn Hydrogen Separation System

The hydrogen can be obtained from coke oven gas, which is a mixture of hydrogen, nitrogen, oxygen, carbon dioxide and other hydrocarbons. All the constituents of coke oven gas are present in small amounts and their boiling points vary greatly. Thus hydrogen separation is not possible by rectification columns and condensation-evaporation process is required. Refer Figure 11.24 for schematic of a Linde-Bronn hydrogen separation system.

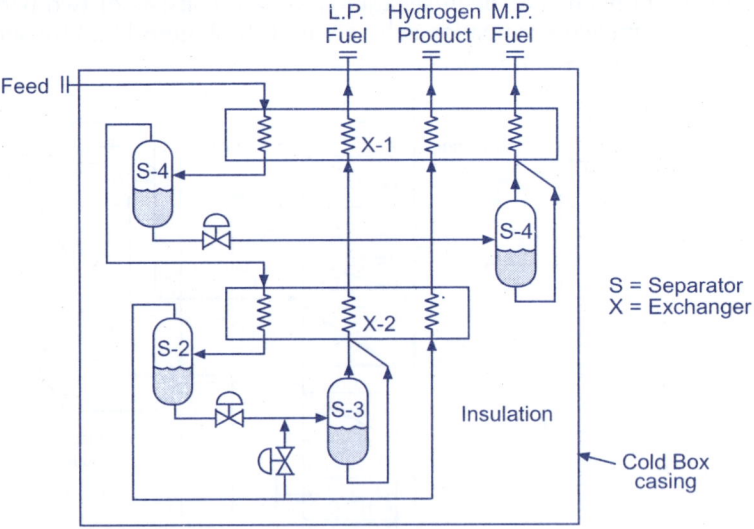

Fig. 11.24 Schematic of Linde-Bronn System

The Linde Bronn system was developed in 1924 to supply hydrogen for ammonia synthesis in which the required mixture was 75% Hydrogen and 25% nitrogen by volume. The nitrogen could be separated from the hydrogen stream by additional condensation and evaporation. An ammonia refrigeration system is used to cool the coke oven gas to –45°C after compression to 1.32 MPa. This removes the impurities such as water vapor and benzene etc. The cooled gas is then passed through a heat exchanger where it is cooled to –100°C and propylene is condensed. The cooled gas is then passed through a second heat exchanger where it is cooled to –145°C and ethylene and ethane are condensed. The cooled gas is then passed through a third heat exchanger where it is cooled to –180°C and some fraction of methane is condensed. The mixture is then passed through a condenser-evaporator where it is cooled to –190°C and all of the remaining methane is condensed. The liquid methane is removed from the system by passing through the heat exchangers where it cools the incoming coke oven gas stream. The main gas stream now contains hydrogen, nitrogen and carbon-monoxide. The carbon monoxide is removed by a scrubber. Liquid nitrogen is added at the top of the scrubber and the liquid collected at the base contains the carbon monoxide. The main gas stream removed from the top of the scrubber now only contains 75% hydrogen and 25% nitrogen. This mixture is then passed to a condenser, where nitrogen is condensed and removed. The hydrogen gas is then passed through a silica gel purifier to remove any traces of nitrogen and exits the system as a product through the three heat exchangers.

L'Air Liquide Hydrogen Separation System

The L'Air Liquide system uses an expansion engine for refrigeration instead of auxiliary liquid nitrogen system. Refer Figure 11.25 for the schematic of the L'Air Liquide system.

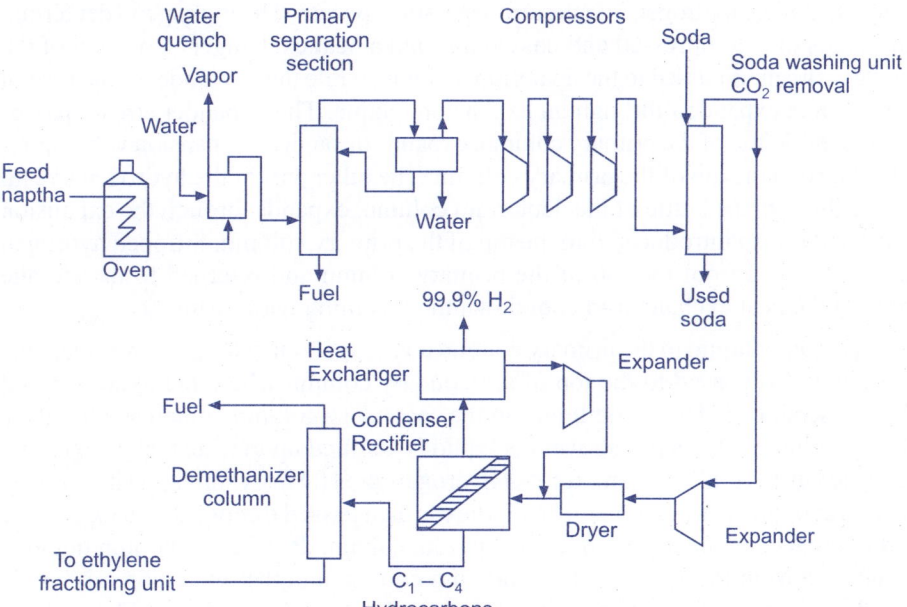

Fig. 11.25 Schematic of L'Air Liquide System

The coke oven gas is compressed to 2.5 MPa and passed through the heat exchanger where it is cooled to –65°C. The heavier hydrocarbons and propylene are condensed in the first heat exchanger and removed from the system. The cool gas is passed through a second heat exchanger in which ethylene, ethane and some methane quantity is condensed. The coke oven gas is then passed into a tubular condenser-evaporator in which methane is condensed. The liquid methane is expanded to atmospheric pressure through an expansion valve and introduced in the shell of the condenser-evaporator. The separated hydrogen gas is removed from the dome of the condenser-evaporator and expanded through an expansion engine to 1 atm and –205°C. The cold hydrogen gas exits the system through the heat exchangers, thereby cooling the incoming gas stream. Expansion engines used in this system are costly and require maintenance and lubrication. The nitrogen that condenses from the hydrogen stream is used as a lubricant. Furthermore bronze piston liners in expanders help to reduce wear and friction. The expander is coupled with a small electric generator which absorbs work done by the expander and produces some power for running the system. In large systems the expander produces work which is utilized by the system compressor.

Hydrogen-Deuterium Separation System

Cryogenic systems for separation of hydrogen and deuterium were proposed in 1959. Natural hydrogen contains 0.032% deuterium by volume. The deuterium combines with hydrogen to form the isotope HD. This isotope has to be concentrated and converted to pure deuterium by removal of hydrogen. Refer Figure 11.26 for the schematic of the hydrogen - deuterium separation system. The hydrogen - deuterium mixture is first compressed and passed through a heat exchanger. A portion of the cool gas stream is passed to the deuterium column, while the remainder of the natural hydrogen is expanded through an expansion engine. The expander stream passes through the kettle of the primary column, expands through an expansion valve and is introduced to the top of the primary column. The other part of the hydrogen stream passes through the kettle of the deuterium column, expands through the expansion valve and is also introduced into the top of the primary column. Stripped hydrogen gas is removed from the top of the primary column and returned to the storage through the heat exchanger to cool down the incoming gas stream.

The kettle liquid in the primary column contains about 3% HD by volume. This kettle liquid is passed to the top of a secondary column where the hydrogen and HD are separated. The kettle liquid in the secondary column contains about 95% HD by volume. The vapor above this liquid is warmed up in a heat exchanger and catalyzed to an equilibrium mixture of hydrogen and HD in a catalyzer. The reaction is rapid at higher temperatures. The mixture is then passed through a heat exchanger and introduced in the middle of the deuterium column. Purified deuterium liquid is withdrawn from the kettle of the deuterium column and the remaining mixture is withdrawn from the dome of the condenser in the deuterium column. The primary column is single column type and deuterium column is of the double column type. Rectification of hydrogen is difficult due to low density and viscosity.

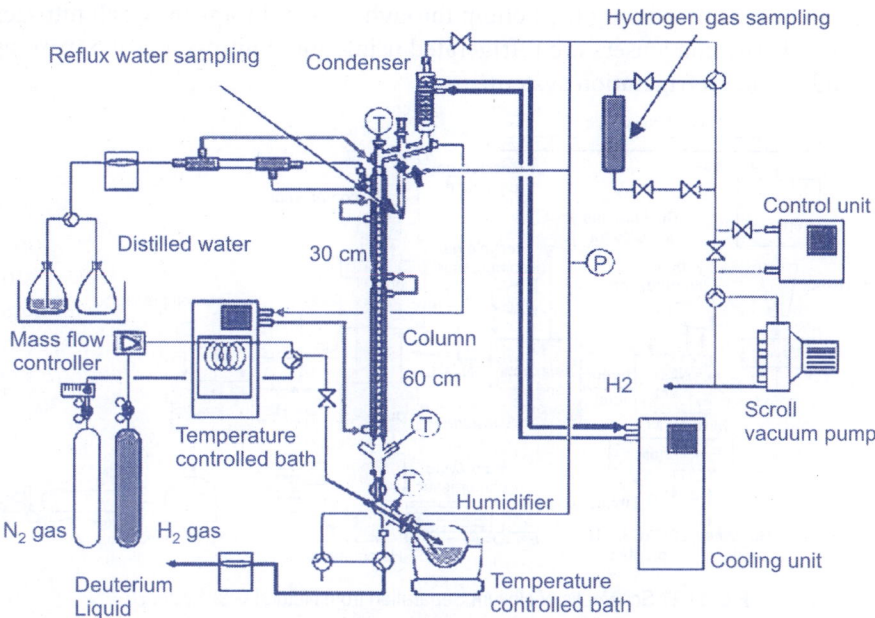

Fig. 11.26 Schematic of Hydrogen-Deuterium System

HELIUM SEPARATION FROM NATURAL GAS

Helium is produced in the US primarily by separation of helium rich natural gas. The natural gas contains about 8% helium and 12% nitrogen. The remaining is methane and other hydrocarbons. Refer Figure 11.27 for the schematic of the helium separation from natural gas system. A rectification column is not required for separation of helium and separation can be achieved with condenser and evaporator system. The natural gas is compressed to 4.26 MPa and passed though a heat exchanger after the carbon dioxide, hydrogen sulfide and water vapor have been removed. Most of the natural gas is condensed in this heat exchanger. The cold gas is then expanded through an expansion valve to 1.82 MPa and −145°C. This fluid is then passed to a crude helium separator which is a condenser in which 98% of the natural gas is liquefied. The vapor from the crude helium separator consists of about 40% nitrogen and 60% helium by volume. This gas mixture is removed from the crude helium separator and warmed to ambient temperature in a heat exchanger. The outgoing liquid from the separator cools the incoming gas in the heat exchanger. This outgoing stream is a good fuel gas as non-condensable impurities have been removed. The crude helium stream is then compressed to 18.75 MPa and passed back through the heat exchanger where it is cooled. The crude helium stream then enters a condenser-evaporator where it is condensed and nitrogen from the mixture is removed from the bottom of the condenser-evaporator. The condensed helium stream is then expanded through an expansion valve to 1.82 MPa into the separator where helium is separated and passed to the condenser. The separated helium stream has purity level of 98.5% with traces of nitrogen. The purity of helium stream is further improved by passing the helium through charcoal traps in which nitrogen is adsorbed. The condensers are refrigerated using liquid nitrogen as the working fluid and Claude refrigeration system.

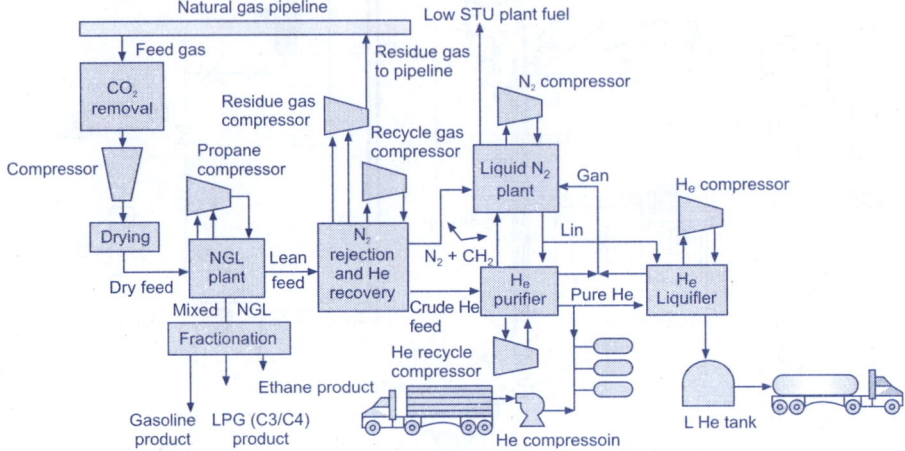

Fig. 11.27 Schematic of Helium Separation from Natural Gas System

REFRIGERATION PURIFICATION

The simplest method of gas purification is to cool the mixture until the impurities either condense or freeze. The effectiveness of this method depends on the vapor pressure of the impurities at the refrigeration temperature. For the system to be effective the gases should have differing vapor pressures. If the gases are assumed to be ideal then mole fraction of the impurity in the vapor phase is equal to ratio of partial pressure of that constituent to the total pressure of the mixture. Thus the following equation holds true

$$y_j = p_j / p_m$$

where y_j = mole fraction of the jth component in the vapor phase, p_j = partial pressure of the jth component in the vapor phase and p_m = pressure of the mixture.

When the impurity is condensed out from the gas mixture, the partial pressure of the impurity in the vapor phase must be equal to the saturation pressure of that component at the temperature of the gas mixture. The saturation pressure can be determined from the slope of the vapor pressure curve as given by the Clapeyron Equation.

$$dp/dT = (s_g - s_f)/(v_g - v_f) = h_{fg} / T(v_g - v_f)$$

where s_g = saturated vapor entropy, s_f = saturated liquid entropy, v_g = saturated vapor specific volume, v_f = saturated liquid specific volume and h_{fg} = heat of vaporization.

For pressures lower than critical pressure, the saturated liquid specific volume (v_f) can be neglected. Also the latent heat of vaporization can be expressed in terms of constants A and B as follows

$$h_{fg} = A - BT$$

At low pressures the vapor obeys the ideal gas equation

$$v_g = RT/p$$

Substituting the above equations in the Clapeyron equation, the following expression is obtained

$$(dp/dT)_{sat} = p(A - BT)/RT^2$$

Integrating the above equation the expression for vapor pressure curve is obtained

$$\ln(p/p_o) = C_1 - C_2/T - C_3 \ln T$$

Where p_o is a reference pressure and C_1, C_2 and C_3 are constants for a particular substance. This equation is not valid for impure gases as they do not behave like ideal gases.

Physical Adsorption

When a gas is brought in contact with a solid, some gas molecules penetrate the solid and this process is known as absorption. If the gas molecules only stick to the surface of the solid, the process is termed as adsorption. If chemical bonds are formed between the solid and the gas the process is known as chemical adsorption. If weak physical bonds are formed between the solid and the gas due to van der Waal's forces the process is known as physical adsorption. The van der Waal's forces are attractive forces between molecules. The quantity of gas absorbed per unit adsorbent in physical adsorption can be determined by the BET equation. The BET equation assumes retention of molecules on the surface of adsorbent in multiple layers till equilibrium is achieved. The expression for BET model is

$$V/m_a = v_m z \, (p/p_{sat})/(1 - p/p_{sat}) \, [1 + (z-1)(p/p_{sat})]$$

Where V = volume of gas adsorbed at 101 kPa and 273 K, m_a is mass of adsorbent, v_m = volume of gas per unit mass adsorbent required to form a molecular layer over the surface, z = parameter for energy of adsorption, p = partial pressure of gas which is adsorbed and p_{sat} = saturation pressure of gas being adsorbed at adsorbent temperature.

For monomolecular adsorption the BET equation becomes

$$V/m_a = v_m z(p/p_{sat}) / [1 + z(p/p_{sat})]$$

The parameter v_m can be calculated as

$$v_m = 2 \, MRT_o(A/m_a) / \sqrt{3} \, N_o p_o D^2$$

Where m_a is mass of adsorbent, v_m = volume of gas per unit mass adsorbent required to form a molecular layer over the surface, M = molecular weight of the gas, R = specific gas constant, T_o = standard temperature, 273 K, D = molecular diameter, A = area of adsorbent, N_o = Avogadro's number = 6.023×10^{23} mol^{-1} and p_o = standard pressure = 101.3 kPa.

11.32 Cryogenics

The ratio A/m_a is termed as effective surface area per nit mass of adsorbent. This quantity should be as large as possible for good adsorbents as is the case with silica gel, charcoal and alumina. Figure 11.28 shows a typical cryogenic adsorption se-tup.

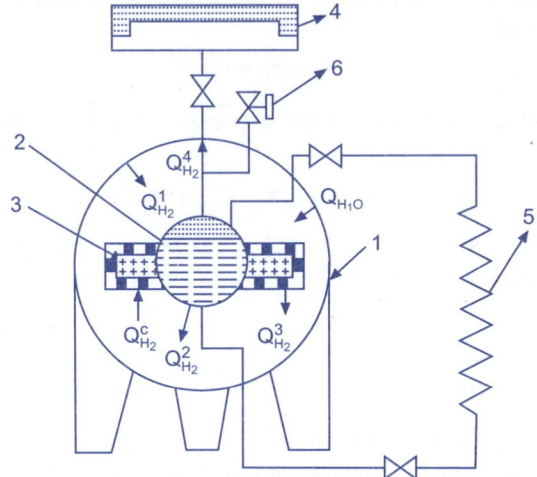

1. Cryogenic tank layout and illustration of gas release and gas absorption processes in a heat-insulating cavity where: 1 — outer casing, 2 — Cryogenic tank with cryoagent, 3 — Built-in cryoadsorption pump, 4 — Chemical patron with gas absorbent, 5 — Cryogenic liquid evaporator, 6 — Drainage gaseous valve

Fig. 11.28 Typical Cryogenic Adsorption Se-tup.

Solved Problems

Problem 1. Determine the ideal work requirements for the separation of a mixture of gases consisting of 79% nitrogen and 21% oxygen (mole fraction). The mixture is at temperature of 300 K and 101.3 kPa. Consider ideal gas assumption for the mixture.

Solution:

The work requirement per unit mole of mixture is given by

$-W_i/n_m = RT_m [y_a \ln(1/y_a) + y_b \ln(1/y_b)]$
$ = (8.314)(300) [0.79 \ln(1/0.79) + 0.21 \ln(1/0.21)]$
$ = (8.314)(300) [0.186 + 0.327] = 1282 \text{ J/mol}$ ...(Ans)

The work requirement in terms of constituents is given as

$W_i/n_{N2} = 1282/0.79 = 1623$ J/mol of Nitrogen

$W_i/n_{O2} = 1282/0.21 = 6105$ J/mol of Oxygen.

Problem 2. Determine the Murphree efficiency for a nitrogen-oxygen rectification plate in which the bubble radii are 3 mm on average. The liquid layer thickness is 15 mm and the liquid temperature is 79 K. The mixture pressure is 101.3 kPa.

Solution:

For air at 79 K and 101.3 kPa, the following values are obtained.

$x = 0.77$, $y = 0.93$, $\rho_g = 4.644$ kg/m³, $\rho_f = 873.2$ kg/m³ and $D = 2.03 \times 10^{-5}$ m²/s

The bubble velocity is calculated as $V_b = C_1 R_b^{0.37}$, Where $C_1 = 1.892$

$V_b = 1.892 (0.003)^{0.37} = 0.221$ m/s

The time required for bubble to pass through liquid layer is given by $t^o = L_f/V_b$

$t^o = 0.015 / 0.221 = 0.068$

The mass transfer coefficient for liquid is calculated as $h_{Mf} = C_2 V_b^{0.5}$, Where $C_2 = 0.441$

$h_{Mf} = 0.441 (0.221)^{0.5} = 0.207$ m/s

The term $\Pi^2 Dt^o / R_b^2$ can be calculated as

$\Pi^2 Dt^o / R_b^2 = (3.142)^2\ 2.03 \times 10^{-5}\ (0.068)/(0.003)^2 = 1.5142$

The mass transfer coefficient for gas is calculated as

$h_{Mg} = - R_b/ 3\ t^o \ln[6/\Pi^2 \Sigma(1/n^2) \exp(-n^2 \Pi^2 Dt^o / R_b^2)]$

$h_{Mg} = -0.003/3\ (0.068) \ln 6/(3.142)^2 [\exp(-1.5142) + (1/4) \exp(-6.0569) + ...]$

$h_{Mg} = 0.0295$ m/s

The ratio of concentrations can be given by

$co_g / co_f = y \rho_g / x \rho_f = (0.932)(4.644)/(0.770)(873.2) = 2910.13$

The overall mass transfer coefficient can be calculated as

$1/ U_M = (1/ h_{Mg}) + (co_g/co_f) / h_{Mf}$

$1/ U_M = (1/ 0.0295) + 2910.13/0.207 = 33.874$

$U_M = 0.029$ m/s

Murphree efficiency can also be expressed by Geddes equation

$\eta_M = 1 - \exp(-3 U_M t^o / R_b)$

$\eta_M = 1 - \exp(-3(0.029)(0.068)/0.003) = 0.866$ ...(Ans)

Problem 3. Determine the number of theoretical plates required to yield 98% nitrogen as top product and 95% oxygen as bottom product. The feed stream has composition of 79% nitrogen and 21% oxygen. The molar fraction of liquid in the feed stream is 0.831 mol liquid / mol mixture. The bottom product and top product leave the column as saturated liquids. The desired flow rate of the bottom feed is 25 mol/s. The heat removed in the condenser at the top of the column is 1071 kW. The mean pressure within the column is 101.3 kPa.

Solution:

Using conservation of mass principle to the column

$$F = B + D = 25 + D$$

It is known that

$$x_F F = x_D D + x_B B$$

11.34 Cryogenics

Substituting values,
$$0.79 F = (0.05)(25) + 0.98 D = 1.25 + 0.98(F - 25)$$
Solving $F = (25 \times 0.98 - 1.25)/(0.98 - 0.79) = 122.37$ mol/s
$$D = F - B = 122.37 - 25 = 97.37 \text{ mol/s}$$
Using first law of thermodynamics for the column
$$Q_B = Q_D + h_d D + h_b B - h_f F$$
Now $h = h_d = h_b = 1050$ J/mol (saturated liquid) and $H = 6916$ J/mol (saturated vapor)

The enthalpy of feed stream is
$h_f = (1 - q)H + qh = (1 - 0.831)(6916) + (0.831)(1050) = 2041$ J/Xmol.
Substituting all the values in the first law equation and solving,
$Q_B = (1071 \times 10^3) + (1050)(25) - (2041)(122.37) = 949731$ $W = 949.731$ kW

The next step is to determine the operating line equations for the enriching and stripping sections of the column. For the top column
$$D/V_n = (H_n - h_{n+1})/(Q_D/D) + h_D - h_{n+1}$$
Substituting values,
$D/V_n = (6916 - 1050)/(1071 \times 10^3/97.37) + 1050 - 1050 = 0.5336$
The reflux ratio for upper column $= L_{n+1}/V_n = 1 - (D/V_n) = 1 - 0.5336 = 0.4664$
The operating line equation for top column is
$$y_n = (L_{n+1}/V_n)x_{n+1} + (D/V_n)x_D$$
Substituting values the equation for operating line for upper column is obtained,
$y_n = (0.4664)x_{n+1} + (0.5336)(0.98) = 0.4464 x_{n+1} + 0.5229$

For the bottom column
$$B/V_m = (H_m - h_{m+1})/(Q_B/B) - h_B + h_{m+1}$$
Substituting values,
$B/V_m = (6916 - 1050)/(949731/25) - 1050 + 1050 = 0.1544$
The reflux ratio for upper column $= L_{m+1}/V_m = 1 + (B/V_m) = 1 + 0.1544 = 1.1544$
The operating line equation for bottom column is
$$y_m = (L_{m+1}/V_m)x_{m+1} - (B/V_m)x_B$$
Substituting values the equation for operating line for upper column is obtained,
$y_n = (1.1544)x_{m+1} - (0.1544)(0.05) = 1.1544 x_{m+1} - 0.00772$
$q = 0.831$ (given)

Form equation 11.9,
$y = [(x_F/1-q) - (q/1-q)] x = [(0.79/1 - 0.831) - (0.831/1 - 0.831)]x$
Thus $y = 4.6746 - 4.9172 x$.

This curve is plotted on the McCabe-Thiele Plot as shown in Figure 11.12. The procedure explained in the earlier discussion is followed. In this case the number of vertical lines is three for the condenser and four for the boiler which is equal to the theoretical plates required. Thus total seven plates are required for the column.

References

1. Cryogenic Engineering – B.A. Hands.
2. Internet Website www.wikipedia.com.
3. Cryogenic Systems – Randall Baron.
4. Cryogenic Engineering – P.K.Bose.
5. Internet Website www.uigi.com.
6. Cryogenic Air Separation Brochure by Linde.
7. Thermo electron corporation brochure on Cryogenic ASU.
8. Internet Website http://www.gasplantsmanufacturer.com/.
9. Journal Paper by W. Street, Journal of Chemical Physics, Vol 40, No.5, 1964.
10. Presentation by Gavin Duffy, DIT, USA.
11. Linde brochure on cryogenic air separation.
12. Paper by Cryogenic Consulting Service Inc.
13. Report By Jeffrey Miller, UOP on Hydrogen separation.
14. Paper on air separation units by Harry Kooijman, Clarkson University.
16. Sugiyama et.al., (2004) Journal of Nuclear Science and Technology, Vol. 41, No. 6, p.696–701.
17. Gusev et.al. Article on Hydrogen storage.

Questions

Q.1. Explain thermodynamically ideal gas separation system.

Q.2. Discuss the temperature-composition and enthalpy composition diagrams in detail.

Q.3. Write a short note on principles of rectification.

Q.4. What are flash calculations?

Q.5. Calculate the minimum number of theoretical plates in a gas separation unit.

Q.6. Explain the working of a Linde-Frankl separation unit with neat sketch.

Q.7. Explain the working of a L'Air Liquide separation unit with neat sketch.

Q.8. What are the advantages of purification of gases?

Q.9. Compare any two gas separation technologies.

Q.10. Describe the process of physical adsorption.

Q.11. Determine the ideal work requirements for the separation of a mixture of gases consisting of 59% hydrogen and 41% oxygen (mole fraction). The mixture is at temperature of 300 K and 101.3 kPa. Consider ideal gas assumption for the mixture.

CHAPTER 12

Air Separation Technologies

INTRODUCTION

Large scale commercial production of industrial gases from the atmosphere generally involves cryogenic processing of air. In addition to desired products such as oxygen, nitrogen, argon and the rare gases, feed air used as a starting material for cryogenic processing also contains impurities or undesirable components such as water vapor, carbon dioxide and one or more hydrocarbon species. These impurities must be removed before processing of feed air can be completed because the impurities interfere with continuous and efficient operation of the cryogenic equipment, or may present hazardous conditions which imperil the safety of operators or damage equipment. A significant portion of the cost of an air separation plant is associated with cleaning or prepurifying air and with cooling the air to cryogenic temperatures.

In May 1895, Carl von Linde performed an experiment in his laboratory in Munich which led to his invention of the first continuous process for the liquefaction of air based on the Joule-Thomson refrigeration effect and the principle of countercurrent heat exchange. This was the breakthrough for cryogenic air separation. Since then various techniques have been used to provide air separation systems with clean, low temperature air streams. Heat exchangers allow simultaneous cooling of feed air and reheating of product streams. Feed air and product streams flow in separate passages through the heat exchanger. Early air separation systems allowed impurities to deposit on the cold heat exchange surfaces in the feed air passages, eventually causing the heat exchanger to become plugged with condensed impurity deposits or to become unable to cool the incoming air to the required low temperature for cryogenic separation. The plant would then be shut down and thawed out. Later plants incorporated chillers for the removal of part of the moisture, and caustic scrubbers to remove carbon dioxide. As the demand for gaseous products grew, devices known as regenerators came into use to accomplish heat exchange between feed air and product streams.

To enable air to be separated into its constituents by means of rectification – the actual separation process – a large part of the air volume used must be liquefied. A gas can only be transformed into a liquid state at temperature and pressure conditions below those of its critical point. The critical point of air is $T_{crit} = -140.7°C$ (= 132.5 K) and $P_{crit} = 37.7$ bar, in other words, air can be liquefied only at temperatures below $-140.7°C$ (132.5 K). The vapor pressure curve shown in

12.2 Cryogenics

Figure 12.1 illustrates the allocation of temperatures and pressures at which a gas condenses or a liquid evaporates. Air below atmospheric pressure (1 bar) must be chilled to –192°C (81.5 K) before condensation sets in. Air below a pressure of 6 bar must be chilled to –172°C (101 K) before condensation begins. The boiling point and condensation conditions of gas mixtures such as air are not identical. There is a condensation line and a boiling point line which delineate the boiling point range.

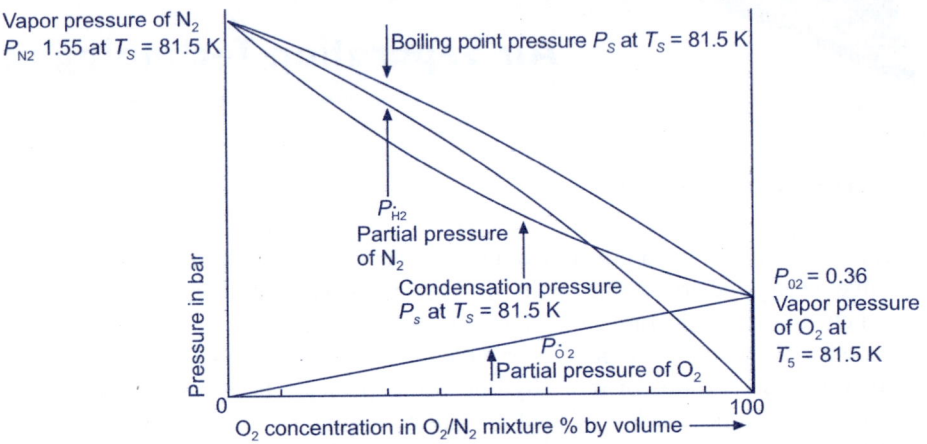

Fig. 12.1 Vapor Pressure Curve

Rectification is a special distillation separation process which enables the individual components of a mixture to be separated with a high purity combined with a good yield, even when their boiling points are relatively close to each other. Due to the different vapor pressures of the individual components of air ($PN_2 > PO_2$) the composition of the vapor differs from that of the liquid mixture. Correspondingly, a higher proportion of the component with the greater pressure vaporizes during the evaporation process. The vapor produced from a boiling liquid mixture of O_2/N_2 will thus have a higher nitrogen concentration than the liquid mixture from which it originates. Accordingly, the condensate produced when an O_2/N_2 vapor mixture is liquefied will display a higher oxygen concentration because the component with the lower partial pressure tends to transform into liquid.

THERMODYNAMICS OF AIR SEPARATION

Refer to a typical air separation column in Figure 12.2. The Incoming stream of air (mixture) M is first cooled by the reboiler and cooled further by the throttling process. The oxygen stream is B and the nitrogen stream is D. The column is assumed to operate adiabatically. For the sake of simplicity, the composition of air in the environment has been assumed to be 0.21 O_2 and 0.79 N_2 fractions by mole.

The thermodynamic work of separation is the reversible work (Ws) and is given by

$$W_s = -T_o \Delta S_1$$

Where ΔS_1 is change in entropy respectively during separation.

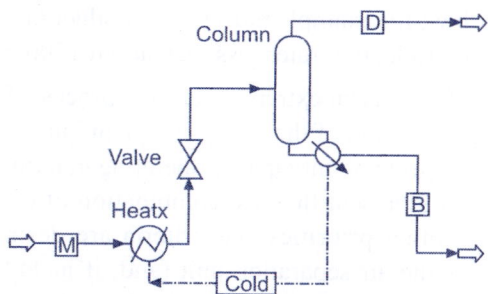

Fig. 12.2 Air Separation Column

The thermodynamic work of liquefaction is the reversible work (W_o) and is given by

$$W_o = \Delta H - T_o \Delta S_2$$

Where ΔH is change in enthalpy and ΔS_2 is change in entropy respectively during liquefaction.

Total reversible work = $W_r = W_s + W_o = \Delta H - T_o(\Delta S_1 + \Delta S_2)$

Considering the irreversibilities in the process, the irreversible work is given by

$$W_i = T_o \Delta S_3$$

Where ΔS_3 is change in entropy due to irreversibility.

Total Work = $W = W_r + W_i = \Delta H - T_o(\Delta S_1 + \Delta S_2) + T_o \Delta S_3$

The thermodynamic efficiency of the process is calculated as

$$\eta = W_r / W$$

The exergy analysis of the air separation column is as follows

$$E_M = E_B + E_D$$

The thermodynamic efficiency of the process can also be defined as

$$\eta = (E_B + E_D) / E_M$$

INTRODUCTION TO CRYOGENIC GAS SEPARATION PLANTS

Cryogenic plants produce nitrogen, oxygen and argon products using very low temperature distillation to separate and purify components of air. Cryogenic plants are most commonly used to produce high purity products at medium to high production rates. Non-cryogenic plants produce gaseous nitrogen or oxygen products using near-ambient-temperature separation processes. There are two major types of non-cryogenic processes, using either selective adsorption or differential permutation through membranes to produce relatively pure oxygen or nitrogen. These processes use differences in properties such as molecular structure, size and mass to achieve the desired degree of separation. Non-cryogenic separation processes are most commonly used when high purify nitrogen or oxygen is not needed (e.g. 99% purity nitrogen or oxygen at about 93% purity) and when product

12.4 Cryogenics

demand is relatively low; for example nitrogen at production rates less than about 500 Nm^3 or oxygen at production rates less than about 1500 Nm^3.

Cryogenic plant design is an extremely critical process. System optimization does not stop with specification of the cryogenic plant but a fully-optimal facility design will reflect all aspects of anticipated operating conditions at the site. Plant supply solutions which represent the best combination of cost and performance, as determined by customer priorities and criteria are desirable. The complete plant design integrates the air separation unit (and, if included, the liquefaction system) with appropriately configured and sized auxiliary equipment such as pipeline compressors, liquid storage tanks for backup and supplementation of plant production, liquid vaporizers (ambient or heated), elevated pressure gas storage, and bulk merchant liquid products storage and trailer loading systems. Well-optimized cryogenic plants provide operational flexibility, high operating reliability, and cost-effective operation. When large volumes of liquid are required to support customer operations and / or bulk merchant liquid product sales, the air separation unit (ASU) can be coupled with a liquefier. Liquefiers provide supplemental refrigeration, which is used to convert the available nitrogen, oxygen and argon to liquefied industrial gas products. Liquefiers can be supplied as stand-alone units, which allows maximum operating flexibility, which can be particularly useful for "piggyback" gas and liquid plants, or they may be fully integrated with the ASU to realize capital cost savings for plants which will produce bulk liquid products (liquid nitrogen, oxygen and argon). Figure 12.3 shows photograph of a typical cryogenic plant.

Fig. 12.3 Photograph of a Typical Cryogenic Gas Separation Plant

CRYOGENIC AIR SEPARATION UNITS (ASU)

Cryogenic air separation processes rely on differences in boiling points to separate and purify products. The basic process was commercialized early in the 20th century. Since then, numerous process configuration variations have emerged, driven by the desire to produce particular gas products and product mixes as efficiently as possible at various required levels of purity and pressure. These air separation process cycles have evolved in parallel with advances in compression machinery, heat exchangers, distillation technology and gas expander technology. The composition of dry air is approximately 78% nitrogen, 21% oxygen, and 1%

argon (by volume) plus small amounts of carbon dioxide, neon, helium, krypton, hydrogen, and xenon. In addition, variable amounts of water vapor will be present (depending upon humidity) plus other gases produced by natural processes and human activities. The composition of air is shown in Table 12.1.

Table 12.1 Composition of Air

Gas	% by Volume	% by Weight	Parts per Million (V)	Chemical Symbol
Nitrogen	78.08	75.47	780805	N_2
Oxygen	20.95	23.20	209450	O_2
Argon	0.93	1.28	9340	Ar
Carbon Dioxide	0.038	0.0590	380	CO_2
Neon	0.0018	0.0012	18.21	Ne
Helium	0.0005	0.00007	5.24	He
Krypton	0.0001	0.0003	1.14	Kr
Hydrogen	0.00005	Negligible	0.50	H_2
Xenon	8.7×10^{-6}	0.00004	0.087	Xe

Cryogenic air separation processes include these steps:
- Filtering and compressing air
- Removing contaminants, including water vapor and carbon dioxide (which would freeze in the process)
- Cooling the air to very low temperature through heat exchange and refrigeration processes
- Distilling the partially-condensed air (at about –300°F / –185°C) to produce desired products
- Warming gaseous products and waste streams in heat exchangers that also cool the incoming air stream

Atmospheric air is sucked in by a multi-stage compressor through a filter and is compressed to the design pressure. The compressed air is then passed through inter-coolers, industrial refrigerator, moisture separators, and then to the molecular sieve battery for removal of Carbon dioxide, hydrocarbons and moisture from the process air. This pure air then passes through the 1st heat exchanger, where it is cooled by the out going nitrogen and oxygen. Part of this cooled air is passed through expansion engine and the other part through the second heat exchanger. Both the expansion engine and second heat exchanger help in further cooling down the air, which is finally released to the bottom of the column through an expansion valve. The air becomes liquid at this stage. The column consists of two parts. In between the lower and upper columns there is a condenser, which acts as a reflux for the lower column and as a re-boiler for the upper column. The liquid air at the bottom of lower column separates through the trays to give crude oxygen at the bottom and approximately 90% pure nitrogen at the top. Crude oxygen termed as rich liquid is then expanded through an expansion valve from the lower column to the middle of the upper column. Crude nitrogen termed as poor liquid is expanded through another expansion valve from the top of the lower column to the top of the upper column. Due to difference in the boiling points, the pure nitrogen boils over and accumulates at the top of the upper column and oxygen, which accumulates at the bottom of the upper column. Both nitrogen and oxygen are removed through separate paths in heat exchangers, for cooling the incoming air. Oxygen is compressed to a prescribed settled pressure by a liquid pump and is directly filled into cylinders. Nitrogen is however available at a pressure of approximately 0.5 kg/cm^2 and the same can be compressed into cylinders with help of an independent high-pressure compressor.

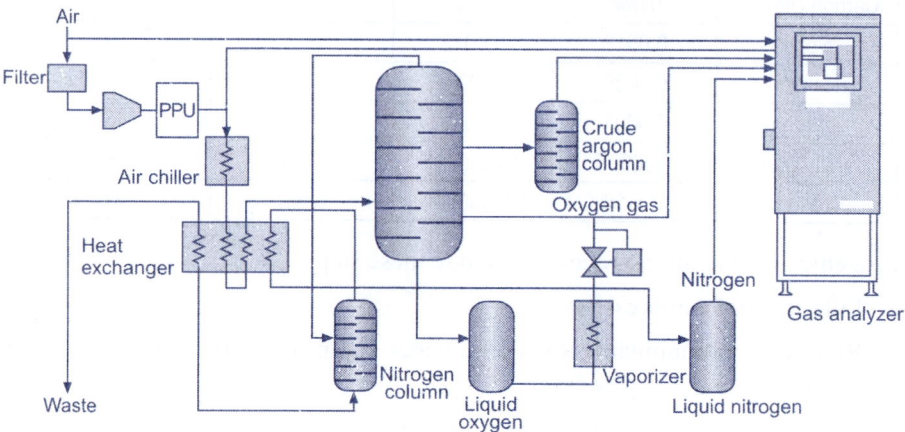

Fig. 12.4 Schematic of an Air Separation Unit

Cryogenic processes are the most cost effective separation process for producing at high production rates and are capable of making the highest purity products. All air separation processes start with compression of air, and additional compressors may be used to boost the pressure of the nitrogen and/or oxygen products leaving the separation and purification process. Even more compression is required to produce large amounts of additional refrigeration when some or all of plant production from an air separation unit (ASU) is desired in liquid form. The cost of electricity is the largest single operating cost incurred in air separation plants. It is usually between one third and two thirds of the operating costs associated with producing gas and liquid products. Electric motors are used to drive the compression equipment, as well as in process heaters, instrumentation systems and cooling systems. Consequently, it is fair to say that electrical power is just as much a raw material as air for the manufacture of atmospheric industrial gas products. Small gaseous product plants may use hundreds of kilowatts (kW). Large liquid plants may have power demands measured in megawatts or tens of megawatts (MW). The various separation technologies that produce commercially useful products from air are based on differences in boiling points (cryogenic distillation); or on differences in molecular weights and molecular size. Refer Figure 12.4 for a schematic of cryogenic air separation unit.

At high production rates, cryogenic processes are the most cost-effective choice. Cryogenic processes can produce very pure end products; and must be used to produce liquid nitrogen, oxygen and argon. Within cryogenic technology choices can be made regarding the specifics of process design, machinery configuration and system control. These all have capital cost / energy cost / and operational flexibility / plant reliability tradeoffs. The portions of the cryogenic air separation process that operate at very low temperatures (the distillation columns, heat exchangers and cold interconnecting piping sections) must be well insulated to minimize energy consumption and avoid operating problems. To accomplish this, these components are located inside insulated, sealed (and nitrogen purged) structures called "cold boxes". Cold boxes may have a rectangular or round footprint and, depending on plant type and capacity, may measure approximately 2 to 4 meters on each side and have a height of 15 to 60 metres. Oxygen and nitrogen gases typically emerge from the air separation process at close to ambient air temperature and at relatively low pressure. Most cryogenic air separation plants can produce a few percent of the plant product as liquid. If large amounts of oxygen or nitrogen are desired in liquid form, additional refrigeration (external to the basic air separation unit) is required. The set of equipment, which accomplishes this is called a Liquefier, and it typically uses nitrogen as the working fluid. Through compression, expansion and heat exchange, liquefiers produce the refrigeration needed to lower the temperature of large amounts of product from ambient to $-185°C$ and sub-cool those products prior to sending them to storage. Sub-cooling is desirable to minimize vaporization of stored product due to heat leak into the storage tank.

CRYOGENIC ASU CYCLES

Cryogenic air separation is currently the most efficient and cost-effective technology for producing large quantities of oxygen, nitrogen, and argon as gaseous or liquid products. Numerous configurations of heat exchange and distillation equipment can separate air into the required product streams. These process alternatives are selected based on the purity and number of product streams required; trade-offs between capital costs and power consumption; and the degree of integration between the ASU and other facility units. Characterization of the types of cryogenic processes used for air separation, hereinafter referred to as cycles, can be based on the method of pressurizing the product streams or on the air feed pressure to the ASU.

1. **Compression Cycles:** Air separation processes typically produce gaseous product streams at slightly above atmospheric pressure and near ambient temperature. Typically the product oxygen leaves the main heat exchangers at low pressure, ranging from 0.5 to 10 psig, and a centrifugal compressor train with relatively high inlet volumetric flow rate delivers the product at the required pressure. Several of these plants operate worldwide.

2. **Pumped Liquid Cycles:** Liquid products can be pumped from the distillation section upstream of the cryogenic heat exchangers for vaporization and warming. These products may be pumped to the required delivery pressure or to an intermediate pressure. However, since producing a liquid product from the distillation system requires two to three times the power of producing a gaseous product, the cycle must efficiently recover the energy spent for refrigeration. This is accomplished by condensing an air or nitrogen feed stream at high pressure against the vaporizing product stream in the cryogenic heat exchangers. The liquefied air returns the refrigeration effect to the distillation section.

3. **Low and Elevated Pressure (EP) Cycles:** Low pressure ASU cycles are based on compressing the feed air only to the pressure required to reject the nitrogen byproduct at atmospheric pressure. As such, feed air pressures will typically vary between 65 and 100 psia, depending on the oxygen purity and the level of energy efficiency desired. Elevated-pressure ASU cycles produce product and byproduct streams at well above atmospheric pressures (above 100 psia) and generally require smaller and more compact cryogenic components that can be cost effective. An EP cycle may be appropriate when all or nearly all of the nitrogen byproduct will be compressed as a product stream. In addition, an EP cycle is often selected for integration of the ASU with other process units such as gas turbines.

NON-CRYOGENIC AIR SEPARATION UNITS (ASU)

Non-cryogenic plants are less energy efficient than cryogenic plants (for comparable product purity) but may cost less to build, in particular when the required production rate is relatively small. They are most suitable when high purity product is not required by end-use applications. Non-cryogenic processes employ membranes or adsorbents (PSA/ VPSA) to remove the unwanted components of air. They produce oxygen, which is typically 90 to 95% pure, or nitrogen, which is typically 95 to 99.5% pure. The basic technology choice between cryogenic or non-cryogenic technology, is largely determined by the number of products that must be supplied (e.g. nitrogen or oxygen or both), the required production rates for each gas and/or liquid product, and required product purities. Refer Figure 12.5 for a chart on selection of different technologies for ASU as per requirement. The need to accommodate user demand patterns (flow rate fluctuations) leads to additional system optimizations. Instantaneous demand is normally satisfied from several sources: gas from the on-site production plant, gas withdrawn from elevated pressure gas storage vessels and distribution lines, and, as needed, additional gas generated by vaporizing stored liquid. When gases are used at a number of distinct pressure levels, with their own usage patterns, defining the optimal gas storage and liquid vaporization system can be complex. In an optimal system, the local gas production plant is sized to cost-effectively provide the maximum possible percentage of gas demand, while the amount of liquid vaporization required to support day-to-day operations is minimized.

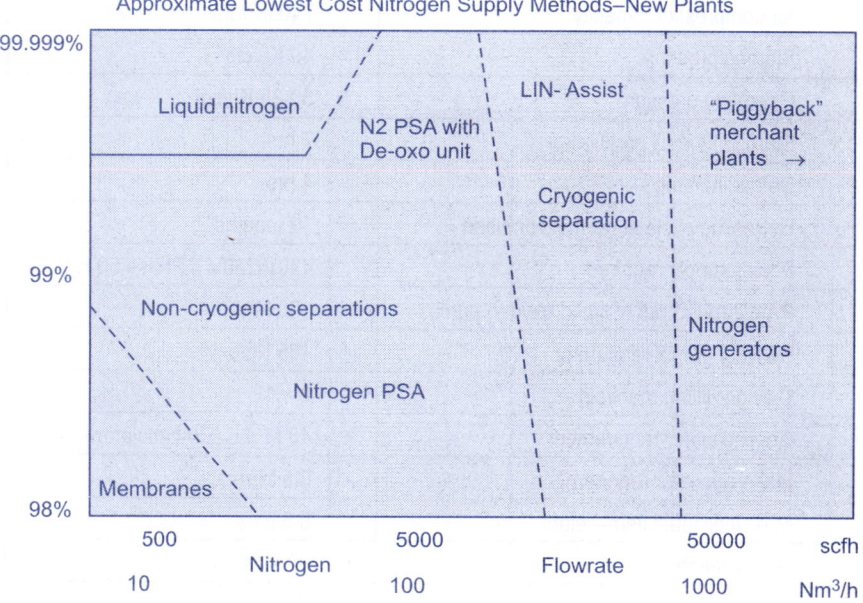

Fig. 12.5 Choice of Technology for Air Separation Units (ASU)

Technical Specifications of a typical ASU plant are given in Table 12.2. Features of ASU plants are:

- Simultaneous production of nitrogen with oxygen. This results in higher economical viability.
- Plants are equipped with a highly efficient Expansion Engine, which helps rapid and adequate cooling. On account of this ASU plant operation is exceedingly stable with a consistency in gas purity.
- High standard of purity of gases.
- Low power consumption.
- A special bypass valve for quick production and better cooling.
- International class pressure differential gauges as level indicators.
- High quality insulation by Perlite powder to prevent cold losses.
- Liquid Nitrogen withdrawal facility.
- Heavy duty air compressor suitable to operate with the power supplied by a diesel generating set or normal power supply.
- Low maintenance cost.

Table 12.2 Typical ASU Plant Technical Specifications

Specifications	Values
Air compressor capacity	740 M^3/hr.
Starting pressure	40 Kg/cm^2
Operating pressure	30-35 Kg/cm^2
Starting time till production	7 hrs.
Defrosting time	4 hrs.
Defrosting cycle at normal condition	12 months
Power supply required	420V 220V 3 Phase 50 Hz
Rate through put of liquid oxygen pump	130 M^3/hr.
Delivery pressure of pump	165 Kg/cm^2
Cylinder filling manifold	2 × 12
Cooling water requirement	40 M^3/hr. (Re-circulated)
Make-up water requirement	0.5 M^3/hr.
Max. Individual item weight	8.5 tons
Gross weight	20 tons
Insulation	Perlite
Assembly height	8.0 Meters
Area required	200 sq.m.
Material of construction	Copper, brass, steel etc
Power consumption	1.2 Units/m^3 of oxygen

AIR SEPARATION BY RECTIFICATION

Air separation by rectification can be carried out in a single or double column. Based on his air liquefaction principle Carl Von Linde constructed the first air separation plant for oxygen production in 1902, using a single column rectification system. In 1910, he set the basis for the cryogenic air separation principle with the development of a double column rectification system. Now, it was possible to produce pure oxygen and pure nitrogen simultaneously. Refer Figure 12.6 for the Linde single and double rectification columns.

Below the low pressure column an additional pressure column is installed. At the top of this column pure nitrogen is drawn off, liquefied in a condenser and fed to the low pressure column serving as reflux. At the top of this low pressure column pure gaseous nitrogen was withdrawn as a product while liquid oxygen evaporated at the bottom of this column to deliver the pure oxygen product. This principle of double column rectification combining the condenser and evaporator to form a heat exchanger unit is still used today.

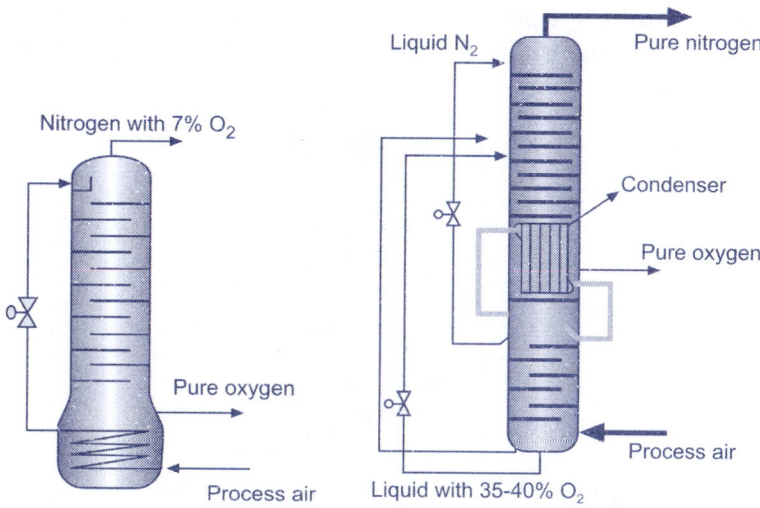

Fig: 12.6 Linde Single and Double Column Rectification Air Separation Units

EXPANSION ENGINE FOR CRYOGENIC ASU PLANTS

The Expansion Engine is vertical, single acting, reciprocating type. This produces the refrigeration effect required for the plant. High- pressure air enters through inlet valve at the start of downward stroke of the piston. On further downward motion, the inlet valve closes and entrapped air expands. During upward stroke, outlet valve remains open and inlet valve remains closed. Therefore, in downward stroke air enters the cylinder and expands. In the upward stroke the expanded air is pushed outside. The electric motor is used to start the machine. Thereafter the engine is moved by the air pressure itself and during which time, the engine motor retains the speed by acting as a brake. Since work is done by air in rotating the flywheel, it loses its heat content (enthalpy), thus the Air gets cooled. This cooling is more than that of an expansion in an expansion valve. An important factor for the function of an expansion engine is to use dry and carbon dioxide free air, as otherwise, ice and dry ice will form on valve seats, causing mal-function. The approx. temperature drop across expansion engine is 50 to 70°C, depending on inlet air pressure, temperature and inlet cam position. Figure 12.7 shows a typical expansion engine.

The expansion engine consists of three major units. The drive unit, the cylinder unit for air expansion and the hydraulic system for operating the valves. The drive unit is similar to any reciprocating machine with usual parts such as, crank case, crankshaft, connecting rod, crosshead, etc. The expansion engine has an extended crankshaft enabling the accommodation of the cam box. The moving parts are lubricated through a hole in crankshaft. Oil scrapper rings are provided to prevent oil escape to cylinder unit. The cylinder unit handling the process air consists of cylinder, piston, inlet and outlet isolation valve assemblies. The ball valves are actuated mechanically by a push rod as per the timing transmitted by the cam. The push rods are housed in a stuffing box to avoid air leakage and are actuated by the hydraulic system.

Fig. 12.7 Photograph of a Typical Expansion Engine

TONNAGE AIR SEPARATION PLANTS

The tonnage series of air separation plants are designed for the applications that require large amount of oxygen and to replace bulk liquid oxygen storage vessels. The output of liquid oxygen can be stored in a liquid oxygen tanks. Liquid nitrogen can be available as a by product simultaneously up to 99.999% purity. The tonnage plant is extremely versatile and all the possible product variants are possible. One can either obtain 100% liquid oxygen output or 50% liquid oxygen output directly into cylinders or 50% liquid oxygen and 50% liquid nitrogen. Tonnage plants can be designed for covered (standard) as well as outdoor installations. These plants are PLC controlled, inclusive of oxygen analyzer. Refer Figure 12.8 for a schematic of the plant.

Operation stages of the plant are as follows:

Compression of Air: The air is first compressed to 150 bar pressure. The compressor is a water-cooled, oil lubricated, heavy duty continuously operating four stage machine. It also composes of horizontal rugged inter-coolers to give normal discharge conditions. The compressor can be single frame primary as well as secondary equipment. The multi-stage composition improves the volumetric efficiency of the compressor and reduces B.H.P. The inter-stage temperature is low and therefore carbon formation is nil.

Purification of Air: This is done by the process skid which comprises of the following components: after cooler with tank, Nitrogen cooler with tank, moisture separator (Purger), chilling unit with Freon unit, oil absorber filled with alumina, molecular sieve battery on skid, defrost heater, gas/air lines as per standard layout on skid/platform, water pump, drain manifold complete with ball valves for draining moisture.

Cooling and Liquefaction of Air: This is done by a highly efficient expansion engine and also by a refrigeration unit (for 100% liquid output). The expansion engine is a single cylinder expansion engine, hydraulically operated with variable cam for automatic control of pressure and temperature, this is also equipped with ball type valve at inlet and outlet which is proof for trouble free operation for years together complete with electric motor.

Air Separation (Rectification): This is done by the special air separation unit consisting of upper and lower column and special exchangers. The separation air is by the cryogenic process. The final product i.e. liquid oxygen/nitrogen goes to the storage tank. The air separation column (cold box) is a leak proof stainless steel column supplied as packed unit consisting of the following components: outer steel casing, main heat exchanger, bottom rectification column, top rectification column, liquid oxygen and nitrogen filter, instrumentation, cold pipe lines and insulation materials.

Liquid Oxygen (LOX) Pumping: There is one liquid pump, namely for oxygen. The oxygen is being pumped in liquid form and thus eliminates unnecessary equipment such as gas holders and oxygen compressor. This pump has special design with stainless steel jacket argon welded for insulation to avoid cold loss during pumping. It has a special seal to prevent any oil carry over during operation. The oxygen purity is high and is bone dry due to usage of special types of seals designed specially for this purpose complete with electric motor vee belts, fly wheel, pulley and guard.

Cryogenic Storage: Cryogenic vessels are designed for storage and transport of liquid gases at sub-zero temperatures. Manufacturing of such cryogenic tanks requires special technical know how and sophisticated fabrication techniques.

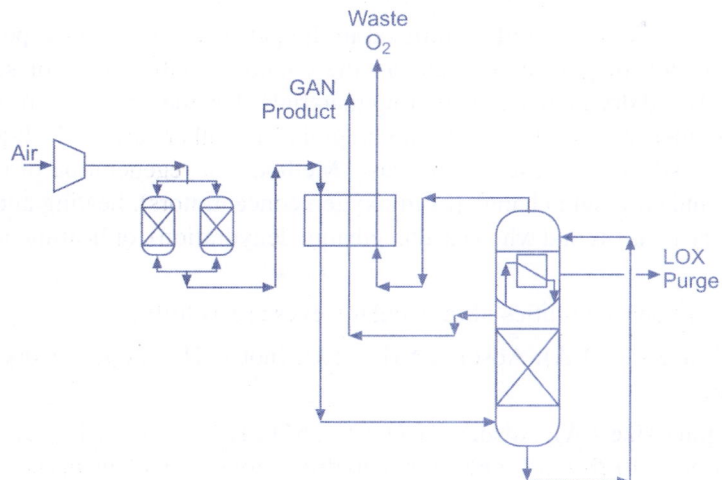

Fig. 12.8 Tonnage Air Separation Plant

Molecular Sieve

A molecular sieve is a material containing tiny pores of a precise and uniform size that is used as an adsorbent for gases and liquids. Molecules small enough to pass through the pores are adsorbed while larger molecules are not. It is different from a common filter in that it operates on a molecular level. For instance, a water molecule may not be small enough to pass through while the smaller molecules in the gas pass through. Because of this, they often function as a desiccant. A molecular sieve can adsorb water up to 22% of its own weight. Often they consist of aluminosilicate minerals, clays, porous glasses, microporous charcoals, zeolites, active carbons, or synthetic compounds that have open structures through which small molecules, such as nitrogen and water can diffuse. Figure 12.9 shows some materials for molecular sieves.

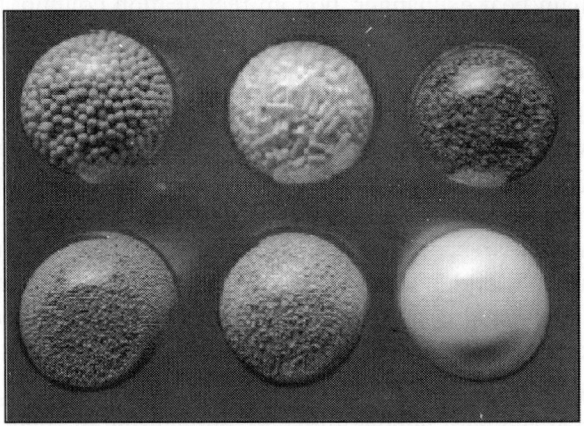

Fig. 12.9 Material for Molecular Sieves

Molecular sieves are often utilized in the petroleum industry, especially for the purification of gas streams and in the chemistry laboratory for separating compounds and drying reaction starting materials. The mercury content of natural gas is extremely harmful to the aluminium piping and other parts of the liquefaction apparatus – silica gel is used in this case. Methods for regeneration of molecular sieves include pressure change (as in oxygen concentrators), heating and purging with a carrier gas (as and when used in ethanol dehydration), or heating under high vacuum.

Adsorption capabilities of molecular sieves are as follows:

- **3A (pore size 3 Å):** Adsorbs NH_3, H_2O, (not C_2H_6), good for drying polar liquids.
- **4A (pore size 4 Å):** Adsorbs H_2O, CO_2, SO_2, H_2S, C_2H_4, C_2H_6, C_3H_6, EtOH. Will not adsorb C_3H_8 and higher hydrocarbons. Good for drying nonpolar liquids and gases.
- **5A (pore size 5 Å):** Adsorbs normal (linear) hydrocarbons to n-C_4H_{10}, alcohols to C_4H_9OH, mercaptans to C_4H_9SH. Will not adsorb isocompounds.
- **10X (pore size 8 Å):** Adsorbs branched hydrocarbons and aromatics. Useful for drying gases.

Pressure Swing Adsorption (PSA) Plants

PSA is a technology used to separate some gas species from a mixture of gases under pressure according to the species' molecular characteristics and affinity for an adsorbent material. It operates at near-ambient temperatures and so differs from cryogenic distillation techniques of gas separation. Special adsorptive materials (e.g., zeolites) are used as a molecular sieve, preferentially adsorbing the target gas species at high pressure. The process then swings to low pressure to desorb the adsorbent material. Pressure swing adsorption processes rely on the fact that under pressure, gases tend to be attracted to solid surfaces, or "adsorbed". The higher the pressure, the more gas is adsorbed; when the pressure is reduced, the gas is released, or desorbed. PSA processes can be used to separate gases in a mixture because different gases tend to be attracted to different solid surfaces more or less strongly. If a gas mixture such as air, for example, is passed under pressure through a vessel containing an adsorbent bed that attracts nitrogen more strongly than it does oxygen, part or all of the nitrogen will stay in the bed, and the gas coming out of the vessel will be enriched in oxygen. When the bed reaches the end of its capacity to adsorb nitrogen, it can be regenerated by reducing the pressure, thereby releasing the adsorbed nitrogen. It is then ready for another cycle of producing oxygen enriched air. Using two adsorbent vessels allows near-continuous production of the target gas. It also permits so-called pressure equalisation, where the gas leaving the vessel being depressured is used to partially pressurise the second vessel. This results in significant energy savings, and is common industrial practice. Refer Figure 12.10 for the schematic of a PSA plant [3].

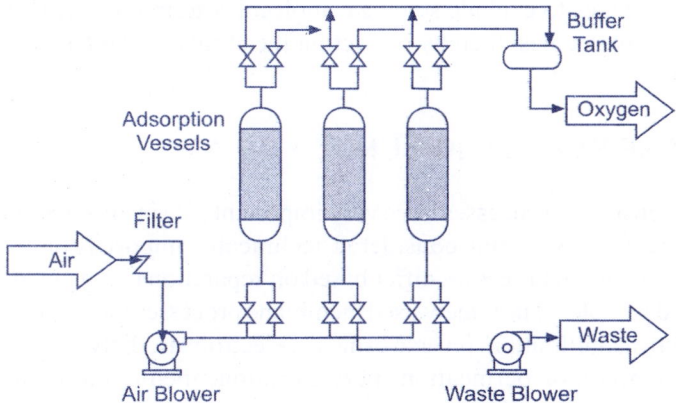

Fig. 12.10 PSA Air Separation Plant

Adsorption processes are based on the ability of some natural and synthetic materials to preferentially adsorb nitrogen. Thus, in air separation, nitrogen molecules are more strongly adsorbed than oxygen or argon molecules. As air is passed through a bed of zeolitic material, nitrogen is retained and an oxygen-rich stream exits the bed. Carbon molecular sieves have pore sizes on the same order of magnitude as the size of air molecules. Since oxygen molecules are slightly smaller than nitrogen molecules, they diffuse more quickly into the cavities of the adsorbent. Thus, carbon molecular sieves are selective for oxygen and zeolites are selective for nitrogen. Pressurized air enters a vessel containing the adsorbent. Nitrogen is adsorbed and an oxygen-rich effluent stream is produced until the bed has been saturated with nitrogen. At this point, the feed air is switched to a fresh vessel and regeneration of the first bed can begin. Regeneration can be accomplished by reducing the pressure in the bed, which reduces the equilibrium nitrogen holding capacity of the adsorbent. Variations in the process that effect operating efficiency include separate pretreatment of the air to remove water and carbon dioxide, multiple beds to permit pressure energy recovery during bed switching, and vacuum operation during depressurization. Optimization of the system is based on product flow, purity and pressure, energy cost and expected operating life. Oxygen purity is typically 93–95 %. Due to the cyclic nature of the adsorption process, bed size is the controlling factor in capital cost. Since production is proportional to bed volume, capital costs increase more rapidly as a function of production rate compared to cryogenic plants.

Aside from their ability to discriminate between different gases, adsorbents for PSA systems are usually very porous materials chosen because of their large surface areas. Typical adsorbents are activated carbon, silica gel, alumina and zeolite. Though the gas adsorbed on these surfaces may consist of a layer only one or at most a few molecules thick, surface areas of several hundred square meters per gram enable the adsorption of a significant portion of the adsorbent's weight in gas. In addition to their selectivity for different gases, zeolites and some types of activated carbon called carbon molecular sieves may utilize their molecular sieve characteristics to exclude some gas molecules from their structure based on the size of the molecules, thereby restricting the ability of the larger molecules to be adsorbed.

MEMBRANE SEPARATION UNITS FOR AIR

Membrane separation processes have very important role in air separation industry. Nevertheless, they were not considered technically important until mid-1970. Membrane separation processes differ based on separation mechanisms and size of the separated particles. The widely used membrane processes include microfiltration, ultrafiltration, nanofiltration, reverse osmosis, electrolysis, dialysis, electrodialysis, gas separation, vapor permeation, pervaporation, membrane distillation, and membrane contactors. All processes except for pervaporation involve no phase change. All processes except dialysis are pressure driven. Microfiltration and ultrafiltration is widely used in food and beverage processing, biotechnological applications and pharmaceutical industry, water purification and wastewater treatment, microelectronics industry, and others. Nanofiltration and reverse osmosis membranes are mainly used for water purification purposes. Dense membranes are utilized for gas separations (removal of CO_2 from natural gas, separating N_2 from air, organic vapor removal from air or nitrogen stream) and sometimes in membrane distillation.

The selection of synthetic membranes for a targeted separation process is usually based on few requirements. Membranes have to provide enough mass transfer area to process large amounts of feed stream. The selected membrane has to have high selectivity (rejection) properties for certain particles; it has to resist fouling and to have high mechanical stability. It also needs to be reproducible and to have low manufacturing costs. The main modeling equation for the dead-end filtration at constant pressure drop is represented by Darcy's law.

$$\frac{dV_p}{dt} = Q = \frac{\Delta p}{\mu} A \left(\frac{1}{R_m + R} \right)$$

Where V_p and Q are the volume of the permeate and its volumetric flow rate respectively (proportional to same characteristics of the feed flow), μ is dynamic viscosity of permeating fluid, A is membrane area, R_m and R are the respective resistances of membrane and growing deposit of the foulants. R_m can be interpreted as a membrane resistance to the solvent (water) permeation. This resistance is a membrane intrinsic property and expected to be fairly constant and independent of the driving force, Δp. R is related to the type of membrane foulant, its concentration in the filtering solution, and the nature of foulant-membrane interactions. Darcy's law allows to calculate the membrane area for a targeted separation at given conditions.

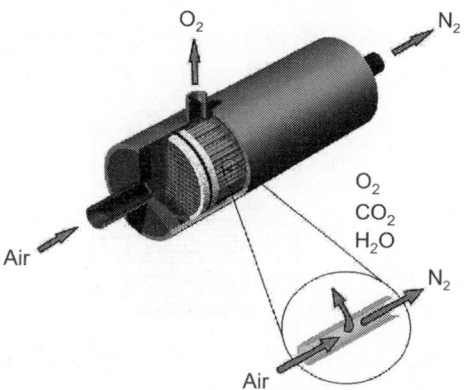

Figure 12.11 Synthetic Membrane for Air Separation

The solute sieving coefficient is defined by the equation.

$$S = \frac{C_p}{C_f}$$

where C_f and C_p are the solute concentrations in feed and permeate respectively. Hydraulic permeability is defined as the inverse of resistance and is represented by the equation

$$L_p = \frac{J}{\Delta p}$$

where J is the permeate flux which is the volumetric flow rate per unit of membrane area. The solute sieving coefficient and hydraulic permeability allow the quick assessment of the synthetic membrane performance. Figure 12.11 shows a typical synthetic membrane.

Membrane air separation processes using polymeric materials are based on the difference in rates of diffusion of oxygen and nitrogen through a membrane which separates high-pressure and low-pressure process streams. Flux and selectivity are the two properties that determine the economics of membrane systems, and both are functions of the specific membrane material. Flux determines the membrane surface area, and is a function of the pressure difference divided by the membrane thickness. A constant of proportionality that varies with the type of membrane is called the permeability. Selectivity is the ratio of the permeabilities of the gases to be separated. Due to the smaller size of the oxygen molecule, most membrane materials are more permeable to oxygen than to nitrogen. Membrane systems are usually limited to the production of oxygen enriched air with 25–50% oxygen. Active or facilitated transport membranes, which incorporate an oxygen complexing agent to increase oxygen selectivity, are a potential means to increase the oxygen purity from membrane systems, assuming oxygen compatible membrane materials are also available. Figure 12.12 shows a typical membrane air separation plant [3].

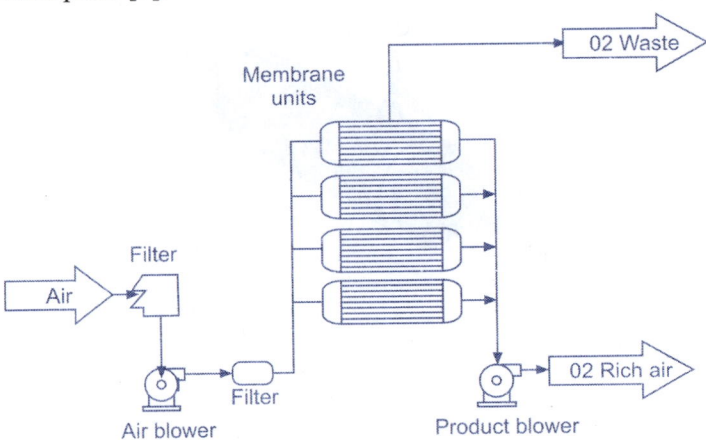

Fig. 12.12 Membrane Air Separation Plant

A major benefit of membrane separation is the simple, continuous nature of the process and operation at near ambient conditions. An air blower supplies enough head pressure to overcome pressure drop through the filters, membrane tubes and piping. Membrane materials are usually assembled into cylindrical modules that are manifolded together to provide the required spiral wound sheets. A vacuum pump typically maintains the pressure difference across the membrane and delivers oxygen at the required pressure. Carbon dioxide and water usually appear in the oxygen enriched air product, since they are more permeable than oxygen for most membrane materials. As with adsorption systems, capital is essentially a linear function of production rate and product backup is typically not available without a separate liquid oxygen storage tank and delivery support system. Membrane systems readily fit applications up to 20 tons per day, where air enrichment purities with water and carbon dioxide contaminants can be tolerated. This technology is newer than adsorption or cryogenics and improvements in materials could make membranes attractive for somewhat larger oxygen requirements. The fast start-up time, due to the near ambient operation, is especially attractive for oxygen-use systems than exhibit discontinuous usage patterns. The passive nature of the process is also appealing.

CHEMICAL SEPARATION UNITS FOR AIR

A number of materials have the ability to absorb oxygen at one set of pressure and temperature conditions, and desorb the oxygen at a different set of conditions. One such process is a molten salt chemical process [3] depicted in Figure 12.13. The process shown is based on absorption of oxygen by a circulating molten salt stream, followed by desorption through a combination of heat and pressure reduction of the salt stream.

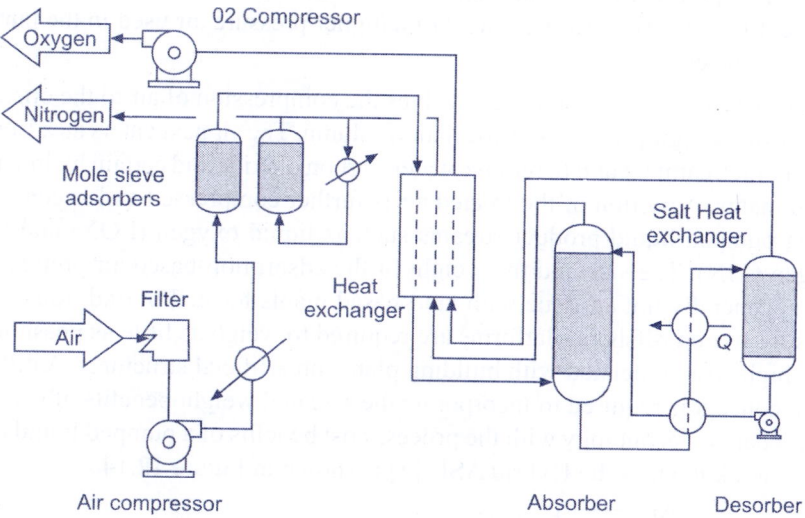

Fig. 12.13 Chemical Air Separation Plant

Air is compressed from 20 to 185 psia and treated to remove water and carbon dioxide in an adsorbent-based system. Air flows through an adsorbent bed until bed saturation is reached. The beds are switched and the saturated bed is regenerated by dry nitrogen from the process. The clean, dry air is heated against returning product streams to between 900 F and 1200 F in the main heat exchangers. The hot air flows to the bottom of the absorber where it contacts molten liquid salt. The oxygen in the air reacts chemically with the salt and is removed with the liquid salt leaving the bottom of the absorber. The oxygen-bearing salt is heat interchanged with oxygen-free salt and further heated before being reduced in pressure and flowing to the desorber. Gaseous oxygen leaves the top of the desorber, while oxygen-lean salt is removed from the bottom of the desorber, heat interchanged and sent to the top of the absorber vessel to close the loop. The hot oxygen and hot nitrogen streams enter the main heat exchanger and are cooled against feed air. The oxygen is compressed to delivery pressure, while a portion of the nitrogen is used to regenerate the air pretreatment system. The major process advantage of this system is that air has only to be compressed to a pressure that overcomes pressure drop through the air pretreatment and heat exchanger, thus reducing the amount of air compression power compared to a cryogenic plant. A source of thermal energy must be available to liberate the salt via heating.

RECENT TRENDS IN AIR SEPARATION UNITS

A small footprint and reduced weight for large ASUs are critical success factors for developing technologies located on offshore platforms, ships or barges, or in remote locations. Pumped liquid process cycles are also of interest in reducing ASU cost. The following concept involves the use of reversing main heat exchangers to purify a low-pressure air stream that enters the distillation system, while using an adsorbent-based system for treatment of a higher pressure air used in the pumped portion of the cycle.

Current ASU design practice involves the compression of air to the operating pressure of the high-pressure HP distillation column. The air next enters an adsorbent based air pretreatment unit to remove water, carbon dioxide and certain hydrocarbon contaminants. A portion of the treated air is further compressed and is condensed against pumped liquid product streams such as liquid oxygen (LOX) and liquid nitrogen (LIN). The size and the weight of the adsorption-based air pretreatment units is generally not an issue with land-based plants located in traditional areas. ASUs for use on offshore platforms are required to weigh as little as possible due to the high cost associated with building plants on artificial structures. A different design concept is required to incorporate the size and weight benefits of reversing heat exchanger technology with the process cost benefits of a pumped liquid cycle. Such a unit known as the Hybrid ASU [3] is shown in Figure 12.14.

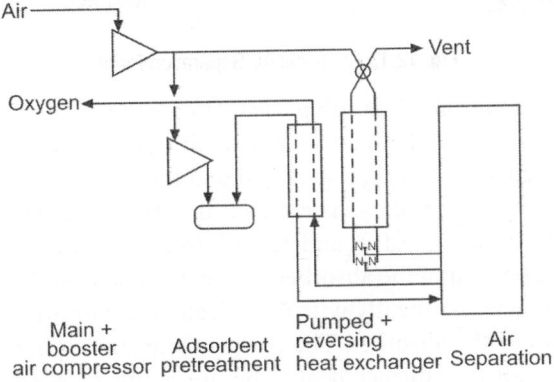

Fig. 12.14 Hybrid Air Separation Plant

In the hybrid unit air is compressed to the operating pressure of the HP distillation column, the majority of the air passes through reversing heat exchangers and into the distillation column. In reversing heat exchangers contaminants in the air stream are captured by desublimation within the heat exchanger passages. Periodically, the flow through the heat exchanger is reversed to allow a dry stream from the distillation system to flow through the air passages to sublimate, liquefy and sweep the contaminants from the system. A reversing heat exchanger and associated valving performs the combined duty of an adsorption pretreatment system and main cryogenic heat exchanger, thus saving weight. A portion of the initially compressed air bypasses the reversing heat exchangers and is further compressed and intercooled before being treated in an adsorption-based system. Since this stream is typically in the 200–1200 psia range, it contains very little water vapor after being cooled.

Another new development in ASU is that extracted air heat can be used to regenerate adsorption-based processes. The adsorption processes can be associated with the air separation system or byproduct recovery systems. Purification recovery systems often use PSA cycles that typically require heat above 200 F for regeneration. Gas turbine extraction air is typically available in the 700–800 F range, making it a suitable regeneration heat source with or without a topping heat recovery step. Refer Figure 12.15 for an integrated turbine and ASU plant.

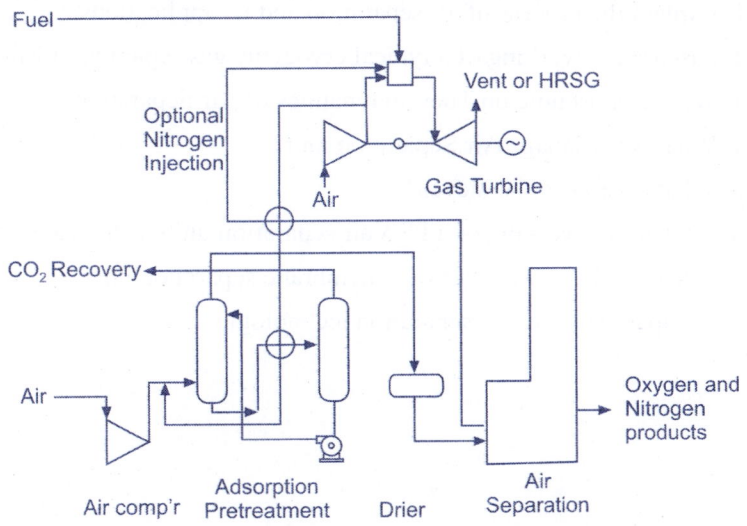

Fig. 12.15 Integrated Gas Turbine and Air Separation Plant

Heat from the extracted air may be recovered by indirect contact of the hot air with a process fluid or by heat transfer from the air to a working fluid such as stream, or an inert gas. High level heat from the extracted air source is transferred to a nitrogen stream returning to the gas turbine. The extracted air is further cooled by contact with adsorption column used to pretreat air feed to the ASU. The extracted air can be combined with supplemental compressed air entering the pretreatment process, or it could be separately treated if the pressure was significantly different than the supplement air source. In this embodiment, carbon dioxide is recovered from the air pretreatment area as a byproduct.

References

1. Cryogenic Engineering – B.A. Hands.
2. Internet Website www.wikipedia.com.
3. A.R.Smith, Fuel Processing technology (2001) v 70, pp 115 – 134.
4. Cryogenic Systems – Randall Baron.
5. Cryogenic Engineering – P.K. Bose.
6. Internet Website www.uigi.com.
7. Website of KVK Corporation, India, www.indiamart.com/kvkcorp.
8. Internet website http://www.freepatentsonline.com/5802872.html.
9. Cryogenic Air Separation Brochure by Linde.
10. Thermo Electron Corporation brochure on Cryogenic ASU.
11. Internet Website http://www.gasplantsmanufacturer.com/.

Questions

- **Q.1.** Explain the process of air separation and its applications.

    Q.2. Discuss the working of a typical cryogenic gas separation plant.

    Q.3. Write a short note on Thermodynamics of Air Separation.

    Q.4. What is a Tonnage Air Separation unit?

    Q.5. What are molecular sieves?

    Q.6. Explain the working of a PSA air separation units with neat sketch.

    Q.7. What are the advantages of a membrane separation unit?

    Q.8. Compare any two air separation technologies.

CHAPTER 13

Cryogenic Refrigeration Systems

INTRODUCTION

Refrigeration and liquefaction of gases are historically at the root of cryogenics, as they constitute the enabling technology which gave access to the low-temperature domain. They have developed over the years along several lines, to become a specialized subject. In this chapter, the basic thermodynamics of refrigeration is described along with the cooling processes at work and the corresponding equipment in the case of refrigerators based on various principles. Cryogenic refrigerators are often required to provide cooling duties at several temperatures or in several temperature ranges, e.g. for thermal shields or continuous heat interception. Cryogenic refrigerators and liquefiers have the same components however the cryogenic liquid is evaporated in the refrigerator, whereas it is collected in the liquefiers for application use.

A block diagram of a typical cryogenic refrigerator is shown in Figure 13.1 below. There is a heat load at the cold temperature (T_c) the application wants to operate at. This heat load can be from the environment, typically radiation, solid/gas conduction and fluid convection from adjacent regions or structures at higher temperatures. The heat load can also be internally generated in the component from sources such as ac losses and eddy currents in conductors. The heat from the load is transferred to the cold region of the refrigerator by solid conduction or fluid convection. Therefore, there has to be a small temperature difference (typically a few degrees Kelvin or less) between the cold region of the refrigerator and the slightly warmer application load. The working fluid (example helium gas) in the cold section of the refrigerator absorbs this heat and flows next to a heat exchanger where this fluid is further heated by exchange with a counter flow gas stream. From there it enters a compressor where external work is done on the fluid compressing it to high pressures and further increasing the fluid temperature. After this the fluid flows to an ambient heat exchanger where the heat of compression is rejected near room temperature (at T_h). Next the fluid goes to the heat exchanger where it is further cooled by rejecting heat to the counter flow gas stream or regenerative matrix. From here the cold fluid can go to the load or be further cooled by expansion of the fluid to remove energy. If the working fluid temperature is below the inversion temperature

13.2 Cryogenics

(~45 K for helium), a Joule-Thomson expansion can reduce the temperature an additional amount. For an ideal process it can be shown that the efficiency (more precisely, this is called the coefficient of performance since it can be >1) of such a refrigerator is:

$$\eta_{Carnot} = \frac{T_c}{T_h - T_c}$$

Several observations can be made. If T_c is in the range 25-80 K compared to 4-10 K typical of low temperature superconductors; this results in higher Carnot efficiencies and means that the cryogenic refrigerators can be smaller. Also heat should be rejected at as low a temperature as practical; usually this is room or ambient temperature. Real refrigerators typically run at a fraction (up to 30%) of Carnot efficiency due to losses in compressors, heat exchanger effectiveness, etc. The inverse of the Carnot efficiency is called the Carnot specific power and is the number of watts required at ambient to provide 1 watt of refrigeration at the lower operating temperature T_c.

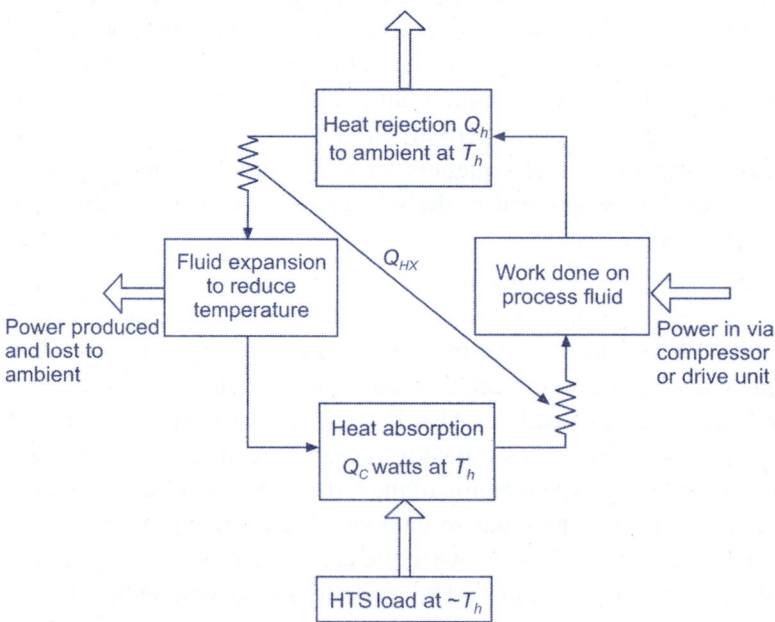

Fig. 13.1 Block Diagram of Cryogenic Refrigerator

Refrigerator Effectiveness

A refrigerator is a heat engine operated in reverse. Figure 13.2 shows a schematic diagram of a refrigerator. The heat sucked out of the cold reservoir is Q_c, while the electrical energy supplied is W. The waste heat, Q_h is dumped into the environment.

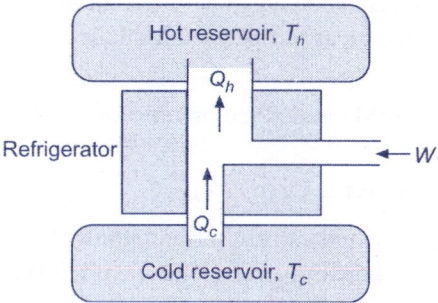

Fig. 13.2 Schematic of Refrigerator

To define the effectiveness or efficiency of the refrigerator we have to refer the benefit/cost ratio, where the benefit is Q_c while the cost is W. This ratio is called the coefficient of performance (COP).

$$\text{COP} = \frac{\text{benefit}}{\text{cost}} = \frac{Q_c}{W}.$$

Coefficient of Performance (COP) which is a measure of performance of the refrigerator can be defined as energy removed from the source divided by work required to remove this amount of energy. Just as for a heat engine, one can use the first and second laws to derive a limit on the COP in terms of the temperatures Th and Tc. The first law can be written as $Q_h = Q_c + W$, so

$$\text{COP} = \frac{Q_c}{Q_h - Q_c} = \frac{1}{Q_h/Q_c - 1}.$$

There is no obvious upper limit on COP; in particular, the first law allows the COP to be greater than 1. Meanwhile, the second law stipulates that the entropy dumped into the hot reservoir must be at least as much as the entropy absorbed from the cold reservoir.

$$\frac{Q_h}{T_h} \geq \frac{Q_e}{T_c}, \quad \text{or} \quad \frac{Q_h}{T_c} \geq \frac{T_h}{T_c},$$

Plugging this inequality into equation for COP gives

$$\text{COP} \leq \frac{1}{T_h/T_c - 1} = \frac{T_c}{T_h - T_c}.$$

To make an ideal refrigerator with the maximum possible COP, one can again use a Carnot cycle, this time operated in reverse. In order to make the heat flow in the opposite direction, the working substance must be slightly hotter than T_h while heat is being expelled and slightly colder than T_c while heat is being absorbed.

The term Q/m is termed as refrigeration effect or the energy absorbed by the refrigerant (working substance) per unit mass of the refrigerant. The COP of an ideal refrigeration system is denoted by COPi. To compare performance of various refrigerators as against the ideal refrigerator we define a term known as Figure of Merit (FOM).

Figure of merit (FOM) is defined as ratio of COP of actual refrigerator to COP of ideal refrigerator.

$$\text{FOM} = \text{COP} / \text{COP}i$$

FOM lies between zero and unity. A refrigerator with FOM near 1 is considered a very good refrigerator, whereas the refrigerator with FOM near zero is considered a very poor refrigerator.

Thermodynamically Ideal Refrigeration System: Isothermal Source

The thermodynamically ideal refrigeration system is important to form the ideal basis of comparison with other practical refrigeration systems. In case of refrigerators, two types of low temperature sources can be used. When the liquid is evaporated energy is added in an isothermal form, whereas if a cold gas like helium is used for liquefaction, then the energy is added in an isobaric form. Thus we have two types of refrigerators, isothermal source refrigerator or isobaric source refrigerator. In both cases the sink is the atmosphere which is isothermal in nature. The source is the application or area to be cooled. The thermodynamically ideal refrigeration system is the reverse Carnot cycle or Carnot refrigeration system as shown in Figure 13.3.

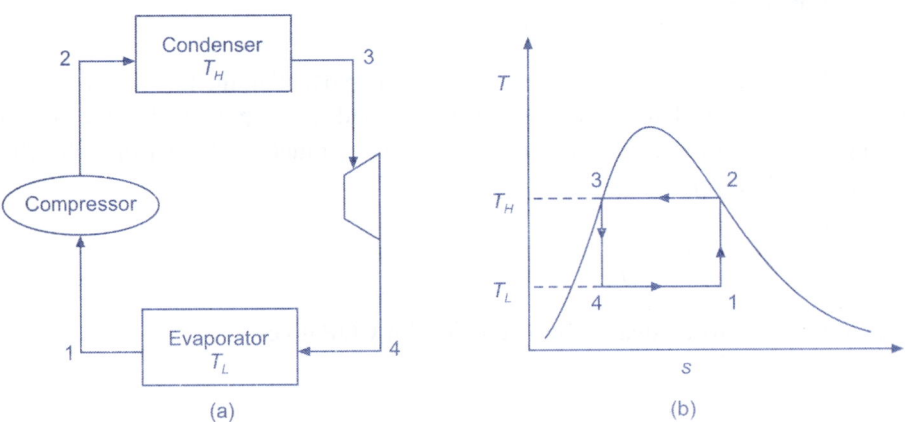

Fig. 13.3 Carnot Refrigeration System and its Representation on T-S Diagram.

The processes in the Carnot Refrigeration System are

1-2: Isentropic compression from state 1(wet vapor) to state 2(saturated vapor). The refrigerant is compressed isentropically from source to sink temperature.

2-3: Heat rejection (Q_H) in the condenser. Here the energy is rejected to sink to maintain the refrigerant temperature constant.

3-4: Isentropic expansion from state 3 (saturated liquid). The refrigerant is expanded isentropically from sink to source temperature.

4-1: Heat absorption (Q_L) in the evaporator. Here the energy is absorbed by the refrigerant from the source to maintain the refrigerant temperature constant.

The Carnot cycle thus contains two isothermal and two isentropic processes all of which are reversible in nature. The net heat transfer in the Carnot cycle is given by

$$Q_{net}/m = -(T_h - T_c)(s_1 - s_2) = W_{net}/m$$

The energy added isothermally is given by

$$Q_a/m = T_c(s_1 - s_2)$$

Thus, the COP of Carnot refrigerator is given by

$$COP = T_c/(T_h - T_c)$$

Thus COP of Carnot cycle is independent of the refrigerant used. Furthermore between two temperature limits T_h and T_c, the Carnot refrigerator is the most efficient. Even for the thermodynamically ideal refrigerator, large amount of work is required to maintain low temperatures.

Thermodynamically Ideal Refrigeration System: Isobaric Source

In refrigerators working on the cold gas, energy is absorbed at varying temperatures instead of a constant temperature as in the Carnot refrigerator. The thermodynamically ideal isobaric source refrigeration system can be represented on the T-S diagram as shown in Figure 13.4. This cycle is a useful benchmark for comparison with real systems which absorb heat at constant pressure.

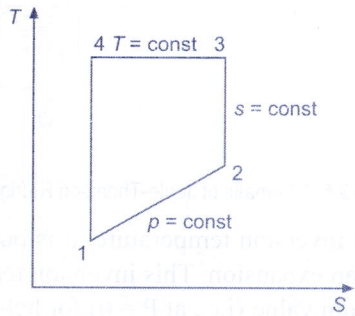

Fig. 13.4 T-S Diagram for Isobaric Source Ideal Refrigeration System

We can denote sink temperature as T_o and lower source and higher source temperatures as T_1 and T_2 respectively. The heat energy is added to the refrigerant at constant pressure whereas the heat rejection is isothermal in nature.

Heat rejected = $Q_r = -mT_o(s_2 - s_1)$

Heat absorbed = $Q_a = \int mT\,ds = m(h_2 - h_1)$

The net work = $W_{net} = Q = Q_r + Q_a = -m[T_o(s_2 - s_1) - (h_2 - h_1)]$

13.6 Cryogenics

For the isobaric source refrigerator COP = $-Q_d/W_{net}$

Thus COP = $(h_2 - h_1) / [T_o(s_2 - s_1) - (h_2 - h_1)]$

$h_2 - h_1 = C_p(T_2 - T_1)$ and $s_2 - s_1 = C_p \ln(T_2/T_1)$

Substituting the above two relations in expression for COP

COP = $(T_2 - T_1)/[T_o \ln(T_2/T_1) - (T_2 - T_1)]$

Thus one can observe that the COP of the isobaric source refrigerator depends only on the temperatures and is independent of the properties of the refrigerant.

Joule-Thomson (JT) Refrigeration

In the early 1850s, two British scientists J.P. Joule and William Thomson (later Lord Kelvin) performed experiments on the expansion of gases. They expanded gases from a high pressure through a porous plug under adiabatic conditions. Under these conditions, the expansion is also isenthalpic and the temperature of the gas increases or decreases depending on the sign of the Joule-Thomson coefficient $\mu = (\partial T/\partial P)_H$.

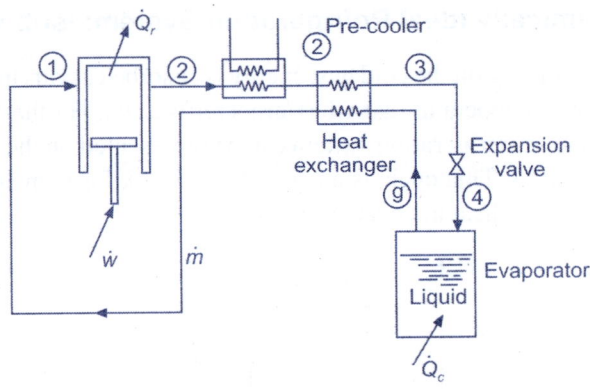

Fig. 13.5 Schematic of Joule-Thomson Refrigerator

Below the so-called inversion temperature, μ is positive and cooling can be obtained as the result of an expansion. This inversion temperature depends on the pressure, and the maximum value (i.e., at P = 0) for helium is 40 K, for hydrogen 205 K, for neon 250 K, and for nitrogen 621 K. A Joule-Thomson Refrigerator utilizes this effect and consists of a compressor, a control and filter unit, a counterflow heat exchanger (CFHX), and a Joule-Thomson expansion stage including a flow restriction (JT-valve). For an ideal CFHX, the cooling power of a JT cooler is given by the product of the mass-flow rate and the difference in specific enthalpy between the high and low-pressure sides at the warm end of the CFHX. Refer Figure 13.5 for a schematic of the JT refrigerator and Figure 13.6 for the T-S diagram of the Joule-Thomson cooling cycle.

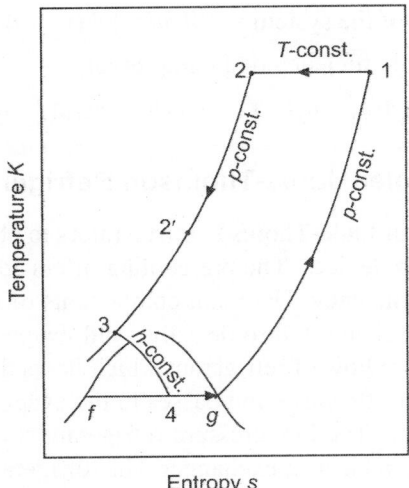

Fig. 13.6 Joule-Thomson Thermodynamic Cycle

Lowering the warm-end temperature i.e., by pre-cooling the fluid before it enters the CFHX can increase this enthalpy difference, and thus the cooling power. Very large enthalpy changes can be obtained by pre-cooling to a level such that the fluid is liquefied before the expansion. In that case, the expansion valve is usually referred to as a throttling valve rather than a JT-valve. The Joule-Thomson refrigerators produce cooling when the high-pressure gas expands through a flow impedance (orifice, valve, capillary, porous plug), often referred to as a JT-valve. The expansion occurs with no heat input or production of work, thus, the process occurs at a constant enthalpy. The heat input occurs after the expansion and is used to warm up the cold gas or to evaporate any liquid formed in the expansion process. In an ideal gas the enthalpy is independent of pressure for a constant temperature, but real gases experience an enthalpy change with pressure. Thus, cooling in a JT expansion occurs only with real gases and at temperatures below the inversion curve. In fact, for a given pressure change the amount of cooling increases as the temperature is lowered and reaches a maximum around the critical point.

Applying first law of thermodynamics we get expression for the heat transfer as $Q_a = m(h_1^* - h_2)$, where h_1^* = actual enthalpy of fluid leaving the heat exchanger.

The heat exchanger effectiveness is given by

$$\epsilon = (h_1^* - hg)/(h_1 - hg)$$

Thus the heat transfer in terms of effectiveness becomes

$$Q_a/m = (h_1 - h_2) - (1 - \epsilon)(h_1 - hg).$$

We can conclude from this expression that for Joule-Thomson refrigerator, the working gases need to be pre-cooled ($h_1 > h_2$) to ensure a positive refrigeration effect. The refrigerator operation is dependent on the heat exchanger effectiveness and below a certain limit of the effectiveness, the refrigerator becomes ineffective.

13.8 Cryogenics

Work requirement of the system $= -W/m = [T_2(s_2 - s) - (h_1 - h_2)]/\eta_c$,
where η_c = Thermal efficiency of a compressor.

$COP = -Q_d/W = \eta_c[(h_1 - h_2) - (1 - \epsilon)(h_1 - h_g)]/[T_2(s_2 - s) - (h_1 - h_2)]$

Cascade or Pre-cooled Joule-Thomson Refrigerator

Pre-cooling is required in Joule-Thomson refrigerators to obtain low temperatures without using expansion devices. The pre-cooling offers advantages of long life, low noise and higher efficiency. The main components of the JT refrigerator are the compressor, heat exchanger, throttle valve and evaporator. The compressor supplies the high pressure flow of refrigerant which flows through heat exchanger and expands through throttle valve and passes to the evaporator where it absorbs heat from the application. The low pressure refrigerant stream is circulated back to the compressor through the heat exchanger. The refrigeration capacity of the JT refrigerator is determined by the enthalpy difference between high pressure and low pressure refrigerant streams. To optimize the refrigeration effect an additional cold source (pre-cooling stage) can be used. Refer Figure 13.7 for the pre-cooled JT refrigerator and its T-h diagram.

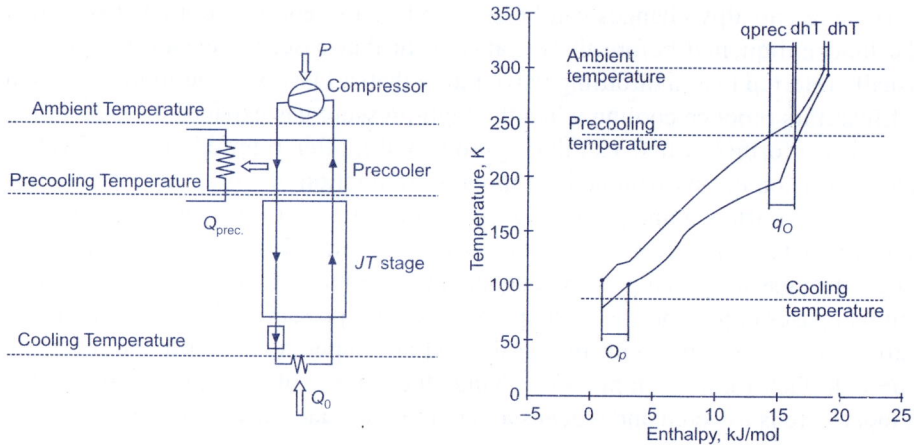

Fig. 13.7 Pre-cooled Joule-Thomson Refrigerator and T-h Diagram

The components of the pre-cooled JT refrigerator are the same as the JT-refrigerator with the addition of the 3 stream heat exchanger (pre-cooler). The high pressure refrigerant stream is cooled by an additional cold stream before entering the JT stage. The additional cooling capacity of the low pressure refrigerant stream after exiting the JT stage is also used in the pre-cooler. The specific cooling capacity q_o of this cycle is given by

$$q_o = q_p + \Delta h$$

where q_p is the cooling capacity of precooler and Δh is the enthalpy difference between high pressure and low pressure refrigerant stream. The T-h diagram shows that the high pressure refrigerant is in gaseous form and value of Δh is normally small. Thus for pre-cooling the refrigeration capacity should be similar to that of the main JT stage. Furthermore power consumption requirements for pre-cooling stage are equal to the cooling capacity of the pre-cooler.

Using first law of thermodynamics, the refrigeration effect (Q) of pre-cooled JT refrigerator can be written as

$$Q/m = (h_1 - h_2) + z(h_a - h_b)$$

Where mass flow rate ratio (z) is written as

$z = mp/m$ = mass flow rate of pre-coolant/mass flow rate of main refrigerant.

$h_1 - h_2$ = enthalpy difference of main refrigerant

$h_a - h_b$ = enthalpy difference of pre-coolant.

Futher low temperatures can be attained by using multiple pre-coolers or cascades.

Refrigerator Optimization

The performance analysis and optimization of a refrigerator is carried out by taking the cooling load density, i.e. the ratio of cooling load to the maximum specific volume in the cycle, as the optimization objective using finite-time thermodynamics or entropy generation minimization. Analytical relationships between cooling load density and pressure ratio, as well as between coefficient of performance (COP) and pressure ratio can be derived. The irreversibilities considered in the analysis include the heat transfer losses in the hot- and cold-side heat exchangers and the regenerator, the non-isentropic compression and expansion losses in the compressor and expander, and the pressure drop losses in the piping. The comparison of the cycle performances under maximum cooling load density and maximum cooling load conditions can also be performed. The optimal performance characteristics of the cycle are obtained by optimizing the pressure ratio of the compressor, and searching the optimum distribution of heat conductances of the hot and cold-side heat exchangers and regenerator for the fixed total heat exchanger inventory. The influences of the effectiveness of the regenerator as well as the hot and cold-side heat exchangers, the efficiencies of the expander and the compressor, the pressure recovery coefficient, and the temperature ratio of the heat reservoirs on the cooling load density and COP are key parameters in refrigerator optimization.

The heat exchanger effectiveness may be determined by selection of heat exchanger surface area. If the effectiveness near unity is chosen, the heat exchanger becomes bulky and expensive. If the effectiveness is low then it results in higher refrigerant flow rates and high compressor costs. It is essential to determine an optimum value for effectiveness of heat exchangers keeping in mind both the constraints explained above.

For a typical refrigerator with refrigerating effect Q, the compressor costs are proportional to the power requirements (W)

$$C_c = \text{compressor cost} = C_1.W = m\, C_1.W/m$$

13.10 Cryogenics

The heat exchanger costs are proportional to surface area of heat exchanger (A)
$$C_e = C_2.A$$
Total cost of system $\quad = C_t = C_c + C_e$
Optimum cost is calculated as $d/d_t(C_t) = 0$
Thus $C_1.W/m.(d_m/d_t) + C_2.(d_A/d_t) = 0$ ...(13.1)
The parameter $\quad i = 1 - \epsilon$ and $H = (h_1 - h_2)/(h_1 - h_g)$
Mass flow rate $\quad = m = Q/(h_1 - h_g)(H - i)$
Thus $\quad d_m/d_t = Q/(h_1 - h_g)(H - i)^2$ ...(13.2)
Also if A = area of heat exchanger
$$d_A/d_t = (m_c/U)/i(1 - C_r - C_r i)$$ ...(13.3)
where c = mean specific heat, U = universal heat transfer coefficient, C_r = capacity ratio

Solving eqn. (13.1) by substituting equn. (13.2) and (13.3) we get optimum effectiveness as
$$i \text{ optimum} = C_r H^2/[2 C_r H + 1 - C_r]$$

Thus the refrigerator optimization can be achieved with selection of optimum heat exchanger effectiveness.

Expansion Engine Refrigeration or Claude Refrigeration

The expansion cycle of refrigeration or Claude closed loop refrigeration system includes a compressor for compressing low pressure gas refrigerant to a high pressure and high temperature gas, a condenser for condensing the high pressure gas refrigerant to a liquid, a reservoir for holding the liquid refrigerant discharged from the condenser, and an evaporator for evaporating the liquid refrigerant at a low pressure to produce cooling. An expansion engine or turbo-expander is disposed between the condenser and the reservoir. The expansion of the liquid leaving the condenser drives the expansion engine. The expansion engine in turn drives, at least partially, a secondary compressor disposed between the reservoir and the condenser for discharging gas from the reservoir to the condenser. The expansion of the liquid refrigerant and the removal of the gas from the reservoir provides a low energy refrigerant to the evaporators, thus improving the overall efficiency of the refrigeration system. Further efficiency improvement is achieved by subcooling the liquid refrigerant in the condenser at all ambient temperatures before it leaves the condenser. A micro-controller control circuit controls the discharge of the liquid refrigerant from the condenser as a function of the temperature difference between the liquid leaving the condenser and the ambient air entering the condenser. The micro-controller control circuit monitors this temperature difference and in response thereto reduces the flow of the liquid refrigerant out of the condenser if this difference is greater than a preselected value and increases the flow if this difference is less than a preselected value. This method achieves the greatest amount of subcooling while minimizing the amount of refrigerant necessary for the subcooling; and, additionally, ensures that the liquid is entering the expansion engine or turbo expander. Refer Figure 13.8 for the schematic of a turbo expander.

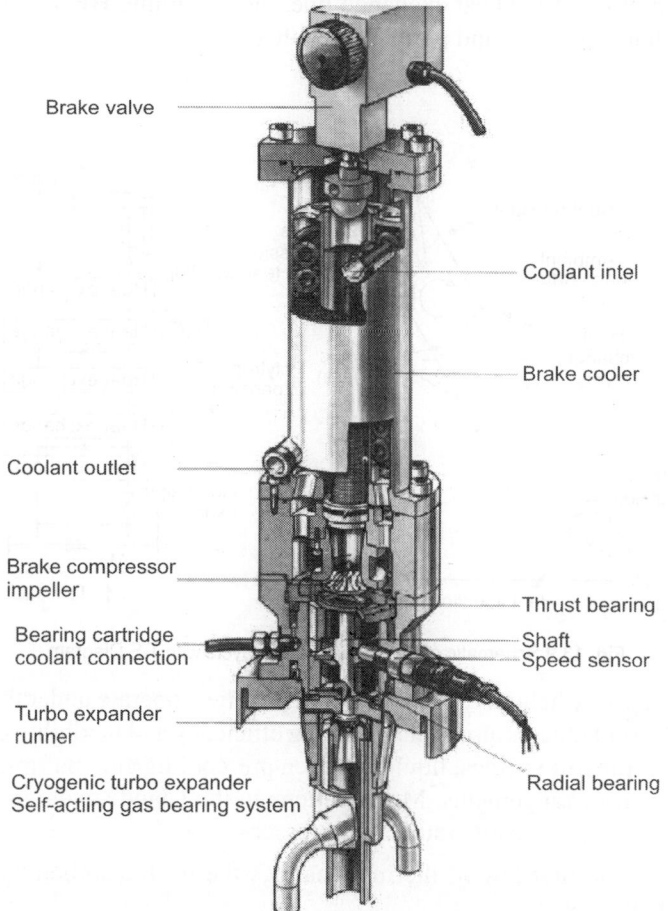

Fig. 13.8 Cryogenic Turbo-expander from Linde

For real gases, enthalpy depends both on temperature and pressure, so that isenthalpic expansion can produce warming or cooling, depending upon the Joule-Thomson coefficient. Joule-Thomson expansion generates substantial entropy. Its relative inefficiency with respect to adiabatic expansion is however accepted in view of the simplicity of its implementation, particularly when it results in partial condensation of the stream entailing two-phase flow conditions which would be difficult to handle in an expansion turbine. These elementary cooling processes are combined in practical cycles, a common example for helium refrigeration is provided by the Claude cycle. A schematic of two-pressure, two-stage Claude cycle along with its T-S diagram is shown in Figure 13.9. The gaseous helium is compressed to high pressure in a lubricated screw compressor and is re-cooled to room temperature in water-coolers and is dried and purified from oil aerosols. It is

13.12 Cryogenics

then sent to the HP side of the heat exchange line where it is refrigerated by heat exchange with the counter-flow of cold gas returning on the LP side. Part of the flow is tapped from the HP line and expanded in the turbines before escaping to the LP line. At the bottom of the heat exchange line, the remaining HP flow is expanded in a Joule-Thomson valve and partially liquefied.

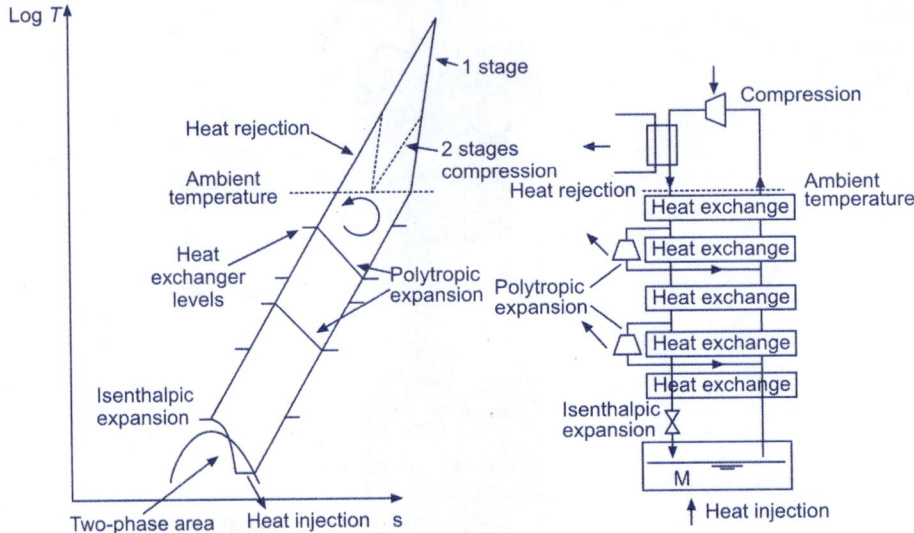

Fig. 13.9 Schematic of two Stage Claude Cycle with T-S Diagram

Large-capacity helium refrigerators and liquefiers operate under this principle, however with refinements aimed at improving efficiency and flexibility of operation, such as three- pressure cycles, liquid nitrogen pre-cooling and multiple turbines in series or parallel arrangements. Modifications in the Claude cycle include use of wet expanders and low temperature compressors.

Applying the first law of thermodynamics the heat absorbed by refrigerant can be calculated as

$$Q/m = (h_1 - h_2) + x\, \eta_a (h_3 - h_e),$$

Where m = mass flow rate, x = expander mass flow rate ratio = m_e/m, η_a = adiabatic efficiency, $(h_3 - h_e)$ = enthalpy drop at end of expansion.

The work requirement of the system is given by

$$W/m = [T_2 (s_1 - s_2) - (h_1 - h_2)]/\eta_c - x\, \eta_c \eta_a (h_3 - h_e),$$

where m = mass flow rate, x = expander mass flow rate ratio = m_e/m, η_a = adiabatic efficiency, η_c = compressor efficiency, $(h_3 - h_e)$ = enthalpy drop at end of expansion.

Regenerators

The regenerator is an internal heat exchanger and temporary heat store placed between the hot and cold spaces such that the working fluid passes through it first in one direction then the other. Its function is to retain within the system that heat which would otherwise be exchanged with the environment at temperatures intermediate to the maximum and minimum cycle temperatures, thus enabling the thermal efficiency of the cycle to approach the limiting Carnot efficiency defined by those maxima and minima. The primary effect of regeneration in a cryogenic refrigerator is to increase the thermal efficiency greatly by 'recycling' internal heat which would otherwise pass through the equipment irreversibly. As a secondary effect, increased thermal efficiency promises a higher refrigeration effect from a given set of hot and cold end heat exchangers. In practice this additional refrigeration effect may not be fully realized as the pumping losses inherent in practical regenerators tend to have the opposite effect. The regenerator works like a thermal capacitor. The ideal regenerator has very high thermal capacity, very low thermal conductivity parallel to fluid flow, very high thermal conductivity perpendicular to fluid flow, almost no volume, and introduces no friction to the working fluid. The design challenge for a regenerator is to provide sufficient heat transfer capacity without introducing too much additional internal volume ('dead space') or flow resistance, both of which tend to reduce efficiency. These inherent design conflicts are one of many factors which limit the COP of refrigerators. A typical design is a stack of fine metal wire meshes, with low porosity to reduce dead space, and with the wire axes perpendicular to the gas flow to reduce conduction in that direction and to maximize convective heat transfer. Refer to Figure 13.10 for the photograph and internal structure of regenerators and Figure 13.11 for the schematic of regenerator.

Fig. 13.10 Regenerator Photograph with Internal Structures

In a regenerator the heat exchange space is occupied alternatively by hot and cold fluid. Energy is stored and released from the regenerator packing material referred to as matrix. The temperature of regenerator is a function of position of fluid in regenerator and time. The steady state operation of the regenerator is known as periodic flow and is achieved after a long operation time. Applying first law to a regenerator gives the equation

$$h(T_s - T)(A/L)d_x - m\, C_p(d_T/d_x)d_x = \rho(V_g/L)\, C_p(d_T/d_t)d_x \qquad ...(13.4)$$

where h = convective heat transfer coefficient, T_s = matrix temperature, T = gas temperature, A = heat transfer area of matrix, L = length of regenerator, m = mass flow rate of gas through regenerator, C_p = specific heat of gas at constant pressure, ρ = density of gas, V_g = void volume within regenerator.

13.14 Cryogenics

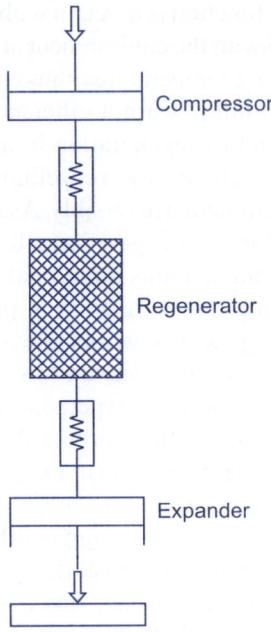

Fig. 13.11 Schematic of Regenerator

The first term on the left side in Equation Z represents heat transfer from the gas, the second term on the left side represents change in enthalpy of gas and the term on the right of the equation represents energy stored in the element of gas.

For solid material within the regenerator the heat transfer equation can be written as

$$h(T_s - T)(A/L)\, d_x = (M_s/L)\, C_s (d_T/d_t) d_x$$

where M_s = mass of matrix material and C_s is specific heat of matrix material. It is assumed that there is negligible heat transfer in longitudinal direction.

For regenerator, matrix capacity ratio (C_m) is defined as

$$C_m = M_s\, C_s / P_o\, C_{min}$$

where P_o = regeneration period and C_{min} is minimum specific heat capacity.

For regenerator, effectiveness is defined as

$$\varepsilon = Q / C_{min}(T_h - T_c).$$

Solved Problems

Problem 1. Determine the ideal COP for an isobaric source refrigerator operating between a sink temperature of 300 K and minimum and maximum source temperatures of 110 K and 180 K. the working fluid is gaseous nitrogen and source pressure is 1.013 MPa

Solution: Given data, T_o = 300 K, T_1 = 110 K and T_2 = 180 K

$$COP = (T_2 - T_1)/[T_o \ln(T_2/T_1) - (T_2 - T_1)]$$
$$COP = (180 - 110)/[(300) \ln(180/110) - (180 - 110)]$$
$$COP = 70/77.78 = 0.9.$$

Thus the COP is 0.9.

Problem 2. Determine the refrigeration effect, work requirement, COP and figure of merit for a Joule-Thomson refrigerator operating from 300K and 101.3 kPa to 10.13 MPa. The overall efficiency of compressor is 75% and heat exchanger effectiveness is 0.96. Assume working fluid as nitrogen.

Solution:

From T-S diagram of nitrogen as shown in Figure 13.12 the following quantities are obtained:

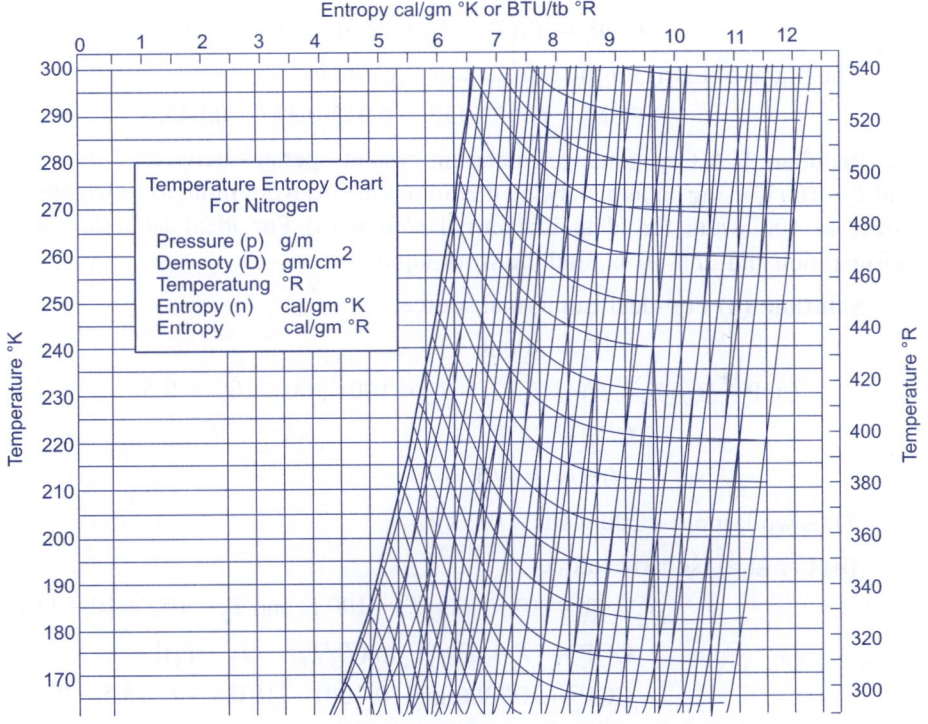

Fig. 13.12 T-S Diagram for Nitrogen

h_1 = 462 J/g at 300 K and 101.3 KPa (1 atm)

h_2 = 444 J/g at 300 K and 10.13 MPa (100 atm)

s_1 = 4.42 J/g K

s_2 = 3.0 J/g K.

hg = 229 J/g at 101.3 KPa(1 atm) and 77.36 K – Saturated vapor

hf = 29 J/g at 101.3 KPa(1 atm) – Saturated liquid.

13.16 Cryogenics

Work requirement for the refrigerator is given by

$$-W/m = [T_2(s_2 - s) - (h_1 - h_2)]/\eta_c$$
$$-W/m = [(300)(4.42 - 3.0) - (462 - 444)]/0.75 = 544 \text{ J/g}.$$

Thus work requirement = 544 J/g.

Refrigeration effect for the refrigerator is given by

$$Q_d/m = (h_1 - h_2) - (1 - \epsilon)(h_1 - h_g)$$
$$Q_d/m = (462 - 444) - (1 - 0.96)(462 - 229) = 8.68 \text{ J/g}$$

Thus refrigeration effect = 8.68 J/g

Coefficient of performance for the refrigerator is given by

$$COP = -Q_d/W = 8.68/544 = 0.015.$$

Coefficient of performance for the ideal carnot refrigerator is given by

$$COP_i = T_c/(T_h - T_c) = 77.36/(300 - 77.36) = 0.347.$$

The figure of merit for the refrigerator is given by

$$FOM = COP/COP_i = 0.015/0.347 = 0.045.$$

Problem 3. A Cryogenic refrigeration system operates between $T_o = 300$ K and $T_r = 100$ K using air as a working medium. Find (a) work input for refrigerating effect of 100 watts for ideal isothermal source (b) For ideal isobaric source refrigeration and $\Delta t = 10°C$, find the work input.

Solution: (a) For ideal isothermal source

$$T_o = 300 \text{ K}$$
$$COP = T_h/(T_h - T_c) = T_o/(T_o - T_r) = 100 / 300 - 100 = 0.5$$
$$COP = Q/W$$
$$0.5 = 100/W$$

Thus work input = 200 watts = 0.2 kW

(b) For ideal isobaric source

$$T_2 - T_1 = 10, \text{ Thus } T_1 = 100 \text{ K and } T_2 = 100 + 10 = 110 \text{ K}$$
$$COP = (T_2 - T_1)/[T_o \ln(T_2/T_1) - (T_2 - T_1)]$$
$$COP = 10/[300 \ln(110/100) - (10)] = 10 / 18.5 = 0.54$$
$$COP = Q/W$$
$$0.54 = 100/W$$

Thus work input = 185 watts.

References

1. Cryogenic Systems – Randall Baron.
2. Report on Cryogenics – Lebrun, CERN (2007).
3. Cryogenic Engineering – B.A. Hands.
4. Cryogenics assessment report by University of Wisconsin (2005).
5. Presentation by CERN on cryogenic engineering.
6. Internet website of Wikipedia on Cryogenics.
7. Paper on Joule-Thomson Cryocoolers – Alexeev et.al. Dresden University.
8. US Patent No. 5157931.

Questions

Q.1. What are cryogenic refrigerators? Describe the basic operating principle for cryogenic refrigerators with a block diagram.

Q.2. Describe a thermodynamically ideal cryogenic refrigeration system.

Q.3. Write a short note on regenerators.

Q.4. Describe the term effectiveness for a cryogenic refrigerator.

Q.5. Explain the process of refrigerator optimization.

Q.6. What is an expansion engine in cryogenics?

Q.7. Describe the Joule-Thomson refrigeration effect with a neat sketch.

Q.8. Determine the refrigeration effect, work requirement, COP and figure of merit for a Joule-Thomson refrigerator operating from 350 K and 1 atm to 120 atm. The overall efficiency of compressor is 85% and heat exchanger effectiveness is 0.9. Assume working fluid as nitrogen.

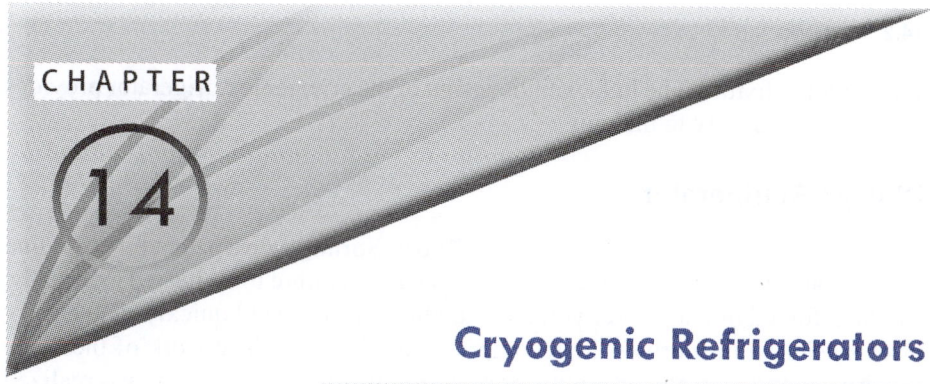

Cryogenic Refrigerators

INTRODUCTION

Cryogenic refrigerators are required to provide cooling duties at several temperatures or in several temperature ranges. Cryogenic refrigerators and liquefiers have the same components however the cryogenic liquid is evaporated in the refrigerator, whereas it is collected in the liquefiers for application use. Typical cryogenic refrigerators and cryogenic storage systems encompass everything from a simple liquid nitrogen dewar to large-capacity cryogenics refrigerators that can store up to 80,000 samples. Cryogenic refrigerators are also designed to safely transport materials at cryogenic temperatures. The controlled-rate freezer systems can be programmed to meet any freezing protocol requirements. Figure 14.1 shows some typical cryogenic refrigerators.

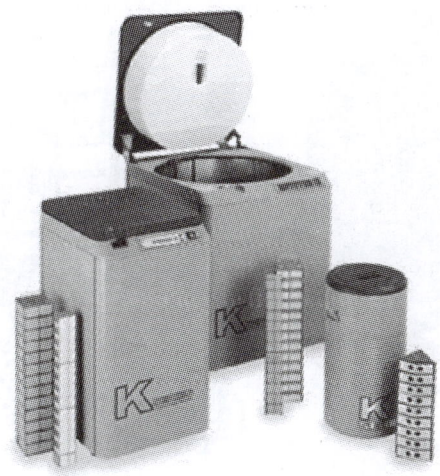

Fig. 14.1 Typical Cryogenic Refrigerators

Modern cryogenic refrigerators use superior insulation materials to extend holding time of the refrigerator. The durability of the refrigerator is increased by providing a ribbed body of high strength aluminum and using a rugged necktube with wear resistant paints. Additional features include a canister indexing system

14.2 Cryogenics

and a guiding fixture. The light weight design of the cryogenic refrigerator provides necessary portability to the unit.

Phillips Refrigerator

In 1816, Scottish engineer and theologian Robert Stirling won a patent for an engine design that called for a volume of gas at a low temperature to be heated externally, and then forced into a greater volume and allowed to expand quickly. The energy of expansion in turn drives the recompression of the gas. The work of the cycle can drive motion, create electricity, or perform other tasks. In 1873, it was realized that one can use a proper coolant that absorbs heat at the compression stage and then that heat can be removed, via an exchanger, from a chamber. The removal of heat regulates the chamber's temperature at a desired low point. This lead to the development of the Stirling cycle refrigerators. Refrigerator designs based on the Stirling cycle offer significant efficiencies, and need no fluorocarbons. Because a Stirling machine needs fewer parts for cooling, these refrigerators can be as small as a few liters. A Stirling refrigerator would weigh about one-third as much as an equivalent rankine system. Free-piston Stirling refrigerators operate efficiently at all levels of demand because they can modulate their capacity to match any requirement.

The Phillips refrigerator works on the Stirling cycle and was first constructed in 1864. The photograph and schematic of the Phillips refrigerator is shown in Figure 14.2.

Fig. 14.2 Phillips Refrigerator

The Phillips refrigerator consists of a cylinder enclosing a piston, a displacer and regenerator. The piston compresses the gas whereas the piston moves the gas from one chamber to another, ideally without changing the volume of the gas. The heat exchange during the constant volume process is carried out in the regenerator. The sequence of operation of the system as shown in Figure 14.3 is as follows.

Process 1-2: Gas is isothermally compressed and rejects heat to the high temperature sink i.e. the surroundings.

Process 2-3: Hot gas is passed through the regenerator by motion of dispacer. The gas is cooled at constant volume and the heat removed is stored in the regenerator matrix.

Process 3-4: Gas is isothermally expanded and absorbs heat from the low temperature source.

Process 4-1: Cold gas is passed through the regenerator by motion of dispacer. The gas is heated at constant volume as the heat stored in the regenerator matrix is transferred back to the gas.

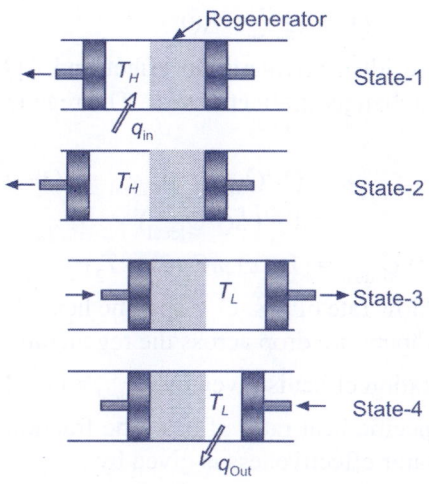

Fig. 14.3 Operation of Phillips Refrigerator

Heat rejected by Phillips refrigerator = $Q_r = mT_1 (s_2 - s_1)$

Heat absorbed by Phillips refrigerator = $Q_a = mT_3 (s_4 - s_3)$

Network required by Phillips refrigerator = $W_{net} = Q_a + Q_r$

$COP = -Q_a/W_{net} = -T_3(s_4 - s_3)/[T_3(s_4 - s_3) + T_1(s_2 - s_1)]$

Thus $COP = T_3/[T_1 (s_1 - s_2)/(s_4 - s_3) - T_3]$

For an ideal gas $(s_1 - s_2) = (s_4 - s_3)$, Thus

COP for ideal Phillips refrigerator = $T_3/(T_1 - T_3)$.

This is same as that for the Carnot refrigerator and thus the FOM = 1 for ideal Phillips refrigerator. The actual Phillips refrigeartor will have a lower COP of around 0.3 due to frictional energy dissipation, pressure drop in regenerator and heat losses. The Phillips refrigerator is marketed by the Phillips company of Netherlands and is used for cooling electronic components, gas separation and in air liquefaction units.

14.4 Cryogenics

Importance of Regenerator for Phillips Refrigerator

The successful performance of the Phillips refrigerator depends on the effectiveness of regenerator used in the system. There are several factors which affect the regenerator effectiveness. The primary requirements of a good regenerator are 1) It should be constructed from a material having large thermal capacity, 2) Period of fluid switching should be small and frequency of regeneration should be large, 3) Heat transfer area and heat transfer coefficient should be large and 4) Mass flow rate of gas should be small. In case of Phillips refrigerator, the matrix material is made up of fine wire having large thermal capacity and heat transfer area. If the regenerator is not fully effective then some of the heat absorbed from the application (refrigeration effect) is lost due to wastage of energy in cooling the gas to the source temperature. Thus actual heat energy absorbed from the source (Q_a) is given by

$$Q_a = Q_{ideal} - \Delta Q$$

where Q_{ideal} is the ideal refrigeration effect and ΔQ is the energy lost in overcoming the heat exchanger ineffectiveness. The regenerator effectiveness can be written as

$$\varepsilon = Q_a/Q_{ideal} = (Q_{ideal} - \Delta Q)/Q_{ideal}$$
$$= 1 - (\Delta Q/Q_{ideal})$$

Then $\Delta Q = (1 - \varepsilon) Q_{ideal} = (1 - \varepsilon) mC_v(T_2 - T_3)$

Where m = mass flow rate of gas, C_v = specific heat of gas at constant volume and $(T_2 - T_3)$ is the temperature drop across the regenerator.

The ideal refrigeration effect is given by $Q_{ideal} = (\gamma - 1) mC_v T_3 \ln (v_4/v_3)$

Where γ is the specific heat ratio of gas. The fraction of refrigeration effect wasted due to regenerator effectiveness is given by

Thus $\Delta Q / Q_{ideal} = [(1 - \varepsilon) (T_2/T_3 - 1)]/[(\gamma - 1) \ln (v_4/v_3)]$.

As a specific case if helium is assumed as a working fluid with $\gamma = 1.67$ and $v_4/v_3 = 1.5$ along with $T_2/T_3 = 300/78$, then the expression for $\Delta Q/Q_{ideal}$ becomes,

Thus $\Delta Q / Q_{ideal} = [(1 - \varepsilon) (300/78 - 1)]/[(1.67 - 1) \ln(1.5)] = 10.48(1 - \varepsilon)$.

Thus one can conclude that entire refrigeration effect will be wasted if the regenerator effectiveness is

$$\varepsilon_{min} = 1 - (1/10.48) = 0.905.$$

Thus regenerator effectiveness is very critical for good performance of Phillips refrigerator.

Vuilleumier (VM) Refrigerator

The Vuilleumier cycle has been around since 1918 and is by definition a combination of a heat engine and a refrigeration cycle. Two distinct features of the cycle are that it is a thermo-compression cycle instead of a mechanical compression cycle and that the engine and refrigerator share the same working fluid. The basic Vuilleumier refrigerator machine is shown in Figure 14.4 and its T-S diagram is shown in Figure 14.5.

Heat is input at the high temperature heat exchanger to increase the pressure of the working fluid, normally helium. Heat is rejected, to the ambient in a cooling application or to the conditioned space for a heat pump, through the two intermediate heat exchangers. Cooling is performed for an air conditioner at the low temperature heat exchanger. The working fluid is moved between hot and cold volumes in the machine through the use of two displacers operated by a crank mechanism 90 degrees out of phase (mechanical energy is input to rotate the crank and move the displacers). The displacers are operating only against the pressure drops across the heat exchangers and regenerators and friction along the tube walls so there is a low pressure ratio. The low pressure ratio results in a low specific volume for the basic Vuilleumier cycle.

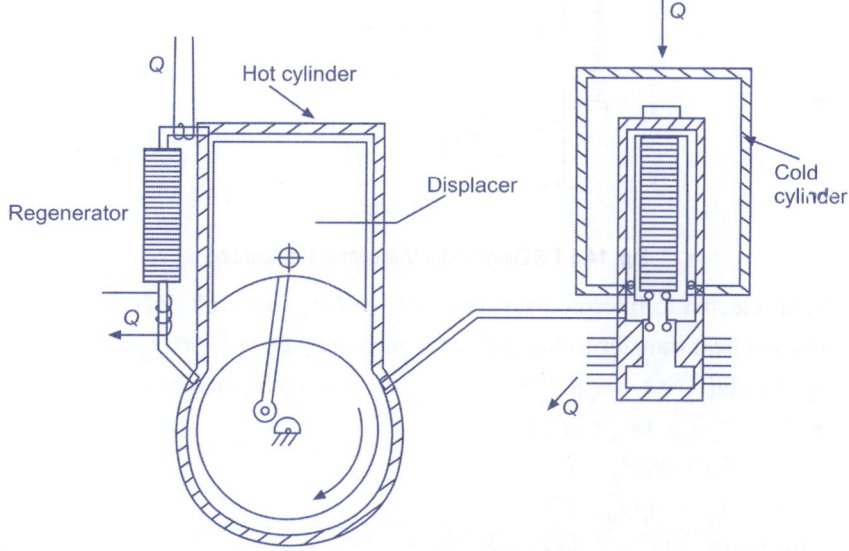

Fig. 14.4 Vuilleumier Refrigerator

While the basic cycle relies on thermal compression, compounded machines may use mechanical compression. These machines closely resemble duplex Stirling cycles, the primary difference being that the working fluid is shared between the engine and refrigerator segments in the Vuilleumier machine while the working fluid in the engine-warm space of the duplex Stirling is separated from the working fluid in the refrigerator warm space. The Vuilleumier cycle transfers work between components through transfers of the working fluid. The duplex Stirling cycle transfers work from the engine to the refrigerator segments through mechanical means (movement of the piston). A Vuilleumier cycle machine can be configured either as a kinematic or a free-piston machine; the free piston Vuilleumier system is expected to have lower costs than kinematic version because of the reduction in equipment size and weight.

14.6 Cryogenics

In the ideal *VM* cycle, heat is added from the high temperature source to the gas in the hot cylinder (T_h). The gas is then displaced to the intermediate volume with temperature (T_a). The heat is hen absorbed by the gas in the cold cylinder at temperature (T_c). Assuming ideal gas and applying first law of thermodynamics,

Heat added from high temp source = $Q_h = m_h R T_h \ln(v_2/v_1)$ and

Heat added from low temp source = $Q_c = m_c R T_c \ln(v_2/v_1)$,

Where m_h and m_c are the hot and cold gas mass flow rates and R is the specific gas constant.

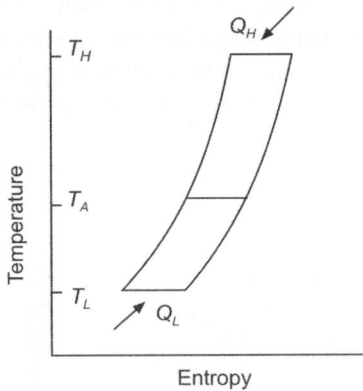

Fig. 14.5 T-S Diagram for Vuilleumier Refrigerator

Heat rejected to Intermediate sink = $Q_a = -(m_h + m_c) R T_a \ln(v_2/v_1)$

The net heat transfer to the system is zero, thus $Q_h + Q_c + Q_a = 0$

$m_h R T_h \ln(v_2/v_1) + m_c R T_c \ln(v_2/v_1) - (m_h + m_c) R T_a \ln(v_2/v_1) = 0$

$m_h T_h + m_c T_c - (m_h + m_c) T_a = 0$

$m_c (T_c - T_a) + m_h(T_h - T_a) = 0$

$m_h (T_h - T_a) = m_c(T_a - T_c)$

Thus $m_h/m_c = (T_a - T_c)/(T_h - T_a)$ or

$m_c/m_h = (T_h - T_a)/(T_a - T_c)$

The COP for the refrigerator is given by

$COP = Q_c/Q_h = m_c T_c/m_h T_h = T_c(T_h - T_a)/T_h(T_a - T_c)$

Thus, the COP of Vuilleumier Refrigerator only depends on the temperatures in the system and is independent of the regenerator ineffectiveness and other thermodynamic losses. The COP of VM refrigerator is lower than Phillips refrigerator due to higher heat rejection. The VM refrigerator has the advantage of using solar energy as an input as well as vibration free operation.

Solvay Refrigerator

The Solvay refrigerator was invented in Germany in 1887 and achieved upto 178 K. Further in 1950 the solvay refrigerator was used in a miniature infrared cooler. Refer Figure 14.6 for the schematic of the Solvay Refrigerator.

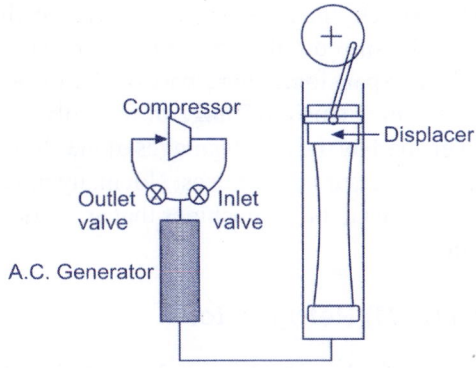

Fig. 14.6 Schematic of Solvay Refrigerator

The expansion stage consists of a cylinder with piston that is connected to a drive mechanism, a separate regenerator and high and low pressure valves. The temperature difference stands over the cylinder and piston so the piston should be made of a low conductivity material. A dynamic gas seal is located at the ambient temperature end of the piston to keep the gas in the cylinder. For the Solvay cycle also five processes can be distinguished in the operation.

Process 1-2: The piston is located at the cold end of the cylinder and the high pressure valve is opened. High pressure gas enters compressing the small amount of gas that remains in the void volumes of the cylinder and the regenerator from pressure P_1 to P_2.

Process 2-3: When the maximum pressure is reached, the piston is raised increasing the cold volume from V_1 to V_2. The pressure remains constant at P_2 since gas is admitted via the opened high pressure valve. All gas reaching the cold end is precooled with the regenerator.

Process 3-4: The high pressure valve is closed and the piston continues to expand from V_2 to V_3, reducing the pressure from P_2 to P_1. In this step, the gas in the cold end performs work on the piston and performs refrigeration. The temperature of the gas decreases.

Process 4-5: The low pressure exhaust valve is opened and the low pressure gas in the cold end is pushed out of the volume by the piston which reduces the volume from V_3 to V_1. When the minimum volume V_1 is reached, the low pressure valve is closed again.

Process 5-1: The gas finally passes out through the regenerator in which the cold gas is warmed back to room temperature.

The work requirement of the system is given by

$$-W_{net}/m = [T_2(s_1 - s_2) - (h_1 - h_2)]/[\eta_{co} - \eta_e \eta_{ad}(h_3 - h_4)]$$

The first term represents compressor work and second term represents work output during expansion process. The energy removed is given by

$$Q_d/m = (h_5 - h_4) - (1 - \eta_{ad})(h_3 - h_4)$$

14.8 Cryogenics

Compared to the G-M cycle, the Solvay cycle has a somewhat higher theoretical efficiency. This is caused by the complete reversible expansion against the piston, in contrast with the G-M expansion, where part of the expansion (over the low pressure valve) is irreversible. This advantage of the Solvay cycle is in practical machines, however, overruled by the disadvantages of the sliding seals of the piston that should withstand high pressure differences. Not many practical examples exist of the Solvay cycle, in contrast to G-M machines that are widely used for a number of different applications.

Gifford Mc-Mohan (G-M) Refrigerator

At the end of the 1950's, Gifford and McMahon developed a refrigerator separating the compressor and expander. In their approach, the compressor was just a high-pressure generator that was connected to the expansion unit by flexible gas lines that can be up to several meters long, and an active-valve unit that generated the pressure wave. Besides the advantage of separation, the compressor was a standard oil-lubricated air-conditioning type, implying much lower cost. Provisions were included for filtering the compressor oil out of the working fluid. The cold head was similar to that of a Stirling refrigerator. The phasing of the displacer movement with respect to the pressure wave was set and controlled by means of the mechanics that drive the valve unit and the displacer/regenerator. Refer Figure 14.7 for the schematic of the G-M refrigerator and Figure 14.8 for the G-M refrigeration cycle. G-M refrigerators with helium gas as the working fluid were primarily developed for cryopumping and are now available from a large number of suppliers.

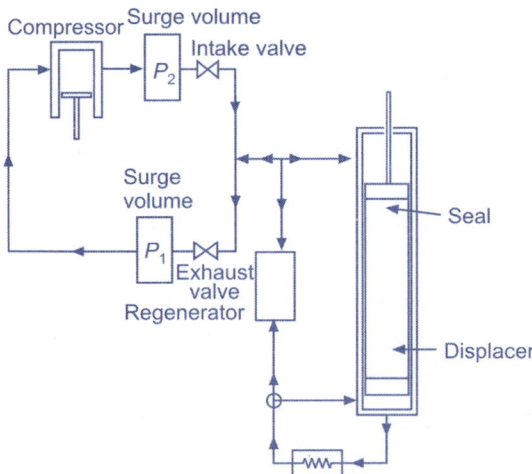

Fig. 14.7 Schematic of G-M Refrigerator

The G-M refrigerator consists of a cylinder, closed at both ends and containing a displacer made of low conductivity material. The displacer is connected to a drive mechanism which can cause the up and down movement of the displacer. As the displacer reciprocates two volumes are created at the top and bottom. These volumes are connected through a thermal regenerator and to a gas supply system consisting of inlet and outlet valves, a compressor and high and low pressure reservoirs. The valves are coupled to the displacer drive mechanism so that their operation is synchronized with the displacer. A heat exchanger is provided to cool the gas to ambient temperature after compression. The cold expansion part of the G-M refrigerator is separated from the hot compressive part. Gas flow to and from the displacer is controlled by the valves. The regenerator consists of fine metallic material which either heats or cools the gas flowing through it. The operation of the system as shown in Figure 14.7 occurs in a cycle with four processes. 1) Compression where there is a pressure build up in the system as the displacer is at the bottom of the cylinder and intake valve is open and exhaust valve is closed. 2) Intake, where the gas flows to the displacer, moving it upward to the top position. The high pressure gas from the space above the displacer (Volume 1) then travels to the space below the displacer (Volume 2) through the regenerator and cools down. 3) Expansion, where there is a pressure release in the system as displacer is at the top, inlet valve closes and exhaust valve opens. The decrease in pressure causes a decrease in temperature of the gas in Volume 2 which causes absorption of surrounding heat producing the refrigeration effect. 4) Exhaust stroke where the gas flows to the displacer, moving it downward to the bottom position. The low pressure gas from the space below the displacer (Volume 2) then travels to the space above the displacer (Volume 1) through the regenerator and heats up to ambient temperature. This completes the cycle.

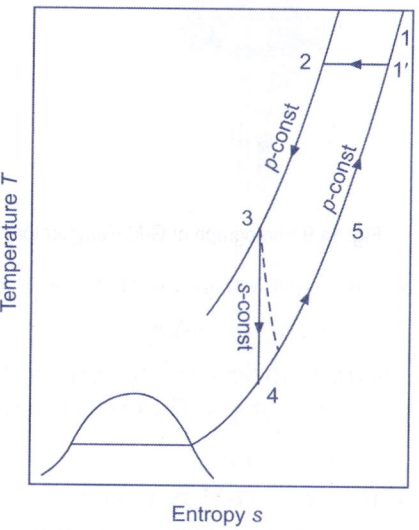

Fig. 14.8 G-M Refrigeration Cycle

The mechanical vibration of the G-M refrigerator becomes a serious problem. In order to eliminate the mechanical vibrations of G-M refrigerator, the bellows flange is usually put between top flange and the G-M refrigerator. Furthermore the cold stages of the G-M refrigerator and cold stage flanges are thermally linked by wires. Gifford-McMahon (G-M) refrigerators were first developed in 1960 and were used in the 1980s for the cooling of charcoal adsorbers to about 15 K in cryopumps. Their use in cryopumps in semiconductor fabrication equipment for producing very clean vacuums provided a large market for these refrigerators and led to many improvements in their reliability and in the reduction of their costs. Maintenance intervals of one to two years are typical for these refrigerators and are usually part of planned maintenance of the entire fabrication equipment. Refer Figure 14.9 for the photograph of the G-M refrigerator.

The network requirement for G-M Refrigerator is given by

$$-W/m = [T_1(s_1 - s_2) - (h_1 - h_2)]/\eta_{co}$$

Where η_{co} is the thermal efficiency of the compressor.

Fig. 14.9 Photograph of G-M Refrigerator

The energy removed by G-M refrigerator from source is given by

$$Q_d/m = \eta_{ad}(m_e/m)(h_3 - h_4)$$

Where η_{ad} is the thermal efficiency of the expander, m_e is the mass of gas expanded and m is the total mass of gas. The expansion ratio is written as

$$m_e/m = \rho_4/\rho_3.$$

The loss of performance in G-M refrigerators is due to factors such as regenerator ineffectiveness, thermal conduction from the displacer and cyclic heat transfer. For improving performance of G-M refrigeration cycle, regenerator effectiveness is the key and the same can be improved by using punched screens of lead separated by coils of stainless steel wire. The advantages of G-M refrigerator include absence of low temperature sealing problems, lower cost due to use of inexpensive regenerators, and tolerance to impure refrigerants.

Comparison between Solvay and Gifford Mc-Mohan Refrigerator

Sr. No	Solvay Refrigerator	GM Refrigerator
1	COP is higher as higher energy is removed from source.	COP is lower as lower energy is removed from source.
2	Expanding gas moves the piston in displacer.	Small motor moves the piston in displacer.
3	There is some leakage of gas past the displacer due to higher pressure difference across the seals	There is no leakage of gas past the displacer due to lower pressure difference across the seals.
4	The motion transmission system i.e. crank arm is complicated	The motion transmission system i.e. crank arm is simple
5	Lesser prone to vibrations	Prone to vibrations
6	Not so adaptable for multistaging	Adaptable to multistaging

Magnetic Cooling or Adiabatic Demagnetization

In 1927 Giaque and Debye independently suggested that a paramagnetic substance and magnetic fields can be used to produce refrigeration instead of expansion of gases or liquids. In absence of external magnetic field, the dipoles of the paramagnetic material are randomly arranged. Application of the magnetic field, allows alignment of magnetic moment of the atoms thereby bringing them in order. For paramagnetic materials, there is a decrease in entropy within the material related to an increase in the ordering of the electron spin state. This is illustrated in the T-S diagram as shown in Figure 14.10. The decreased entropy is associated with a decrease in internal energy and an increase in temperature of the material for an adiabatic process. If the process occurs isothermally, heat must be rejected to the surroundings through a secondary loop. The process is nearly reversible with very small internal losses, at least for small temperature rises, and can be used in a heat pump. The maximum temperature rise is proportional to the magnetic field.

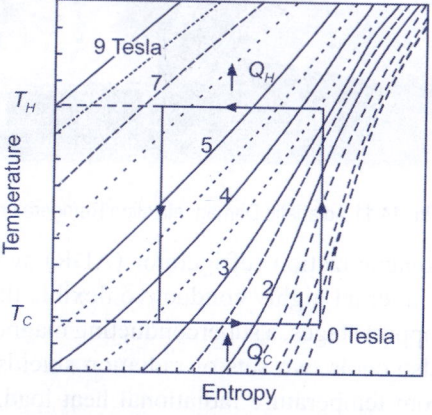

Fig. 14.10 T-S Diagram for Paramagnetic Materials

If the magnetic field is removed reversibly and adiabatically, entropy is constant however the alignment of dipole moments is disturbed. To maintain the alignment and order the temperature of the paramagnetic material decreases. This process is called adiabatic demagnetization and it is successful in producing temperatures as low as 0.6 K.

Adiabatic demagnetization refrigerators (ADR) are ultra low temperature systems, which offer a simple method of achieving temperatures of 50 mK. These systems make use of two built in pills that operate at temperatures of 1 K and 50 mK, supported below a 4 K surface that is cooled with either liquid helium or a mechanical (pulse tube) cooler. These systems offer a typical hold time of two to three days below 100 mK, and require no active pumping on the cryostat. A recycling period of two to three hours is needed at the end of the three-day period, to reduce the temperature back to 50 mK. These systems include a conductively cooled superconducting magnet, which is used to magnetize the two pills, and a mechanical heat switch that links and isolates the two stages from the 4 K surface. After magnetizing the pills at 4 K, the pills are isolated from the 4 K surface and demagnetized to achieve the lowest temperature. The cold stage may then be allowed to drift up in temperature or controlled to operate at any higher temperature between 50 mK and 150 mK. Refer Figure 14.11 for a photograph of an ADR developed by Janis Corporation.

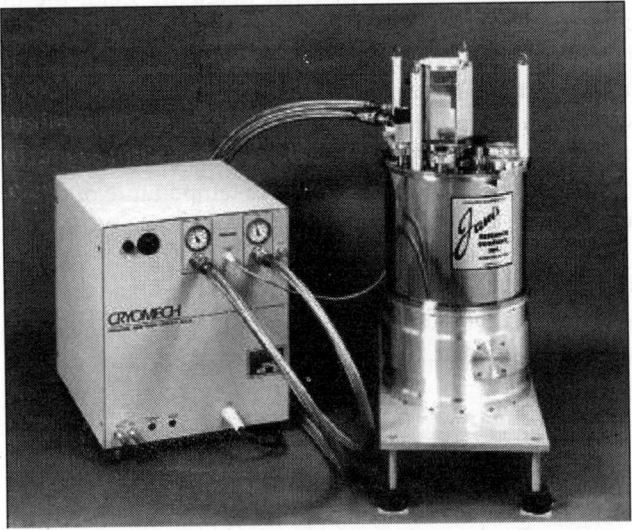

Fig. 14.11 Adiabatic Demagnetization Refrigerator

The adiabatic demagnetization refrigerator (ADR) system includes a Pulse Tube refrigerator with internal highly conductive flexible thermal linkage to cool two stages, the ADR support stages, a superconducting magnet and associated leads and thermometry. It also cools two sets of radiation shields that protect the cold assembly from the room temperature radiational heat load. The two-stage ADR system contains a gadolinium-gallium garnet (GGG) crystal and a ferric ammonium alum (FAA) salt pill. Each pill has its own ultra low thermal conducting support structure isolating it from the 4 K flange and the intermediate stage. The two stages operate at a temperature of ~1 K (GGG) and 50-100 mK (FAA), with a typical hold time of three days below 100 mK. A mechanical thermal heat switch assembly is provided for attaching and detaching the two stages of the ADR during initial cool down, and during its magnetization and demagnetization cycles. Two ruthenium oxide thermometers with appropriate heat sunk wiring at the various intermediate stages are provided to monitor the temperature of the two stages of the ADR. The two thermometers are supplied with a standard calibration data between 50 mK and 20 K. A 4-Tesla conductively cooled multi-filamentary Niobium-Titanium superconducting magnet surrounds the GGG and FAA pills. The magnet is wired with an optimized arrangement of low and high temperature superconducting leads, and is provided with voltage taps and two standard thermometers to monitor its cool down and operating temperature. The magnet reaches its rated field of 4-Tesla at a nominal 10 amperes. A high permeability magnetic shield surrounds the magnet and the two pills, to reduce the stray magnetic field in the vicinity of the cold stages. Refer Figure 14.12 for a schematic of ADR.

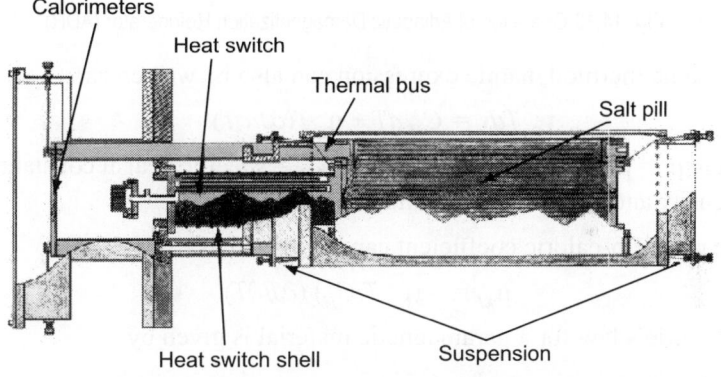

Fig. 14.12 Schematic of Adiabatic Demagnetization Refrigerator (ADR)

The paramagnetic salt pill is suspended by nylon threads in a chamber. The chamber is filled with gaseous helium and is immersed in a liquid helium bath. The liquid helium bath is covered by liquid nitrogen shield. The temperature of the salt is around 1 K. The entire assembly is placed between poles of an electromagnet as shown in Figure 14.13. The magnetic field is first switched on by a switch and the salt pill becomes hot. The heat of magnetization is removed by conduction through the gaseous helium and liquid helium bath. The salt pill then returns to its original temperature. Once the thermal equilibrium is established, the gaseous helium is then pumped away maintaining a vacuum around the paramagnetic pill. The magnetic field is then switched off and the salt is cooled to very low temperatures, i.e. around 0.001 K. The upper limiting temperature (T_o) achieved in adiabatic demagnetization can be calculated from the Debye temperature (θd).

$$T_o = 0.0446\ \theta d.$$

Applying thermodynamic relations to the ADR process we get

$$T_{ds} = d_u - \mu_o H \, dJ,$$

where μ_o = permeability of free space, H = magnetic field intensity and J = magnetic moment per unit mass.

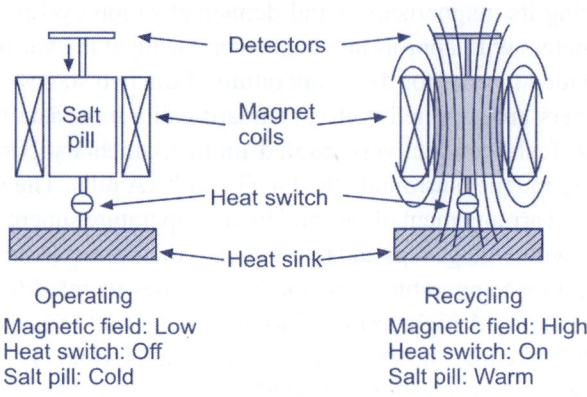

Fig. 14.13 Operation of Adiabatic Demagnetization Refrigerator (ADR)

The same thermodynamic expression can also be written as

$$Tds = C_H \, dT + \mu_o T(\partial J/\partial T)$$

where μ_o = permeability of free space, C_H = specific heat at constant magnetic field intensity and J = magnetic moment per unit mass.

The magneto-caloric coefficient can be written as

$$\mu_M = -(\mu_o T/C_H)(\partial J/\partial T)$$

The Curie's law for a paramagnetic material is given by

$$J = CH/T$$

where C = Curie's constant permeability of free space, H = magnetic field intensity and J = magnetic moment per unit mass.

Magnetic or Magneto-Caloric Refrigerator

The magnetic refrigeration system was developed by Daunt in 1954. Magnetic refrigeration technology takes advantage of the magneto-caloric effect, the remarkable ability of a magnetic material to heat up in the presence of a magnetic field and cool when the field is removed. Magneto-caloric materials store heat energy in the way the atoms vibrate and in the way in which electrons spin within each atom. More heat energy increases the vibrations and also makes the spins more random. This disorder causes an increase in entropy. When a strong magnetic field is applied to the coolant material, the magnetic moments of its atoms become aligned, making the system more ordered. The more ordered material has a lower entropy and compensates for the loss by heating up. But when the strong magnetic field is removed, the material is forced to cool down. The magnetic moments return to their random directions, entropy increases and the material cools. Typically, the temperature of a material can drop by about 10 to 15°C depending on the magnetic field strength. Refer Figure 14.14 for the magnetic refrigeration effect and Figure 14.15 for schematic of magnetic refrigerator.

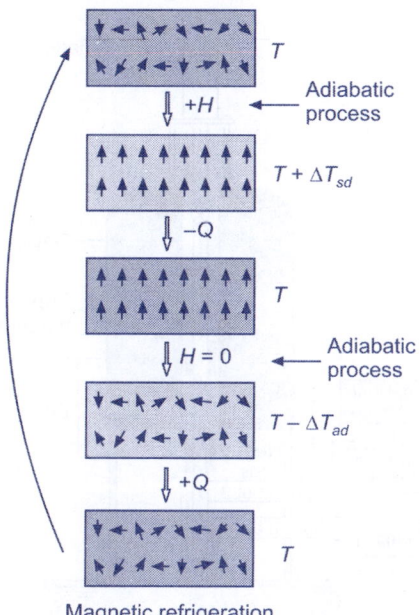

Fig. 14.14 Representation of Magnetic Refrigeration Effect

The temperature at which most of the change in magnetic entropy occurs is known as the material's ordering temperature or its Curie point. This is the point where the material changes from being ferromagnetic to paramagnetic, and the farther away from this point the weaker the magneto-caloric effect. The useful portion of the magneto-caloric effect usually spans about 25°C on either side of the material's Curie temperature. Therefore, in order to span a wide temperature range, a refrigerator must contain several different coolants arranged according to their differing ordering temperatures.

The working medium for the refrigerator is a paramagnetic material (iron ammonium alum). The operation of the magnetic refrigerator is given as follows:

Process 1-2: The magnetic field is applied to the working salt while the upper thermal valve is open and the lower thermal valve is closed. When upper thermal valve is open, the heat is transferred from the working salt to the helium bath.

Process 2-3: Both the thermal valves are closed and magnetic field around the working salt is reduced adiabatically thereby reducing the temperature of the salt.

Process 3-4: The thermal valve between working salt and reservoir is opened and heat is absorbed by the working salt.

Process 4-1: Both thermal valves are closed and the magnetic field applied to the salt is increased to its original value.

Refer Figure 14.15 for the schematic of the operation of magnetic refrigerator.

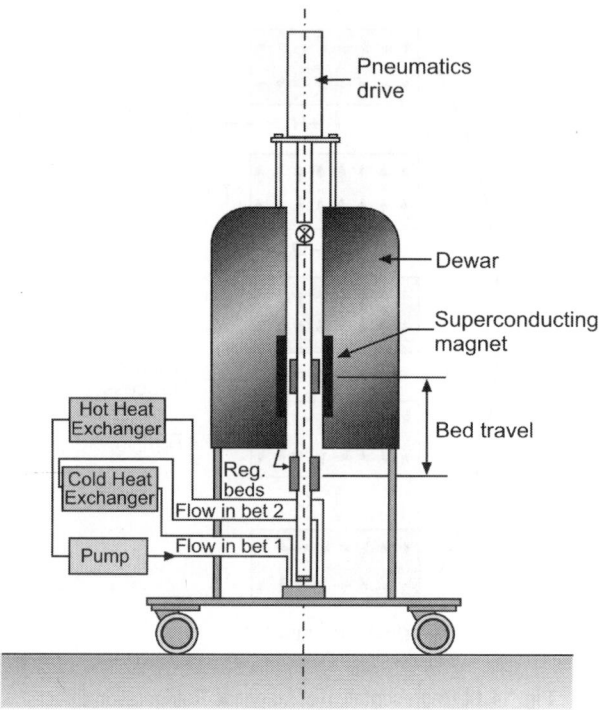

Fig: 14.15 Schematic of Magnetic Refrigerator

The Magnetocaloric effect (MCE, from magnet and calorie) is a magneto-thermodynamic phenomenon in which a reversible change in temperature of a suitable material is caused by exposing the material to a changing magnetic field. Refer Figure 14.16 for the curve of magneto-calorific effect. The heating and cooling that takes place in magnetic refrigeration is proportional to the size of the applied magnetic field and the magnetic moments, which are generally largest in rare-earth elements. One such material, a compound based on gadolinium, has previously been shown to work as a magnetic refrigerant, but in a modest magnetic field its entropy only changes significantly at low temperatures.

Applying first law of thermodynamics to magnetic refrigerator,

Heat absorbed = $Q_a = mT_3 (s_4 - s_3)$

Heat rejected = $Q_r = -mT_2 (s_4 - s_3)$

Work required = $W_{net} = -m(T_2 - T_3)(s_4 - s_3)$.

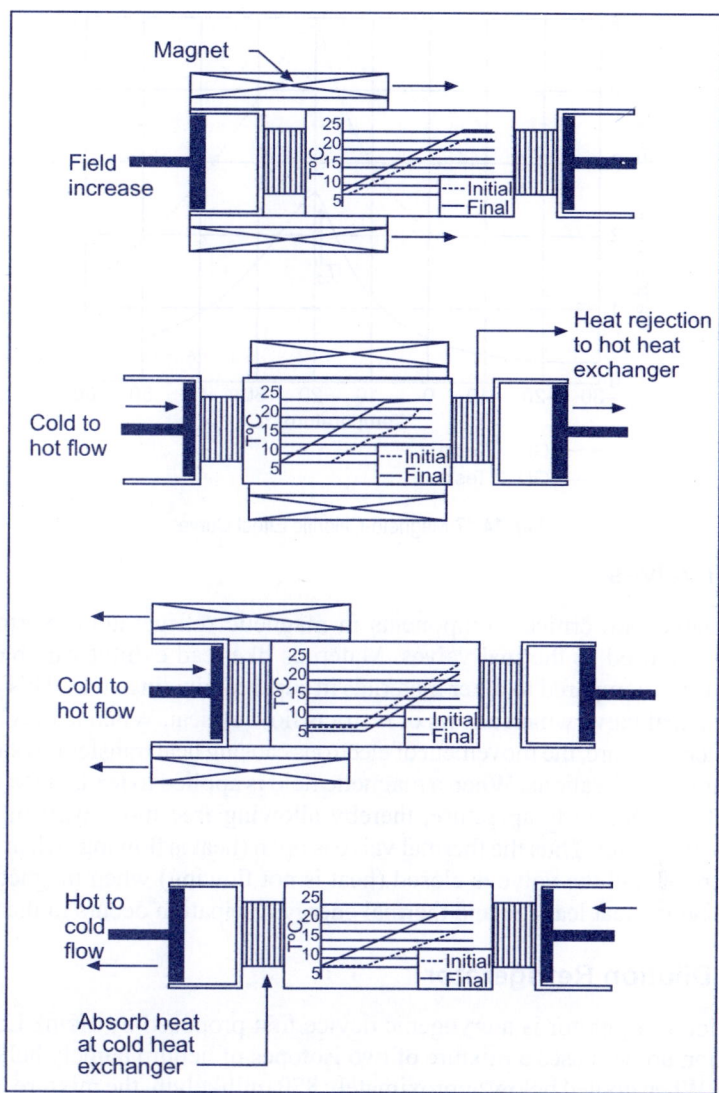

Fig. 14.16 Operation of Magnetic Refrigerator

COP of ideal magnetic refrigerator is same as that of carnot refrigerator, however due to irreversibilites, the actual performance is lower. Magnetic refrigerators have two main advantages over today's commercial devices, which extract heat from a vapour using a compressor: they do not use hazardous or environmentally damaging chemicals, such as chlorofluorocarbons, and they are up to 60% efficient. In contrast, the best gas-compression refrigerators achieve a maximum efficiency of about 40%. Further more magnetocaloric refrigerators can work in zero gravity which makes them suitable for space applications.

14.18 Cryogenics

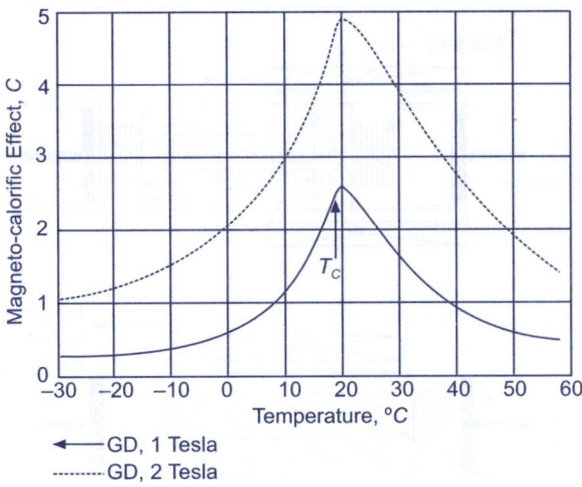

Fig. 14.17 Magneto-Calorific Effect Curve

Thermal Valves

Thermal valves are critical components in magnetic refrigerators. Normally thin lead strips are used as thermal valves. Materials like lead exhibit variable thermal conductivity as compared to other materials in superconducting state. This allows its use as a thermal valve which allows or restricts flow of heat. When lead is below the transition temperature, the movement of electrons causing heat transfer is restricted due to Quantum considerations. When a magnetic field is applied to lead, the temperature exceeds the transition temperature, thereby allowing free movement of electrons causing heat transfer. Thus the thermal valve is open (heat is flowing) when magnetic field is applied and the valve is closed (heat is not flowing) when magnetic field is removed. Some heat leakage and thermal energy dissipation occurs in the valves.

Helium Dilution Refrigerator

The dilution refrigerator is a cryogenic device first proposed by Heinz London. Its refrigeration process uses a mixture of two isotopes of helium namely helium-3 and helium-4. When cooled below approximately 870 millikelvin, the mixture undergoes spontaneous phase separation to form a ^{3}He-rich phase and a ^{3}He-poor phase. As with evaporative cooling, energy is required to transport ^{3}He atoms from the ^{3}He-rich phase into the ^{3}He-poor phase. If the atoms can be made to continuously cross this boundary, they effectively cool the mixture. Because the ^{3}He-poor phase cannot have less than 6% helium-3 at equilibrium, even at absolute zero, dilution refrigeration can be effective at very low temperatures. The volume in which this takes place is known as the mixing chamber. Refer Figure 14.18 for a schematic and Figure 14.9 for a photograph of helium dilution refrigerator.

The simplest application is a "single-shot" dilution refrigerator. In single-shot mode, a large initial reservoir of helium-3 is gradually moved across the boundary into the ^{3}He-poor phase. Once the ^{3}He is all in the ^{3}He-poor phase, the refrigerator cannot continue to operate. More commonly, dilution refrigerators run in a continuous cycle. The ^{3}He/^{4}He mixture is liquified in a condenser, which is

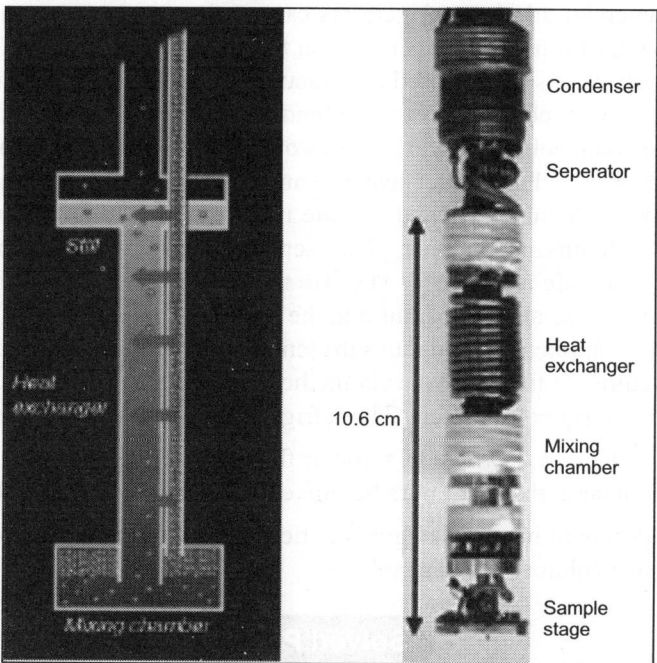

Fig. 14.18 Helium Dilution Refrigerator

connected through an impedance to the ^{3}He-rich area of the mixing chamber. Atoms of ^{3}He migrate across into the ^{3}He-poor phase, providing cooling power, and then into a still where the liquid ^{3}He evaporates. Outside the refrigerator, this gas is pumped up to a higher pressure and usually purified, and finally returns to the condenser to start the cycle again. Continuous-cycle dilution refrigerators are commonly used for low-temperature physics experiments. Temperatures below 2 millime Kelvins can be achieved with the best systems.

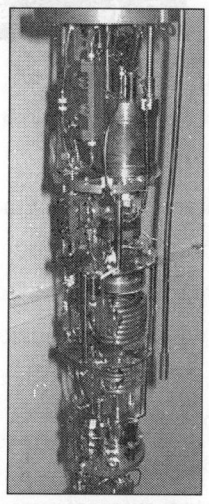

Fig. 14.19 Photograph of Helium Dilution Refrigerator

14.20 Cryogenics

Commercial dilution refrigerators can reach as low temperatures as 0.05 K. The working of dilution refrigerator can be described as follows. The gas which is essentially ^{3}He is compressed in a vacuum pump to 4 kPa pressure and is then cooled by a heat exchanger and to 4.2 K and then condensed in a liquid helium bath at 1.2 K. The liquid helium then expands through a capillary tube at around 0.6 K. The liquid is then cooled in another heat exchanger before entering the mixing chamber. In the mixing chamber ^{3}He and ^{4}He are mixed at temperatures between 0.005 K and 0.05 K. In mixing chamber, phase separation occurs and two phases of dense ^{3}He and dilute ^{3}He are formed. The ^{3}He molecules expand from the dense phase to the dilute phase and temperature in the mixing chamber reduces. The absorbed heat from the application maintains the temperature to its original value. The dilute mixture returns to the reservoir via the heat exchanger and the ^{3}He is evaporated causing the refrigeration effect. The refrigeration effect can be calculated as

$Q = n_3 (h_m - h_1)$, where n_3 = molar flow rate of ^{3}He, h_m = molar enthalpy of ^{3}He, h_1 = molar enthalpy of pure helium entering the mixing chamber.

The design of heat exchanger is critical for the helium dilution refrigerator as large surface volumes are desirable.

Solved Problem

Problem 1. Determine COP and FOM for Phillips refrigerator to maintain 80 K. Heat is rejected at 305 K. Assume all processes are ideal.

Solution:

The operating temperatures $T_1 = 305$ K and $T_3 = 80$ K.

$COPi = COP = T_3/(T_1 - T_3) = 80/(305 - 80) = 80/225 = 0.35$

$FOM = COP/COPi = 1$

References

1. Cryogenic Systems – Randall Baron.
2. Cryogenics assessment report by University of Wisconsin (2005).
3. Cryogenic Engineering – B.A. Hands.
4. Presentation by CERN on cryogenic engineering.
5. Internet website http://www.midsci.com/lp/cryogenic.
6. Taylor Wharton Brochure on Cryogenic Refrigerators.
7. Internet website http://www.memagazine.org/backissues/may99.
8. Report by Steve Fischer (2000) Oak Ridge National Laboratory.
9. Internet website www.nasa.gov.
10. Practical Cryogenics – Balshaw.
11. Internet Website http://www.external.ameslab.gov/.

Questions

Q.1. Enlist atleast three cryogenic refrigerators and describe one in detail.

Q.2. Write short notes on G-M refrigerator, Phillips Refrigerator and Solvay Refrigerator.

Q.3. Write a short note on Vuilleumier Cryogenic refrigerator.

Q.4. Explain the importance of regenerator for Phillips refrigerator.

Q.6. What is a Helium dilution refrigerator? How does it work?

Q.7. Explain the concept of Thermal Valve.

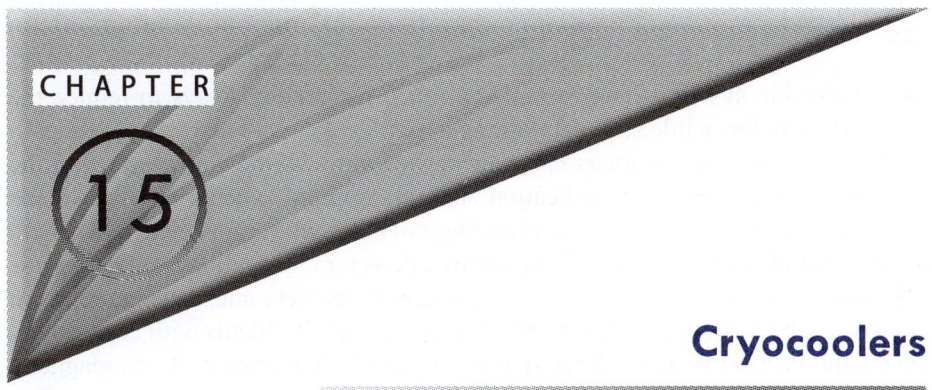

CHAPTER 15

Cryocoolers

INTRODUCTION

Cryocoolers are the devices used to attain cryogenic temperatures by cycling certain gases. These are also known as small refrigerators operating in the cryogenic temperature range. A cryostat is used to maintain similar conditions or keep some environment in cryogenic state. Conventional coolers for the cryogenic temperature range (i.e., < 120 K), all consist of a mechanical compressor that compresses a working gas at about ambient temperature and an expansion space at a low temperature. In the expansion space, the compressed fluid is allowed to expand, and by doing so it establishes refrigeration. The cryocoolers serving this market have predominantly been Gifford-McMahon cryocoolers that use large, steady-flow, oil-lubricated compressors from the refrigeration industry. The input power for these systems is typically in the range of a few kilowatts. Some emerging applications that need cooling in this regime, such as superconducting electronics for communications, require small and potentially mobile packages that cannot tolerate large size, inefficiency and vibration.

Small cryogenic refrigerators or cryocoolers are required for a wide variety of applications, and the number of applications keeps expanding as improvements to cryocoolers are made. One of the earliest applications, and one that appeared about 50 years ago, was for cooling infrared sensors to about 80 K for night vision capability of the military. Over 125,000 Stirling cryocoolers for this tactical military application have been produced with refrigeration power varying from 0.15 W to 1.75 W. In the last decade or so the desire for night vision surveillance and missile detection from satellites has prompted much research on improved cryocoolers to meet the very stringent requirements for space use. Each application has a particular set of requirements that have led to many recent improvements in cryocoolers. Still, some of the problems associated with cryocoolers have hampered the marketing of many potential applications. For satellite applications a space-qualified cryocooler can cost $1.0M or more. Lifetimes of 10 years are now standard requirements for most space applications. Until recently efficiencies of 3 to 10% of Carnot were typical of these small cryocoolers, whereas now efficiencies of 15 to 25% of Carnot are now possible. New commercial applications, particularly the use of high temperature superconductors as microwave filters in base stations for wireless

15.2 Cryogenics

communication systems, have stimulated much research on ways to reduce the cost of cryocoolers while still maintaining a lifetime goal of 3 to 5 years. Research on various types of cryocoolers has been carried out with the goal of meeting the requirements for a particular application. As a result, improvements have been made in all the types of cryocoolers. The advantages of pulse tube cryocoolers make them natural candidates for the cooling of infrared detectors for both tactical and space applications. This chapter focuses on various cryocoolers and their comparison along with discussion on recent trends in cryocooling. Problems with cryocoolers are reliability, low efficiency, large size and weight, high vibration, electromagnetic interference (EMI), heat rejection and high cost.

Classification of Cryocoolers by Walker's Chart

Cryocoolers can be classified as shown using Walker's chart as shown in Figure 15.1.

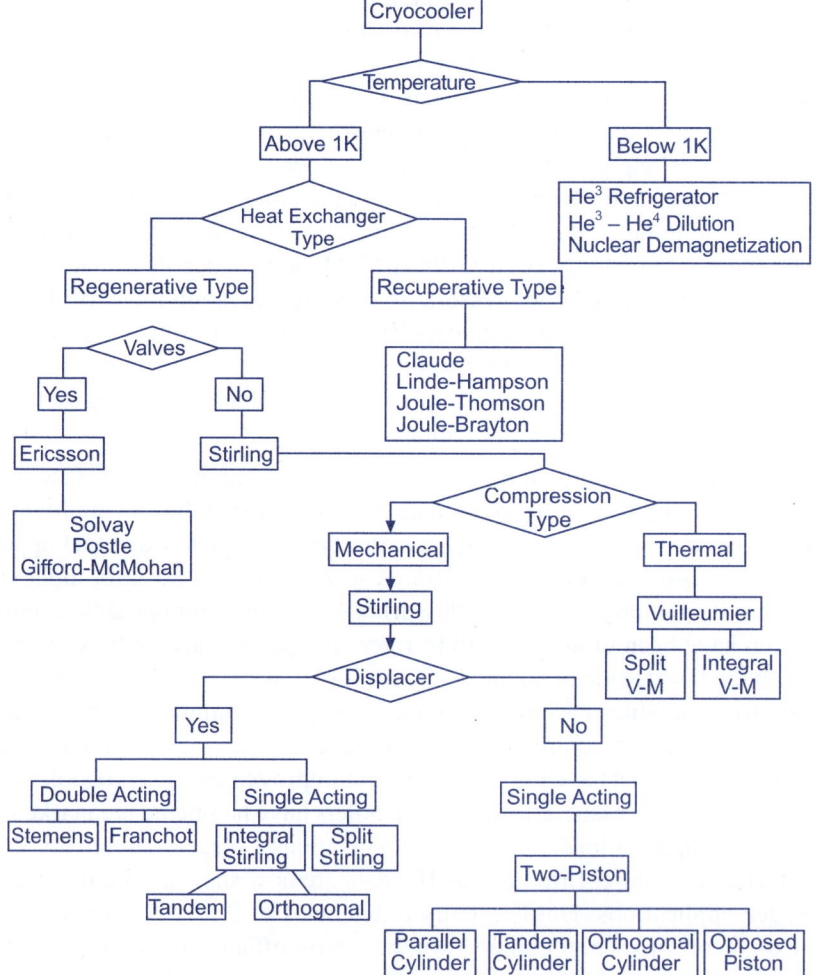

Fig. 15.1 Cryocooler Classification by Walker's Chart

Mechanical Cryocoolers

Consumable cryogenics systems have problems of difficulties of liquid supplies, perturbations during refilling, operation in all orientation and high cost of operation. These constraints can be overcome by mechanical cryocoolers which have high reliability, high thermodynamic efficiency, low investment and maintenance cost and low vibration levels. Mechanical cryocoolers have demonstrated their long term reliability, with several coolers having logged over 10 years of space flight operation. There are active coolers currently in development that are applicable for the various configurations, though their heat lift capability and temperature stages need modifications. Cooler build cycles are on the order of 3 years, but each new generation cooler leads towards improvements that lower mass or increase efficiency. The need to oversize the cooler cannot be stressed enough as thermal loads inevitably increase. The cooler must be designed for end-of life conditions when parasitic loads are highest. The first cryocooler for space applications was developed by Lockheed Martin's Advanced Technology Center in 1971. These developments have continued on with major emphasis on stored cryogen systems, including solid hydrogen systems, and superfluid helium systems. In 1987, a Stirling cryocooler was developed by Oxford University in England and proved to be a major breakthrough in mechanical cryocooler technology. Also, several pulse tube cryocoolers have been developed. including nine single-stage coolers which operate at as low as 3.8 K while rejecting heat at 295 K. A mechanical cryocooler consists of three modules (compressor, coldhead, and electronics).

Mechanical cryocoolers have characteristics of reliability, simplicity, and robustness in the order of priorities. Toward this end several generations of cryocoolers have been developed using moving magnet linear motors, which have a number of advantages over the moving coil motors. While moving magnet motors typically have lower efficiency than moving coil motors, recent evolutionary developments have resulted in improvements in motor efficiency so they are now equal to or better than moving coil motors. Some of the major advantages of the moving magnet design are the placement of the epoxy-potted coil outside the working gas space which reduces outgassing and gas contamination and the elimination of the flexing current carrying leads to the moving coil. In addition, all electrical penetrations into the working gas space are eliminated to improve the reliability and simplify the construction of mechanical cryocoolers. Additional features, which improve the reliability, are the utilization of the pulse tube cold head, which eliminates the moving parts and linear motor associated with the Stirling coldhead, and reduces the complexity and weight of the electronic controller by removing the functions required to control the displacer. The third component of a mechanical cryocooler system is the electronic controller. Reliability of this module is improved principally through the reduction of the number of electronic parts. This reduction results in a direct increase in the reliability, and numerous system studies indicate that the electronic controller, rather than the compressor or the coldhead, limits the reliability of a mechanical cryocooler. The reduction in electronic parts is

15.4 Cryogenics

achieved in part by the elimination of the displacer of the Stirling cycle but also by innovative circuit design with emphasis on reduced parts. The electronics includes a vibration reduction control loop, a temperature control loop, and an active ripple suppression circuit. We shall now study some important cryocoolers in detail.

Cryocooler Requirements

(A) **High Reliability**: The most important issue in cooler development is reliability. Available coolers have specified lifetimes of approximately 5 years. Considerable developmental work was done on the compressor and on the gas-expansion unit. Concerning the latter, the main attention has been on eliminating the moving parts in the cold. Typical examples are the pulse-tube refrigerator and the developments in Joule-Thomson cooling. In the case of the compressor, the key issue was the rubbing contact between the compressor piston and the cylinder. Wearing of the rubbing seals was the limiting factor in compressor lifetime. Now-a-days, using flexure bearings that support the piston and the displacer inside their respective cylinders without any contact can eliminate these rubbing contacts. By very accurate machining, the gap between piston and cylinder can be reduced to a few µm. The flow impedance of this clearance gap is so high that it acts as a dynamic seal for the helium gas. The "mean-time-to-failure" (MTTF) of these compressors can exceed 50000 hours.

(B) **Low Cost**: Another challenge is to manufacture highly reliable cryocoolers at a sufficiently low price. Cooler manufacturers are continuously improving their designs and manufacturing processes, with cost reduction being the main driver. The US DARPA program that sponsored research aiming at $ 1,000 coolers largely stimulated these efforts towards lower cryocooler cost. Nowadays, the cost of a single unit 1W cryocooler roughly ranges from $ 20,000 to $ 40,000 for low temperature ($\sim$ 4 K), and from $ 5000 to 15000 for high temperature ($\sim$ 70 K). The production volume is an important cost determining parameter. As a rough indication, the price per cooler goes down by a factor of two when the number of coolers produced is increased by one order of magnitude.

(C) **Small Size**: Increasing attention is paid to the size of the cooler. Since electronic devices have become smaller in terms of size and dissipation, there is a genuine need for extremely small coolers (with dimensions of centimeters) with small cold heads (order of mm) and low cooling powers (order of mWs). Extreme miniaturization is possible by using micromechanical techniques. Already in the mid-90's, patents were filed on a micro-pulse-tube cooler and on a micro-Stirling-cooler.

Cryocooler vs Cryostat

In comparing cryostats with cryocoolers, it is important to note that the cryostat with the liquid cryogen is only a small part of a rather complex cryogenic system.

It just serves as a thermal buffer in the total system. First, gas has to be liquefied in a large cooler that produces relatively large amounts of liquid in big storage dewars. This liquefaction is performed industrially. The liquid cryogen is then shipped in large transport/storage dewars to a customer, who transfers the liquid into smaller SQUID system dewars and vents it to air. This is the standard procedure with liquid nitrogen, but helium is so expensive that it is often worthwhile to recover the vented gas. In this case, it is captured in large inflatable gas-bags or gas tanks, and, after filtering and purification, re-pressurized by a compressor to be re-liquefied through expansion. The main drawback of liquid cryogens is that periodical replenishment of the cryostat is required for long-term operation. The most attractive feature of a cryocooler is, in general terms, its turn-key operation. No expertise or cryogenic facility is required. This lowers the barrier preventing applications of superconducting devices outside the cryogenic laboratories. A very important requirement in this respect is reliability. The cooler should be a black box to the user, who need not worry about it.

The operating temperature of any liquid cryogen is fixed by its boiling point and a temperature variation is only possible by changing the vapor pressure, e.g. by pumping the space above the bath. Practical cryocoolers are available for cryogenic temperatures down to well below 4 K. Within some range, changing the input power to the cooler can simply vary the temperature. Temperature stability and low interference are clear advantages of cryogens although methods are available to control the cryocooler temperature fluctuations and reduce interference. As mentioned above, cryocoolers are very "friendly" to the user because of their turn-key operation. On the other hand, they are rather "unfriendly" to the manufacturer because of their constructional complexity. The latter aspect directly translates into higher costs. Liquid-cryogen cooling operates for a predictable period of time and is not vulnerable to unscheduled interruptions of the electric power. However, availability may be the problem of cryogens. A further point related to the reliability of cryostats is the availability of a good clean vacuum. The necessary periodical re-pumping may be problematic at some places. A big advantage of cryocoolers compared to liquid cryogens is the weight required for long-term operation, which is especially relevant in space applications. The possibility to operate the cryogenic systems in all orientations is also a clear advantage of cryocoolers. For earth applications, a liquid cryogen clearly has its limitations, although special constructions can be used to keep the liquid inside the cryostat at all orientations.

Joule-Thomson (JT) Cryocooler

In the early 1850s, two British scientists J.P. Joule and William Thomson (later Lord Kelvin) performed experiments on the expansion of gases. They expanded gases from a high pressure through a porous plug under adiabatic conditions. Under these conditions, the expansion is also isenthalpic and the temperature of the gas increases or decreases depending on the sign of the Joule-Thomson coefficient $\mu = (\partial T/\partial P)_H$.

15.6 Cryogenics

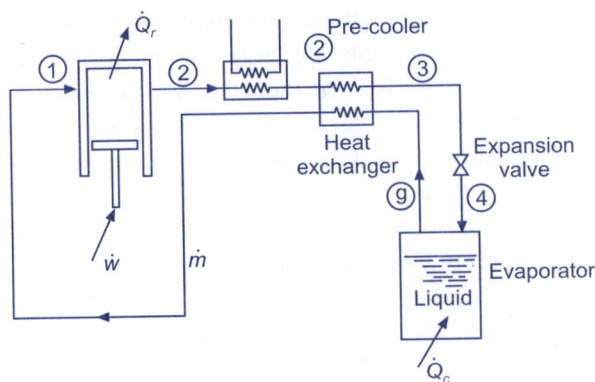

Fig. 15.2 Schematic of Joule-Thomson Cryocooler

Below the so-called inversion temperature, μ is positive and cooling can be obtained as the result of an expansion. This inversion temperature depends on the pressure, and the maximum value (i.e., at P = 0) for helium is 40 K, for hydrogen 205 K, for neon 250 K, and for nitrogen 621 K. A Joule-Thomson cryocooler utilizes this effect and consists of a compressor, a control and filter unit, a counter-flow heat exchanger (CFHX), and a Joule-Thomson expansion stage including a flow restriction (JT-valve). For an ideal CFHX, the cooling power of a JT cooler is given by the product of the mass-flow rate and the difference in specific enthalpy between the high and low-pressure sides at the warm end of the CFHX. Refer Figure 15.2 for a schematic of the JT cryocooler and Figure 15.3 for the T-S diagram of the Joule-Thomson cooling cycle.

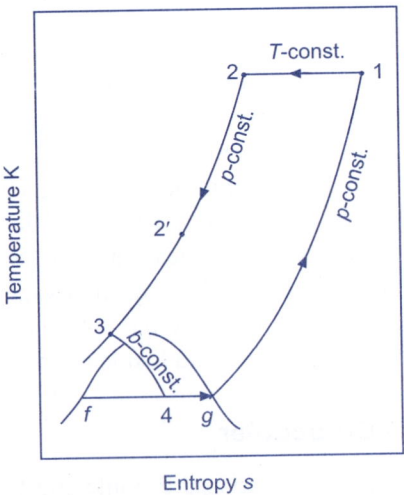

Fig. 15.3 Joule-Thomson Cycle

Lowering the warm-end temperature, i.e., by pre-cooling the fluid before it enters the CFHX can increase this enthalpy difference, and thus the cooling power.

Very large enthalpy changes can be obtained by pre-cooling to a level such that the fluid is liquefied before the expansion. In that case, the expansion valve is usually referred to as a throttling valve rather than a JT-valve. The Joule-Thomson cryocoolers produce cooling when the high-pressure gas expands through a flow impedance (orifice, valve, capillary, porous plug), often referred to as a JT valve. The expansion occurs with no heat input or production of work, thus, the process occurs at a constant enthalpy. The heat input occurs after the expansion and is used to warm up the cold gas or to evaporate any liquid formed in the expansion process. In an ideal gas the enthalpy is independent of pressure for a constant temperature, but real gases experience an enthalpy change with pressure. Thus, cooling in a JT expansion occurs only with real gases and at temperatures below the inversion curve. In fact, for a given pressure change the amount of cooling increases as the temperature is lowered and reaches a maximum around the critical point.

Typically, nitrogen or argon is used in JT coolers, but pressures of 20 MPa (200 bar) or more on the high pressure side are needed to achieve reasonable cooling. Such high pressures are difficult to achieve and require special compressors with limited lifetime. The main advantage of JT cryocoolers is the fact that there are no moving parts at the cold end. The cold end can be miniaturized and provide a very rapid cool-down. This rapid cool-down (around 77 K) has made them the cooler of choice for cooling infrared sensors used in missile guidance systems. These coolers utilize a small cylinder pressurized to about 45 MPa with nitrogen or argon as the source of high-pressure gas. In this open-cycle mode, cooling lasts for only a few minutes until the gas is depleted. Figure 15.4 shows a typical JT cryocooler used for missile guidance. Miniature finned tubing is used for the heat exchanger. An explosive valve is used to start the flow of gas from the high-pressure bottle, and after flowing through the system, the gas is vented to the atmosphere. A disadvantage of the JT cryocooler is the susceptibility to plugging by moisture of the very small orifice. Another disadvantage is the low efficiency when used in a closed cycle mode. Compressor efficiencies are very low when compressing to such high pressures.

Fig. 15.4 Photograph of a JT Cryocooler

Recent advances in JT cryocoolers have been associated with the use of mixed-gases as the working fluid rather than pure gases. It is commonly referred to as the mixed refrigerant cascade (MRC) cycle. The use of small JT coolers with mixed gases for cooling infrared sensors was first developed under during the 1970s. Typically, higher boiling-point components, such as methane, ethane, and propane can be added to nitrogen to make the mixture behave more like a real gas over the entire temperature range. Enthalpy changes as much as 5 times that with pure nitrogen are possible with pressures of only 2.5 MPa. Such large enthalpy changes then lead to greatly increased efficiencies with the JT cryocooler at pressures that can be achieved in conventional compressors used for domestic or commercial refrigeration. In general, the freezing point of a mixture is less than that of the pure fluids, so temperatures of 77 K are possible with the nitrogen-hydrocarbon mixtures. The advantage of these JT-type coolers with regard to cooling is the absence of cold moving parts, which helps attaining a low level of interference. Furthermore, the cold stage can be constructed in a very simple and straightforward way and is well suitable for miniaturization. The biggest problem with these JT-type coolers is clogging of the cold stage caused by moisture and contamination. In the case of 4 K cooling, a further disadvantage is that a helium-based JT cooler has to be pre-cooled to typically 20 K using a pre-cooler, because of the inversion temperature.

Brayton Cryocooler

The Brayton cycle was developed in the nineteenth century by George Brayton, a pioneer in the development of internal combustion engines and the continuous ignition combustion (turbine) engine. There are many forms of the Brayton-cycle, ranging from the simple open cycle used in gas turbine and jet engines to the closed cycle with external combustion; the reverse Brayton-cycle is used to provide cooling, having the potential for excellent thermodynamic efficiency. Development of reverse Brayton-cycle technology for cryogenic cooling in space was funded by NASA. Brayton-cycle cryocoolers are being developed to provide efficient cooling in the 6 K to 70 K temperature range. The cryocoolers are being developed for use in space and in terrestrial applications where combinations of long lifetime, high efficiency, compactness, low mass, low vibration, flexible interfacing, load variability, and reliability are essential. The key enabling technologies for these systems are a scale expander and an advanced oil-free scroll compressor. The emphasis on the component and system development has been on invoking fabrication processes and techniques that can be evolved to further reduction in scale tending toward cryocooler miniaturization. Refer Figure 15.5 for the schematic of the Brayton cryocooler

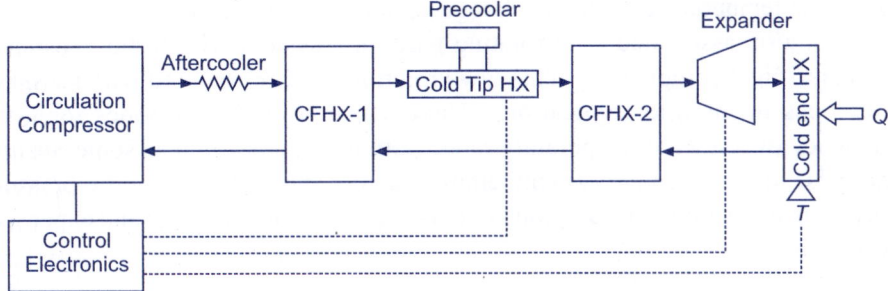

Fig. 15.5 Schematic of Brayton Cryocooler

The Brayton-cycle cryocooler is a recuperative system that derives its thermodynamic performance from efficient compression, expansion, and heat exchange processes. The key components in a system include the compressor, expander, counter flow heat exchangers, and a cold-end heat exchanger. Newer Brayton-cycle cryocooler design incorporates technology including a DC-flow oil-free scroll compressor, a reciprocating expander, and compact high-effectiveness counter flow heat exchangers. An intermediate stage (precooler) is used in lower temperature cooling applications to compensate for imperfect heat transfer in the warm recuperative heat exchanger and to intercept parasitic heat load into the cold end. This stage can be another expander that is integral with the fluid system. Second stage cooling at an intermediate load temperature may also be provided with proper component sizing. The Brayton-cycle cryocooler has many attractive features that can benefit actively cooled systems. Its modular configuration allows system operation flexibility; e.g., changes can be made during integration with the sensor package since components can be independently integrated. Refer Figure 15.6 for the construction of a typical Brayton cryocooler.

In Brayton cryocoolers cooling occurs as the expanding gas does work. An expansion turbine supported on gas bearings is more commonly used to give high reliability. According to the First Law of Thermodynamics the heat absorbed with an ideal gas in the Brayton cycle is equal to the work produced. This process is then more efficient than the JT cycle and it does not require as high a pressure ratio. The Brayton cycle is commonly used in large liquefaction plants. For small Brayton cryocoolers the challenge is fabricating miniature turbo expanders that maintain high expansion efficiency. Turbine diameters of about 6 mm on shafts of 3 mm diameter are typical in systems for use in space applications of cooled infrared sensors. Turbine speeds of 2000 to 5000 rev/s are typical. Centrifugal compressors providing a pressure ratio of about 1.6 with a low side pressure of 0.1 MPa are used with these systems. The working fluid used in the Brayton cryocoolers is usually neon when operating above 35 K, but helium is required for lower temperatures. The advantage of the Brayton cryocooler is the very low vibration associated with rotating parts in a system with turbo expanders and centrifugal compressors. This low vibration is often required with sensitive telescopes in satellite applications. The expansion engine provides for good efficiency over a wide temperature range, although not as high as some Stirling and pulse tube cryocoolers at temperatures above about 50 K. The low-pressure operation of the miniature Brayton systems requires relatively large and expensive heat exchangers.

Fig. 15.6 Construction of Brayton Cryocooler

Plank Sorption Cryocooler

Adsorption coolers rely on the capability of porous materials to adsorb or release a gas when cyclically cooled or heated. Using this physical process one can design a compressor/pump which by managing the gas pressure in a closed system, can condense liquid at some appropriate location and then perform an evaporative pumping on the liquid bath to reduce its temperature. These self-contained coolers require only electrical connections and thermal contact in order to operate from a cold heat sink. During ground testing they can be recycled in a wide range of orientations. Moreover the absence of moving parts makes them reliable and vibration free. They can be recycled indefinitely with over 95% duty cycle efficiency using simple control electronics. Adsorption is the physical mechanism upon which a gas can be trapped onto a material surface. Depending on the magnitude and nature of the attractive forces the effect is described as chemisorption or physisorption. In the first case a chemical bond is formed and the process involves a transfer of electric charges. Physisorption relies on the relatively weak van der Walls forces. Evidently material with high specific area, such as activated charcoal are required. The amount of gas adsorbed is then a strong function of pressure and temperature; it increases as the temperature decreases and the pressure increases. Thus by varying the temperature it is possible to provide either a compression or a pumping effect. The former can be used for instance to drive a Joule Thomson loop, the latter can be used to perform an evaporative pumping on a liquid bath and thus reduce its temperature. In addition physisorption, or the ability to vary the pressure by varying the temperature, can also be used to design very efficient gas gap heat switches. Refer Figure 15.7 for a schematic of the absorption cryocooler.

Plank Sorption cryocoolers provide continuous operation, vibration free, long-life operation. 20 K Hydrogen sorption cryocoolers were built using hydrogen as the refrigerant and $LaNi_{4.8}Sn_{0.2}$ as the hydride sorbent. The materials, components, design margins, and assembly procedures were entirely consistent with space-flight qualification requirements. In addition to being vibration-free, the mass and power requirements are significantly less as compared to Stirling cycle cryocoolers.

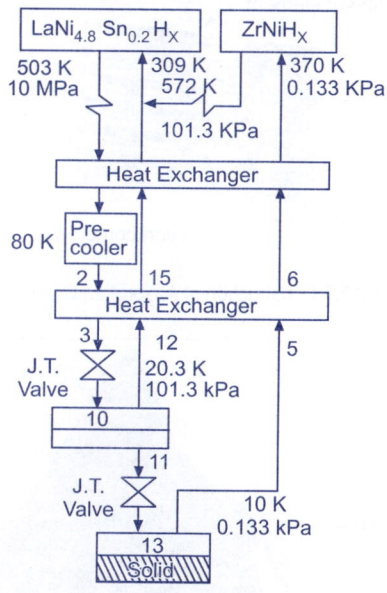

Fig. 15.7 Schematic of Absorption Cryocooler

Plank sorption cryocoolers have no moving parts, are vibration less and can be designed to be self contained and compact with a high duty cycle efficiency. These features and the expected reliability that follows make them very attractive for space applications. In addition the thermal and mechanical interfaces are fairly simple. The links to ambient temperature are limited to the heater wires used to drive the sorption pump and the heat switches. It should be noted that this type of cooler is the last stage of a cooling system. It requires a pre-cooling stage at a temperature lower than the vapor transition (3-5 K) with enough cooling power. This can either be a helium bath or a mechanical. Concerning the drawbacks relatively poor thermodynamic efficiency with regards to Carnot due to the heat of adsorption, and one shot operations can be noted. Refer Figure 15.8 for construction and Figure 15.9 for the photograph of the absorption cryocooler.

15.12 Cryogenics

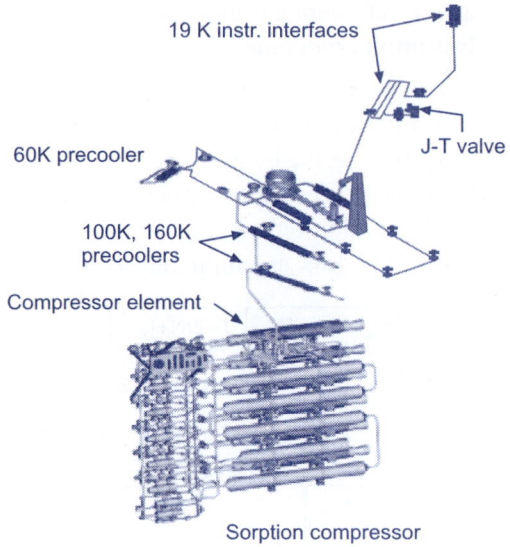

Fig. 15.8 Construction of Absorption Cryocooler

Fig. 15.9 Photograph of Absorption Cryocooler

For operation in a zero-gravity environment liquid is condensed directly into the evaporator by establishing the necessary temperature gradients using a heat switch, and held there by capillary attraction inside some porous material. Both the surface tension and the vapor pressure provide forces that drive and hold the liquid at the coldest point. Like any other coolers a number of specifications are required for the design of a sorption cooler. Three main requirements are the ultimate temperature, the hold time and the average power dissipated on the main cold stage.

Gifford Mc-Mahon (G-M) Cryocooler

One disadvantage of the stirling cooler was the inability to separate the compression and expansion sections, thus limiting the flexibility of the cooler. At the end of the 1950's, Gifford and Mc-Mahon developed a cooler separating the compressor and expander. In their approach, the compressor was just a high-pressure generator that was connected to the expansion unit by flexible gas lines that can be up to several meters long, and an active-valve unit that generated the pressure wave. Besides the advantage of separation, the compressor was a standard oil-lubricated air-conditioning type, implying much lower cost. Provisions were included for filtering the compressor oil out of the working fluid. The cold head was similar to that of a stirling cooler. The phasing of the displacer movement with respect to the pressure wave was set and controlled by means of the mechanics that drive the valve unit and the displacer/regenerator. Refer Figure 15.10 for the schematic of the G-M cryocooler and Figure 15.11 for the G-M refrigeration cycle. G-M coolers with helium gas as the working fluid were primarily developed for cryopumping and are now available from a large number of suppliers. Cryopumps are the largest commercial application of G-M cryocoolers.

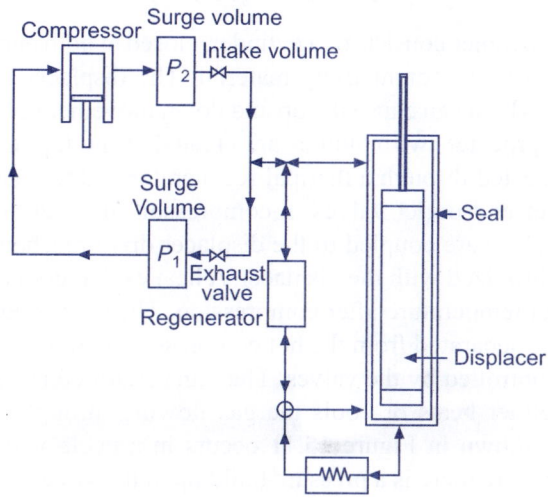

Fig. 15.10 Schematic of G-M Cryocooler

In addition, G-M coolers are installed in MRI systems for reducing the boil-off rate of the liquid helium bath. These coolers are commonly manufactured in single-stage and double-stage versions. Refer Figure 15.12 for the photograph of various sizes of G-M cryocoolers. The double stage type has cooling powers of

15.14 Cryogenics

several watts at the coldest stage (typically at 20 K) and several tens of watts at the other stage (at 80 K) with an input of a few kilowatts. Two-stage G-M coolers are also available that can cool to 4 K by using rare-earth regenerator materials. These materials have magnetic phase transitions in the temperature range of 4 K to 10 K, and thus a high latent heat in that temperature range. The operating frequency of the displacer and the valves is relatively low, in the range of a few Hz.

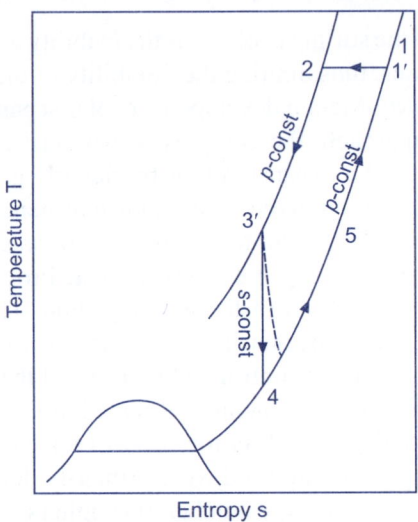

Fig. 15.11 G-M Refrigeration Cycle

The G-M cryocooler consists of a cylinder, closed at both ends and containing a displacer made of low conductivity material. The displacer is connected to a drive mechanism which can cause the up and down movement of the displacer. As the displacer reciprocates two volumes are created at the top and bottom. These volumes are connected through a thermal regenerator and to a gas supply system consisting of inlet and outlet valves, a compressor and high and low pressure reservoirs. The valves are coupled to the displacer drive mechanism so that their operation is synchronized with the displacer. A heat exchanger is provided to cool the gas to ambient temperature after compression. The cold expansion part of the G-M cryocooler is separated from the hot compressive part. Gas flow to and from the displacer is controlled by the valves. The regenerator consists of fine metallic material which either heats or cools the gas flowing through it. The operation of the system as shown in Figure 15.11 occurs in a cycle with four processes.
1) Compression where there is a pressure build up in the system as the displacer is at the bottom of the cylinder and intake valve is open and exhaust valve is closed.
2) Intake, where the gas flows to the displacer, moving it upward to the top position. The high pressure gas from the space above the displacer (Volume 1) then travels to the space below the displacer (Volume 2) through the regenerator and cools down. 3) Expansion, where there is a pressure release in the system as displacer is at the top, inlet valve closes and exhaust valve opens. The decrease in pressure causes a decrease in temperature of the gas in Volume 2 which causes absorption

of surrounding heat producing the refrigeration effect. 4) Exhaust stroke where the gas flows to the displacer, moving it downward to the bottom position. The low pressure gas from the space below the displacer (Volume 2) then travels to the space above the displacer (Volume 1) through the regenerator and heats up to ambient temperature. This completes the cycle.

Fig. 15.12 Photographs of G-M Cryocoolers

The mechanical vibration of the G-M cryocooler becomes the serious problems In order to eliminate the mechanical vibrations of G-M cryocooler, the bellows flange is usually put between top flange and the G-M cryocooler. Furthermore the cold stages of the G-M cryocooler and cold stage flanges are thermally linked by wires. Gifford-McMahon (G-M) cryocoolers were first developed in 1960 and were used in the 1980s for the cooling of charcoal adsorbers to about 15 K in cryopumps. Their use in cryopumps in semiconductor fabrication equipment for producing very clean vacuums provided a large market for these cryocoolers and led to many improvements in their reliability and in the reduction of their costs. Maintenance intervals of one to two years are typical for these cryocoolers and are usually part of planned maintenance of the entire fabrication equipment. In the late 1980s the G-M cryocooler also began to be used for the cooling of radiation shields in MRI equipment for reducing the boil-off rate of the liquid helium that was used to maintain the superconducting magnet at 4.2 K. The use in 1990 of regenerators made with rare earth materials with high heat capacities in the range of 4-20 K allowed the G-M cryocooler to achieve temperatures of 4 K. These new coolers could then eliminate the helium boil off, and some compact MRI systems have been developed recently that use no liquid helium (cryogen-free systems). The commercial availability of the G-M cryocoolers, both one and two-stage systems, has led them to be used for many development projects involving new applications.

Pulse Tube Cryocooler

The moving displacer in the Stirling and Gifford-McMahon cryocoolers has several disadvantages. It is a source of vibration, has a limited lifetime, and contributes to

15.16 Cryogenics

axial heat conduction as well as to a shuttle heat loss. In the pulse tube cryocooler, shown in Figure 15.13, the displacer is eliminated. The proper gas motion in phase with the pressure is achieved by the use of an orifice, along with a reservoir volume to store the gas during a half cycle. The reservoir volume is large enough that negligible pressure oscillation occurs in it during the oscillating flow. The oscillating flow through the orifice separates the heating and cooling effects just as the displacer does for the Stirling and Gifford-McMahon refrigerators. The orifice pulse tube refrigerator (OPTR) operates ideally with adiabatic compression and expansion in the pulse tube. Thus, for a given frequency there is a lower limit on the diameter of the pulse tube in order to maintain adiabatic processes.

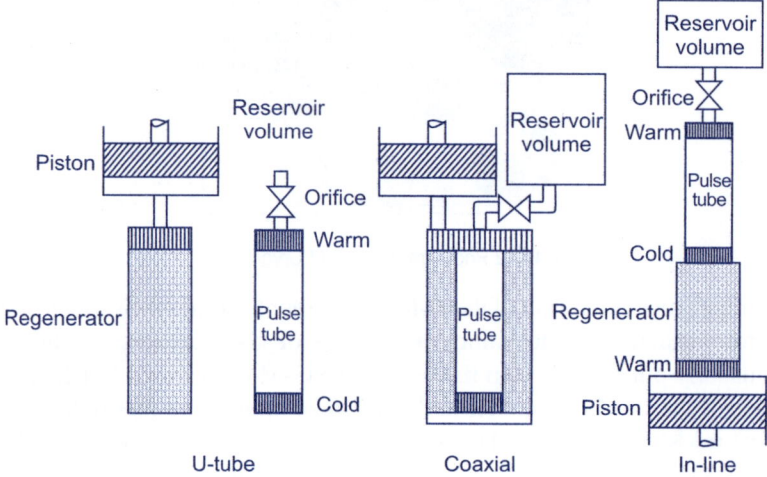

Fig. 15.13 Pulse Tube Cryocoolers

The four steps in the cycle are as follows. 1) The piston moves down to compress the gas (helium) in the pulse tube. 2) Because this heated, compressed gas is at a higher pressure than the average in the reservoir, it flows through the orifice into the reservoir and exchanges heat with the ambient through the heat exchanger at the warm end of the pulse tube. The flow stops when the pressure in the pulse tube is reduced to the average pressure. 3) The piston moves up and expands the gas adiabatically in the pulse tube. 4) This cold, low-pressure gas in the pulse tube is forced toward the cold end by the gas flow from the reservoir into the pulse tube through the orifice. As the cold gas flows through the heat exchanger at the cold end of the pulse tube it picks up heat from the object being cooled. The flow stops when the pressure in the pulse tube increases to the average pressure. The cycle then repeats. The function of the regenerator is that it precools the incoming high-pressure gas before it reaches the cold end. Refer Figure 15.14 for a Photograph of the Pulse Tube Cryocooler developed by NIST.

One function of the pulse tube is to insulate the processes at its two ends. That is, it must be large enough that gas flowing from the warm end traverses only part

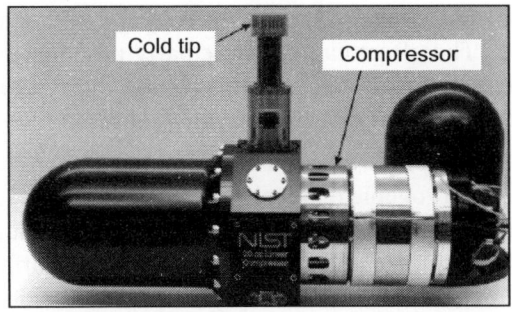

Fig. 15.14 Photograph of a Pulse Tube Cryocooler

way through the pulse tube before flow is reversed. Likewise, flow in from the cold end never reaches the warm end. Gas in the middle portion of the pulse tube never leaves the pulse tube and forms a temperature gradient that insulates the two ends. Roughly speaking, the gas in the pulse tube is divided into three segments, with the middle segment acting like a displacer but consisting of gas rather than a solid material. For this gas plug to effectively insulate the two ends of the pulse tube, turbulence in the pulse tube must be minimized. Thus, flow straightening at the two ends is crucial to the successful operation of the pulse tube refrigerator. The pulse tube is the unique component in this refrigerator that appears not to have been used previously in any other system. It could not be any simpler from a mechanical standpoint. It is simply an open tube. But the thermodynamics of the processes involved in it are extremely complex and still not well understood or modeled. The overall function of the pulse tube is to transmit hydrodynamic or acoustic power in an oscillating gas system from one end to the other across a temperature gradient with a minimum of power dissipation and entropy generation.

Pulse tube refrigerators were invented by Gifford and Longsworth in the mid 1960s, and only reached a low temperature of 124 K. In 1984 Mikulin introduced the concept of an orifice to the original pulse tube concept and reached 105 K. In 1985 Radebaugh changed the location of the orifice and reached 60 K. Further improvements since then have led to a low temperature limit of about 20 K with one stage and 2 K with two stages. There are three different geometries that have been used with pulse tube cryocoolers as shown in Figure 15.13 The inline arrangement is the most efficient because it requires no void space at the cold end to reverse the flow direction nor does it introduce turbulence into the pulse tube from the flow reversal. The disadvantage is the possible awkwardness associated with having the cold plate located between the two warm ends. The most compact arrangement and the one most like the geometry of the Stirling cryocooler is the coaxial arrangement. That geometry has the potential problem of a mismatch of temperature profiles in the regenerator and in the pulse tube that would lead to steady heat flow between the two components and a reduced efficiency. The absence of a moving displacer in pulse tube cryocoolers gives them many potential advantages over Stirling cryocoolers for the cooling of infrared sensors. One of the few disadvantages of pulse tube cryocoolers is the potential for gravitationally induced convective instabilities inside the pulse tube whenever the cold end of the pulse tube is raised above the warm end. The effect can be particularly pronounced in the off state, but the oscillating flow in the on state prevents the instability except for pulse tube diameters greater than about 10 mm.

In the last few years, pulse-tube coolers have been improved to such a level that their efficiencies at cryogenic temperatures are comparable to those of Stirling coolers. The most important improvement was a so-called double inlet that was introduced by Zhu in 1990. He added a by-pass from the warm end of the pulse tube to the inlet of the regenerator. In that by-pass a second orifice was placed. The benefit of this step is two-fold: first, it provides an extra parameter in optimizing the phasing between mass flow in the tube and pressure wave at the inlet; second, part of the gas flow that is required for expansion and compression at the warm end of the pulse tube is taken directly from the inlet, instead of passing through the regenerator and the pulse tube. A reduction in mass flow through the regenerator means lower regenerator losses, and thus a higher efficiency, especially at higher operating frequencies. Pulse-tube coolers may exhibit temperature stability problems because of unwanted gas flows. This may be a dc flow through the double inlet or gravity-induced secondary streaming in the tube. Apart from these problems, it is clear that the pulse-tube cooler has great advantages over the Stirling or the GM cooler: the absence of moving parts in the cold results in longer life, lower interference, and lower cost. Moreover, a pulse-tube cooler is less sensitive to side loads (i.e. forces in the radial direction) and less sensitive to contamination coming from the compressor. Two types of pulse-tube cryocoolers can be distinguished: A Stirling-type pulse tube is directly connected to the pressure-wave generating compressor, whereas a GM-type pulse tube is connected to an active-valve unit connected to a compressor. In terms of cooling power, input power, size, operating temperature and frequency, these pulse-tube coolers resemble their Stirling and GM-type counterparts.

Stirling Cryocooler

The Stirling cycle, as invented and patented by Robert Stirling in 1815, was first used as a prime mover. Refer Figure 15.15 for the Stirling cycle. In 1834 John Herschel proposed its use as a refrigerator in producing ice. It was not until about 1861 that Alexander Kirk reduced the concept to practice. Air was used as the working fluid in these early regenerative systems. Very little development of Stirling refrigerators occurred until 1946 when a Stirling engine at a Dutch company was run in reverse with a motor and was found to liquefy air on the cold tip. The engine used helium as the working fluid.

Stirling cryocoolers have been used for about last 40 years in cooling infrared sensors for tactical military applications in such equipment as tanks and airplanes. They cannot provide the very fast cool down times of JT cryocoolers, so they are not used on missiles for guidance. The long history of the Stirling cryocooler in cooling infrared equipment has resulted in many specifications being tailored to the geometry characteristics of the stirling cryocooler. As a result, newer cryocoolers,

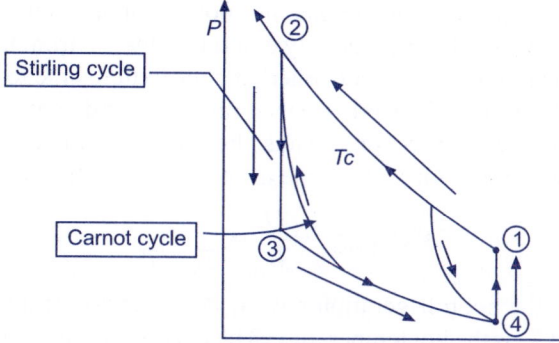

Fig. 15.15 Stirling Cycle

like the pulse tube cryocooler, with different geometries are difficult to adapt to the geometry specifications. A pressure oscillation by itself in a system would simply cause the temperature to oscillate and produce no refrigeration. In the Stirling cryocooler the second moving component, the displacer, is required to separate the heating and cooling effects by causing motion of the gas in the proper phase relationship with the pressure oscillation. Refer Figure 15.16 for the schematic of Stirling Cycle Cryocooler.

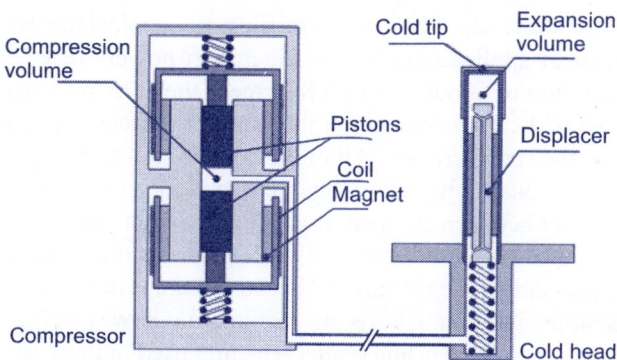

Fig. 15.16 Schematic of Stirling Cryocooler

When the displacer in Figure 15.16 is moved downward, the helium gas is displaced to the warm end of the system through the regenerator. The piston in the compressor then compresses the gas, and the heat of compression is removed by heat exchange with the ambient. Next the displacer is moved up to displace the gas through the regenerator to the cold end of the system. The piston then expands the gas, now located at the cold end, and the cooled gas absorbs heat from the system it is cooling before the displacer forces the gas back to the warm end through the regenerator. There is little pressure difference across the displacer (only enough to overcome the pressure drop in the regenerator) but there is a large temperature difference. Most actual Stirling cryocoolers have the regenerator inside the displacer. With this condition the mass flow or volume flow through the regenerator is approximately in phase with the pressure. Though the moving piston causes both compression and expansion of the gas, net power input is required to drive the system. The moving displacer reversibly extracts net work from the gas at the cold end and transmits it to the warm end where it contributes some to the compression work. In an ideal system, with isothermal compression and expansion and a perfect regenerator, the process is reversible. Thus, the coefficient of performance COP for the ideal Stirling refrigerator is the same as the Carnot COP given by

$$\text{COP}_{\text{Carnot}} = \frac{Q_c}{W_o} = \frac{T_c}{T_h - T_c},$$

where Q_c is the net refrigeration power, W_o is the power input, T_c is the cold temperature, and T_h is the hot temperature. The occurrence of T_c in the denominator arises from the PV power (proportional to T_c) recovered by the expansion process and used to help with the compression. Practical cryocoolers have COP values that range from about 1 to 25% of the Carnot value. The refrigeration effect for a stirling cryocooler depends upon regenerator effectiveness and any deviation from ideal behavior causes a loss in refrigeration effect. For the stirling cycle cryocooler the loss in refrigeration effect (*RE*) is

$$\Delta Q/Q_{\text{ideal}} = (1 - \varepsilon)(T_1/T_2 - 1)/(\gamma - 1) \ln(v_2/v_1)$$

where ε = regenerator effectiveness, v_2/v_1 = expansion ratio, T_1/T_2 = operating temperature ratio, γ = specific heat ratio.

Figure 15.17 shows the four sizes of Stirling cryocoolers that are currently used for military tactical applications. The refrigeration powers listed for each cooler are for a temperature of about 77 to 80 K, except the 1.75 W system, which is for a temperature of 67 K. Their specified minimum efficiencies range from about 3 to 6% of Carnot as the size increases. All of the coolers shown in Fig. 15.17 use linear drive motors with a dual-opposed arrangement to reduce vibration. The linear drive reduces side forces between the piston and the cylinder and the Mean-Time-To-Failure (MTTF) is at least 4000 hours. The displacer is driven pneumatically with the oscillating pressure in the system and because there is only one displacer it gives rise to considerable vibration. Efforts are currently underway to increase the MTTF of Stirling cryocoolers The Stirling cooler was first used in these space applications after flexure bearings were developed for supporting the piston and displacers in their respective cylinders with little or no contact in a clearance gap of about 15 µm.

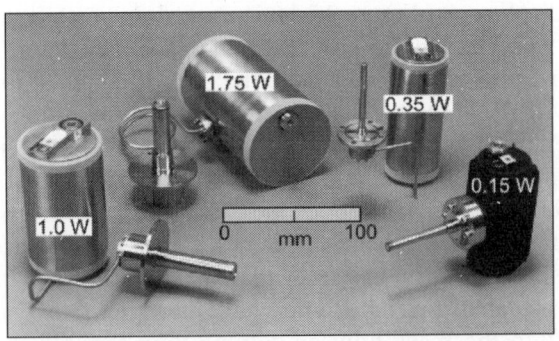

Fig. 15.17 Photograph of Stirling Cryocoolers

In the compressor, two pistons are driven by linear motors. The motion of each piston is produced by a time-varying magnetic field of a coil that interacts with a permanent magnetic field around the coil. The permanent magnets are fixed to the compressor housing. Because the coil is connected to the piston, a pressure wave is established in the gas that acts on the spring-loaded displacer in the cold head. In order to establish high compressor efficiency, the pistons are operated close to mechanical resonance (50/60 Hz). First, by moving the compressor pistons towards one another, the gas is compressed isothermally and the heat of compression is rejected to the environment. After the compression, the displacer moves down and displaces gas from the compression space to the expansion space. In this process, the gas is forced to flow through the interior of the displacer filled with a regenerator material. This is a material with a large specific heat (e.g., copper screens or lead spheres). Along the regenerator, a temperature gradient is established from the compression space at ambient temperature to the expansion space at low temperature. The gas exchanges heat with the regenerator material and is thus cooled to a temperature close to that of the cold tip. Then, the compressor pistons move apart and the gas is allowed to expand at the cold tip. Because the gas is in good thermal contact with the environment, the heat of expansion is absorbed from the environment. This is the actual cooling step in the cycle. After the expansion, the gas is returned to the compression space by upward motion of the displacer. Passing the regenerator the gas warms up to the compression space temperature. In this cooling cycle, the phase of the displacer motion with respect to that of the compressor pistons is very important. In general, the required phase difference can be obtained by attaching the compressor piston(s) and the displacer to a kinetic drive mechanism resulting in a disciplined piston/displacer Stirling cooler. Here, the dynamic gas forces acting on the displacer and the mechanical spring cause the movement of the displacer. The effective spring constant and the mass of the displacer determine the resonance frequency of this displacer motion. By adjusting the latter two parameters, the phase of the displacer motion can be tuned to that of the compressor pistons. Thus, at a given compressor operating frequency, the cooling performance can be optimized. A wide variety of smaller size Stirling cryocoolers are commercially available for tasks such as the cooling of electronic sensors and sometimes microprocessors. For this application, Stirling cryocoolers are the highest performance technology available, due to their ability to lift heat efficiently at very low temperatures. They are silent, vibration-free, and can be scaled down to small sizes, and have very high reliability and low maintenance.

Since the late 1950's, small Stirling coolers with helium as the working gas were especially developed for cooling infrared detectors in military applications. At present, thousands of units a month are produced worldwide for cooling these detectors in night vision equipment and missile guidance systems. Coolers for this market are manufactured with more or less standard dimensions: cold tip diameter 5 to 10 mm, length of the cold finger typically 6 cm, compressor diameter 55 mm and length 120 mm. Their cooling powers range from 0.2 W to 2 W at 80 K and an input power of typically 50 W. They are, usually operated at 50/60 Hz. Larger Stirling coolers are, for instance, applied for cooling superconducting filters in telecommunication systems. Custom-made multi-stage plastic Stirling coolers were investigated by Zimmerman at NIST in USA. The first Stirling cycle cryocooler was developed at Philips in the 1950s and commercialized in such places as liquid air production plants. The Philips Cryogenics business evolved until it was split off in 1990 to form the Stirling Cryogenics BV, The Netherlands. This company is still active in the development and manufacturing of Stirling cryocoolers and cryogenic cooling systems.

Comparison of Cryocoolers

The low operating temperature (below 4 K) can be attained with G-M coolers and pulse-tube coolers. Joule-Thomson coolers with two-stage precooling are available but are more complex and more expensive. G-M coolers have higher cooling capacities than pulse-tube coolers and correspondingly higher efficiencies; G-M coolers are available in the range 0.5 W to 1.5 W at 4 K with efficiencies of typically 1% of Carnot whereas GM-type pulse-tube coolers have cooling powers below 1 W at a Carnot efficiency of about 0.7%. Despite the lower efficiency, the pulse-tube cooler is more attractive for cooling then a G-M cooler because of the much lower level of vibration. In this respect, also a pulse-tube pre-cooled Joule-Thomson cooler may be attractive. In the high temperature range (above 4K) a large variety of coolers can be used. Below 1 W of cooling power, Stirling coolers are the most efficient with Carnot efficiencies up to 15%. Between 1 W and 10 W, pulse-tube coolers perform almost as well as Stirling coolers with typical Carnot efficiencies between 10% and 20%. Cooling powers well above 10 W can be attained with G-M coolers and pulse tubes; Carnot efficiencies about 5% at a few tens of watts, approaching 20% for cooling powers around 300 W. If the system shows hardly any power dissipation, the required cooling power will be below 10 W and, therefore, Stirling-type coolers are the most attractive in terms of cooling performance. However, the compressor in these coolers usually is close to the cold head and generates a lot of noise (electromagnetic noise and vibrations). Special measures are, therefore, required to reduce the noise. Joule-Thomson coolers are much less noisy but have a far lower efficiency, and as a result require relatively big compressors. Figure 15.18 shows the comparison of various cryocoolers in terms of Carnot efficiency and compressor input power.

The efficiencies reported here refer to the input electrical power to the compressor. The majority of pulse tube refrigerators have not achieved high efficiencies. Careful attention to details of the design are required with experimental optimization and computer modeling. In most cases these detailed designs remain proprietary information. Figure 15.18 shows the comparison of data for high efficiency Stirling, Gifford-McMahon, Brayton, and mixed refrigerant Joule-Thomson cryocoolers, all operating at temperatures near 80 K. This graph shows a general trend that efficiency increases with increasing size. The graph also indicates that pulse tube refrigerators have equaled or exceeded the efficiency of the best Stirling refrigerators. As a result, pulse tube refrigerators have now become the most efficient cryocoolers for a given size. Efficiencies as high as 24% of Carnot have now been achieved with pulse tube cryocoolers. The use of a valved compressor reduces the efficiency of the pulse tube cryocooler to that of Gifford-McMahon cryocooler.

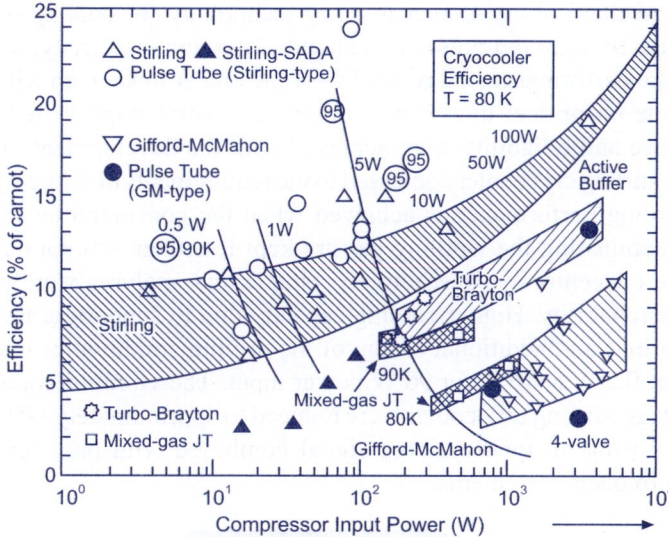

Fig. 15.18 Comparison of Cryooolers

Applications of Cryocoolers in Space

An advanced two-stage Stirling (2ST) cryocooler is used for the cryogenic systems of astronomical and earth-observation missions. These cryocoolers are used to cool infrared space telescopes with optical bench to be maintained at 4.5 K. The temperature of 4.5 K is achieved using a Joule-Thomson (JT) circuit combined with the cryocooler for precooling to 15-20 K; this precooling contributes strongly to the JT-circuit cooling performance. Therefore, the cooling performance and reliability of the cryocooler is the key to the success of the space mission. Reduction of vibration induced by the cryocoolers is another important technical issue to avoid deterioration of spatial resolution and pointing stability of optical devices. Cryocooler for space require higher cooling performance, better reliability, and less vibration. Refer Figure 15.19 for a space cryocooler.

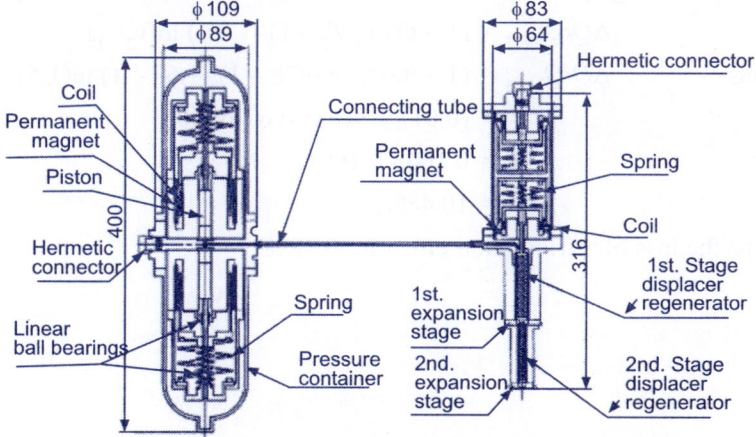

Fig. 15.19 Space Cryocooler

15.24 Cryogenics

The mechanical cryocooler is a key technology for cooling optical devices to minimize background noise and enhance detection sensitivity; it is required to improve the performance and reliability of the cooler to keep up with the progress being made on optical device performance. Several improvements in cooling performance and reliability were achieved with the development of an advanced two-stage stirling cryocooler addressed to the requirements of future space missions. Stable cooling performance is achieved when the concentration of CO_2 and N_2 gases contaminating the working gas are kept less than 500 ppm and 1000 ppm of volume concentration, respectively. Outgassing analyses were used to verify the higher reliability. Higher cooling capacity due to supporting the displacer on flexure springs and additional tuning of the driving condition provided a cooling capacity of 0.2 W at 16 K for 90 W power input. The vibration forces induced by the two-stage Stirling cryocooler were reduced to approximately 1/50th the original amplitude using an input voltage signal combined with predetermined higher harmonic frequency elements.

Solved Problem

Problem 1. In the case of stirling cycle cryocooler for 1% deviation from ideal value of effectiveness of regenerator, calculate% loss of refrigeration effect (RE) when volume expansion ratio is 1.5 and helium is the working fluid. Assume suitable operating temperatures.

Solution:

We assume operating temperatures as 300 K and 78 K.

Thus $T_1 = 300$ K and $T_2 = 78$ K

Effectiveness = ε = 0.99 (Given as 1% deviation from ideal value of 100%)

v_2/v_1 = Volume expansion ratio = 1.5 (given)

$\gamma = 1.67$ for helium

For the Stirling cycle cryocooler the loss in refrigeration effect (RE) is

$$\Delta Q/Q_{ideal} = (1-\varepsilon)(T_1/T_2 - 1)/(\gamma - 1) \ln(v_2/v_1)$$

Thus $\Delta Q/Q_{ideal} = (1 - 0.99)(300/78 - 1)/(1.67 - 1) \ln(1.5)$

$= (0.01)(2.846)/(0.67)(0.405)$

$= 0.1048 \times 100\%$

$= 10.48\%.$

Thus the loss of refrigeration effect is 10.48%.

References

1. Cryogenic Systems – Randall Baron.
2. Research paper on Brayton cycle cryocoolers by Nieczkoski et.al. (2003).
3. Cryogenic Engineering – B.A. Hands.
4. Presentation on Cryocoolers by Budapest University.
5. Cryogenics assessment report by University of Wisconsin (2005).
6. Presentation on Cryocoolers by NASA (2008).
7. Research paper on Stirling cycle cryocoolers by Sato et.al., JAEA, Japan.
8. Internet website of Wikipedia on cryocoolers.
9. Research paper on Small Cryocoolers by Baek et.al., Korea Institute of Energy & Resources, Korea (1990).
10. Ray Radebaugh, paper on Pulse Tube Cryocoolers, Proceedings of Infrared Technology and Applications XXVI, Vol. 4130, pp. 363-379 (2000).
11. Presentation by CERN on cryogenic engineering.
12. Paper on Squid Systems by Foley et. al.
13. Overview of Lockheed Martin Cryocoolers by Nast et.al. Cryogenics v 46 (2006) pp164–168.
14. Internet website http://www.jlab.org/hydrogen.

Questions

Q.1. What are cryocoolers? Enlist at least three cryocoolers.

Q.2. Describe the construction of G-M cryocooler with a neat sketch.

Q.3. Write a short note on pulse tube cryocooler.

Q.4. Describe the classification of cryocoolers.

Q.5. Explain the Stirling cryocooler with a neat sketch.

Q.6. What is a sorption cryocooler?

Q.7. Describe the Brayton cryocooler with a neat sketch.

Q.8. Discuss the space applications of cryocoolers.

Q.9. In the case of Stirling cycle cryocooler for 1% deviation from ideal value of effectiveness of regenerator, calculate% loss of refrigeration effect (RE) when volume expansion ratio is 1.5 and helium is the working fluid. Assume operating

References

1. Cryogenic Systems – Randall Barron.
2. Research paper on Stirling Cycle cryocoolers by Atrey et al. (2001).
3. Cryogenic Engineering – R. B. Timms.
4. Presentation on Cryocoolers by Dartmouth University.
5. Cryocooler assessment report by University of Wisconsin (2009).
6. Presentation on Cryocoolers by NASA GODR.
7. Research paper on Stirling cycle cryocoolers by Sato et al., AIAA, Japan.
8. Internet website of Wikipedia on cryocoolers.
9. Research paper on Small Cryocoolers by Choi et al., Korea Institute of Energy & Resources, Korea (1991).
10. Radebaugh, paper on Pulse Tube Cryocoolers for coolings of infrared detectors, and appearance AXVI AIAA 130, pp 363-379 (2007).
11. A presentation by NASA on Cryocoolers cooling.
12. Paper on Stirling Systems, by Pohl et al.
13. Overview of Lockheed Martin Cryocoolers by Nast et al., Cryogenics v 46 (2006) pp164-168.
14. Internet website http://www.nist.org by design.

Questions

1. Q. What are cryocoolers? Further at least the four questions.
2. Q. Describe the operation of a GM cryocooler with a neat sketch.
3. Q. Write a short note on pulse tube cryocooler.
4. Q. Describe the classification of cryocoolers.
5. Q. Explain the Stirling cryocooler with a neat sketch.
6. Q. What is a sorption cryocooler?
7. Q. Describe the Joule-Thomson cryocooler with a neat sketch.
8. Q. Discuss the space applications of cryocoolers.
9. Q. In a J-T type of Stirling cycle cryocooler for the evaluation the technical value of effectiveness of regeneration. Influence of loss of refrigeration effect (RE) when cooling expansion unit is T_e and fraction of the working fluid. Assume operating conditions.

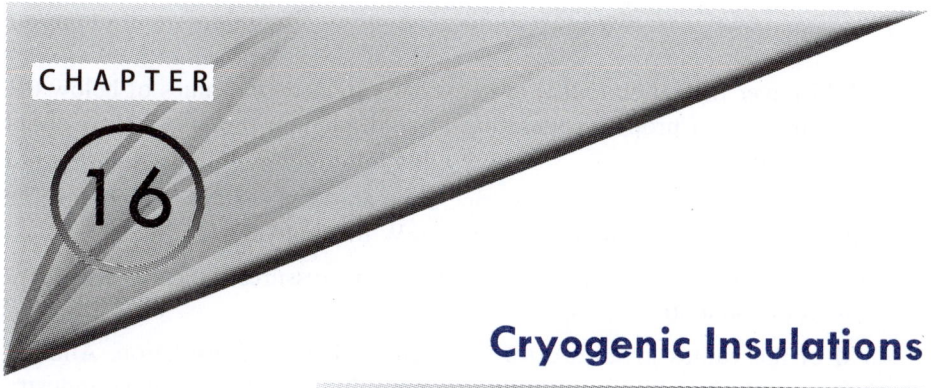

CHAPTER 16

Cryogenic Insulations

INTRODUCTION

Cryogenic vessels that require a high level of thermal isolation are typically enclosed in an outer vessel with a separating space that is vacuum evacuated. With an ambient vacuum 24-hour settle pressure in the 10^{-4} torr range, *convective heat transfer* across this space is virtually eliminated. Creating a very small heat path from the outer to the inner vessel typically controls *conductive heat transfer*. *Radiated heat transfer* is typically controlled by the barrier placed around the inner vessel. Its mission is to prevent heat from radiating into the inner vessel. Thus selecting a material for the heat path which blocks all three modes of heat transfer and has very low thermal conductivity properties normally works well. Such materials which have poor conductivity characteristics are called insulators.

Insulation may be categorized by its composition (material), by its form (structural or non-structural), or by its functional mode (conductive, radiative, convective). Non-structural forms include batts, blankets, loose-fill, spray foam, and panels. Structural forms include insulating concrete forms, structured panels, and straw bales. Sometimes a thermally reflective surface called a radiant barrier is added to a material to reduce the transfer of heat through radiation as well as conduction.

Several types of insulations can be used in cryogenic equipment. These include 1) expanded foam, 2) solid foam, 3) gas filled powders and fibers, 4) vacuum insulation, 5) evacuated porous and fibre materials, 6) opacified powder, 7) multilayer insulation 8) composite layer insulation 9) glass microspheres and 10) vapour shielding. These insulations are used in various applications based on their cost, effectiveness, ruggedness, ease of application, weight, availability and other factors. We shall now study each type of insulation in detail.

Insulation Performance Regarding Heat Transfer

Cryogenic insulation system performance is often reported for large temperature differences in terms of an apparent thermal conductivity, or "k" value. Boundary temperatures of 77 K (liquid nitrogen) and 295 K (room temperature) are common. The following "k" values discussed apply generally to these boundary conditions.

16.2 Cryogenics

- Multi Layer Insulation (MLI) systems can produce "k" values of below 0.1 mW/m-K when properly operating at cold vacuum pressure below about 1×10^{-4} torr.
- For bulk filled insulation systems operating at a cold vacuum pressure below about 1×10^{-3} torr, k values of about 2 mW/m-K is typical.
- Foam and other similar materials at ambient pressures typically produce k values of about 30.

All of the values given are for 1.0-inch thickness of insulation. Another characteristic of cryogenic insulations is the R value. The R value is an industry standard unit of thermal resistance for comparing insulation values of different materials. The R value is a measure of resistance to heat flow in units of degree F-hour-square foot/BTU-in.

The relationship between "k" and "R" values is described as follows:

$k = d/R$ where d is the heat flow distance.

Comparative assessment of performance of insulations can be done using effective conductivity as a tool. At a vacuum pressure of 0.02 torr many of the basic insulation designs give about the same thermal performance. Tests were conducted with 1" of insulating material (60 layers of MLI) in a test dewar to compare various insulating materials as a function of vacuum pressure. The results are shown as a plot of effective conductivity in Figure 16.1 below [3]. The tests indicate effectiveness of MLI and Opacified bead insulation vis a vis some commonly used commercial insulators.

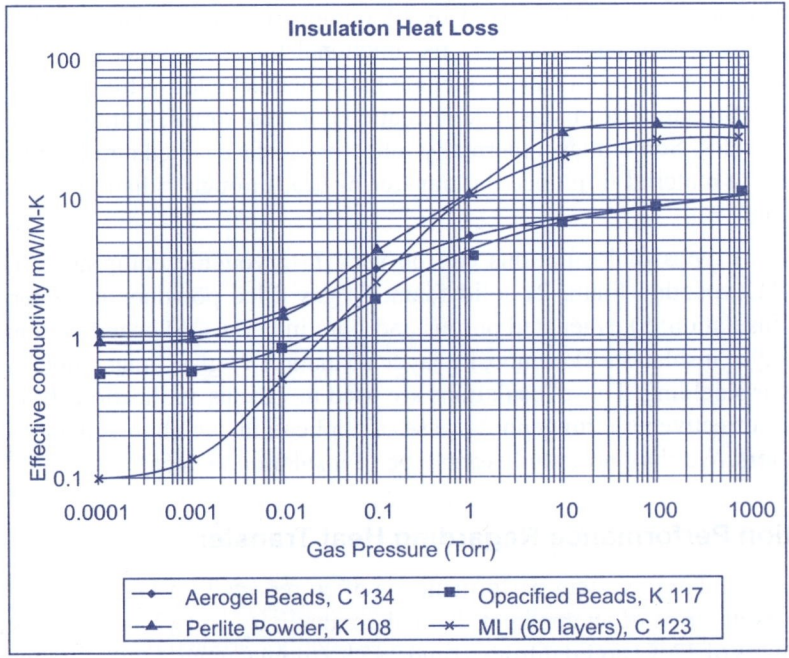

Fig. 16.1 Insulation Performance Chart

EXPANDED FOAM INSULATION

Expanded foam insulation is a spray applied cellular plastic that forms a continuous, monolithic barrier on walls, around corners, and on all contoured surfaces, and protrusions. It adheres directly to the substrate forming a seamless permanent coverage over and around even the most difficult shapes. Styrofoam is one such example of expanded foam, which was developed by the Dow chemical company. Other examples are expanded polystyrene, silica foam etc. Expanded foam is a rigid white material made by expanding plastic beads like polystyrene with steam and bonding the beads together under pressure in a block or mold shape. The properties of a typical expanded foam insulation are given in Table 16.1 and Figure 16.2 shows some expanded foams. Expanded foam comes in several varieties that can meet cryogenic insulating needs. Spray foam insulation has water and fire-retardant properties as well as resistance to mold and mildew.

Table 16.1 Expanded Foam Properties

Method	Description	Value
ASTM D 1622	Density	32-35 Kg/m^3 (2.0 – 2.2 lb/ft^3)
ASTM C-518	Thermal Resistance 90 days @ 23°C	1.22 m^2.°C/W (6.9 ft^2.h.°F/BTU.in)
ASTM D-2856	Open Cell Content (%)	6.02%
ASTM D-1621	Compressive Strength	174 kPa (25 psi)
ASTM D-1623	Tensile Strength	212 kPa (31 psi)
ASTM D-2126	Dimension Stability (% Volume change @ 28 Days) −20°C, ambient R.H. 100°C 70°C	0.47 5.89 2.58
ASTM D-2842	Water Absorption (% volume)	0.62
ASTM E-96	Water Vapour Permeance	86.6 ng/Pa.s.m^2
CCMC 07273	Air Barrier Material Test	0.00014 L/ms^2 @ 75 pa
ASTM E-283	Air Leakage (75 Pa = 40 Km./h)	0.000 L/s.m^2
ASTM E-330	Gust Wind (3000 Pa = 255 Km/h)	No delamination
CAN/ULC	Flame spread classification (2 inches thick)	335
S-102 M&S-127	Smoke develped (2 inches thick)	315
MVSS 302	Extinction	B (Self-extinguishing)

Expanded foam insulations have a cellular structure formed by evolving gas during the manufacture of the foam. The thermal conductivity of the foam depends on the gas used in manufacture of the foam. Normally Carbon dioxide due to its low vapor pressure is preferred. When using expanded foam for cryogenic applications,

16.4 Cryogenics

the thermal conductivity of the foam reduces when in contact with low temperatures due to condensation of the trapped gas in the foam. However when exposed to ambient the conductivity again rises.

Advantages of expanded foam include blockage of airflow by expanding and sealing off leaks, gaps and penetrations. It can serve as a vapor barrier with a better permeability rating than plastic sheeting and consequently reduce the build up of moisture, which can cause mold growth. It works well in tight spaces and provides acoustical insulation. It expands while curing thereby filling bypasses and increaseing structural stability. Disadvantages are the high cost and lower life. One of the major limitations of these kinds of foam are large thermal contractions when exposed to varying temperature conditions. Due to unusually high coefficient of expansion (Approx seven times that of steel), the foam may crack during uneven expansion of the cryogenic vessel. The cracked foam can also allow air and water to enter through the cracks further deteriorating the performance.

Fig. 16.2 Expanded Polystyrene Foam

SOLID FOAM INSULATION

This foam is a multiple purpose, two-component polyurethane froth foam designed within the international guidelines for protection of the ozone layer, with respect to the Montreal Protocol, 1987 and other environmental guidelines, utilizing a non-flammable, non-ozone depleting blowing agent to assist in the safety of the end user and the environment. This foam is any polymer consisting of a chain of organic units joined by carbon links. Polyurethane foams are examples of such foams. These polymers are formed through step-growth polymerization by reacting a monomer containing at least two isocyanate functional groups with another monomer containing at least two hydroxyl (alcohol) groups in the presence of a catalyst. The R value ranges from 11 to 22.

Polyurethane formulations cover an extremely wide range of stiffness and hardness and are ideally suited for cryogenic applications. The pioneering work on polyurethane polymers was conducted by Otto Bayer and his coworkers in 1937 in Germany. They observed rigid foams based on polymeric MDI offered better thermal stability and combustion characteristics. In 1967, urethane modified polyisocyanurate rigid foams were introduced, offering even better thermal stability and flammability resistance to low-density insulation products. Polyurethane foam (including foam rubber) is often made by adding small amounts of volatile materials called blowing agents, to the reaction mixture. These simple volatile chemicals yield important performance characteristics, primarily thermal insulation. In the early 1990s, because of their impact on ozone depletion, the Montreal Protocol led to the greatly reduced use of many chlorine-containing blowing agents, such as trichlorofluoromethane (CFC-11). By the late 1990s, the use of blowing agents such as carbon dioxide, pentane, and 1,1,1,2-tetrafluoroethane (HFC-134a) became more widespread.

Building on existing polyurethane spray coating technology and polyetheramine chemistry, extensive development of two-component polyurea spray elastomers took place in the 1990s. Their fast reactivity and relative insensitivity to moisture make them useful coatings for large surface area projects, such as secondary containment, manhole and tunnel coatings, and tank liners. Excellent adhesion to steel is obtained with the proper primer and surface treatment.

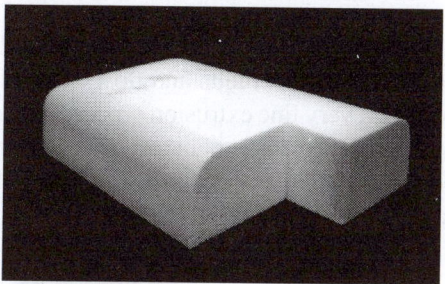

Fig. 16.3 Solid Polyurethane Foam

Solid foams can be made in a variety of densities and hardnesses by varying the type of monomer(s) used and adding other substances to modify their characteristics, notably density, or enhance their performance. Other additives can be used to improve the fire performance and stability in difficult chemical environments. Though the properties of the polyurethane are determined mainly by the choice of polyol, the isocyanate exerts some influence, and must be suited to the application. The cure rate is influenced by the functional group reactivity and the number of functional isocyanate groups. The mechanical properties are influenced by the functionality and the molecular shape. There are then two main foam variants: one in which most of the foam bubbles (cells) remain closed and the other having mostly open cells. The retention of the cell gas is desired since this gas gives the foams their characteristic high thermal insulation performance.

GAS FILLED POWDERS AND FIBRE INSULATION

Gas filled powder insulations and fibre insulations are available in a range of forms from light weight rolled products to heavy rigid slabs including preformed pipe sections. In addition to good thermal insulation properties, they exhibit excellent acoustic insulation and fire retardation properties. Examples are rock wool, Perlite, Vermiculite etc. When this porous insulation is used, a vapor barrier must be applied to the outer surface of the insulation to prevent water vapor diffusion into the insulation. The primary mechanism for insulation in gas filled powders is the reduction in convection due to lower void size in the material. Furthermore fine powders also alter the conduction mechanism from continuum to free molecular type.

Fiberglass or fiber glass which consists of extremely fine glass fibers–is one of the most commonly used insulation materials. It is used in two different insulation forms: blanket (batts and rolls) and loose-fill. Fiberglass insulation is naturally non-combustible because it is made from sand and recycled glass. The insulation requires no additional fire-retardant chemical treatments. Rock wool is made from basaltic rock and recycled material from steel mill wastes. The batts or rolls typically come at a density of 0.5 pcf with an R-value of 3.1. Fiberglass is a common choice for new construction and there is considerable flexibility in its application. It is fireproof but is an irritant to the applicator's lungs and skin. Fiberglass insulation resembles cotton candy, and the manufacturing process even resembles a large cotton candy machine. However, instead of sugar, the main ingredient is silica sand. The sand is heated to a high temperature where it melts and flows as liquid glass. The glass is then spun into fibers through a very fine extrusion process. It is collected and sprayed with binders (glue) and shaped into large blankets of various thicknesses. It is then cut into strips for rolled product or left large for batt products. Fiber glass insulation products come in R values ranging from R-11 to R-38 for fiber glass batts and rolls.

Fig. 16.4 Fibre Glass Insulation

Newer fibrous insulation products combine two types of glass, which are fused together. As the two materials cool during manufacturing, they form random curls of material. This material is less irritating and possibly safer to work with. It also requires no chemical binder to hold the batts together, and the material even comes in a perforated plastic sleeve to assist in handling. Advantages of fibre insulation are 1) relatively inexpensive, 2) very light-weight, 3) dries out if it gets wet and retains R-value, 4) good R-value per inch thickness. Disadvantages of fibre insulation are 1) can be a skin irritant when handling, 2) does not block air penetration and 3) does not provide an integral vapor barrier.

VACUUM INSULATION

The use of vacuum insulation essentially eliminates two components of heat transfer namely solid conduction and gaseous convection. Heat is transferred across the annular space of a vacuum insulated line by radiation from the hot outer jacket to the cold inner line and by gaseous conduction through the residual gas within the annular space. The radiant heat transfer rate between the two surfaces is given by the modified Stefan Boltzman's equation.

$$Q = F_0 F_{1-2} \sigma A_1 (T_2^4 - T_1^4)$$

where
F_e = Emmisivity Factor
F_{1-2} = Configuration Factor
σ = Stephen Boltzman's constant
A_1 and A_2 = Area of surface in m^2
T = Absolute temperature K.

In addition the emmisivity factor for the diffused radiation for concentric cylinder is given by

$$\frac{1}{F_e} = \left[\frac{1}{e_1} + \frac{1}{e_2} - 1\right] + \left[\frac{2}{e_3} - 1\right] + \left[\frac{1}{e_2} + \frac{1}{e_3} - 1\right]$$

where
e_1 and e_2 = Exissivity of surface 1 and 2 respectivety.

For cryogenic transfer line where inner line is completely enclosed by the outer line, $F_{1-2} = 1$, where 1 and 2 refers to the enclosed inner line and the enclosing outer line respectively. In addition to heat transfer by radiation, energy transmitted by conduction through residual gases is also to be considered. If the pressure of the gas is low enough the mean free path of the gas molecules is greater than the distance between the two surfaces, then the conduction differs from usual continuum type at atmospheric pressure and in such a case free molecular conduction may be determined from the following equations.

$$\lambda = \frac{\mu}{p}\left[\frac{\pi RT}{2ge}\right]^{1/2}$$

$$Q = GpA(T_2 - T_1)$$

$$G = \frac{\gamma+1}{\gamma-1}\left[\frac{g.R}{8\pi T}\right]^{1/2} F_a$$

where
λ = mean free path m
μ = Gas viscosity Pa-s
T = Gas temperature K
P = Absolute pressure of the gas Pa
g_c = Conversion factor 1 kg-m/N-s^2
γ = Specific heat ration
R = Specific gas constant
F_a = Accommodation coefficient given by the equation

$$\frac{1}{F_a} = \frac{1}{e_1} + \frac{A_1}{A_2}\left[\frac{1}{e_2} - 1\right].$$

Vacuum insulation is used for laboratory size dewar vessels. Radiant heat transfer can be reduced by interposing floating thermally isolated radiation shields between hot and cold surfaces. For the gas molecules between hot and cold surfaces, the degree of approach to thermal equilibrium is given by accommodation coefficient "a" which is a ratio of the actual energy transfer to the maximum energy transfer.

EVACUATED PORUS AND FIBRE INSULATION

To significantly improve the performance of insulating systems, vacuum or evacuated porus technology is currently being used by the cryogenic industry. The absence of air or gas in panels affords the possibility of substantial enhancement of insulating performance. A useful evacuated insulation system is one employing a panel core stock of an open-cell foam. The open-cell structure of the foam allows rapid and substantially complete withdrawal of gases from within the foam structure and the panel. The rigid foam matrix provides a core stock of substantial mechanical strength and performance. Refer Figure 16.5 for the evacuated foam matrix.

Fig. 16.5 Evacuated Foam Matrix

Though evacuated insulation panels are hermetically sealed, gases and vapors such as air and water vapor will seep or permeate into the evacuated interior of the panel over a long period of time. The presence of gases or vapors in the evacuated interior of the insulation panel denudes the insulation capability of the panel. It would be particularly desirable to use such an evacuated insulation panel employing an open-cell foam of an alkenyl aromatic polymer. Normally a getter material is provided in the interior of the panel to adsorb gases and vapors which may permeate into the panel over an extended period of time. The interior of the core stock is evacuated to an absolute pressure of about 10 torr or less, the polymer foam is having an open cell content of about 90% or more and a density of from about 16 to about 250 kilograms per cubic meter. Other types of useful getter materials include conventional desiccants, which are useful for adsorbing water vapor or moisture. Such materials are advantageously incorporated into the evacuated insulation panel in the form of a packet having a porous or permeable wrapper or receptacle containing the material therein. Useful materials include silica gel, activated alumina, aluminum-rich zeolites, calcium chloride, calcium oxide, and calcium sulfate. A preferred material is calcium oxide.

The variation in heat flux and total heat transferred by an evacuated powder insulation in relation to residual pressure is shown in Figure 16.6. The heat flux is directly proportional to the conductivity as per the Fourier's law.

$$Q = -kA\, d/d_x = -kA(T_1 - T_2)/\Delta x$$

Note that as the pressure of the gas within insulation is reduced by evacuation to 15000 millitorr or 15 torr there is practically no change in heat flux and the conductivity. However as the pressure is reduced to 100 millitorr, one can see a large reduction in the heat flux and the conductivity. The heat flux exhibits linear variation with the pressure. The magnitude of gaseous conduction diminishes and radiation effects become dominant. If the pressure of gas is further decreased to 10 millitor or less, the heat flux and thermal conductivity levels off. Evacuated powder insulations are superior to vacuum insulation near room temperature due to higher radiant contribution in total heat transferred.

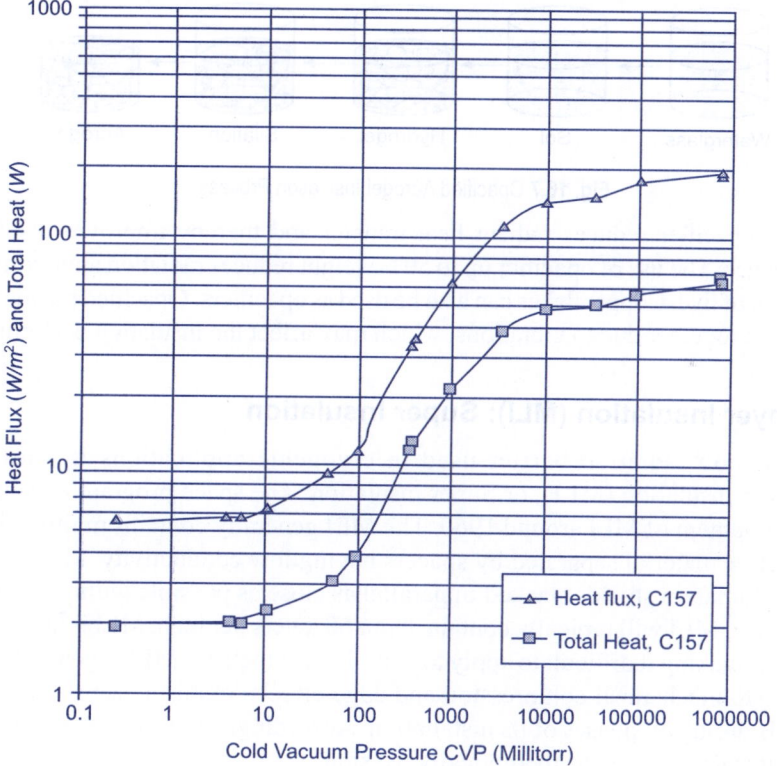

Fig. 16.6 Heat Flux vs Residual Pressure

OPACIFIED POWDER INSULATION

Insulators with reduced high temperature thermal conductivities such as the opacified aero gel silica composite insulators are now being used. The advantage of an opacified aero gel-matrix composite material lies in formulations and processes

16.10 Cryogenics

that result in superior properties, which include 1) much less shrinkage during a supercritical-drying process employed in producing a typical aero gel, 2) much less shrinkage during exposure to high temperature, and 3) as a result of the reduction in shrinkage, much less or even no cracking. An opacified aero gel-silica-fiber composite is synthesized by means of a sol-gel process. Except for the addition of the TiO_2 powder, the process is almost identical to that used to make the prior, non-opacified version. Fumed silica mesh powder and TiO_2 powder (in particle sizes between 1 and 2 μm) are suspended in acetonitrile solution and then the silica gel is added. After thus preparing the aero gel casting solution, a piece of silica fiber felt is placed in a mold. Then the aero gel casting solution is poured into the mold, where it permeates the silica fiber felt. It is necessary to consider that as the concentration of TiO_2 increases, the opacity increases (and thus the radiative contribution to heat transfer decreases) while the conductive contribution to heat transfer increases. Refer Figure 16.7 for the opacified aero gel insulation process.

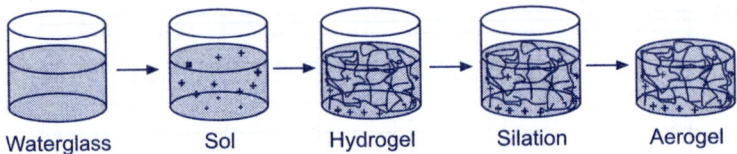

Fig. 16.7 Opacified Aerogel Insulation Process

The opacifier reduces radiant heat transfer and thereby improves insulation performance. The use of opacifier up to 50% weight reduces radiation approximately by a factor of five. Copper flakes can also be used as opacifiers. Opacified powders can get packed together due to vibrations, which may affect the insulation performance.

Multilayer Insulation (MLI): Super Insulation

One common radiation barrier used in cryogenic applications is known as Multilayer Insulation (MLI), or Super insulation. The space program encouraged the development of MLI around 1960. The MLI generally contains multiple layers of reflective material separated by spacers having low conductivity. MLI consists of many radiation shields stacked in parallel as close as possible without touching one another. MLI will typically contain about 60 layers per inch. MLI is anisotropic by nature, making it difficult to apply to complex geometries. MLI is generally very sensitive to mechanical compression and edge effects, requiring careful attention to details during all phases of its installation. Accordingly, performance in practice is not typically as good as theoretically possible.

Fig. 16.8 Multilayer Insulation

Each layer is isolated from the other by spacer material such as polyester, nylon, or mylar. Refer Figure 16.8 for a photograph of multilayer insulation. The aluminum foil is carefully wrapped around the container such that it covers the entire surface of the inner vessel. Spacer material, as described, is placed between the layers to completely prevent the separate coverings of foil from contacting. Should they touch, a thermal short circuit will occur and increase the heat transfer. The layers can be applied manually as blankets. These are hand cut to fit and wrapped over the vessel and vessel ends. Tape that has low out-gassing properties is then used to hold the blanket layers in place. Another method of applying the layers is by "orbital wrapping". This method is used where high-volume vessels are being manufactured. Special equipment is required that wraps the alternating layers much like the wrapping of a spool of string.

As the number of layers increase the insulation capability is also increased. Refer Figure 16.9 for the graph of layer effectiveness of MLI. Typically layers adding up to about one inch in total thickness is applied in the liquid nitrogen temperature range described. MLI is designed to work under high order vacuum, i.e., pressure below about 1×10^{-4} torr. To obtain this vacuum generally requires lengthy pumping along with heating and purging cycles. Chemical materials are required to absorb the out-gassed molecules to maintain the vacuum over extended periods. Figure 16.8 shows a typical multilayer insulation. Polyimide and/or polyester layers (from 5 to 30 layers) that are vapor deposited with 99.99% aluminum, on one or both sides, assist in the MLI material's heat management. The terms absorption and emittance are two critical factors in the design and effectiveness of an MLI blanket.

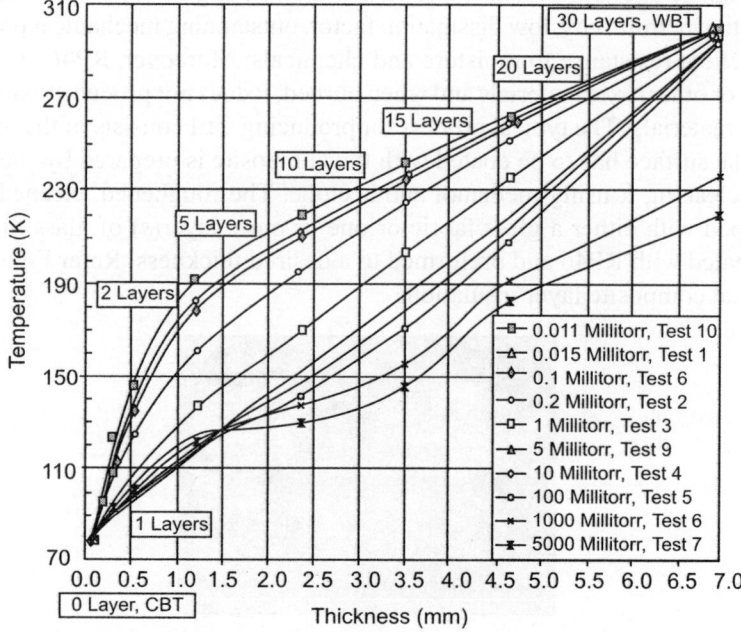

Fig. 16.9 Layer Effectiveness of MLI

Heat transfer through a multi-layer insulation involves combined modes of heat transfer including solid conduction through fibers, gas conduction and natural convection in spaces between fibers, and radiation interchange through participating media in the fibrous insulation spacers between reflective foils. The overall heat transfer from MLI is reduced to a minimum. Radiation is reduced by use of reflective metal foils, convection is reduced by use of lower residual gas pressure and conduction is reduced by crinkling shield material to minimize contact.

Thermal conductivity (k) of multilayer insulation is given by

$$K = (N/\Delta x)^{-1} [h_c + \sigma e(T_h^2 + T_c^2)(T_h + T_c)/(2-e)]$$

Where $N/\Delta x$ = layer density, σ is the Stefan Boltzmann constant, e is the emissivity of shield material, h_c is the solid conductance, T_h and T_c are the boundary temperatures. The thermal conductivity of MLI in direction parallel to the sheets is higher than in the normal direction.

The drawback of MLI is the problems associated with evacuation of residual gas from spaces in between insulation layers. This may be solved by providing small vent holes for outgassing the trapped gas.

Composite Layer Insulation (CLI)

Lightweight composites of RP46 polyimide and glass fibers have been found to be useful as extraordinarily fire-resistant insulation materials. RP46 is a polyimide of the polymerization of monomeric reactants (PMR) type, developed by NASA Langley Research Center. Properties of CLI insulators make them attractive for use at high temperatures. These properties include high-temperature resistance, low relative permittivity, low dissipation factor, outstanding mechanical properties, and excellent resistance to moisture and chemicals. Moreover, RP46 contains no halogen or other toxic materials and when burned, it does not produce toxic fume or gaseous materials. The typical process for producing CLI consists of the following steps. The surface has to be coated with the composite is prepared by roughening it, then cleaning it using methanol and acetone. The roughened, cleaned surface is wrapped with either a glass fabric or one or more layer(s) of glass fibers pre-impregnated with RP46 and preformed to a desired thickness. Refer Figure 16.10 for typical composite layer insulations

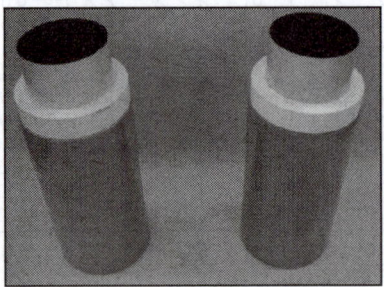

Fig. 16.10 Composite Layer Insulations

Glass Microsphere Insulation

Glass microsphere insulation, typically consisting of hollow glass bubbles, combines in a single material the desirable properties that other insulations only have individually. The material has high crush strength, low density, is noncombustible, and performs well in soft vacuum. Microspheres provide robust, low-maintenance insulation systems for cryogenic transfer lines and dewars. They also do not suffer from compaction problems typical of perlite that result in the necessity to reinsulate dewars because of degraded thermal performance and potential damage to its support system. Since microspheres are load bearing, autonomous insulation panels enveloped with lightweight vacuum-barrier materials can be created. Refer Figure 16.11 for a micrograph of glass microspheres.

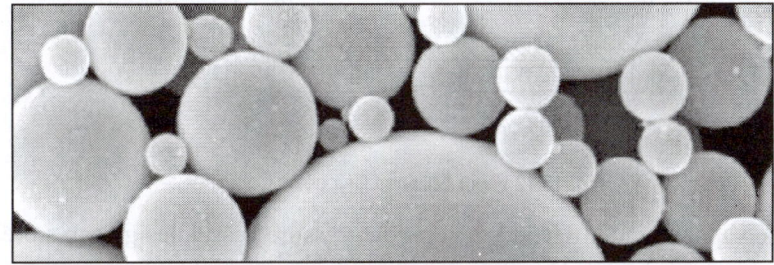

Fig. 16.11 Glass Microspheres

Cryogenic insulation systems that employ glass microspheres in evacuated powder form offer significant advantages over traditional materials for many practical applications. The best insulation material for a cryogenic system is the one that offers the optimal combination of thermal performance, low cost, light weight, durability, and minimal or no maintenance. While the thermal performance of microspheres falls considerably short of MLI, it is important to note that the vacuum space of a commercial vacuum-jacketed cryogenic tank or transfer line is usually only partially filled with MLI in order to enable nesting of the insulated inner vessel within the vacuum shell. The complete filling of the vacuum space with microspheres can improve the comparative thermal performance of microspheres to MLI to within a factor of two.

Vapor Shielding

The storage of cryogenic liquid is an efficient storage method due to high volumetric energy density. The main problem of cryogenic liquid storage is the boil-off losses due to heat leakage. To minimize the boil-off losses, multi layer insulation with Vapor Cooled Shield (VCS) is used in cryogenic vessels. In a cryogenic tank with VCS, the evaporated cryogenic liquid flows circularly through a spiral pipe which surrounds the insulator of the tank and absorbs some part of the heat leakage. The spiral pipe acts as a VCS since it decreases the heat leakage and boil-off losses significantly. To design a cryogenic tank with VCS, the main issue is to find the optimum location of VCS to minimize the heat leakage. One has to consider both radial and axial heat transfers as well as the temperature dependence of heat capacity and heat transfer coefficient of the cryogenic liquid. Figure 16.12 shows a vapour shielded cryogenic container.

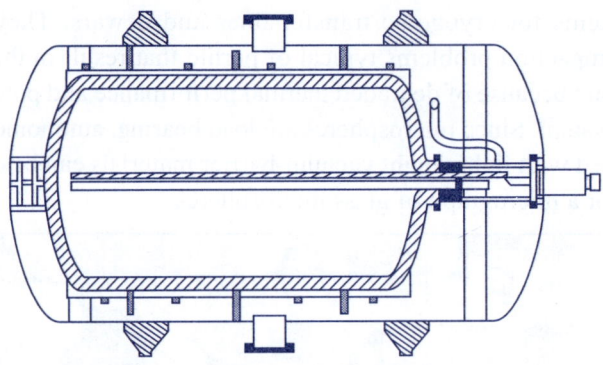

Fig. 16.12 Vapor Shielded Cryogenic Container

The vapor shielded second-generation pressure vessel design can store about 6 kg of cryogenic liquid. This design includes a vapor shield to reduce evaporative losses in addition to the instrumentation and safety devices that exist in the first generation vessel. The vessel contains a vacuum space, for obtaining high thermal performance from the multilayer insulation, instrumentation for pressure, temperature and level. These vessels are currently being used for DOT and SAE tests, and for incorporation into demonstration vehicles.

The vapor shield consists of the cold vent gas which intercepts some heat loss. The effectiveness of the shield depends upon ratio of sensible heat absorbed by vent gas to the latent heat of cryogenic fluid. The heat transfer rate (Q) is calculated as

$$Q = U \Delta T = (kA/\Delta x) \Delta T = mhf = m \, cp \, \Delta T,$$

where U = overall heat transfer co efficient, k = thermal conductivity, A = Heat transfer cross section area, ΔT = temperature gradient, Δx = length of conduction path, m = mass flow rate of boil off vapour, hf = latent heat of vaporization of cryogenic fluid, cp = specific heat of fluid.

Comparison of Insulations

All the insulations discussed so far can be compared on the basis of their advantages and disadvantages as shown in Table 16.2. Figure 16.13 shows the comparison of various insulations on the basis of heat conductivity parameter (λ).

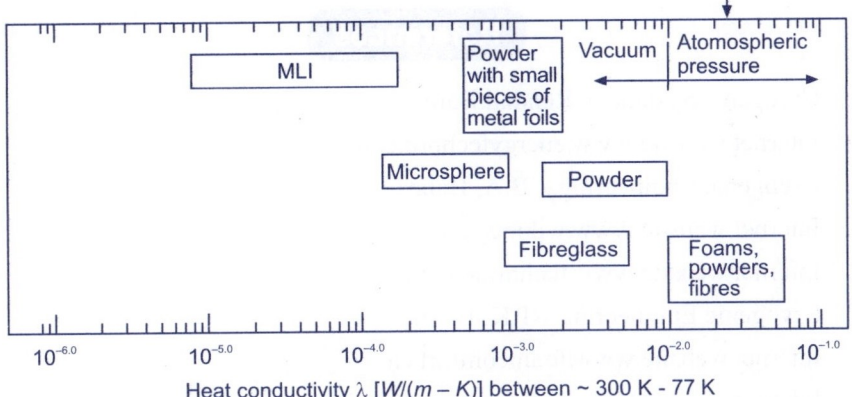

Fig. 16.13 Comparison of Cryogenic Insulations

Table 16.2 Comparison of Insulations : Advantages and Disadvantages

S.N.	Type of Insulation	Advantages	Disadvantages
1	Expanded foam	Low cost, good strength, flexible vacuum jacket acceptable.	High thermal contraction, deterioration in thermal conductivity over time.
2	Solid foam	Good strength, low thermal contraction.	Rigid vacuum jacket required.
3	Gas filled powder	Low cost, non-flammable, applied to irregular shapes	Vapor barrier required, conductivity higher
4	Vacuum	Low heat flux, small cooling loss, irregular geometry is ok	High vacuum required, low emissivity surface needed.
5	Evacuated powder	Vacuum requirements are less, ease of evacuation	Prone to vibratory packing, retains moisture
6	Opacified powder	Vacuum requirements are less, better performance	Higher cost and explosion hazard
7	Multilayer	Best performance, low weight, lower cooldown loss	Higher cost and simple shapes required
8	Composite layer	Best stability, low weight, lower cooldown loss	Higher cost and simple shapes required and stringent vacuum conditions
9	Glass microsphere	Good stability, applied to irregular shapes	Higher cost, high weight
10	Vapor shielding	Lower boiling losses	High vacuum required.

References

1. Cryogenic Systems – Randall Baron.
2. Internet website www.energytechpro.com.
3. Cryogenic Engineering – B.A. Hands.
4. Internet website www.wikipedia.com.
5. Internet website www.technifab.com.
6. Cryogenic Engineering – P.K. Bose.
7. Internet website www.foamcomfort.ca.
8. Internet website www.univfoam.com.
9. US patent 20060101 – Dow Chemical Company.
10. NASA Publication on Glass Microsphere Insulation (2003).
11. Proceedings of the USDOE Hydrogen Program Review (2001).

Questions

Q.1. Explain the importance of cryogenic insulation.

Q.2. Discuss the performance parameters for any insulation.

Q.3. Write a short notes on (a) vacuum insulation, (b) expanded foam insulation and (c) opacifed powder insulation.

Q.4. What are the advantages of multilayer insulation?

Q.5. Explain the working of vapor shielded cryogenic vessels.

CHAPTER 17

Cryogenic Instrumentation and Measurement

INTRODUCTION

It is desirable to measure properties of cryogens at low temperatures and same can be done by cryogenic instruments. The four most common properties measured at cryogenic conditions are the temperature, pressure, liquid level and flow. In this chapter we shall focus on some cryogenic equipment and their range of usage. Cryogenic measurement field has grown over the years and newer, highly accurate equipments are continually being developed. Today's cryogenic measurement systems are specially designed and tested to thrive in the harsh conditions of the cryogenic industry. The selection of electronic and mechanical registers allows assembly of cryogenic measurement systems suited to customer needs. The space saving construction and multiple mounting options are made for easy installations. Cryogenic measurement systems are designed and certified to meet or exceed the requirements of NIST.

Cryogenic Liquid Level Measurement

For a cryogenic liquid storage vessel, it is important to know the exact level of liquid remaining in the container at any given point of time. The liquid quantity is directly proportional to the liquid level in the cryogenic container. Liquid level measurement has come a long way from early means such as dipstick to recent electronic level measurement techniques. In this section we will study various level measurement techniques.

Hydrostatic Gauges

The pressure at any given point in a liquid is proportional to the density and height of the liquid above that point. The reading given by a hydrostatic gauge showing the pressure at a fixed point near the bottom of the tank may therefore be used to determine the depth of liquid in the tank if the density is known and the weight of the liquid above the fixed point if the horizontal area of the tank is also known. By assuming average densities or areas, gauges can be calibrated to read in dips or units or weight. Hydrostatic gauges (such as the mercury column manometer) compare pressure to the hydrostatic force per unit area at the base of a column of

17.2 Cryogenics

fluid. Hydrostatic gauge measurements are independent of the type of gas being measured, and can be designed to have a very linear calibration. They have poor dynamic response.

The manometer which is a level measurement device of the hydrostatic type could also be referred as a pressure measuring instrument, usually limited to measuring pressures near to atmospheric. The term manometer is often used to refer specifically to liquid column hydrostatic instruments. Liquid column gauges consist of a vertical column of liquid in a tube whose ends are exposed to different pressures. The column will rise or fall until its weight is in equilibrium with the pressure differential between the two ends of the tube. A very simple version is a U-shaped tube half-full of liquid, one side of which is connected to the region of interest while the reference atmospheric pressure is applied to the other end. The difference in liquid level represents the applied pressure. The pressure exerted by a column of fluid of height h and density ρ is given by the hydrostatic pressure equation, $P = hg\rho$. Therefore the pressure difference between the applied pressure P_a and the reference pressure P_0 in a U-tube manometer can be found by solving $P_a - P_0 = hg\rho$. The difference in fluid height in a liquid column manometer is proportional to the pressure difference.

$$H = \frac{P_a - P_o}{g\rho}.$$

If the fluid being measured is significantly dense, hydrostatic corrections may have to be made for the height between the moving surface of the manometer working fluid and the location where the pressure measurement is desired. Refer Figure 17.1 for the construction of a U-Tube Manometer.

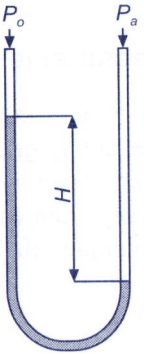

Fig. 17.1 U-Tube Manometer

Although any fluid can be used in the U-Tube manometer, mercury is preferred for its high density (13.534 g/cm^3) and low vapor pressure. For low pressure differences water is commonly used and "inches of water" is a common pressure unit. Liquid-column pressure gauges are independent of the type of gas being measured and have a highly linear calibration. They have poor dynamic response. When measuring vacuum, the working liquid may evaporate and contaminate the vacuum if its vapor pressure is too high. When measuring liquid level, a loop filled with gas or a light fluid must isolate the liquids to prevent them from mixing. Simple hydrostatic gauges can measure pressures ranging from a few Torr (a few 100 Pa) to a few atmospheres (1,000,000 Pa).

Problems associated with hydrostatic gauges are the pressure oscillations and low sensitivity. The sensitivity of hydrostatic liquid level gauge is given by the equation

$$S = d(\Delta p)/dL_f = (\rho_f - \rho_g) g/g_c.$$

Where Δp is the pressure difference, L_f is the height of liquid column, ρ_f is the liquid density, ρ_g is the vapor density, g is the gravitational acceleration, g_c is Newton's conversion factor. Thus one can conclude that the sensitivity of hydrostatic liquid level gauge is directly proportional to difference between liquid and vapor densities.

Electric Resistance Gauges

The electrical resistance of an object is a measure of its opposition to the passage of a steady electric current. An object of uniform cross section will have a resistance proportional to its length and inversely proportional to its cross-sectional area, and proportional to the resistivity of the material. By heating a wire with small electrical current, its temperature can be increased above the fluid temperature in which it is immersed. The temperature of wire will be higher if it is immersed in liquid as compared to vapor. By measuring the electrical resistance of heated wire as it is raised through the liquid allows determination of liquid level. The heat transfer coefficient (h) for the heated wire can be determined from the Langmuir equation.

$$h = (2k/D)/\ln(1 + \varphi \, G_r^{1/4})$$

where k = thermal conductivity, D = wire diameter, φ = Function of Prandtl number, and G_r = Grashoff's number.

The heat wire can also be in the form of strips immersed in the liquid. The resistance of the strips is calibrated to the liquid level in the tank. The advantage of these level indicators is the continuous indication capability without movement whereas the disadvantage is the higher energy dissipation losses.

Capacitance Liquid Level Probes

Today, industrial refrigeration systems use capacitance level probes to accurately measure liquid refrigerant levels. Capacitance level probes have many characteristics that make them ideal for refrigeration applications, and have become the choice of many engineers. There are no moving parts in the probe so it does not wear or jam. They are also not affected by reasonable changes in refrigerant temperature and pressure. There are two basic types of capacitance level probes namely continuous and single-point. Continuous level probes can provide measurements along the entire length of the probe. Single-point level sensors indicate the presence of liquid at a single point or at their installed location only. Capacitance type level switches

17.4 Cryogenics

provide accurate and reliable level detection of conducting as well as non-conducting liquids, solids or slurries that are stored in open or closed containers or are flowing in closed pipes, open channels or on conveyor belts. The performance of these instruments is neither affected by properties of the material such as conductivity and corrosiveness nor by process parameters such as temperature and pressure. Figure 17.2 shows some capacitance level probes.

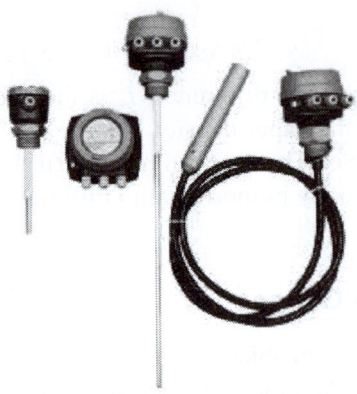

Fig. 17.2 Capacitance Level Probes

Capacitance is the ratio of the electric charge on one of a pair of conductors to the potential difference between the conductors. A capacitance level probe determines the level of liquid in a column or receiver by measuring the combined capacitance of the liquid and gas (vapor) in the column. As the liquid level rises in the column, the total capacitance value increases. This increase is measured by the controlling electronic system and an output control signal is created. The following formula can be used; Capacitance = Dielectric constant × (Area of the plates ÷ Distance between plates)

The dielectric constant is the ability of a material to store electrostatic energy for a voltage potential, or difference. Since it is a constant, it does not change. Liquid refrigerant will always have one constant, and vapor will always have different constant. Looking at the above formula, the distance between the plates also does not change. This is because the rod and the column (the plates) do not move. Therefore, the only variable in the above formula is the area of the plates in the liquid. (The capacitance of vapor is very small compared to the capacitance of the liquid.) This area changes as the liquid rises or falls in the column. Therefore, the total capacitance changes approximately proportional to the liquid rise or fall in the column. The liquid level can then be determined electronically by the change in capacitance.

The capacitance of the level gauge "C" can be written as

$$C = 2 \Pi L \varepsilon_g / \ln(D_o/D_i) [\varepsilon_g + (\varepsilon_f - \varepsilon_g) L_f / L]$$

where L = gauge length, L_f = length of gauge immersed; D_o and D_i = outer and inner diameter of gauge, ε_g and ε_f = dielectric constants for liquid and vapor.

The sensitivity of the capacitance gauge = $S = 2 \Pi L \varepsilon_g (\varepsilon_f - \varepsilon_g)/\ln(D_o/D_i)$

The sensitivity of gauge is independent of the liquid level and can be increased by reducing the annular space between the capacitance elements.

Magnetic Bond Type Level Measurement

The magnetic bond method was developed to overcome the problems of cages and stuffing boxes. The magnetic bond mechanism consists of a magnetic float which rises and falls with changes in level. The float travels outside of a non-magnetic tube which houses an inner magnet connected to a level indicator. When the float rises and falls, the outer magnet will attract the inner magnet, causing the inner magnet to follow the level within the vessel. Refer Figure 17.3 for the construction of the magnetic bond level measurement.

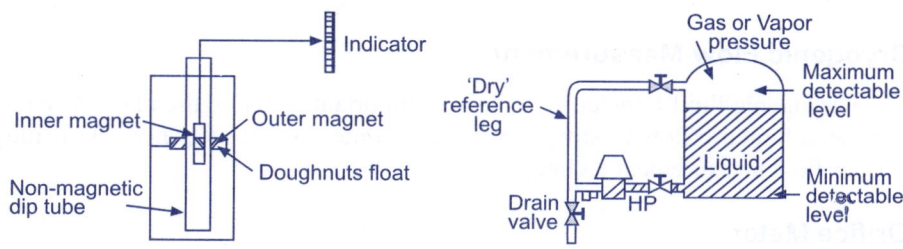

Fig. 17.3 Magnetic Bond Level Measurement

Thermodynamic Liquid Level Gauge

Thermodynamic liquid level probes are based on the principle that a liquid undergoes a large change in volume when it is evaporated. Refer Figure 17.4 for the schematic of the system.

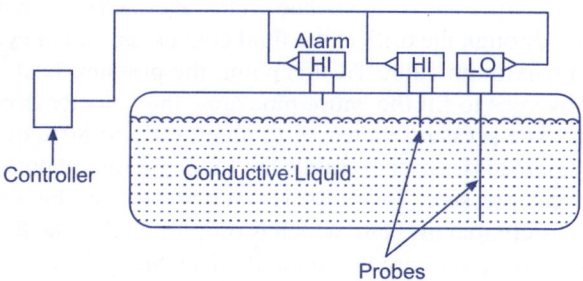

Fig. 17.4 Thermodynamic Liquid Level Gauge

The probe consists of a thin capillary tube that is heated by an electric current. The capillary is connected to the dead volume at ambient temperature. The gauge is charged with vapors of the liquid whose level is to be measured. As soon as the capillary tube is immersed in the cryogenic liquid, the stored gas is condensed thereby reducing the dead volume and gas pressure. This reduced gas pressure is calibrated for the liquid level. The liquid level is obtained by the equation

17.6 Cryogenics

$$L_f/L = [(m - \rho_o V_o/AL) - \rho_g]/[\rho_f - \rho_g]$$

Where, L = length of capillary tube, L_f = liquid level, m = mass of liquid, ρ_f = density of liquid, ρ_g = density of vapor, ρ_o = dead volume density, V_o = volume of liquid, A = area of cross section.

Thus the liquid level measurement depends upon the densities, which in turn depends on temperature. One can also consider the super compressibility factor Z of the liquid for a more realistic analysis. The compressibility factor is defined as $P = ZRT$. The super compressibility correction can be applied to the liquid level calculation and total mass of gas in the gauge can be calculated as

$$m = (P_o V_o/ZR_o T_o)(1 + A_L/V_o).$$

The initial charging pressure of gas should be above saturation pressure of the liquid whose level is to be measured. The sensitivity of the probe can be improved by increasing the parameter A_L/V_o.

Cryogenic Flow Measurement

For a cryogenic fluid transfer system, it is important to know the mass flow rate and the volumetric flow at any given point of time. In this section we will study various flow measurement techniques.

Orifice Meter

The orifice plate is the simplest of the flow path restrictions used in flow detection, as well as the most economical. Orifice plates are flat plates 1/16 to 1/4 inch thick. They are normally mounted between a pair of flanges and are installed in a straight run of smooth pipe to avoid disturbance of flow patterns from fittings and valves. Three kinds of orifice plates are used: concentric, eccentric, and segmental as shown in Figure 17.5. The concentric orifice plate is the most common of the three types. As shown, the orifice is equidistant (concentric) to the inside diameter of the pipe. Flow through a sharp-edged orifice plate is characterized by a change in velocity. As the fluid passes through the orifice, the fluid converges, and the velocity of the fluid increases to a maximum value. At this point, the pressure is at a minimum value. As the fluid diverges to fill the entire pipe area, the velocity decreases back to the original value. The pressure increases to about 60% to 80% of the original input value. The pressure loss is irrecoverable; therefore, the output pressure will always be less than the input pressure. The pressures on both sides of the orifice are measured, resulting in a differential pressure which is proportional to the flow rate. Segmental and eccentric orifice plates are functionally identical to the concentric orifice.

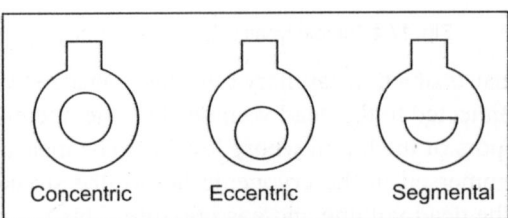

Fig. 17.5 Orifice Plates

The circular section of the segmental orifice is concentric with the pipe. The segmental portion of the orifice eliminates damming of foreign materials on the upstream side of the orifice when mounted in a horizontal pipe. Depending on the type of fluid, the segmental section is placed on either the top or bottom of the horizontal pipe to increase the accuracy of the measurement. Eccentric orifice plates shift the edge of the orifice to the inside of the pipe wall. This design also prevents upstream damming and is used in the same way as the segmental orifice plate. Orifice plates have two distinct disadvantages; they cause a high permanent pressure drop (outlet pressure will be 60% to 80% of inlet pressure), and they are subject to erosion, which will eventually cause inaccuracies in the measured differential pressure.

Earlier there was reluctance to use orifice plates due to the problem of flashing (bubble formation) of liquid due to pressure drop across the orifice. However, this problem is not instantaneous and can be ignored. The accuracy of the meter is within 1%. Normally a length of at least ten times of tube diameter is placed upstream of the orifice. The orifice meter is a thin plate with holes and the mass flow rate (m) is calculated as

$$m = C_d C_a A (2 g \rho \Delta p)^{1/2}$$

where C_d = discharge coefficient, C_a = velocity of approach coefficient, A = Area of orifice, g = gravitational acceleration, ρ = fluid density, Δp = pressure drop across orifice.

The discharge coefficient accounts for irreversibilities in the orifice flow, whereas the velocity of approach coefficient accounts for kinetic energy of fluid upstream of the orifice. The pressure taps are normally located at corners or in flanges or diametrically mounted. The advantage of this method of measurement is the simplicity, however the disadvantage of this type of measurement is the large pressure drop.

Venturi Meter

The venturi tube, illustrated in Figure 17.6, is the most accurate flow-sensing element when properly calibrated. The venturi tube has a converging conical inlet, a cylindrical throat, and a diverging recovery cone. It has no projections into the fluid, no sharp corners, and no sudden changes in contour. The inlet section decreases the area of the fluid stream, causing the velocity to increase and the pressure to decrease. The low pressure is measured in the center of the cylindrical throat since the pressure will be at its lowest value, and neither the pressure nor the velocity is changing. The recovery cone allows for the recovery of pressure such that total pressure loss is only 10% to 25%. The high pressure is measured upstream of the entrance cone. The major disadvantages of this type of flow detection are the high initial costs for installation and difficulty in installation and inspection.

17.8 Cryogenics

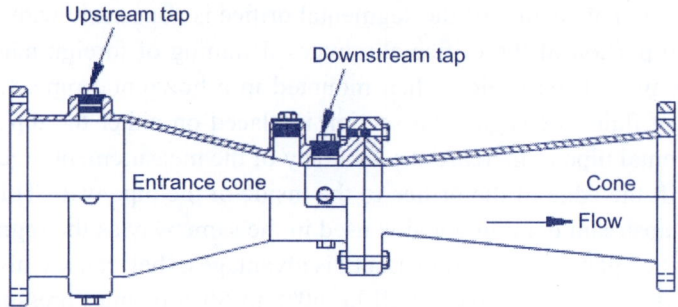

Fig. 17.6 Venturi Flowmeter

The inlet cone angle of the venture meter is 20 to 22 degrees and exit cone angle is about 5 to 7 degrees. The throat diameter is approximately half of the tube diameter. The discharge coefficient for a venturi Flowmeter is

$$C_d = \log R_e/(0.6 + 0.9 \log R_e),$$

where R_e = Reynolds number.

The mass flow rate equation for venture Flowmeter is the same as that of orifice meter. The advantage of this meter is reduced pressure drop as compared to orifice meter however disadvantage is the increased chances of cavitation.

Turbine Flowmeter

Turbine Flowmeters use the mechanical energy of the fluid to rotate a "pinwheel" (rotor) in the flow stream. Blades on the rotor transform energy from the flow stream into rotational energy. The rotor shaft spins on bearings. When the fluid moves faster, the rotor spins proportionally faster. Shaft rotation can be sensed mechanically or by detecting the movement of the blades. Blade movement is often detected magnetically, with each blade generating a pulse. When the fluid moves faster, more pulses are generated. The electronic transmitter processes the pulse signal to determine the flow of the fluid. Refer Figure 17.7 for the construction of turbine Flowmeter and Figure 17.8 for the photograph of the turbine Flowmeter.

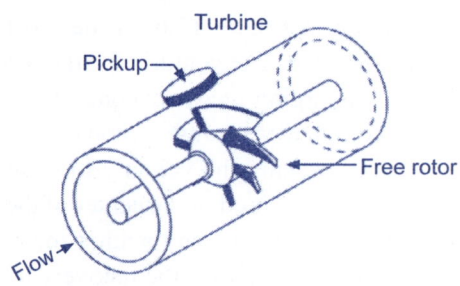

Fig. 17.7 Turbine Flowmeter Construction

Turbine Flowmeters measure the velocity of liquids, gases and vapors in pipes, such as hydrocarbons, chemicals, water, cryogenic liquids, air, and industrial gases. High accuracy turbine Flowmeters are available for custody transfer of hydrocarbons and natural gas. Turbine Flowmeters are not suitable for fluids that are non-lubricating, because the Flowmeter can become inaccurate and fail if the bearings prematurely wear. Some turbine Flowmeters have grease fittings for use with non-lubricating fluids. The flow of corrosive liquids can be measured with proper attention to the materials of construction of all wetted parts, such as the body, rotor, bearings, and fittings. Turbine Flowmeters are less accurate at low flow rates due to rotor/bearing drag that slows the rotor. Turbine Flowmeters should not be operated at high velocity because premature bearing wear and/or damage can occur. Abrupt transitions from gas flow to liquid flow should be avoided because they can mechanically stress the Flowmeter, degrade accuracy, and/or damage the Flowmeter. These conditions generally occur when filling the pipe and under slug flow conditions.

The volumetric flow rate (V) of the turbine meter is calculated as

$$V = \Pi D An/\tan \theta,$$

where D = diameter of the rotor, A = free flow area of turbine, n = speed of turbine, θ = angle between blade and meter center line.

Turbine Flowmeters are sensitive to viscosity changes of the fluid however are not affected by density of single phase or two phase fluid flow. The turbine Flowmeter has the advantage of good dynamic response to transient flows and fast response time of few milliseconds. The Flowmeter can be damaged due to severe flow oscillations and surges.

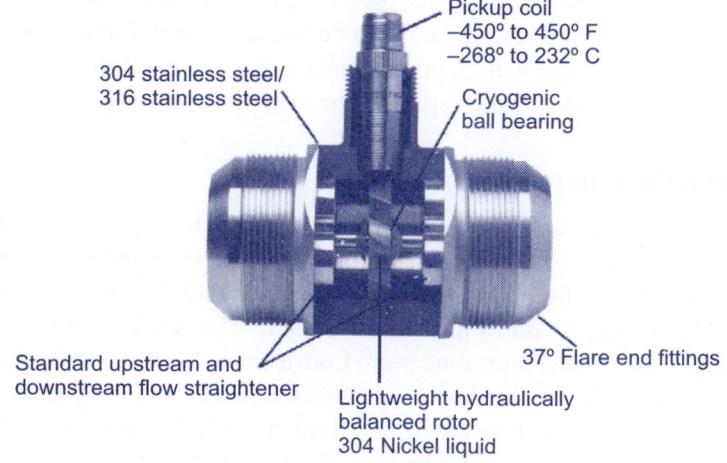

Fig. 17.8 Turbine Flowmeter Photograph

Capacitance Flowmeter

Based on the RF capacitance principal, this instrument is ideally suited for metering flow in cryogenic applications. The rate of flow indicator is a capacitance based solution for reliable detection of rate of flow of liquids in an open channel. The

17.10 Cryogenics

level change is measured using the capacitance principle and converted to rate of flow utilizing the appropriate formula. Galvanically isolated 4 to 20 mA analog output is configured via user interface to be either proportional to level or rate of flow. Three relays can be configured to either operate as per level change or as per flow rate change, as desired. All other functions such as calibration, fail safe mode selection, switching time delay and delay modes can be configured via the user interface. A very stable and noise free signal communication, over a simple 2-wire interconnection ensures a high degree reliability.

When the liquid flows through the measuring device, the level of the liquid is directly proportional to the rate of flow which can be calculated by the formula: $Q = K \times L_n$, where Q is the rate of flow in cubic meters per hour, L is the level in meters, k and n are constants. The capacitance meter is usually in the form of a concentric cylinder whose capacitance "C" can be calculated as $C = 2 \Pi L \varepsilon / \ln(D_o/D_i)$, Where L = length of cylinder, ε = dielectric constant of the material, D_o is the outer diameter and D_i is the inner diameter of the cylinder.

Electromagnetic Flowmeter

The electromagnetic Flowmeter is similar in principle to the generator. The rotor of the generator is replaced by a pipe placed between the poles of a magnet so that the flow of the fluid in the pipe is normal to the magnetic field. As the fluid flows through this magnetic field, an electromotive force is induced in it that will be mutually normal (perpendicular) to both the magnetic field and the motion of the fluid. This electromotive force may be measured with the aid of electrodes attached to the pipe and connected to a galvanometer or an equivalent. For a given magnetic field, the induced voltage will be proportional to the average velocity of the fluid. However, the fluid should have some degree of electrical conductivity.

Ultrasonic Flowmeter

Devices such as ultrasonic flow equipment use the Doppler frequency shift of ultrasonic signals reflected from discontinuities in the fluid stream to obtain flow measurements. These discontinuities can be suspended solids, bubbles, or interfaces generated by turbulent eddies in the flow stream. The sensor is mounted on the outside of the pipe, and an ultrasonic beam from a piezoelectric crystal is transmitted through the pipe wall into the fluid at an angle to the flow stream. Signals reflected off flow disturbances are detected by a second piezoelectric crystal located in the same sensor. Transmitted and reflected signals are compared in an electrical circuit, and the corresponding frequency shift is proportional to the flow velocity.

Cryogenic Pressure Measurement

Pressure measurement for cryogenic systems such as for liquid oxygen tanks has always presented a challenge to instrumentation engineers. These cryogenic liquids, possess many properties which cause problems for designers. Primarily, these problems are associated with large temperature variations and highly volatile nature of the liquid. One solution is to connect the pressure measurement transducer away from the extreme temperature section thereby providing a thermal buffer for the transducer. This is however unsuitable for dynamic pressure environments. Currently available pressure measurement techniques solve these problems by use of low temperature compatible materials. We shall now study some cryogenic pressure measurement techniques.

Bourdon Gauge

A Bourdon gauge uses a coiled tube, which, as it expands due to pressure increase causes a rotation of an arm connected to the tube. In 1849 the Bourdon tube pressure gauge was patented in France by Eugene Bourdon. The pressure sensing element is a closed coiled tube connected to the chamber or pipe in which pressure is to be sensed. As the gauge pressure increases the tube will tend to uncoil, while a reduced gauge pressure will cause the tube to coil more tightly. This motion is transferred through a linkage to a gear train connected to an indicating needle. The needle is presented in front of a card face inscribed with the pressure indications associated with particular needle deflections. In a barometer, the Bourdon tube is sealed at both ends and the absolute pressure of the ambient atmosphere is sensed. Differential Bourdon gauges use two Bourdon tubes and a mechanical linkage that compares the readings. Refer Figure 17.9 for the bourdon pressure gauge.

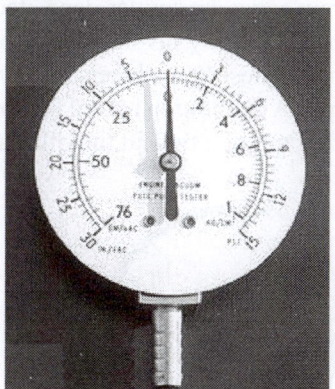 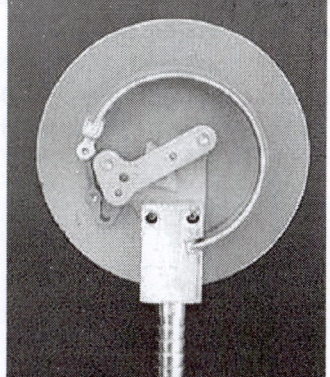

Fig. 17.9 Bourdon Pressure Gauge

Electronic Scanned Pressure (ESP) Modules

Electronically scanned pressure modules have been developed that can operate in ambient and in cryogenic environments. Because they can operate directly in a cryogenic environment, their use eliminates many of the operational problems associated with using conventional modules at low temperatures. For over two decades, conventional pressure modules have been used for low temperature applications because of their high accuracy and measurement of facility pressures at rates of 500 samples per second. However, at cryogenic temperatures, their use presents obstacles that contribute to the overall expense of testing and the loss of data quality. One obstacle is the need for thermal protection of the module because of charge carrier freeze-out. Another major obstacle is the requirement that a test be interrupted periodically for line calibration, consuming as much as 25% of the total test time. Electronically scanned modules are differential pressure measurement units consisting of an array of silicon pressure sensors which are electronically multiplexed at rates of 20,000 Hz through an on board multiplexer and instrumentation amplifier. Found today in widespread practice throughout the world to measure surface pressure, these state-of-the-art ESP modules typically contain 16-64 measurement channels connected to pressure ports through small diameter tubing.

Cryogenic Pressure Transducers

Sputtered Thin Film Pressure Transducer are designed for cryogenic service; it can operate in temperatures from $-160°C$ to $+149°C$. Yet, even in these difficult temperatures, it provides outstanding accuracy, long-term calibration stability and reliability. Static accuracy is $\pm 0.25\%$, and thermal zero and sensitivity shifts over the compensated range of $-160°C$ to $+27°C$ are less than $\pm 0.01\%$. The all-welded stainless-steel pressure cavity and double-isolated case ensures reliability in the tough environments normal to cryogenic service. The thin film technology makes this premium performance possible. The strain gages are sputter-deposited, forming a molecular bond with the substrate. There is virtually no shift, drift, or creep to cause the transducer's calibration to change. These transducers are available in many standard ranges from 15 psi to 1,000 psi. A calibration record is supplied with every unit. Refer Figure 17.10 for the photograph of a typical thin film transducer.

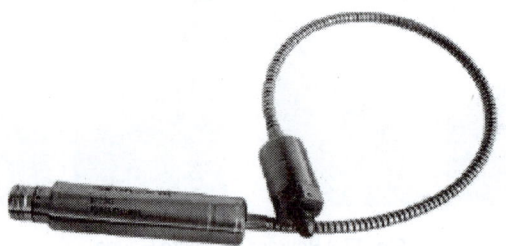

Fig. 17.10 Thin Film Pressure Transducer

Cryogenic Strain Gauge Sensors

Silicon strain gauge sensors for measurements of pressure are characterized by high sensitivity, reliability and the possibility of mass production with relatively high accuracy. However, the peculiarity of piezo-resistance in semiconductors and *p-n* junction isolation leads to significant non-linearity, temperature dependence of the sensitivity and restricted operating temperature range of silicon sensors. Many different types of circuitry are used to compensate the errors that arise. As a result, the best silicon transducers have an accuracy of about 0.1-0.2%, but to achieve this individual adjustment is necessary, which makes their production more expensive. A good illustration of silicon's possibilities as a material for integrated pressure transducers is sensing elements made of 'silicon-on-sapphire' (SOS) structures. The evident advantage of SOS usage is a significant extension of the operating temperature range due to the absence of a p-n junction. Using SOS structures, pressure sensors can operate in a wide temperature range (from -272 to $+350°C$) with high accuracy.

Cryogenic Pressure Sensitive Paint (PSP)

Novel pressure sensitive paint suitable for use in a cryogenic wind tunnel has been developed. This new PSP uses a high polymer as a binder. Unlike other polymer-based paints, the PSP maintains high oxygen sensitivity even at cryogenic temperatures. The luminescent intensity of PSP is related to air pressure. PSP sprayed on a model surface is illuminated by excitation light, and the luminescent image is detected by a CCD camera. PSP measurement provides a simple and inexpensive way to obtain a full-field image of the pressure distribution on aerodynamic model surface with high spatial resolution. Considering that visual recognition of detailed flow field is rather easy, and quantitative detailed pressure image can also be obtained, the PSP measurement technique would be applicable to the research and development of aerospace vehicles.

Cryogenic Temperature Measurement

Temperature is a fundamental quantity to be measured in cryogenics and can be measured by several means but not with comparison with a standard. The temperature measurement systems need to exhibit qualities such as accuracy, sensitivity, reproducibility and stability over time. The selection of a particular measurement system takes into account the application and requirements of these four parameters. In this section we shall study cryogenic temperature measurement systems which have evolved over time.

Thermodynamic Temperature Scale

Thermodynamic temperature is the absolute measure of temperature and is one of the principal parameters of thermodynamics. Thermodynamic temperature is an "absolute" scale because it is the measure of the fundamental property underlying temperature: its *null* or zero point, absolute zero, is the temperature at which the particle constituents of matter have minimal motion and can be no colder. Temperature arises from the random submicroscopic vibrations of the particle constituents of matter. These motions comprise the kinetic energy in a substance.

17.14 Cryogenics

More specifically, the thermodynamic temperature of any bulk quantity of matter is the measure of the average kinetic energy of a certain kind of vibrational motion of its constituent particles called translational motions. Translational motions are ordinary, whole-body movements in three-dimensional space whereby particles move about and exchange energy in collisions. Thermodynamic temperature's null point, absolute zero, is the temperature at which the particle constituents of matter are as close as possible to complete rest; that is, they have minimal motion. Thermodynamic temperature is measured in Kelvin (K). Many engineering fields in the U.S. however, measure thermodynamic temperature using the Rankine scale. Temperatures expressed in kelvins are converted to degrees Rankine simply by multiplying by 1.8 as follows: $T_{°R} = 1.8 \, T_K$, where T_K and $T_{°R}$ are temperatures in kelvin and degrees Rankine respectively.

By international agreement, the unit *kelvin* and its scale are defined by two points: absolute zero, and the triple point. Absolute zero, the lowest possible temperature, is defined as being precisely 0 K or −273.15°C. The triple point of water is defined as being precisely 273.16 K or 0.01°C. Strictly speaking, the temperature of a system is well-defined only if its particles are at equilibrium, so that their energies obey a Boltzmann distribution. There are many possible scales of temperature, derived from a variety of observations of physical phenomena. The thermodynamic temperature can be shown to have special properties, and in particular can be seen to be uniquely defined (up to some constant multiplicative factor) by considering the efficiency of idealized heat engines. Thus the ratio T_2/T_1 of two temperatures T_1 and T_2 is the same in all absolute scales. Loosely stated, temperature controls the flow of heat between two systems, and the universe as a whole, as with any natural system, tends to progress so as to maximize entropy.

One way to study this is to analyze a heat engine, which is a device for converting heat into mechanical work, such as the Carnot heat engine. Carnot's theorem states that all reversible engines operating between the same heat reservoirs are equally efficient. Thus, a heat engine operating between temperatures T_1 and T_3 must have the same efficiency as one consisting of two cycles, one between T_1 and another (intermediate) temperature T_2, and the second between T_2 and T_3. Refer Figure 17.11.

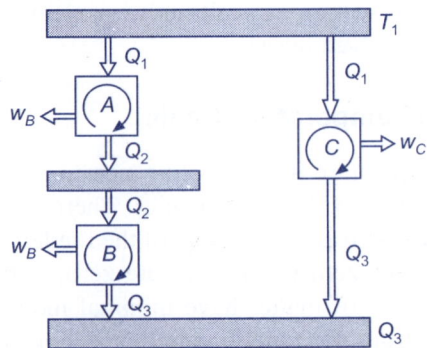

Fig. 17.11 Carnot Heat Engine for Temperature Scale

The working medium of Carnot cycle is not limited to an ideal gas but any medium. One can define a thermodynamic temperature scale that is independent of the working medium. There is a high temperature heat reservoir at T_1 and a low temperature heat reservoir at T_3. For any two temperatures, the ratio of the magnitudes of the heat absorbed and rejected in a Carnot cycle has the same value for all systems. The heat supplied for the Carnot engine is Q_1 and heat rejected is Q_3.

For a Carnot cycle

$$\eta = 1 + \frac{Q_L}{Q_H} = F(T_L, T_H); \; \eta \text{ is only a function of temperature.}$$

Also

$$\frac{Q_1}{Q_2} = F(T_1, T_2)$$

$$\frac{Q_2}{Q_3} = F(T_2, T_3)$$

$$\frac{Q_1}{Q_3} = F(T_1, T_3)$$

But

$$\frac{Q_1}{Q_3} = \frac{Q_1}{Q_2} \frac{Q_2}{Q_3}.$$

Hence

$$\underbrace{F(T_1, T_3)}_{\text{Not a function of } T_2} = \underbrace{F(T_1, T_2) \times F(T_2, T_3)}_{\text{Cannot be a function of } T_2}.$$

The ratio of the heat exchanged is therefore

$$\frac{Q_1}{Q_3} = F(T_1, T_3) = \frac{f(T_1)}{f(T_3)}.$$

In general,

$$\frac{Q_H}{Q_L} = \frac{f(T_H)}{f(T_H)},$$

so that the ratio of the heat exchanged is a function of the temperature.

We could choose any function that is monotonic, and one choice is the simplest

$$f(T) = T$$
$$Q_H/Q_L = T_H/T_L$$

This is the thermodynamic scale of temperature. The temperature defined in this manner is the same as that for the ideal gas; the thermodynamic temperature scale and the ideal gas scale are equivalent.

Thermocouples

A thermocouple is constructed of two dissimilar metal wires joined at one end. When one end of each wire is connected to a measuring instrument, the thermocouple becomes a sensitive and highly accurate measuring device. Thermocouples may be constructed of several different combinations of materials. The performance of a thermocouple material is generally determined by using that material with platinum. The most important factor to be considered when selecting a pair of materials is the "thermoelectric difference" between the two materials. A significant difference between the two materials will result in better thermocouple performance. Figure 17.12 illustrates the characteristics of the more commonly used materials when used with platinum. For example: Chromel-Constantan is excellent for temperatures up to 2000°F and Tungsten-Rhenium is used for temperatures up to 5000°F.

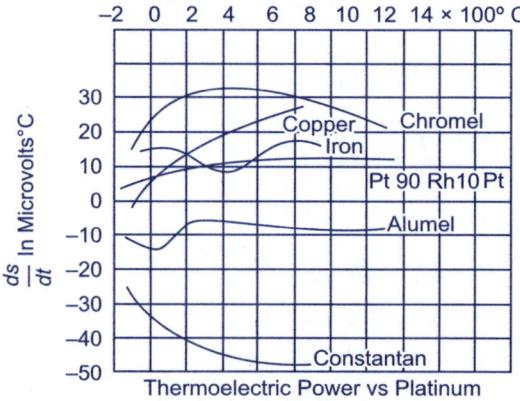

Fig. 17.12 Thermocouple Materials

Figure 17.13 shows the internal construction of a typical thermocouple. The leads of the thermocouple are encased in a rigid metal sheath. The measuring junction is normally formed at the bottom of the thermocouple housing. Magnesium oxide surrounds the thermocouple wires to prevent vibration that could damage the fine wires and to enhance heat transfer between the measuring junction and the medium surrounding the thermocouple.

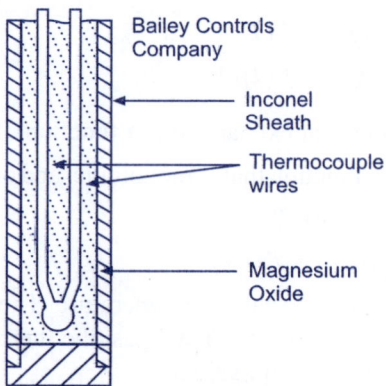

Fig. 17.13 Thermocouple Construction

Thermocouples will cause an electric current to flow in the attached circuit when subjected to changes in temperature. The amount of current that will be produced is dependent on the temperature difference between the measurement and reference junction; the characteristics of the two metals used; and the characteristics of the attached circuit. Figure 17.14 illustrates a simple thermocouple circuit. Heating the measuring junction of the thermocouple produces a voltage which is greater than the voltage across the reference junction. The difference between the two voltages is proportional to the difference in temperature and can be measured on the voltmeter in millivolts. For ease of operator use, some voltmeters are calibrated directly in terms of temperature through use of electronic circuitry.

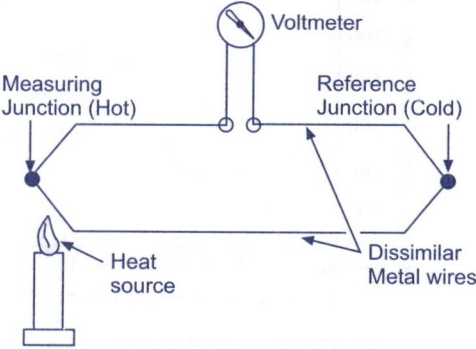

Fig. 17.14 Thermocouple Circuit

The temperature indication is also determined by measuring emf of the thermocouple and using a calibration curve of the form $e = a_1 t + a_2 t^2 + a_3 t^3 + a_4 t^4$, where e = emf, t = temperature difference and a = constant. As output of individual thermocouple is small, sometimes thermocouples are joined to form a thermopile. Thermocouples suffer from thermal conduction errors which can be reduced by insulating the leads.

Metallic Resistance Thermometer

The metallic resistance thermometer or resistance temperature detector (RTD) incorporates pure metals or certain alloys that increase in resistance as temperature increases and, conversely, decrease in resistance as temperature decreases. RTDs act somewhat like an electrical transducer, converting changes in temperature to voltage signals by the measurement of resistance. The metals that are best suited for use as RTD sensors are pure, of uniform quality, stable within a given range of temperature, and able to give reproducible resistance-temperature readings. Only a few metals have the properties necessary for use in RTD elements. RTD elements are normally constructed of platinum, copper, or nickel. These metals are best suited for RTD applications because of their linear resistance-temperature characteristics as shown in Figure 17.15. Their high coefficient of resistance, and their ability to withstand repeated temperature cycles make them suitable for RTDs. The coefficient of resistance is the change in resistance per degree change in temperature, usually expressed as a percentage per degree of temperature. The material used must be capable of being drawn into fine wire so that the element can be easily constructed. Metallic resistance thermometers are calibrated by the equation

$$R/R_o = 1 + At + Bt^2 + Ct^3(t - 100)$$

and the sensitivity is calculated as

$$S = R_o(A + 2Bt + Ct^2(4t - 300))$$

Where R = Resistance of thermometer, R_o = Platinum resistance, t = temperature, A, B, C = constants which vary with material.

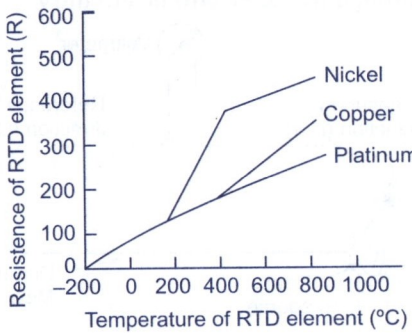

Fig. 17.15 Electrical Resistance Curves

RTD elements are usually long, spring-like wires surrounded by an insulator and enclosed in a sheath of metal. Figure 17.16 shows the internal construction of an RTD.

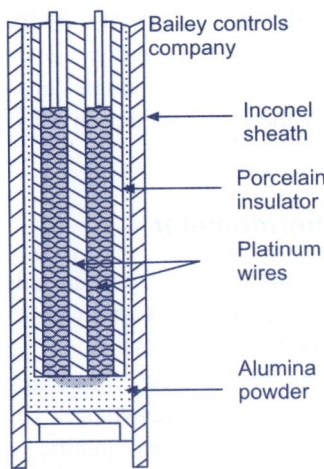

Fig. 17.16 Metallic Resistance Thermometer Construction

This particular design has a platinum element that is surrounded by a porcelain insulator. The insulator prevents a short circuit between the wire and the metal sheath. Inconel, a nickel-iron-chromium alloy, is normally used in manufacturing the RTD sheath because of its inherent corrosion resistance. When placed in a liquid or gas medium, the Inconel sheath quickly reaches the temperature of the medium. The change in temperature will cause the platinum wire to heat or cool, resulting in a proportional change in resistance. This change in resistance is then measured by a precision resistance measuring device that is calibrated to give the proper temperature reading. This device is normally a bridge circuit. It is important to mount the thermocouple properly to prevent mechanical and thermal strains.

Semiconductor Resistance Thermometer

Semiconductor Resistance Thermometer or Thermistors differ from resistance temperature detectors (RTD) in that the material used in a thermistor is generally a ceramic or polymer, while RTDs use pure metals. The temperature response is also different; RTDs are useful over larger temperature ranges, while thermistors typically achieve a higher precision within a limited temperature range [usually −90°C to 130 °C]. A thermistor is a type of resistor whose resistance varies with temperature. Thermistors are widely used as inrush current limiters, temperature sensors, self-resetting over current protectors, and self-regulating heating elements. Figure 17.17 shows a typical thermistor.

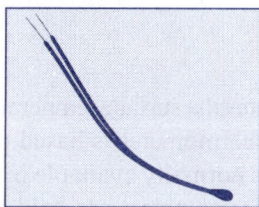

Fig. 17.17 Semiconductor Resistance Thermometer

Assuming, as a first-order approximation, that the relationship between resistance and temperature is linear, then:

$$\Delta R = k \Delta T$$

where ΔR = change in resistance, ΔT = change in temperature and k = first-order temperature coefficient of resistance.

Thermistors can be classified into two types, depending on the sign of k. If k is positive, the resistance increases with increasing temperature, and the device is called a positive temperature coefficient (PTC) thermistor. If k is negative, the resistance decreases with increasing temperature, and the device is called a negative temperature coefficient (NTC) thermistor. Resistors that are not thermistors are designed to have a k as close to zero as possible, so that their resistance remains nearly constant over a wide temperature range. Instead of the temperature coefficient k, sometimes the temperature coefficient of resistance α (alpha) or α_T is used. It is defined as

$$\alpha_T = \frac{1}{R(T)} \frac{dR}{dT}.$$

For example, for the common PT100 sensor, $\alpha = 0.00385$.

Many NTC thermistors are made from a pressed disc or cast chip of a semiconductor such as a sintered metal oxide. They work because raising the temperature of a semiconductor increases the number of electrons able to move about and carry charge – it promotes them into the conduction band. The more charge carriers that are available, the more current a material can conduct. This is described in the formula:

$$I = n \cdot A \cdot v \cdot e,$$

where I = electric current (amperes), n = density of charge carriers (count/m^3), A = cross-sectional area of the material (m^2), v = velocity of charge carriers (m/s), e = charge of an electron (coulomb).

The current is measured using an ammeter. Over large changes in temperature, calibration is necessary. Over small changes in temperature, if the right semiconductor is used, the resistance of the material is linearly proportional to the temperature. There are many different semiconducting thermistors with a range from about 0.01 kelvin to 2,000 Kelvins (−273.14°C to 1,700°C). Germanium resistance thermometers are common. Advantages of semiconductor resistance thermometers are their low cost, high sensitivity, small size and simple temperature-resistance curve.

Magnetic Thermometer

Magnetic Thermometer indicates the surface temperature of steel and other magnetic materials continuously. The thermometer is based on a bimetallic strip, and so it does not require batteries. It is normally available in two scale ranges: −35 to 55°C and 0 to 120°C. Magnetic thermometers are held magnetically to any magnetic surface. These instruments feature two high temperature alnico magnets that act as the thermometers base and hold the instrument in place on magnetic surfaces. The bimetallic sensor is located in a draft shield and is in virtual thermal contact with the surface to be measured. This gives the instrument a relatively quick response time. The thermometer reaches sensing equilibrium (stability) within three minutes. Refer Figure 17.18 for the photograph of magnetic thermometer.

Fig. 17.18 Magnetic Thermometer

Magnetic thermometry is based upon measurement of paramagnetic susceptibility. For an ideal paramagnet the zero-field susceptibility is related to temperature through the Curie law $\chi = C/T$ where C is the Curie constant. Although in magnetic thermometry one approximates to this by using dilute paramagnetic salts, it is generally necessary to take account of interactions and other effects and write

$\chi = C/(T + \Delta + \gamma/T)$, where Δ includes first-order dipole-dipole and exchange couplings and also a shape factor, while γ is due primarily to crystal field splitting of the ground state and second-order interaction effects. The susceptibility measurement is usually made by the ac mutual inductance method in which the salt sample is situated in a set of coils whose mutual inductance M is balanced against a reference. The bridge balance X is linearly related to M and hence to χ. The working equation for a magnetic thermometer becomes

$$X = A + B/(T + \Delta + \gamma/T).$$

Unless Δ or γ is obtainable from theory, a minimum of four fixed points is needed to calibrate the thermometer. Salt crystals are hydrates and almost all of these tend to lose water of crystallization if kept at room temperature. This tendency is considerably diminished if the enclosing volume is small and is filled with an inert gas; but it is catastrophically increased if the enclosure is evacuated. Simple refrigeration, however, avoids most such problems. Cerous magnesium nitrate (CMN) is the closest approximation to an ideal paramagnet in common use: $\gamma = 0$ and for a sphere Δ is about 0.3 mK. It is, however, highly anisotropic and its usefulness is limited to temperatures below 3 K because of its low Curie constant of 0.011 K. Deviations from Curie law occur at low temperatures for all paramagnetic materials.

Constant Volume Gas Thermometer

Constant volume gas thermometer measures temperature by the variation in volume or pressure of a gas. It consists of a bulb filled with a fixed amount of dilute gas that is connected by a capillary tube to a mercury manometer. A manometer is a device used to measure pressure and has a column partially filled with mercury that is connected to another partially filled column of mercury, called a reservoir with a flexible tube. The height of the mercury in the first column is set to a reference point or pressure P that it must stay at, while the mercury in the reservoir is allowed to move up and down. The bulb is filled with a gas such that the volume of the gas in the bulb remains constant. The volume is related to temperature by Charles's Law. The pressure of the gas in the bulb can be obtained by measuring the level difference in the two arms of the manometer. Refer Figure 17.19 for the photograph and Figure 17.20 for the construction of the constant volume gas thermometer.

Figure 17.19 Constant Volume Gas Thermometer

17.22 Cryogenics

As per Gay-Lussac Law, when the temperature of an ideal gas increases, there is a corresponding increase in pressure. Conversely, when the temperature decreases, so does the pressure. Constant volume gas thermometers operate on the same principle, with the restrictions that the gas is at low pressure and the temperature of it is well above that of liquefaction. For a pressure P, the equation used to find the temperature is $T = aP + b$, where a and b are constants determined from two fixed points, such as ice at 0°C and steam at 100°C. Gas thermometers are often used to calibrate other thermometers using values from the triple point of water. To recall, 273.16 K (Kelvin) is the temperature where water exists in an equilibrium state as a gas, liquid, and solid. For equation of temperature, where $a = 273.16$ K, $b = 0$, P_{tp} is the pressure of the gas at the triple point of water, and P is the pressure of the gas at the temperature to be measured, we have $T = 273.16$ K (P/P_{tp}). For low pressure and high temperatures, where real gases behave like ideal gases, equation becomes $T = 273.16$ K $\lim P/P_{tp}$ as $P_{tp} \to 0$

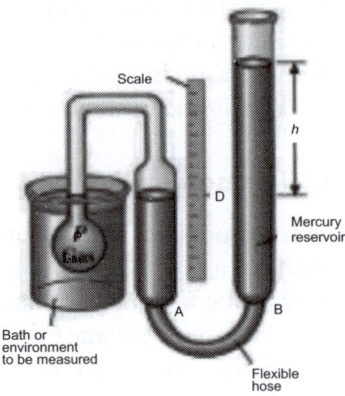

Fig. 17.20 Constant Volume Gas Thermometer Construction

The gas bulb is inserted into a bath to find the temperature. When the temperature increases or decreases, the volume also increases or decreases as well as the pressure, as we have seen in Charles' Law and the aforementioned Gay-Lussac's Law. The pressure of the mercury also changes such that it begins to move up or down and thus away from the reference point. To stop this movement, which will also stop the gas from expanding, the reservoir at the other end is physically lifted up or down, and the ensuing new pressure at this height is measured. The difference between the reference and reservoir heights gives the final pressure P, which is then used to calculate the temperature. The temperature obtained by extrapolating the measured temperatures to the case of no gas in the bulb is a temperature that does not depend on the specific properties of the materials involved. This temperature is a truly fundamental physical quantity, whose definition is independent of the properties of the specific materials. The sensitivity equation for this thermometer can be written as $S = P/T$ and the corrections for volume change of thermometer bulb, gas adsorption, variation of capillary pressures need to be accounted for. The sensitivity of the thermometer can be increased by increasing the pressure.

Vapor Pressure Thermometer

Vapor pressure thermometers are based on the saturated vapor pressure in a two-phase system in an enclosure. The techniques of vapor pressure thermometry apply also to the special case of a boiling or a triple point determination. The behavior of a liquid-vapor system in equilibrium, for example, is describable by an equation $P = f(T)$. Along the curve for a pure substance, the pressure depends on the temperature and not on the quantity of substance enclosed or vaporized. Refer Figure 17.21. The temperature range associated with vaporization is limited to temperatures between the critical point and the triple point of the substance. The range is even further reduced if the extreme pressures to be measured are outside the usable range of the pressure sensor.

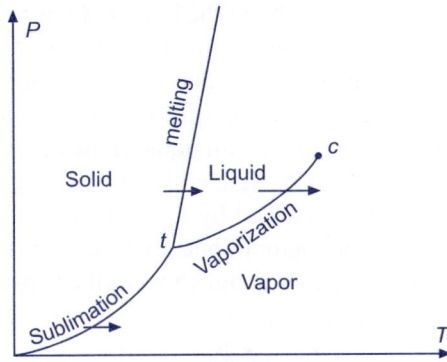

Fig. 17.21 Vaporization Curve

For a given substance, the sensitivity of the thermometer increases approximately inversely with the temperature since u_v varies roughly as 1/P. According to the Clausius Clapeyron equation, we have

$$\frac{dP}{dT} = \frac{L}{T(u_V - u_L)}$$

where L is the molar heat of vaporization which is temperature dependent, and u_v and u_L are the molar volumes of the saturated vapor and liquid respectively. Experimental tables giving $P = f(T)$ have existed for a long time for commonly-used fluids and interpolation formulae have been internationally agreed upon for many of them. It is therefore easy to obtain temperature from measurement of pressure. The sublimation curve can be similarly used, but the range of measurable temperatures is then much smaller, limited on the high side by the triple point temperature and on the low side by the pressure becoming too low to be measured accurately enough. Here, the discussion emphasizes thermometers based on liquid-to-vapor.

A vapor pressure thermometer essentially consists of: a bulb containing the pure measuring substance in thermal equilibrium with the sample whose temperature is to be measured; a pressure sensor; a connecting tube long enough to connect the bulb to the pressure sensor; and a valve for filling. The bulb can be very small depending upon the quantity of substance necessary to cover the required temperature range. It is necessary to ensure that at no time during temperature cycling the substance reassembles into a single phase. The maximum pressure that will occur determines the thickness of the walls of the bulb. The bulb is formed from a material that is a good heat conductor, is chemically neutral, and neither absorbs nor desorbs gases. It is thermally anchored into a copper block which also contains the sample whose temperature is to be measured. If the thermometer is intended to measure the temperature of a large body, the bulb is most often an elongated cylinder (for an industrial thermometer, typically 120 mm long by 15 mm diameter), but its shape can vary as required.

The quantity of filling substance is determined by the necessity to ensure that the surface between the two phases be in the bulb and not in the connecting tube. Moreover, as we have seen, the temperature changes during boiling according to the nature of the impurities. Referring to Figure 17.22, we note that the bulb should contain relatively little substance if the impurities are more volatile than the solvent, and should be almost completely filled with liquid if the impurities are less volatile than the solvent. The bulb's form is dictated by the necessity to take account of eventual concentration gradients caused by impurities in the liquid phase. If the enclosed substance is a poor conductor of heat in the condensed phase, the interior of the bulb is so structured that no large thickness of the substance can become the centre of a temperature gradient.

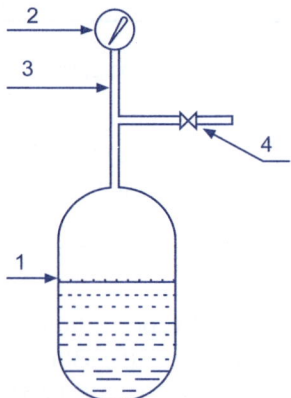

Schematic construction of a saturated vapor pressure thermometer: 1. bulb containing two phases of a substance; 2. pressure sensor; 3. connecting tube; 4. filling system.

Fig. 17.22 Vapor Pressure Thermometer

Many substances could be used as a vapor pressure thermometer. The only constraint is to obtain sufficient purity. The reproducibility of a vapor pressure thermometer is essentially determined by the pressure sensor. The actual magnitude of the imprecision is dependent upon the particular manometer; as typical examples, for a mercury manometer between 233 000 Pa and 6 600 Pa, the imprecision can be ≤ 1.3 Pa. The vapor pressure thermometer can be used as a thermodynamic thermometer only within the limits of application of the Clausius-Clapeyron equation. Otherwise it is an excellent practical thermometer based upon a physical property of a pure substance. Simple and practical, it allows a high measurement precision once the pressure-temperature relationship has been established. The bulb is simple to construct and can be very small. There are not many corrections to apply (a few for impurities possibly, but no dead space corrections as in the gas thermometer). The major inconveniences are its small working range (no pure substance covers a large temperature range) and its nonlinearity of response. The development of new pressure sensors have renewed interest in vapor pressure thermometers. The sensitivity of vapor pressure thermometer is $S = dP/dt$. Precautions such as minimisation of dead volume and removal of cold spots is essential for smooth functioning of the thermometer.

Cryostat

A Cryostat (cryo=cold and stat=stable) is a vessel, similar in construction to a vacuum flask, or Dewar used to maintain cold cryogenic temperatures. Typically cryostats are manufactured with two vessels, one inside the other. The outer vessel is evacuated with the vacuum acting as a thermal insulator. The inner vessel contains the cryogen and is supported within the outer vessel by structures made from low-conductivity materials. An intermediate shield between the outer and inner vessels intercepts the heat radiated from the outer vessel. This heat is removed by a cryocooler. Older helium cryostats used a liquid nitrogen vessel as this radiation shield and had the liquid helium in an inner, third, vessel. Refer Figure 17.23 for the construction of a typical cryostat. Nowadays few units using multiple cryogens are made with the trend being towards 'cryogen-free' cryostats in which all heat loads are removed by cryocoolers. There are four cryostats sizes, the Standard 50 mm, 70 mm, the Mini, and the Maxi. These cryostats offer a range of sample sizes, hold times and physical dimensions which will cover most of the applications likely to be encountered. The cryostat design has been developed and refined over many years, and provides the following unique features:

1. The complete separation of the variable temperature insert and the helium container by the liquid nitrogen shield which allows the insert to be heated to 300 K with no increase in the consumption of liquid helium.
2. The separation of the cooling system and the sample well, thus considerably reducing the risk of blockage in the cooling system.
3. A design of cold valve which allows an 'all or little' operation. The valve can be opened for rapid cooling and when closed maintains a flow, thus eliminating regulation problems. Furthermore, the valve can be removed from the cold cryostat should any maintenance be required.
4. An overpressure facility is provided on the helium can so that temperatures above 5 K can be maintained without liquid helium circulation pumps.

17.26 Cryogenics

5. All separate volumes are fitted with pressure relief valves with the exception of the liquid nitrogen container which is open to the atmosphere.
6. A carefully designed sample holder is provided on which samples can be mounted in close contact with a temperature sensor. The baffles on this sample holder ensure correct thermal coupling between the sample holder and the cooling gas.

All cryostats in the range are manufactured with alluminum alloy casing and stainless steel SS 304 cryogen vessels.

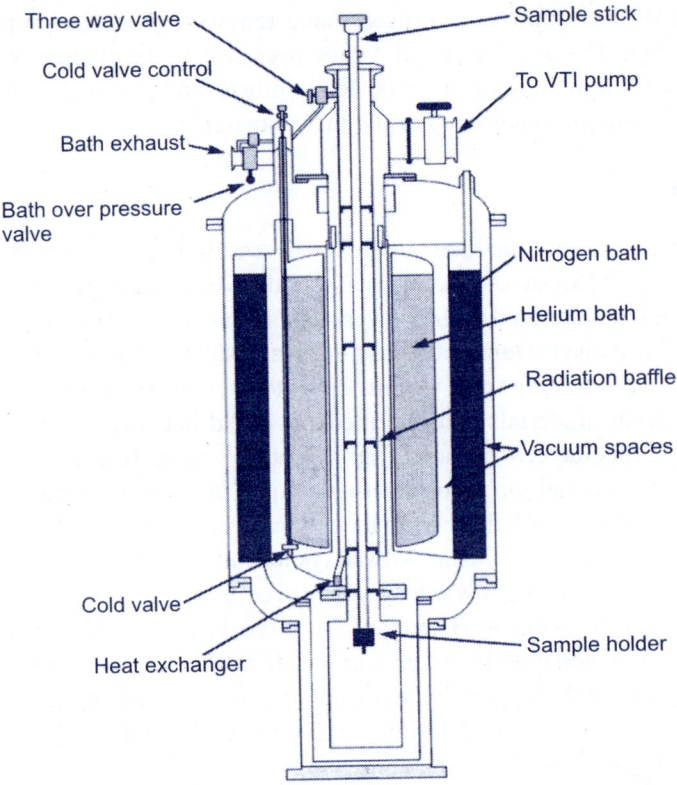

Fig: 17.23 Construction of a Cryostat

References

1. Cryogenic Engineering – B.A. Hands.
2. Internet website www.wikipedia.com.
3. Cryogenic Systems – Randall Baron.
4. Cryogenic Engineering – P.K. Bose.
5. Internet website http://www.asscientific.com/products/cryostats/index.html.
6. Brochure from Liquid Controls Sponsler Inc.
7. NASA report TM-2001-211031 on cryogenic ESP measurement.
8. Brochure of GP-50 Pressure Transducer.
9. Internet website http://www.nivocontrols.com.
10. USDOE Instrumentation Fundamentals Handbook.
11. Internet website http://www.Flowmeters.com.
12. Internet website www.bipm.org.

Questions

Q.1. Write a short note on cryogenic instrumentation.

Q.2. Discuss the operation of a U-Tube Manometer with a neat sketch.

Q.3. Compare Orifice meter and venturi meter.

Q.4. What is a cryostat ? Where is it used?

Q.5. Discuss any two methods of cryogenic temperature measurement.

Q.6. How is cryogenic pressure measured?

Q.7. Discuss the working of capacitance level probes.

Q.8. Explain the working of Magnetic Thermometer.

CHAPTER 18

Cryogenic Storage and Transfer Systems

INTRODUCTION

Liquefied gases, such as liquid nitrogen and liquid helium, are used in many cryogenic applications. Liquid nitrogen is the most commonly used element in cryogenics and is legally purchasable around the world. Liquid helium is also commonly used and allows for the lowest attainable temperatures to be reached. These liquids are held in either special containers known as Dewar flasks, which are generally about six feet tall (1.8 m) and three feet (91.5 cm) in diameter, or giant tanks in larger commercial operations. Dewar flasks are named after their inventor, James Dewar, the man who first liquefied hydrogen. Museums typically display smaller vacuum flasks fitted in a protective casing. The cryogenic transfer pumps are the pumps used on LNG piers to transfer fuel from Carriers to LNG storage tanks, as are cryogenic valves. Cryogenic storage tanks must also be well insulated to prevent boil off. Insulation by design for cryogen tanks adds to the costs. The cryogenic storage industry has progressed a lot from the early years. The range of cryogenic storage vessels is from 1 litre Dewar flasks to approximately 100,000 litre tanks for space and industrial applications. Storage vessels also vary with low tech insulation of foam or fibres to the state of the art multilayer insulation systems which nearly prevent boiloff. Developments in this field have been reported by Hallett and other scientists from the year 1960. In this chapter we shall focus on the design aspect of cryogenic storage and transfer systems.

Basic Cryogenic Storage or Dewar Vessel

A Dewar flask is a vessel designed to provide very good thermal insulation and is named after its inventor, the Scottish physicist Sir James Dewar (1842–1923). When James Dewar was appointed Professor of Chemistry at the Royal Institution, London, in 1877, he set about tackling the two obstacles to progress in cryogenics at that time. The first was the complete lack of understanding of heat transfer processes and of how to achieve thermal insulation; the second was the lack of basic data on the properties of fluids for producing low temperatures. For 15 years progress was slow until, in 1892, he was able to employ his invention of the silvered, double walled, glass vacuum vessel to contain cryogenic liquids for the first time,

18.2 Cryogenics

for relatively long periods, before they evaporated. The idea of vacuum insulation had been used by Dewar and others as early as 1873 to show significant reduction (up to 6 times) in heat influx by introducing into the vacuum space powders such as charcoal, lamp black, silica, alumina and bismuth oxide powder insulations. He also found that three turns of aluminum sheet was not as good as silvered surfaces. His discovery of silvering as an effective means of reducing the radiated heat flux component was a breakthrough. From 1802, the glass dewar flask quickly became the standard container for cryogenic liquids, leading to the successful liquefaction of hydrogen and helium in later years. Dewar had considerable difficulty in finding competent glass blowers willing to undertake the construction of his double-walled vessels, and was forced to get them made in Germany. Dewar never patented his silvered vacuum flask and never benefited financially from his invention. Following Dewar's invention in 1892, the design of Dewars for containing cryogenic liquids did not change for over 60 years, until the growing availability of helium in the 1950s gave rise to its use in open cryostats without liquid hydrogen shielding. Refer figure 18.1 for the construction of a Dewar vessel. Note the red area is the vacuum.

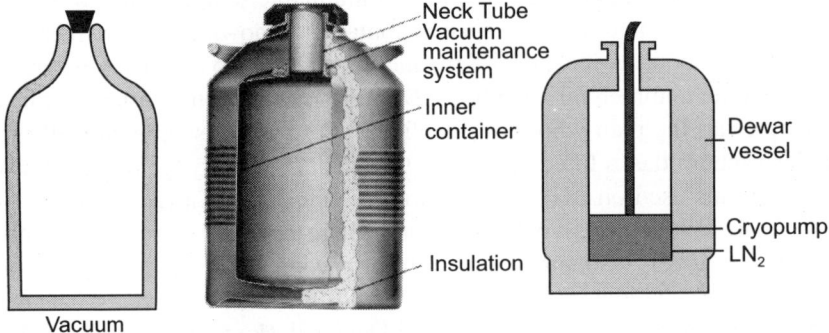

Fig. 18.1 Construction of the Dewar Vessel

In form, a Dewar flask is a glass or metal bottle, with a double-layer construction; in fact, it can be considered to be two thin-walled bottles nested one inside the other, and sealed together at the neck. The narrow space between is evacuated almost entirely of air; the near vacuum minimizes conduction and convection of heat. The inner surfaces of the outer bottle, and the outer surface of the inner bottle, have a metallic or similar reflective coating to prevent heat from being transmitted via radiation. Dewar himself used silver for this purpose. This is so effective that what little heat transport does occur is almost entirely via the neck and stopper; insulating materials such as cork are often used for the stopper. A very common use of the Dewar flask in laboratories is the storage of liquid nitrogen; in this case, the leakage of heat into the extremely cold interior of the bottle results in a slow "boiling-off" of the liquid (a pressure relief valve is provided to prevent pressure from building up). The excellent insulation of the Dewar flask results in a very slow "boil" and thus the nitrogen lasts a long time without the need for expensive refrigeration equipment. The Dewar flask consists of two flasks, one inside the other, separated by a vacuum. The vacuum greatly reduces the transfer of heat, preventing a temperature change. The walls are usually made of glass because it is a poor conductor of heat; its surfaces are usually lined with a reflective metal to reduce the transfer of heat by radiation. The inner vessel of the Dewar is known as the product container storing the cryogenic liquid. The inner vessel is enclosed by the vacuum jacket and the outer vessel. The space between the vessels is filled with vacuum or multilayered insulation. Fill line and drain lines are provided for larger Dewar vessels. A vapor vent is also provided along with vapor diffusers for safe venting of the cryogenic liquid. Antislosh baffles are provided in transportable vessels and a suspension system is provided to support the product container within the vacuum jacket. The design capacity and storage pressure is decided based on user requirements, however standard sizes are available. Dewars are not fully filled due to heat inleaks and rapid boil off phenomena. Dewars can be constructed in cylindrical, conical and spherical shapes. Generally cylindrical shapes are preferred.

Refer Figure 18.2 for a typical cryogenic Dewar vessel. The whole fragile flask rests on a shock-absorbing spring within a metal or plastic container, and the air between the flask and the container provides further insulation. A "Thermos bottle" (trademarked), or vacuum flask (generic) used to keep coffee or other beverages hot, is usually based on the design of a dewar flask; since glass dewar flasks are fragile and somewhat expensive to fabricate, other constructions based on plastic or insulated metal are also popular. Thermos is a proprietary name or trademark applied to a type of dewar flask protected by a metal casing.

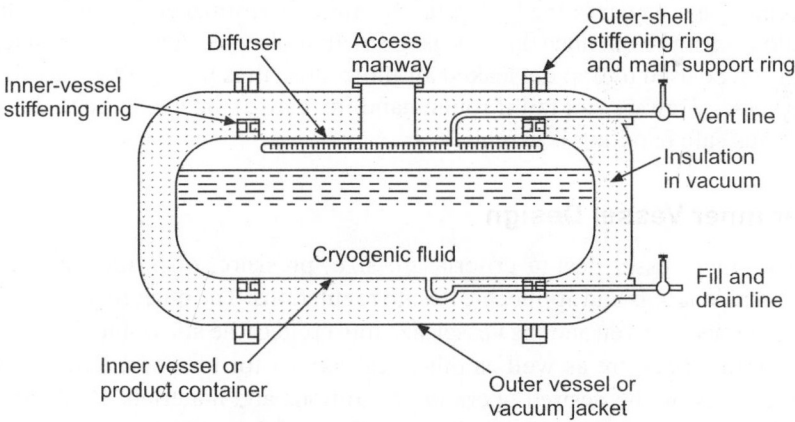

Fig. 18.2 Cryogenic Dewar Vessel

Most of the laboratory dewars are designed for storing and dispensing small quantities of liquid nitrogen. Easy to operate, the snap on cap and necktube core assure positive closure and easy access without unnecessary exposures to the cryogen. Constructed from materials of the highest performance and quality including an aluminum exterior, these containers are not only rugged, robust, and dependable, but also have very high thermal efficiencies. Refer Figure 18.3 for the photograph of Laboratory dewars.

Fig. 18.3 Photographs of Laboratory Dewar Vessels

Most dewars are of all metal construction and are glass-free. Those dewars that contain glass are much more fragile and less robust than the metallic dewars. dewars are of two types, namely pressurised and non pressurised. Non pressurised type are vessels that are open to the atmosphere so that the cryogenic liquid boils at a pressure of one atmosphere. The pressurized dewars is a closed vessel, where liquid nitrogen becomes a boiling liquid at a temperature determined by the pressure within the vessel. This type of dewar is regulated as pressure vessels, because there is the potential for a substantial pressure to develop within them, and generally they are protected by a pressure relief valve. One should note however that there is no need to utilize a pressurized vessel for the local moving of the liquid, such as from the master supply tank (or dewar) to some laboratory application. Integrity of the vacuum in a dewar is the key to its sustained performance. A perfect vacuum insulation cannot be obtained due to leakage of air molecules. A vacuum maintenance system can be used to trap the leaked air molecules, thus prolonging the life of the dewar vessel. Dewars can fail if rough handling causes the necktube to deform or break, thus safe handling is essential.

Dewar Inner Vessel Design

Primary inner vessel design criteria are size, pressure, operating temperature, mechanical loadings and applicable design regulations. The quantity of the cryogen to be contained is given and the vessel size must reflect the allowable filling density and operating pressure as well as other restrictions for transport dewars. Design pressure is set by the desired operating conditions and materials of construction. Operating temperature guides the selection of materials and dictates necessary weld qualifications, particularly the need for impact testing. Operating conditions govern the design and selection of supports and shell structure consistent with support loads. Most cryogenic vessels are designed in accordance with one or more of the following criteria:

1. Section VIII of the ASME (American Society of Mechanical Engineers) Code for Unfired Pressure Vessels.
2. Various parts of 49 CFR (Code of Federal Regulations) particularly 173.316, 173.318, 173.319, 178.57 and 178.338.

3. European Agreement concerning the International Carriage of Dangerous Goods by Road (ADR) and "Recommendations on the Transport of Dangerous Goods".
4. Compressed Gas Association: CGA S-1.2 and CGA S-1.3: Pressure Relief Device Standards for Cargo and Portable Tanks and Compressed Gas Storage containers.
5. National Fire Protection Association: NFPA 50- Bulk Oxygen Systems at Consumer Sites and NFPA 50B- Liquefied Hydrogen Systems at Consumer sites.

The product container must withstand the design internal pressure, weight of fluid in the vessel and the bending stresses due to beam bending action. The inner vessel must be constructed of cryogenic compatible materials like steel, aluminum, monel and copper. Stiffening rings are used to support weight of fluid in lower vessel. The minimum thickness of vessel (t) is given by

$$t = pD/(2 S_a Ew - 1.2 p)$$

where p = internal design pressure, S_a = allowable stress, D = diameter of shell.

The bending moment (M max) for the inner shell stiffening rings is given by

$$Z = M \max / S_a$$

where Z = section modulus and S_a = allowable stress.

Dewar Outer Vessel Design

Dewar vacuum jackets (outer shells) are normally warm and material selection is primarily driven by cost and fabrication concerns, not low temperature compatibility. Carbon steel is the leading material for industrial Dewars. It is easy to weld and has an elastic modulus (the governing factor in external pressure calculations) of approximately 207 GPa (30E06 psi). Outgassing and hydrogen diffusion from vacuum jacket material is important and the warm hydrogen getters and cold outgassing (mostly water) getters must be sized to reflect these loads for long term vacuum performance. Stainless steel and aluminum vacuum jackets are generally cleaner and are often preferred for this and other reasons. The vacuum jacket is a vital element in the Dewar structure in addition to reacting the external pressure vacuum loads. Inner vessel support loads are typically anchored to the outer shell and then distributed to external load points. These may be legs or skirts for stationary dewars, the fifth wheel and multiple axles in transport dewars and ISO corners in container dewars for combined land and sea service. Careful stress analyses are required to properly distribute the loads to avoid stress concentrations which may lead to fatigue failures. These may be catastrophic because a vacuum shell crack usually causes an instant failure in the insulating vacuum. Design to take care of contingencies is desirable.

The external shell of a dewar should not be thin and fail under externally applied stress. Failure by elastic instability is given by the expression of critical pressure (P_c).

$$P_c = 2E\,(t/D_o)^3/(1-v^2)$$

where E = Young's modulus, t = shell thickness, D_o = outside diameter of shell and v = Poisson's ratio.

The outer shell stiffening rings are used to support weight of inner container and fluid and hold the outer shell. The moment of inertia (I) for the stiffening rings is given by

$$I = P_c D_o^3 L/24\,E$$

where P_c = Critical pressure, L = Distance between stiffening rings, D_o = outside diameter of shell and E = Young's modulus.

The bending moment (M max) for the outer shell stiffening rings is given by

$$I = M\max c/S_a.$$

where I = Moment of Inertia, c = distance from axis to outer fibre of the stiffening ring and S_a = allowable stress.

Dewar Vessel Support Design

Supports for storage dewars are fairly straight forward. The dewar is never moved while it is in service and seismic and wind loads are the only on-site factors to be considered and temporary shipping supports may be utilized to handle loads not encountered in service. Supports for transport dewars are more of a challenge than for storage units. These dewars are subject to three dimensional loads, vibration and smaller units may be designed to withstand dropping while being loaded or unloaded. Regardless of the configuration cryogenic vessel supports have a common problem: they must occasionally resist high stress loads while continually contributing minimum heat leak to the cryogen. Optimizing support materials is an important part of cryogenic technology. Desired qualities include low thermal conductivity, high strength, long fatigue life, ease of fabrication, low temperature and vacuum compatibility and acceptable cost. Generally, tensile configurations are most effective. 300 series stainless steel (as rods, tubes and cables) supports are most commonly used. Better performance is obtained with Inconel 625 and Titanium members at higher costs. However, composite materials are superior to any of the metals. Filament wound S-glass/epoxy is at the top of the conventional list with NEMA Grade G-11 and G-10 widely used in less demanding applications.

The suspension system design is very critical and a wrong design can nullify the effect of using high performance insulation. Some commonly used suspension systems are steel tension rods, metal saddle bands, plastic compression blocks, stacked discs, chains, wire cables or compression tubes. The force (F) on the suspension is given by

$$F = (1 + Ng)\,W/2$$

where Ng = acceleration load, W = weight of the inner vessel.

The heat transfer through the support (Q) is given by the Fourier's equation

$$Q = k\,A\,\Delta T/L$$

where k = Thermal conductivity, A = Area of cross section, ΔT = Temperature gradient, L = Length.

Cryogenic Industrial Storage Tanks

Industrial gases such as liquid oxygen, nitrogen and argon are delivered to customers in liquid form at cryogenic temperatures and stored in tanks before further use. The pressure ratings and sizes of these cryogenic tanks are standardized in accordance with the requirements of distribution logistics and economical series production. Typical cryogenic industrial storage tanks are the vacuum insulated double wall tanks. The vacuum-insulated double wall tanks consist of two concentric vessels, an austenitic steel inner tank and an outer jacket in carbon steel with an anti-corrosion primer and a special environmentally friendly top coat. The interspace between inner and outer tank is filled with insulating powder (Perlite) and evacuated. An adsorbent is also added to maintain the vacuum in the insulation interspace. The standard tanks come in gross nominal water capacities from 3,160 litre to 61,620 litre. The maximum allowable working pressure for these tanks is 18, 22 or 36 bar gauge for design temperatures ranking from –196°C up to 20°C. All standard tanks have vertical configuration, requiring little space for installation.

Refer Figure 18.4 for the construction of an industrial cryogenic tank and Figure 18.5 for the photograph of a cryogenic tank.

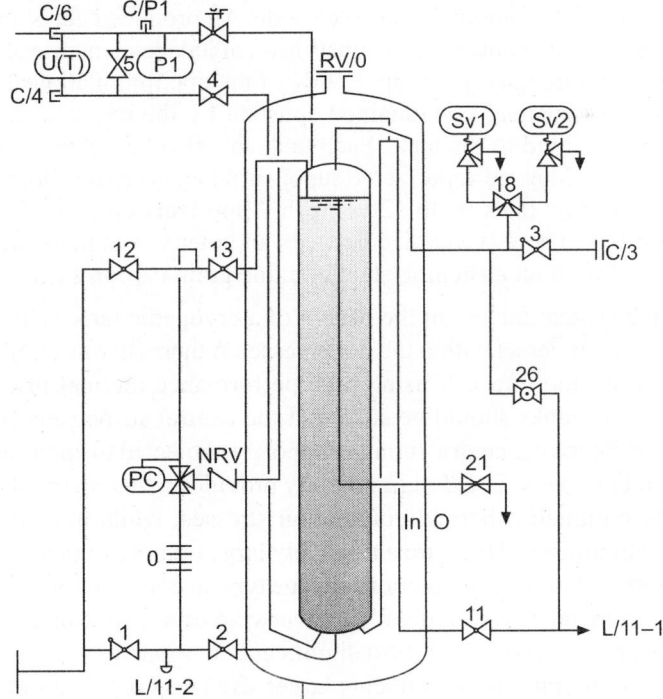

Fig. 18.4 Construction of a Cryogenic Tank

Fig. 18.5 Photograph of a Linde Cryogenic Tank

The pressure vessels are typically manufactured and tested in accordance with EN 13458 which is internationally accepted code. All produced tanks are subjected to inspection and quality control under supervision of independent inspection bodies. The operating pressure may be set up to 90% of the maximum allowable working pressure and is automatically maintained constant by the regulator and pressure build-up vaporizer fitted to the tank. Each tank can also be equipped with a tank mounted (clip-on) air-heated vaporizer to supply product in gaseous form at ambient temperatures and flow rates up to 120 Nm^3/h. Vaporizers up to 1,000 Nm^3/h are installed separately. There is a choice between an inner vessel manufactured from low temperature resistant austenitic steel or a fine grain carbon steel.

One of the critical factors in the design of a cryogenic tank is the method of suspending the inner vessel within the outer vessel. A thermally inferior suspension system can nullify the effect of using high performance thermal insulation. The support system for tanks should be such that the central suspension rod is under tensile stress while the concentric tubular support is subjected to compressive stress due to acceleration loads. The design not only provides a long thermal conduction path, but also eliminates thermal contraction stresses, while allowing marginal assembly misalignments. There are a relatively large number of patents relating to cryogenic tanks of various constructions. In one typical construction, an inner liner, usually made of metal, is covered with an overwrap of a matrix of an epoxy resin and carbon fibers. In general, the most significant component with respect to cost and schedule considerations in producing larger size tanks or vessels of this type is the introduction of the traditional metallic liner. In this regard, larger metallic tank liners require significant tool set-up and the weld joints required for the liners are troublesome sites for leakage.

Storage of cryogenic fluids requires tanks withstanding operating temperatures lower than −161°C. Cryogenic fluids are usually stored in double-walled tanks at atmospheric pressure. The inner tank serves to keep the fluid content enclosed, whereas the outer tank keeps the insulating material in position, protects the inner tank and the insulation against external influence, and provides increased security in case of leakage in the inner tank. For small to medium size tanks it is not unusual to use pressure tanks or isolation provided by arranging a vacuum between the tank walls. Such arrangements are inadequate for large tanks because the walls must be made unduly solid. In order to limit the wall thickness a circular cross section of the tank is the most usual. The outer tank is usually constructed to keep gas enclosed if a leakage should occur in the inner tank, especially if the tank is situated at a location having bad ventilation or in a populated area. The cryogenic tanks may have sizes of up to about 160,000 m^3, or even 200,000 m^3, which may necessitate a tank diameter of about 70 m and a tank height of about 60 m. For large cryogenic tanks, mainly two types of constructions are used. The first type of construction is a cylindrical, self-standing tank of which the inner tank is made of a suitable steel and the outer tank is built of steel or reinforced / prestressed concrete. The second type of construction is a membrane tank in which a thin metal membrane, for instance of a thickness of 1.2 mm, is installed in a cylindrical concrete structure which is built either below or above the ground level. An insulating layer is positioned between the metallic membrane of stainless steel and the load-bearing concrete structure. For a double walled cryogenic tank, there are problems connected with temperature expansion, which are substantially reduced by using materials having equal thermal expansion coefficient.

Nowadays linerless composite cryogenic tanks are providing cost-effective, lighter weight storage solutions for a variety of markets and applications. Weight savings of 15% to 25% over metal lined can be realized with this new composite tank technology. Using a unique set of micromechanics models, a new family of toughened, microcrack resistant resin based cryogenic tanks have now been developed. When properly manufactured into a fiber reinforced composite laminate, these novel resin systems and optimized engineering designs enable the production of cost-effective, microcrack resistant, light weight linerless composite cryogenic tanks. Markets such as space and aerospace, where weight is paramount, are embracing composite tanks as an exciting new alternative. In addition composite technology has been shown to be an enabling technology for lightweight composite cryogenic tanks for launch vehicle and other applications. Based on a new family of toughened resins developed for filament winding and fiber placement manufacturing processes, the microcrack resistance performance of composite materials far exceeds that of industry-standard materials. Test results show that composite materials microcrack at significantly higher strain levels, at both room and cryogenic temperatures, than do industry standard materials. This increased toughness enables linerless tanks made with composite resins to achieve higher design-operating strains, which results in lower tank dry mass. The progress in mechanics of composites, development of novel materials, and innovative manufacturing technologies has led the way for advancements in linerless composite cryogenic tanks. Analysis packages have been developed to predict microcrack accumulation and identification of potential leakage paths in linerless tanks. In addition, analytical studies of thermal-fatigue behavior in composite materials have developed fatigue-failure models to help guide the design of linerless composite tanks subjected to cyclic mechanical or thermal loading. Predictions from the analysis have been validated against actual test results and have been shown to be effective in improving cryogenic tank reliability and minimizing weight. Refer Figure 18.6 for the finite element analysis of a Cryogenic Tank.

Fig. 18.6 Finite Element Analysis of a Cryogenic Tank

Novel manufacturing technologies have been developed that enable the superior performance of composite materials to be realized in cryogenic tank performance. Key techniques include non-autoclave curing of the composite tanks, hermetic-interface technology to join the composite shell to metallic fitting, and the ability to cost-effectively attach composite tanks to other structures. Results of new developments have shown that these technologies can be applied in synergy with composite resin materials to provide highly attractive storage solutions for a variety of cryogenic tank applications.

Cryogenic Storage in Space

The propellant combination of liquid hydrogen and liquid oxygen has a high exhaust velocity, and is thus favored for orbital transfer and interplanetary missions. In such missions, there can be waiting periods of days to months between when the propellant is launched from earth, and when the propellant is used for propulsion. Since both liquid oxygen and liquid hydrogen are cryogenic and only remain liquids at very low temperatures, the issue of the practical boiloff rate in space is important in mission planning. A tank set for on-orbit storage of propellants for a reusable orbital transfer vehicle contains approx 60,000 kg of liquid oxygen and hydrogen, at a mixture ratio of 6 to 1. Normally space cryogenic tanks use multilayer insulation consisting of 50 layers of double aluminized Kapton, separated by filler material. Heat leakage includes that through the insulation, as well as through tank/shell struts and fill, feed and vent lines. The heat leakage causes a boiloff in this specific tank set of 0.45 kg/hour of liquid hydrogen, and 0.36 kg/hr of liquid oxygen.

It should be noted that boiloff is governed by heat leakage, and the rate in kilograms per hour does not depend on the amount of propellant in the tanks. With partly filled tanks, the percentage loss per day or month would be higher. Also note that heat leakage is driven by surface area, while the original mass of propellant in the tanks is governed by volume. Thus, the smaller the tank, the faster the liquids will boil off (square/cube law). Boiloff rates were estimated for tank sets in low earth orbit. Lower boiloff rates would be expected for tanks at a further distance from the sun. Lower boiloff rates can also be expected if tanks can be maintained at an angle which minimizes their cross-sectional area exposed to the sun, or tanks protected by a sunshade. The boiloff rates in low earth orbit thus most probably represent a worst case scenario, with better results to be expected for spacecraft on Mars trajectories etc. Taking into account the enthalpy of vaporization, the oxygen tank heat flux across the walls is $0.36 \times 213.1 = 76.7$ kJ per hour. For the larger and colder hydrogen tank, the figure is $0.45 \times 445 = 200.25$ kJ per hour. The hydrogen tank is venting 0.45 kg/hour of very cold vapor, at just above the boiling point of liquid hydrogen. This vapor has a very high heat capacity, 12.24 kJ/kg K. If the hydrogen vapor were conducted through a heat exchanger in thermal contact with the liquid oxygen tank, it would be warmed from approximately 20 K to approximately 90 K. This has the capability of providing as much as $(90-20) \times 12.24 = 857$ kJ/hour of refrigeration. In other words, if desired the boiloff vapor from the hydrogen tank can be used to keep the liquid oxygen from boiling off entirely.

Cryogenic tanks having internal insulation and a relatively thin sealing membrane within the insulation have been found suitable for space applications. Generally such tanks have a relatively thick outer shell which provides the desired strength to the tank and an internal thermal insulation such as plastic foam. The innermost surface of the tank is a low strength wall of either thin metal or synthetic resinous material such as polyester film or sheeting which prevents a liquid contained within the tank from directly contacting the insulation or outer tank wall. Such an insulated tank may be readily prepared. However, on filling of the tank with cryogenic liquid the thin inner membrane contracts with the foam insulation. As a result of the thermal contraction, stresses are set up which tend to separate the insulation from the outer tank wall. Oftentimes such a separation occurs in a tank which is supported on a fixed base; such tanks, however, are still useable. Separation of the insulation from the outer tank wall can present serious problems when the tank is employed in rockets. This was the reason of failure of the ill fated Columbia shuttle of the USA. Oftentimes minor leakage or permeation occurs of the cryogenic liquid through the inner sealing membrane or inner tank wall and the location of the general area of the leak is extremely difficult to establish. Refer Figure 18.7 for the photographs of cryogenic space tanks used in rockets. Cryogenic propulsion enables a significant increase of space launchers performance, when compared to storable propellant solutions. Cryogenic tanks used to store liquid oxygen, liquid hydrogen and liquid helium. On some scientific satellites, cryogenic tanks are also used to store the cryogen that provides the cooling power required by the scientific instruments.

18.12 Cryogenics

Fig. 18.7 Photographs of Cryogenic Space Tanks

Cryogenic Piping

Cryogenic piping systems have evolved over the past 150 years. During that time thermodynamic engineers have developed amazing new technologies that have greatly increased the efficiency of the cryogenic piping supply systems. The first and the oldest type of supply system used simple copper or stainless steel pipe wrapped with wood insulation. Wood was later replaced by mica, then shortly after polyurethane foam. Foam insulation is highly inefficient and cumbersome; much of the liquid cryogen traveling through the pipe turns to gas and pipe diameters are often over 8 inches. These large, bulky, inefficient piping systems were relatively cheap to install, but the loss of liquid cryogen make them very expensive in the long run. Vacuums are one of the best ways to eliminate heat conduction, so finally a dynamic vacuum piping system was developed. This piping system uses a pipe inside a pipe. The outer jacket pipe is constructed of upper bellow tubing and surrounds the copper inner bellow tube with an airtight-as-possible seal that technology could afford. To maintain a thermo-arresting vacuum, the ultimate insulator, between the two layers a vacuum pump is attached to the line. The vacuum pump works continuously. The dynamic vacuum piping system dramatically decreased heat loss, and therefore was a much more efficient system. While the dynamic system is much more effective than foam, materials and installation cost much more, and the time and maintenance required for the pumping system is highly cost ineffective due to electricity consumption. Refer Figure 18.8 for the photographs of cryogenic pipes.

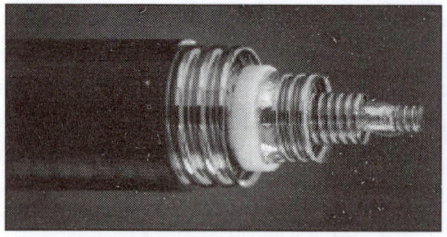

Fig. 18.8 Photographs of Cryogenic Pipes

The most recent and best engineered cryogenic piping system available is the Static Vacuum Jacketed system. The Static system contains all the benefits of the Dynamic System without all the maintenance headaches and optional cost issues. Pieces of vacuum jacketed pipe (VJ sections) are designed according to specifications and are connected with very low energy loss bayoner style fittings. This system is less complex to install than a Dynamic System and extremely efficient. While it generally is the most expensive system to install, it easily makes up the cost in performance, efficiency and long term duration; some well-engineered systems built 35 to 40 years ago are still functioning with good vacuums. Refer figure 18.9 for the construction of the cryogenic pipe.

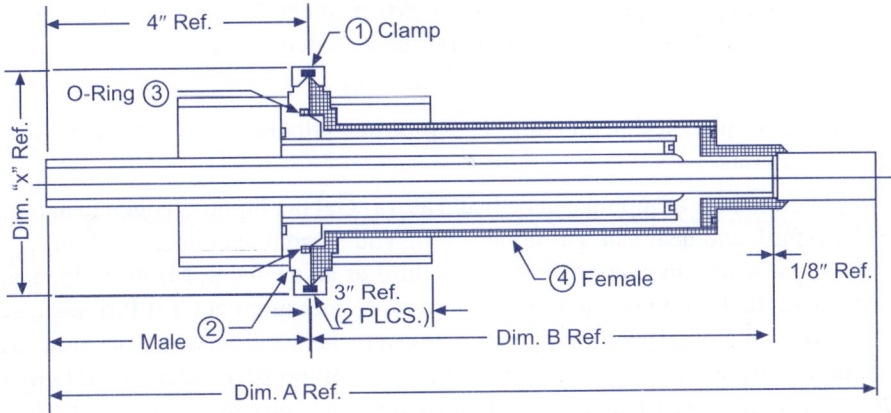

Fig. 18.9 Construction of a Cryogenic Pipe

Liquid cryogens normally exist as a two-phase fluid — a mix of liquid and gas. Liquid cryogen quality refers to the ratio of liquid cryogen within the two-phase cryogenic fluid. A higher percentage of liquid results in a higher quality cryogenic fluid as well as higher cooling capacity per pound of fluid. Consequently, cryogenic liquid quality is an important factor when designing a cryogenic piping system for process cooling. One way to maintain high quality is to minimize heat leak into the piping system. This is best accomplished with vacuum-insulated, multilayer insulation. However, certain applications require a higher quality liquid than that delivered by a vacuum-insulated piping system. Cryogenic piping system design is dependent upon the system's application and operating characteristics. When selecting a pipe insulating system for cryogenic applications, the decision should be based upon the application's operating characteristics and the usage pattern. Key operating factors to consider include: cryogenic fluid flow rate, fluid type, usage pattern or cycle time of the cooling application, required liquid quality, distance from the liquid storage point to the use point. Cryogenic applications can be segmented into four general categories: high flow continuous operation, low flow continuous operation, high flow intermittent operation and low flow intermittent operation. To determine each category's most efficient insulating system, the designer must calculate the system's total cooling losses. This total includes the initial cool down losses to chill the process pipe to cryogenic temperature plus the steady-state heat leak into the piping system.

Many of the problems with underground refrigeration piping is refrigerant leaks due to stress or corrosion and underground freezing. The result is costly repairs. Leaks can occur from corrosion due to soil conditions or stress caused by expansion and contraction due to the varying temperatures of the piping. To avoid this problem, every cryogenic pipe is helium leak tested and cold shocked to ensure its vacuum integrity. A properly designed pipe will restrict the heat transfer across the pipe to conduction only. Piping runs should be as long as possible and walls should be as thin as possible. The thermal contraction of piping should be considered in cryogenic pipe design and provision of expansion bellows should be provided. The minimum wall thickness of piping (t) can be calculated as

$$t = pD_o/(2S_a + 0.8p),$$

where p = pipe design pressure, D_o = pipe outer diameter, S_a = allowable stress of pipe material.

The functions of piping can be listed as: (a) Get the liquid cryogen in and out, (b) Vent flash and heat leak gas under filling and normal operating conditions, (c) Vent all gas safely in event of vacuum failure or an accident, (d) Provide means for hydrostatic liquid level and internal pressure measurement, (e) Provide a self pressurization loop. These Cryogenic pipes must be sized to safely perform their function with attention to flexibility and heat leak. Improperly oriented pipes in the Dewar vacuum space (including external vacuum jackets) can have a huge heat leak impact. In all cryogenic piping runs, the warm end must be at least one diameter higher than the cold end. Any piping segment not meeting this criteria must be considered a thermal short circuit. Cryogen piping and tubes should be made of 300 series (but not 303) stainless steel for purposes of heat leak and ease of fabrication.

Draining of Cryogenic Tanks

Cryogenic liquids are substances that normally exist as gasses, but are liquids at cold temperatures. Special vessels and systems must be used to store and transfer cryogenic liquids because of difficulty in maintaining the extremely cold temperatures. Such vessels typically include a double walled vessel having a vacuum in the annulus or plenum. While the vacuum provides an effective insulation, the insulation is not perfect and, as such heat penetrates the vessel. When heat is added to the cryogenic liquid, a portion returns to the gaseous state. The gas within the vessel increases the internal pressure. Eventually, to prevent over pressurization of the vessel, the gas must be vented. It is desirable to prevent, or at least delay, the venting of the gas. One means of delaying the over pressurization of the vessel is to not completely fill the vessel with the cryogenic liquid. That is, vessels are filled by spraying the cryogenic liquid into the top of the tank via a nozzle. This spray condenses the gas in the tank and collapses the pressure head therein. When the pressure head collapses, the pressure in the tank is substantially reduced allowing the flow of the cryogenic liquid into the tank until it submerges the nozzle. Once the nozzle is submerged, the pressure in the tank is gradually increased. A pressure gage monitors the fill line and the filling procedure is automatically shut-off when the pressure in the tank reaches a predetermined value. At this point, the vessel is not completely full and an ullage space exists above the level of the cryogenic liquid. Refer Figure 18.10 for a cryogenic tank with a drain line.

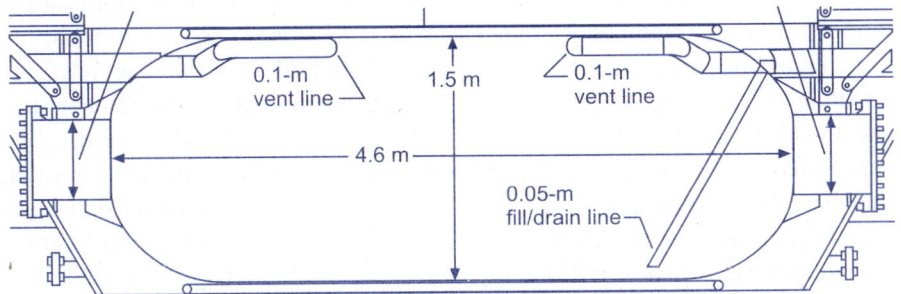

Fig. 18.10 Cryogenic Tank with a Drain Line

The ullage space accommodates any cryogenic liquid that is subsequently vaporized due to heat penetration. Because the newly vaporized gas has a space to fill, the hold time of the tank is increased and venting is delayed. Because monitoring pressure in the fill line to determine when the tank was filled to the proper point does not always result in the vessel being filled to the same level this system was improved by the addition of an ullage space vessel. The ullage space vessel was essentially another vessel having a small opening therein, and which was disposed within the cryogenic vessel. During the fill procedure, the small opening in the ullage space vessel prevented the ullage space vessel from being filled as rapidly as the cryogenic vessel. That is, a small amount of cryogenic fluid would seep into the ullage space vessel at a slow rate until the fill levels are essentially even. There is, therefore, a need for a cryogenic vessel with an ullage vessel structured to drain any residual liquid while resisting filling with a liquid during the filling procedure.

The cryogenic tank draining can be done by three methods. (A) Self pressurization of inner vessel (B) External pressurization of inner vessel or (C) Pump transfer. Self pressurization involves removing some of the liquid from the product container and boiling the liquid in an external evaporator or pressurization coil. The vapor generated is then passed back to the ullage space of the vessel through a diffuser. External pressurization involves introducing high pressure gas from an external source into the inner vessel to increase pressure inside the tank. The high pressure gas is obtained from compressed gas cylinders. Estimation of pressurizing gas required (Δm) is made by the saturation rule given by

$$\Delta m = V_2 \rho_2 - V_1 \rho_1$$

where V_1 = initial ullage volume, V_2 = final ullage volume, ρ_1 = initial pressure saturated vapor density, ρ_2 = final pressure saturated vapor density. In case of pump transfer, a cryogenic pump in the liquid drain line is used. This system is suitable for high flow rate applications.

Cryogenic Safety Devices for Tanks

The basic cryogenic safety devices used on tanks are the inner vessel pressure relief valve, inner shell burst disc assembly, annular space burst disc assembly. The inner vessel pressure relief valve is a spring loaded valve set such that inner vessel pressure should not exceed design pressure by ten percent and for relieving the pressure in extreme condition. The inner shell rupture disc is an additional safety device that operates if the main pressure relief valve of the inner shell fails to operate. The rupture disc has to be replaced after each use. The annular space burst disc operates to protect the outer vessel from excessive internal pressure. Even the most basic dewar needs a pressure gauge and liquid level indicator. For safety, it must have a combination of relief valves or a relief valve and rupture disc with the combined capacity to safely vent the dewar in the event of vacuum failure. A vacuum gauge tube is desirable but not absolutely essential. Cryogenic tanks have a very efficient thermal insulation. When gas is withdrawn from such a tank, the pressure, which is typically a few bar relative, drops because the heat influx is too low to compensate for the loss of fluid. Consequently, when gas is withdrawn, the pressure in the tank may drop excessively with respect to the requirements of the user network. In order to keep the pressure in the tank constant, heat has to be supplied to the tank during withdrawal. For this purpose, pressure-regulating devices for cryogenic tanks are used. These safety devices use an electrical resistor as heating element, in combination with electrical safety means in case of electrical power outage. Inexpensive pressure-regulating devices which can provide a cryogenic tank with heat over a long period are desirable. These devices include a feed pipe suitable for feeding the heating chamber with a heating fluid having a temperature above the temperature of the said cryogenic fluid, and an exhaust pipe intended for discharging the heating fluid.

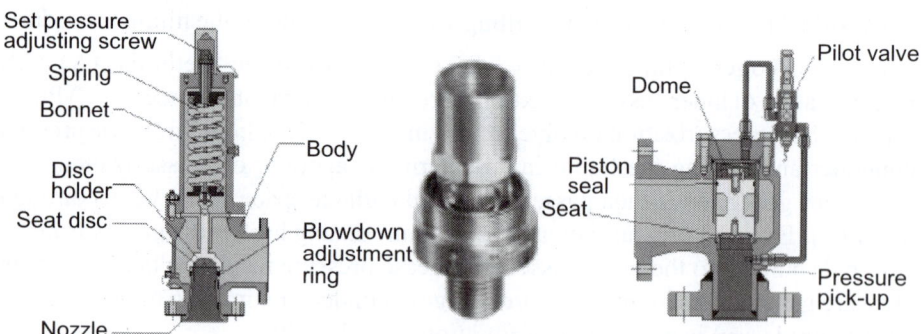

Fig. 18.11 Cryogenic Tank Relief Valves

Most probable malfunction of a cryogenic tank is over pressurization of inner vessel or transfer line. Every dewar that can be sealed off with valves, or somehow be isolated, must be equipped with at least one if not two pressure relief devices. Safety devices provide necessary insurance against catastrophe. The inner vessel should be hydrostatically or pneumatically tested to 1.5 -2.0 times working pressure. The ASME code requires two safety devices namely the pressure relief set to 1.1 times working pressure and a rupture or burst disk system set to burst at 1.2 times working pressure. Relief valves and burst disks are often used in combination. Each device must be sized to handle the flow requirements of the entire vessel. Relief valves are of various types such as conventional relief valve, rupture burst disk type or rupture pin type. Refer Figure 18.11 for two cryogenic tank relief valves.

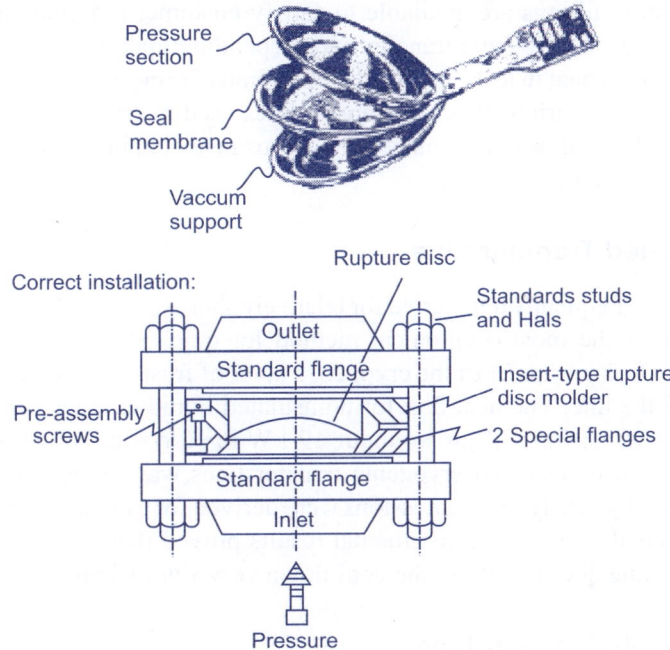

Fig. 18.12 Cryogenic Tank Rupture Disk

Cryogenic vessels or piping subject to thermal expansion must be protected by a thermal relief device. Multiple vessels may be protected by a single relief device provided there is a clear, unobstructed path to the device. At least one pressure relief device must be set at or below the maximum allowable working pressure (MAWP). Refer Figure 18.12 for a typical rupture disk used on cryogenic tanks.

Cryogenic Transfer Systems

Cryogenic fluid transfer through tubes in varying quantities has been achieved for several years. However requirements of space applications has made necessary the development of high flow rate cryogenic transfer lines of the capacity 2000 gallons per minute. The length of the lines is usually 2000 feet or more. For larger lines, there are problems such as thermal contraction, cool down with two phase flow and bowing of the line under partial fill conditions. To overcome these problems cryogenic transfer lines were constructed in sections with low heat in leak joints at periodic intervals. In this section various types of transfer lines are explained with joints and the two phase flow phenomenon.

For the most demanding applications vacuum jacketed hose provides efficient low-loss transfer of cryogenic liquids. These lines are particularly suitable where large liquid losses cannot be tolerated or for continuous flow systems. When only intermittent low volume transfers are required, the un-insulated flexible hose is the ideal choice for providing convenience and economy. This particular type of hose is offered with an option of foam, which adds insulating properties. Cryogenic hoses are furnished with a protective braid covering for extended life. A wide variety of termination fillings are available to satisfy customer requirements. There are different types of cryogenic transfer lines depending upon the insulation used for the reduction of heat in leak. Types of specific transfer line depend upon the specific application. The various types of transfer lines used in cryogenic application can be classified as follows: uninsulated line, porus insulated line, expandable transfer line and vacuum insulated line.

Uninsulated Transfer Line

Cryogens are frequently transferred for relatively short distances through uninsulated lines. This is the most economical method for quick short distance transfer of cryogens. During transfer of the cryogen, a layer of frost builds on the line thereby insulating the line. The heat flux for uninsulated liquid cryogen line ranges from about 11 kW/m^2 for still air to about 19 kW/m^2 for wind velocity of 6.7 m/s. Cool-down of uninsulated cryogenic transfer lines was studied analytically and experimentally. Analytical expressions were derived for estimating cool-down time of uninsulated lines. The experimental results proved that for short sections, the mass flow-rate does not affect the cool-down very significantly.

Expandable Transfer Line

The expandable transfer line is the proposed technique for rapidly and economically erecting a vacuum insulated cryogenic transfer line especially for space applications. The concept consist of a rigid foam – insulated line which, before its initial use, can be packaged in a very small volume. The expandable pipe consists of 5 separate layers. The innermost layer is a thin membrane of Mylar 0.025 mm to 0.125 mm thick. To provide additional bursting strength the Mylar tube is surrounded by a fiberglass fabric with polyurethane resin. At the final destination the pipe is unfolded and placed in its intended location. The interior of the pipe is pressurized and the inner mylar fiber glass tube takes a cylindrical shape. At this stage, the uncapsulated foam is heated throughout the length of the pipe. This heating completes the curing of the fiber glass so that it becomes hard and rigid.

Porous Insulated Transfer Line

Insulations such as fiber glass, polystyrene foam, and polyurethane foam may be applied to the bare line to reduce the heat in leak obtain an inexpensive transfer line. Such transfer lines are known as porous insulated transfer lines. When these porous insulation are used, a vapor barrier must be applied to the outer surface of the insulation to prevent water vapor diffusion into the insulation. Condensation of air within the insulation is hazardous as it poses an explosion risk due to its oxygen content. However the amount of liquid wasted in cool down for relatively short line is greater for a porous insulated line than for a bare line. Normally porous insulated transfer lines include an inner hose and an outer hose arranged around the inner hose in a concentric manner and a spacer member bridging an annular gap between the inner and outer hoses. A fibrous insulation material forms a three dimensional matrix of fibres which is included in the annular gap. The outer hose includes an elastomeric and/or a plastic material. The heat transfer (Q) from a porous insulated line is given by

$$Q = (A \cdot \Delta T)/((D_o \ln (D_o/D_i)/2k) + (1/h))$$

Where A = area of cross section, ΔT = temperature gradient, D_o = outer diameter, D_i = inner diameter, k = thermal conductivity, h = convective heat transfer coefficient.

Vacuum Insulated Transfer Line

Vacuum insulated transfer line consists of an inner line in which the liquid cryogen flows, concentric to an outer vacuum jacket. The annular space may contain a multilayer insulation or evacuated powder alone or vacuum alone. The vacuum insulated line may be used with any cryogenic fluid like liquid oxygen to attain fluid transfer with low losses. For long distance long transfer, the vacuum insulated line is usually much more advantageous than the other types of transfer lines. Refer Figure 18.13 for the photograph of a vacuum insulated transfer line.

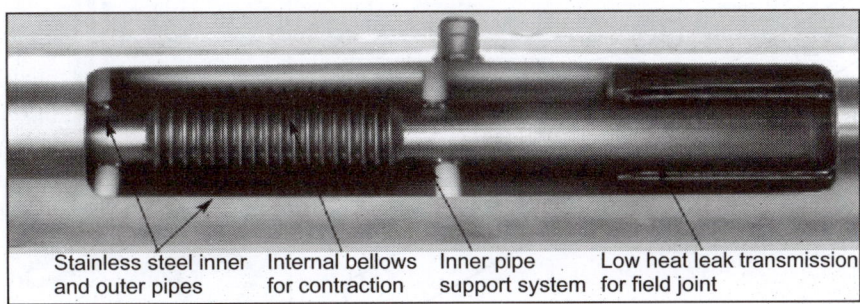

Fig. 18.13 Vacuum Insulated Cryogenic Transfer Line

Vacuum insulated lines are designed according to pressure piping codes. The thermal contraction problem can be solved by using expansion internal bellows and U-bends. With internal bellows, the pressure thrust must be absorbed by the external supports of the transfer line. Internal bellows introduce additional pressure drop and are inaccessible for repair. Longer sections of vacuum jacketed transfer lines require spacers to support the inner line within the outer line. These spacers should minimize the heat in leak (Q) and boil off rate (m) which is given by the equation.

$$\dot{m} = \frac{Q}{h_{fg} \times c.f},$$

where h_{fg} = latent heat of vaporization and $c.f$ = correction factor. The spacer should have low thermal conductivity, high mechanical strength and low outgassing rate. Spacer configuration should be such that it does not block the annular space. Spacers are constructed of plexiglass, plastic or stainless steel and are of square, triangular or spherical shape.

Cryogenic Transfer Line Joints

For cryogenic lines with lengths more than 40 feet it is difficult to construct them in one piece. Normally joints are required to be given at predetermined intervals. These joints could be vacuum jacketed such as the bayonet joint with low heat in leak levels. The bayonet joint as shown in Figure 18.14 consists of a male piece which telescopes within the female piece. The small clearance between the male and female piece restricts flow of liquid in the space and suppresses convection. Due to a close tolerance design, bayonets utilize a metal to metal in-line seal with a silicon O-ring placed between the flanges. These combine to make a reliable heavy duty cryogenic seal between the two vacuum jacketed pipe sections, and simultaneously preserve the low heat in leak of the system. Bayonet joint is suitable for transfer lines which are frequently opened, whereas field coupling (welded) joints as shown in Figure 18.15 are suitable for transfer lines which are seldom taken apart.

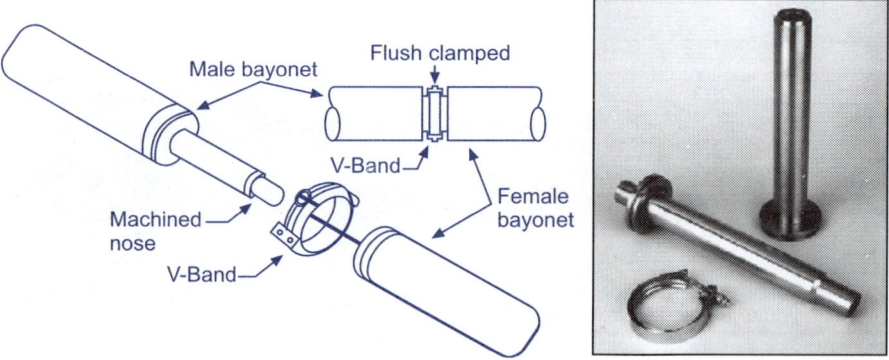

Fig. 18.14 Bayonet Joint

Field joint couplings are vacuum insulated field welded connections between two sections of vacuum jacketed pipe. They have a long heat leak transition between the outer jacket and the inner pipeline to reduce the heat input into the system. After the weld is made, the joint is insulated and a coupling is moved into place over the section. The coupling is then field welded to the collars on the ends of the two piping spools. This coupling is then evacuated to a low vacuum to insure a low heat leak.

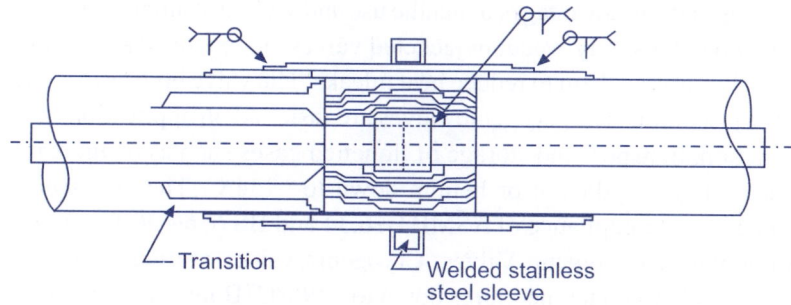

Fig. 18.15 Field Coupling (Welded) Joint

Cryogenic Valves

A cryogenic valve includes a main body which has an opening, a bonnet which is coupled to the main body, a disc which opens or closes the opening, a stem which is placed through the bonnet, and a drive unit which actuates the disc and the stem. A first seat is provided at the position at which the disc comes into contact with the opening, and a second seat is provided at the position at which the disc comes into contact with the bonnet, so that either when the valve is closed or when the valve is completely open, fluid is prevented from being drawn into the bonnet, thus making it possible to replace a packing with a new one even when the system is in operation. Refer Figure 18.16 for the cross-section of a cryogenic valve.

No.	Discription	Material
1	Body	ASTMA 182 F316/A351 CF8M
2	Non Rotating Disc	ASTM A479 TP316
3	Bonnet Gasket	Graphite & SS 316
4	Bolted Bonnet	ASTM A479 TP 316
5	Bonnet Bolt	ASTM A320 BBM2
6	Extension Bonnet	ASTM A182 F316/A351 CF BM
7	Extension Stem	ASTM A 479 TP 316
8	Stem Packing	Graphite
9	Position Indicator	SS 316
10	Hand Wheel	ASTM A240 TP304

Cryogenic valves are of two types, namely the extended stem valve and the vacuum jacketed valve. The extended stem valve uses an extended stem with thin walled tubing and has advantages of handle use and sealing at ambient temperatures, increasing operator safety. Vacuum jacketed valves are extended stem valves with vacuum jacket around them to reduce heat inleaks. The valve body is insulated with multilayer insulation. The cryogenic valves are installed on applications involving the production, transport and storage of liquefied gases such as oxygen, nitrogen, argon, natural gas, hydrogen or helium down to –254°C. The cryogenic valves are designed with compliance of ASME B16.34 and BS 6364. Refer Figure 18.17 for photograph of Cryogenic Valves. Cryogenic valves are designed for use at pressures to 52 bar and temperatures down to –196°C. Both cryogenic ball valves and cryogenic needle valves are constructed from austenite stainless steel for high corrosion resistance in marine environments. Cryogenic valves feature a top entry bolted bonnet design for easy in-line maintenance. Available end connections include flanged, butt welds, and NPT. Reduced and full-bore options are available from 1/2" to 1 1/2".

Fig. 18.17 Cryogenic Valves

Generally, cryogenic valves are used in LNG carriers or the aerospace industry, and are specially designed such that they can be used in cryogenic conditions. In other words, elements constituting the cryogenic valves must meet conditions that sufficient toughness is ensured at cryogenic temperature, weldability and workability are superior, and corrosion resistance to fluid is ensured. Furthermore, parts that open and valve bodies must be able to ensure the airtightness at cryogenic temperature. In addition, drive units must be thermally isolated at cryogenic temperatures to ensure the normal operation thereof. Cryogenic valves are most often employed to deliver liquid coolant directly into a closed chamber. Chamber sizes can vary from small 10cc quartz vessels for cryo-focusing in gas chromatography, to large cargo trailers employed in food transportation.

Cryogenic Two Phase Flow

Cryogenic flow systems that include two-phase flow are widely used in LNG (Liquefied Natural Gas) plants, aerospace technology, superconductivity technology, and many other engineering applications. Thus, the investigation of the two-phase flow characteristics of cryogenic fluids such as liquid helium is very interesting and important not only in the basic study of the hydrodynamics of cryogenic fluid but also for providing solutions to problems related to practical engineering applications of cryogenic two-phase flow. Cryogenic fluids are characterized by large compressibility as compared with fluids, such as water, at room temperature, by a small difference in density between vapor and liquid phases and by a small latent heat of vaporization. These unique characteristics of cryogenic fluids can be utilized to realize high performance in fluid apparatuses, such as the two-phase operation of inducers for liquid rocket turbo pumps. Cryogenic flow has been investigated for many years; however, little information has been obtained which is useful for clarifying the fundamental characteristics of two phase flow. The primary difficulty in the study of two phase flows is the difficult nature of techniques for visualizing two-phase transient flow. Flow visualization of two-phase cryogenic flow passing through a convergent-divergent nozzle or an orifice nozzle installed in a horizontal pipe can be carried out to clarify the fundamental characteristics of the nucleation and transient growth process. Furthermore, flow instability can be discussed based on instable pressure oscillations measured in experiments. Figure 18.18 shows the representation of a cryogenic two phase flow.

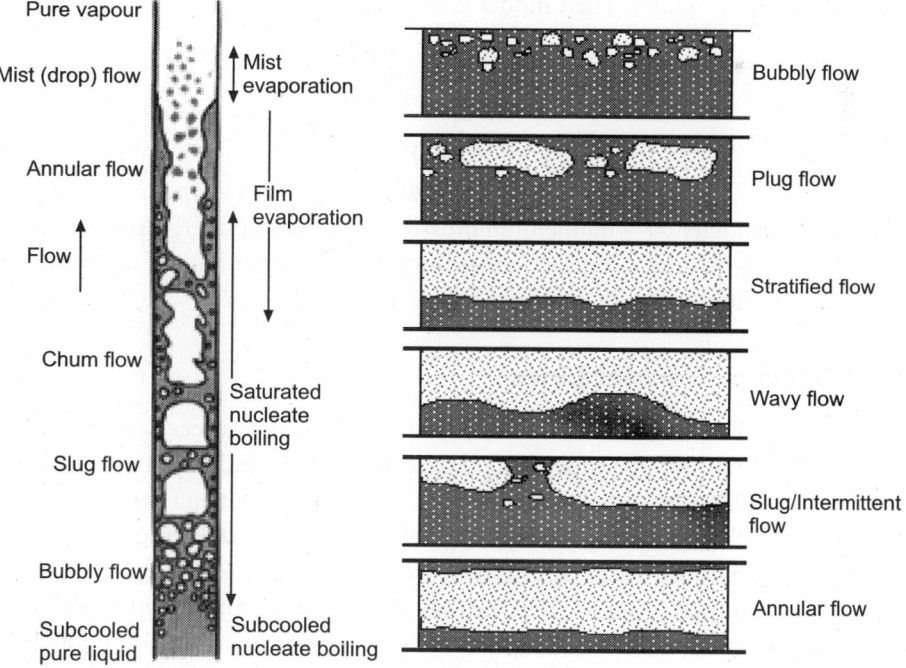

Fig. 18.18 Cryogenic Two Phase Flow–Baker Diagram

18.24 Cryogenics

Two phase flows occur due to heat inleaks and complicate the prediction of pressure drop of the moving cryogenic fluid. The flow pattern could be horizontal or vertical, laminar or turbulent as characterized by the Baker diagram shown in Figure 18.18. The flow pattern may change due to heat transfer across the pipe. In stratified flow, liquid flows at the bottom of the tube and vapor flows at the top without interference. Wave flow results when the vapor velocity is increased thereby resulting in generation of shear forces between two phases. Slug flow is formed when the high shear forces form slugs of liquids throughout the flow. For very high liquid and vapor flow rates, dense liquid is forced along the tube wall and the vapor flows along the central core forming the annular flow. At high vapor velocities, the shear forces tear up the annular layer forming a mist or spray resulting in mist flow. For low quality flow, bubbles are formed in the liquid resulting in bubble flow. If the vapor content of the stream increases, the bubbles agglomerate to form plug flow.

Pressure Drop across two Phase Flow

Two models have been developed to study the pressure drop in two phase flow. The homogenous model assumes two phases as merged with average fluid properties. The separated flow model assumes two phases to be artificially segregated. Lockhart and Martinelli developed a correlation for pressure drop across two phase flow using the separated flow model. Their model is as follows

$$\Phi = (X^2 + CX + 1)^{1/2}/X$$
$$X^2 = (\Delta p/L)\text{liquid}/(\Delta p/L)\text{gas}$$
$$(\Delta p/L) = f(m/A)^2/2g\rho D$$

where C = constant, Δp = pressure drop, L = length of pipe, X = Lockhart Martinelli Parameter, Φ = momentum pressure drop parameter, f = friction factor, m = mass flow rate, A = area of cross section, g = gravitational acceleration, ρ = density of liquid, D = diameter of tube.

Cool Down of Transfer Lines

Cryogenic cool down process is a complicated interaction process among liquid, vapor and solid pipe wall. Analytical and experimental studies on the cool down of vacuum insulated cryogenic transfer lines have been carried out in the literature. An expression for estimating the cool down history is derived taking into account the gas conduction effects. Experiments were carried out for cryogenic flow in glass lines and it was observed that the cool down time decreases with increasing vacuum up to 10^{-2} torr. Beyond this there is no significant change in cool down time. The outer jacket size has no appreciable influence on the cool down process. Higher mass flow rates result in faster cooling of the pipe line. The experimental and theoretical cool down curves agree well, particularly at higher jacket vacuums and at higher mass flow rates as shown in Figure 18.19.

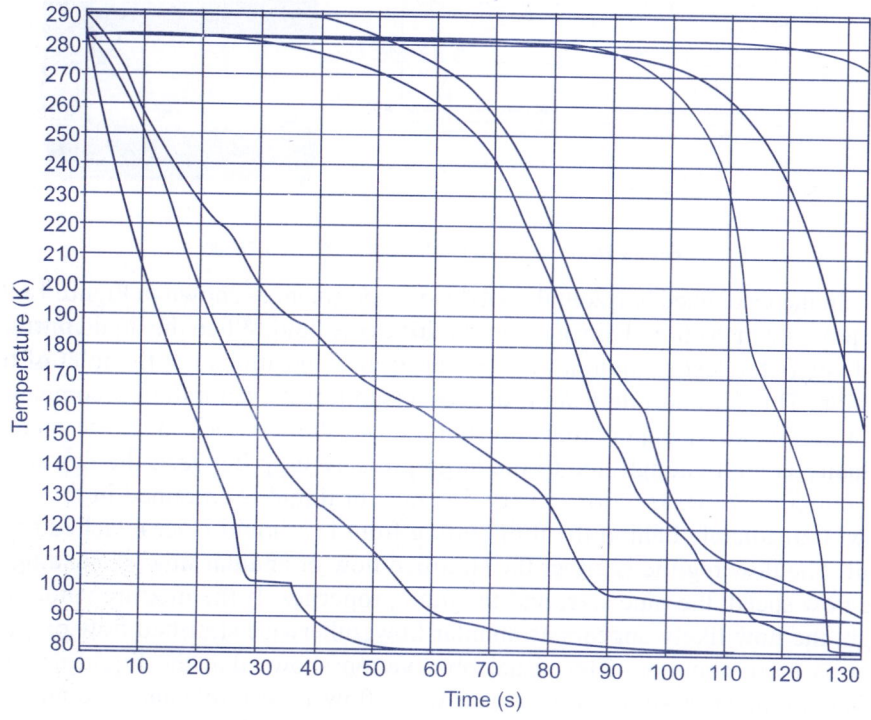

Fig. 18.19 Experimental and Theoretical Cool Down Curves for Transfer Lines

The amount of liquid cryogen required to cooldown transfer lines from room temperature is often of significant importance. The physical phenomena involved can be complex, including film boiling and transition boiling, and two phase pressure drops. Pressure surges can also be experienced when liquid, insulated by a layer of vapor film in the "inverted annular" regime, strikes a bend or obstruction downstream and boils explosively. In the study of cool down, one of the most important aspects of analysis is to determine the type of flow regime in the given region of the pipe. The flow in cryogenic cool down is typically a two-phase flow, because liquid evaporates after a significant amount of heat is transferred from the wall to the fluid during cool down. The two-phase flow regime is determined by many factors, such as fluid velocity, fluid density, vapor quality, gravity, and pipe size. The cryogenic two-phase flow is characterized by low viscosity, small density ratio of the liquid to the vapor, low latent heat of vaporization, and large wall superheat. Furthermore, film boiling, which is prevalent during cool down, causes low wall friction. These factors combined with the complex interaction between the momentum and the thermal transportation make the two-phase flow during the cool down process to distinguish itself from ordinary two-phase flows.

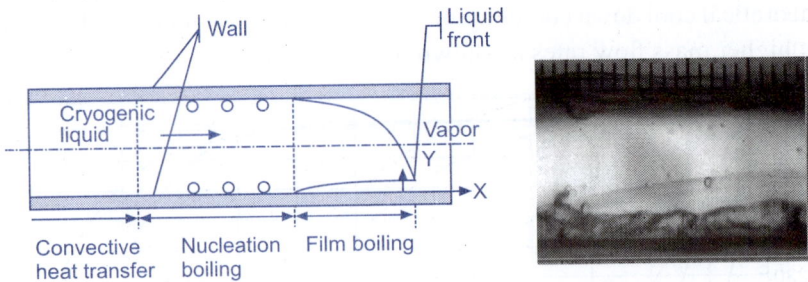

Figure 18.20 Cool Down Process and its Visualization

In the visualized horizontal cool down experiment as shown in Figure 18.20, the pressure in the liquid nitrogen dewar drives the fluid. When the liquid nitrogen first enters the test section, a film boiling front is positioned at the inlet of test section. This film boiling front produces a significant evaporation accompanied by a high velocity vapor front traversing down the test section. If the mixture velocity is high enough due to the large pressure drop between the dewar and the outlet of the test section, a very fine mist of liquid is entrained in the vapor flow. Immediately behind the film boiling front is a liquid layer attached to the wall. The flow regime is either the stratified flow or annular flow, depending on the flow speed, the pipe size, and the fluid properties. If the mixture velocity is high, the flow likely appears as annular flow, otherwise stratified flow or wavy flow is more common. The visual observation shows that the liquid droplets being entrained in stratified flow and wavy flow is insignificant. The nucleate boiling front follows the film boiling, indicating the end of film boiling and the cryogenic liquid starts contacting the wall. The position where the liquid starts contacting the wall is affected by the wall super heat, the liquid layer velocity and the thickness of the liquid layer. It is a complex hydrodynamic and heat transfer phenomenon. Usually Leiden frost temperature indicates the transition from film boiling to nucleate boiling. If the wall temperature is lower than the Leiden frost temperature, the vapor film cannot sustain the weight of liquid layer and becomes unstable. Therefore, the liquid starts contacting the wall, and film boiling ceases.

Once the liquid contacts the wall, the nucleate boiling starts. In the nucleate boiling regime the heat transfer from the wall to the liquid is significantly larger than that in the film boiling regime, and the wall is chilled down much faster. If the nucleation sites are not completely suppressed, a region of rapid nucleate boiling is seen at the quenching front. If most of nucleate sites are suppressed by the subcooled liquid, the flow directly transforms to the forced convection heat transfer, and nucleate boiling stage is not visible. After the nucleate boiling stage, the cool down process dramatically slows down as the convection heat transfer dominates. The wall superheat is relatively low at this stage but the heat leaking from the test section to the environment emerges. These factors lead to a lower cool down rate. In the meantime, the liquid gradually builds up in the pipe due to less vapor generation and the friction between the liquid and the wall. The increase of the liquid layer thickness eventually leads to the transition of the flow regimes. When the liquid layer is thick enough, the stratified flow or wavy flow becomes unstable. Eventually slugs are formed and the flow transforms to the slug flow. In the final stage of cool down, the flow is almost a single-phase liquid flow, occasionally with some small slugs. In this stage, the cool down is almost completed, and the pipe wall temperature gradually reaches the liquid saturated temperature. Figure 18.21 shows the temperature gradient across the pipe during cool down.

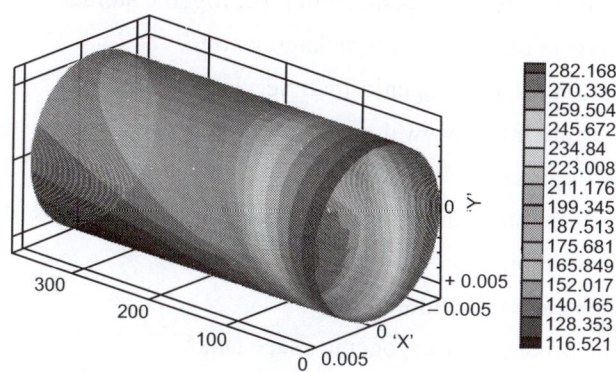

Fig. 18.21 Cool Down Process Temperature gradient on Pipe

The heat transfer in cool down can be given as $Q = \int q\, dt + \int m\, C_p\, dT$

Where q = heat transfer rate, m = mass flow rate and C_p = specific heat at constant pressure.

The cool down process leads to choking of flow. The minimum mass flow rate of vapor (mg) for cool down is given by

$$mg > Q/2\, (hg_2 - hf_1)$$

where Q = total heat transfer, hg_2 = enthalpy of vapor leaving the system and hf_1 = enthalpy of liquid entering the line

$$\text{Mass of liquid} = mf = V(\rho f - \rho g)$$

where V = Volume of fluid, ρf = density of liquid, ρg = density of vapor.

The total system cool down requirement is given by

$$C_p \Delta T(hg_1 - hg_2 + hfg) < mf/M$$

where C_p = specific heat at constant pressure, ΔT = temperature gradient during cool down, hg_1 = enthalpy of vapor entering the system, hg_2 = enthalpy of vapor leaving the system, hfg = enthalpy of vaporization, mf = mass of liquid, M = mass of transfer line.

References

1. Cryogenic Systems – Randall Baron.
2. Internet website http://www.princetoncryotech.com/pd.htm.
3. Linde Brochure on Cryogenic Storage Tanks.
4. Internet website www.wikipedia.com on cryogenic storage.
5. Internet website http://www.idspackaging.com/.
6. Brochure from Eden Cryogenic Tech Services, USA.
7. Internet website http://www.dunnspace.com/cryogen_space_storage.htm.
8. Cryogenic Engineering – B.A. Hands.
9. Internet website http://inventors.about.com/library/inventors/blthermos.htm.
10. Taylor Wharton brochure on LD series Dewar.
11. Internet website www.wipo.int.
12. Air Liquide brochure on cryogenic space tanks.
13. Internet website http://www.ctd-materials.com/products/tanks.htm.
14. US Patent 6336332.
15. Internet website http://www.dunnspace.com/cryogen_space_storage.htm.
16. Presentation from Cryoco Inc.
17. Conference paper by R.M. Shah, LDCE on Cryogenic Transfer Line.
18. Brochure from Quality Cryogenics on vacuum jacketed pipes.
19. Internet website www.hylokusa.com.
20. Ph.D Thesis of Jun Lao, University of Florida, 2005.

Questions

Q.1. What is a Dewar vessel? Describe its construction.

Q.2. Write a short note on cryogenic valves.

Q.3. Discuss two phase cryogenic flow.

Q.4. Explain cryogenic transfer systems in detail.

Q.5. How is cool down of cryogenic storage achieved?

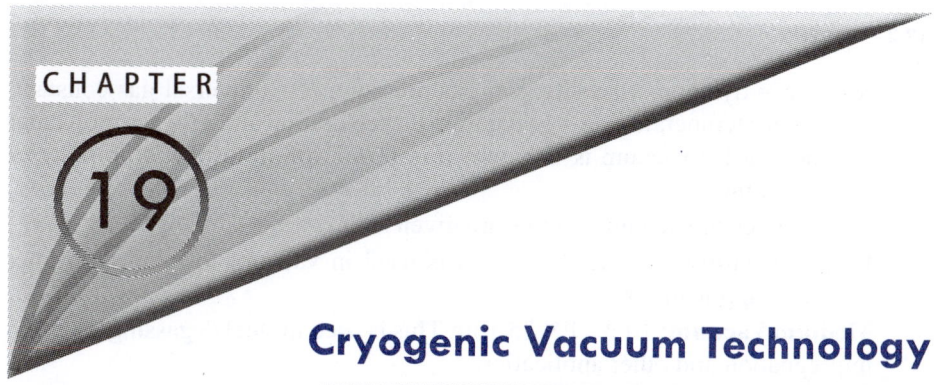

CHAPTER 19

Cryogenic Vacuum Technology

INTRODUCTION

In everyday usage, vacuum is a volume of space that is essentially empty of matter, such that its gaseous pressure is much less than atmospheric pressure. The quality of a vacuum refers to how closely it approaches a perfect vacuum. Residual gas pressure is a primary indicator of vacuum quality, and is most commonly measured in units called torr, even in metric contexts. The quality of a vacuum is indicated by the amount of matter remaining in the system, so that a high quality vacuum is one with very little matter left in it. Vacuum is primarily measured by its absolute pressure, but a complete characterization requires further parameters, such as temperature and chemical composition. One of the most important parameters is the mean free path (MFP) of residual gases, which indicates the average distance that molecules will travel between collisions with each other. Vacuum technology is important for cryogenic applications. Vacuum is used in cryogenic multilayer insulations along with reflective shields and low conductivity spacers. The performance of liquefaction systems can be improved by using low pressures generated by vacuum pumps. Vacuum pumps are also used in helium dilution refrigerators. Cryogenic applications are also used in vacuum systems. In high vacuum systems, cold traps are used to eliminate impurities from gases. Cryopumping technique which is very popular in environmental simulation chambers involves using helium cooled surface to freeze out residual gas. Details of vacuum technology are discussed in this chapter.

Flow Regimes in Vacuum Systems

Flow at ordinary pressures is termed as continuum flow as the fluid is assumed to be continuous media. The fluid flow is laminar if Reynolds number is less than 2300 and transition for Reynolds number between 2300 and 3000 and turbulent for Reynolds number above 3000. At low pressure we have to assume free molecular flow which is not continuous and the mean free path of the molecules is larger than dimensions of the flow channel. The Knudsen number is defined as ratio of mean free path of molecules (λ) to the characteristic dimension (D) of the flow channel. The fluid flow is continuous if Knudsen number is less than 0.01 and mixed flow for Knudsen number between 0.01 and 0.3 and turbulent for Knudsen number greater than 0.3. The Knudsen number (Nk) can also be written as

$$Nk = \lambda/D = (\mu/pD)(\Pi RT/2gM)^{1/2}$$

19.2 Cryogenics

where μ = dynamic viscosity, p = gas pressure, R = universal gas constant, T = absolute gas temperature, g = gravitational acceleration and M = gas molecular weight. The unit for vacuum is Torr which is 1/760th of atmospheric pressure or 1. mm Hg column.

The flow regimes in vacuum systems are given as
1. **Rough Vacuum:** P < 25 Torr. This is used in vacuum forming and other industrial applications.
2. **Medium Vacuum:** 10^{-3} < P < 25 Torr. This is used in steel degassing, vacuum impregnation and other applications.
3. **High Vacuum:** 10^{-6} < P < 10^{-3} Torr. This is used in vacuum coating, particle accelerators and other applications.
4. **Very High Vacuum:** 10^{-9} < P < 10^{-6} Torr. This is used in environmental chambers.
5. **Ultra High Vacuum:** P < 10^{-9} Torr. This is used for semiconductor applications.

Conductance in Vacuum Systems

For laminar continuum flow in tube, the pressure drop and mass flow rate are related by the Poisseuille equation

$$\Delta p = 128\ \mu L m / \Pi\ D^4 \rho g$$

where Δp = pressure drop, μ = dynamic viscosity, L = tube length, m = mass flow rate through tube, D = tube diameter, ρ = gas density and g = gravitational acceleration. Thus the mass flow rate is directly proportional to pressure drop and fourth power of tube diameter.

For mixed flow in tube, the pressure drop and mass flow rate are related by the Kennard equation

$$m = \Pi\ D^4 g\ p\ \Delta p / 128\ \mu L R T [1 + 8\mu/pD\ (\Pi R T / 2 g M)^{1/2}]$$

where Δp = pressure drop, μ = dynamic viscosity, L = tube length, m = mass flow rate through tube, D = tube diameter, p = average gas pressure and g = gravitational acceleration, M = gas molecular weight, T = gas temperature and R = universal gas constant. Thus the mass flow rate is directly proportional to pressure drop.

For free molecular flow in tube, the pressure drop and mass flow rate are related by the Poisseuille equation

$$m = (\Pi g M / 18 R T)^{1/2}\ \mu L m / (D^3 \Delta p / L)$$

where Δp = pressure drop, μ = dynamic viscosity, L = tube length, m = mass flow rate through tube, D = tube diameter, M = gas molecular weight and g = gravitational acceleration. Thus the mass flow rate is directly proportional to pressure drop and third power of tube diameter.

The throughput for a vacuum system is given by

$$Q = pV = mRT/M$$

where p = pressure at the point where flow is measured, V = volumetric flow rate, m = mass flow rate through tube, M = gas molecular weight, T = gas temperature and R = universal gas constant. The units of Q are Pa $-m^3/s$.

In analogy with linear transport relationships in heat transfer (Fourier's equation) and electric circuits (Ohm's Law), a conductance C for a vacuum system can be defined as.

For laminar continuum flow
$$C = \Pi D^4 g p / 128 \mu L$$
For mixed flow
$$C = \Pi D^4 g p / 128 \mu L [1 + 8\mu/pD (\Pi RT/2gM)^{1/2}]$$
For Free Molecular flow
$$C = (\Pi g RT / 18 M)^{1/2} (D^3/L).$$

The conductance concept is important in analyzing complex vacuum piping systems. An equivalent circuit can be drawn and an overall conductance (C_o) can be derived as

For series circuit $\quad 1/C_o = \sum 1/C_i$
For parallel circuit $\quad C_o = \sum C_i$

The conductances can also be combined for free molecular flow through a short tube ($L/D < 30$). The overall conductance is the sum of conductance of the tube and the orifice.
$$1/C_o = 1/C_1 + 1/C_2.$$
The expression for overall conductance is as follows:
$$C_o = (\Pi g R T / 18 M)^{1/2} D^2/(L/D + 4/3).$$

Calculation of Pump Down Time for a Vacuum System

Determination of pump down time is an important design task in vacuum systems. Pump down time is the time required to reduce the pressure of the system from ambient to desired operating pressure. The capacity of the vacuum pump in terms of pump speed (S_p) is given by
$$S_p = Q/p_i,$$
where Q = throughput of the pump and p_i = pressure at inlet of the pump.

The system pumping speed S_s is given by
$$S_s = Q/p,$$
where Q = throughput of the pump and p = pressure within the vacuum space. The system pumping speed has a great effect on pump down time of the vacuum system.

The overall conductance of the piping system between the vacuum space and vacuum pump is related to the throughput as follows.
$$C_o = Q/(p - p_i).$$
Combining all the earlier equations leads to the following equation
$$1/S_s = 1/S_p + 1/C_o$$

Refer Figure 19.1 for a simple vacuum pump setup including a tank and a vacuum pump. The mass flow rate (m) from the vacuum system is given by
$$m = pS_s/RT,$$

19.4 Cryogenics

where p = pressure of space to be evacuated, R = universal gas constant and T = absolute temperature of the gas.

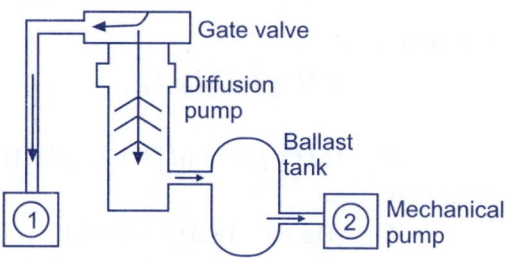

Fig. 19.1 Vacuum Pump Set-up

The leakage of air into the system is denoted by m_1. This mass influx may be due to actual leaks through the pressure wall, virtual leaks due to gas trapped in the system and outgassing from metal seals. The outgassing is a phenomena in which a material releases gases from its surface and interiors, when exposed to a vacuum. Applying law of conservation of mass to the vacuum system, the following equation is obtained

$$m_1 - m = dm/dt = V/RT \, dp/dt.$$

here V = volume of the system and t = overall time of the system.

$$\text{Outgassing rate} = Q_1 = m_1 RT.$$

The expressions for outgassing rate and system pumping speed can be substituted in the equation for the mass balance as follows

$$dp/dt = (Q_1 - S_s p)/V.$$

This is the governing equation to evaluate pump down time for any vacuum system. The ultimate pressure of the vacuum system is calculated as

$$p_u = Q_1/S_s.$$

The system pumping speed is independent of pressure and the pump down time can be calculated as follows

$$t_p = (V/S_s) \ln(p_1 - p_u/p_2 - p_u).$$

The system pumping speed is linked to the pump down time by the following equation

$$S_s = (FV/t_p) \ln(p_1/p_2),$$

where F is the system allowance factor which allows for the outgassing within the vacuum vessel.

Components of Vacuum Systems

The typical vacuum system, illustrated in Figure 19.2, contains the essential elements required to obtain high vacuum. The most common and reliable systems utilize three pumping devices: The rotating mechanical pump, the diffusion pump, and the cold trap. Other system components, such as valves and baffles, aid or control the action of these pumps. The mechanical pump is used as a fore pump or roughing pump to reduce the system pressure to 1 Pa through a bypass line. Once the pressure reaches 1 Pa, the diffusion pump is turned on and the valve in the bypass line is closed. The diffusion pump cannot operate above 10 Pa, hence mechanical pump is required to reduce the pressure at the inlet of diffusion pump. Vacuum valves are used in the system to isolate the diffusion pump from rest of the system to avoid contact of atmospheric air and hot oil of the pump. A cold trap or baffle is provided near the inlet of the diffusion pump to prevent back streaming of oil vapor and to freeze out condensable impurity gases like water vapor. A diffusion pump obtains pressures between 0.1 Mpa and 0.01 Mpa. In order to obtain still lower pressures an ion pump or cryopumping system is used. Table 19.1 shows the operating ranges for various vacuum pumps.

Table 19.1 Vacuum Pump Operating Ranges

Sr. No	Name of Pump	Operating Range (Torr)
1.	Mechanical Pump	10^2 to 10^{-3}
2.	Diffusion Pump	10^{-2} to 10^{-7}
3.	Ion Pump	10^{-4} to 10^{-10}
4.	Sorption Pump	10^2 to 10^{-4}
5.	Cryopump	10^{-3} to 10^{-12}

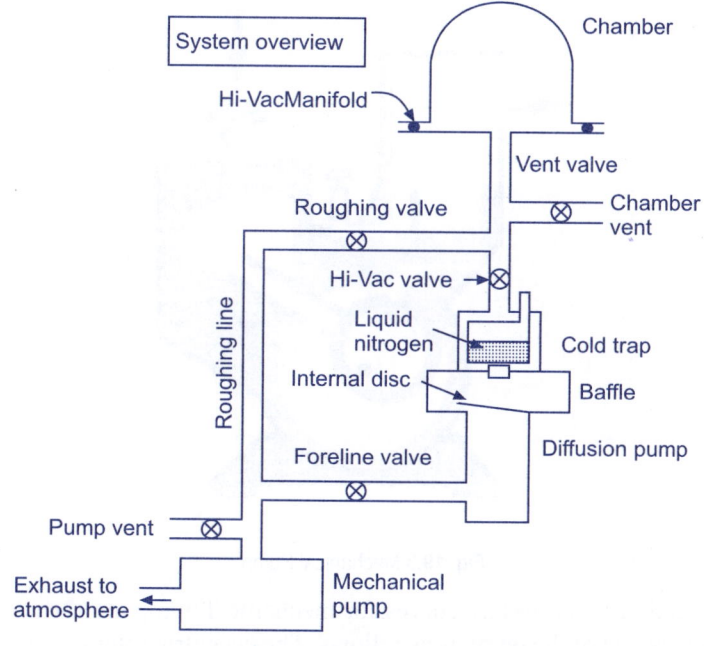

Fig. 19.2 Typical Vacuum System

Mechanical Vacuum Pumps

Mechanical positive gas displacement pumps of one type or another have been used since 1640. Almost all of the very early pumps used liquid mercury within glass tubes and vessels to create a vacuum. Modern mechanical pumps may well be considered the workhorses of vacuum technology; they are simple in design, require little maintenance, are relatively inexpensive, and can operate for long periods of time without failure. Mechanical vacuum pumps can operate continuously for fifteen years with only occasional oil changes. The range of pumping speeds for commercially available pumps runs from about 0.5 liters per second to over 300 liters per second. Mechanical vacuum pumps fall into two basic categories: reciprocating pumps, and rotary pumps. Further distinctions for mechanical pumps include: the number of stages (single stage or compound), the use of oil in a pump (pumps may be oil sealed or "dry"), and the means of driving the mechanics of a pump (direct drive or belt drive). Rotary piston (or rotary plunger) mechanical pumps as shown in Figure 19.3 operate on the principle of positive displacement of gas. On each cycle the rotating eccentric piston and the sliding valve work together to suck gas into the stator, compress it, and expel the gas to atmosphere. As with rotary vane pumps, rotary piston type pumps may be single stage or compound. Rotational speed is typically 600 to 800 rpm. Dimensional tolerances between the stator and piston in pumps of this design are usually 0.003 to 0.004". Because of this, piston pumps are more tolerant of particulate contamination than rotary vane pumps. Higher viscosity oil is used in rotary piston pumps due to the larger dimensional tolerances. Large rotary piston pumps are often water cooled to increase pump life and performance.

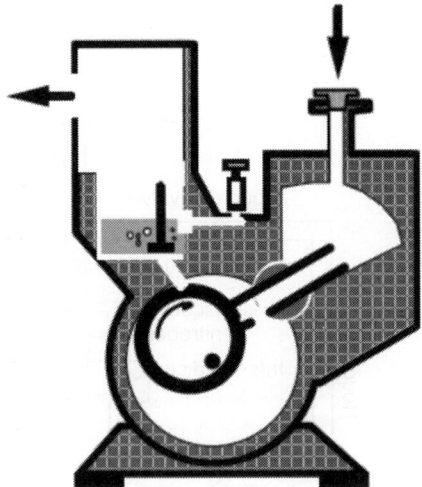

Fig. 19.3 Mechanical Pump

The speed versus pressure curves for mechanical pump are shown in Figure 19.4. The operation of the pump is as follows. The eccentric rotor i.e. piston rotates within the cylindrical jacket. The gas enters the space between two cylinders and is compressed to higher pressure as the piston rotates and is discharged through a check valve that prevents a backflow. The piston is sealed within the housing by a thin film of oil. The oil which is discharged with the gas is trapped by an oil separator and is returned back to the pump. The pump speed of mechanical pumps is nearly constant and they operate in the range of 10^2 to 10^{-3} torr.

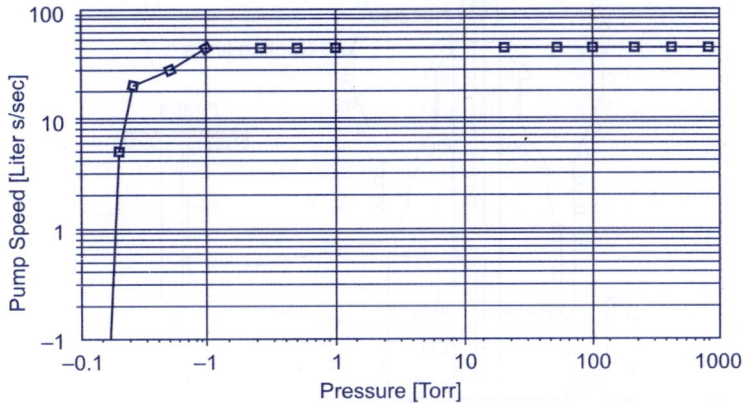

Fig. 19.4 Speed vs Pressure Curves for Mechanical Pump

To prevent moisture condensation within the pump volume, some of the mechanical vacuum pumps are equipped with a gas ballast valve. The gas ballast valve admits atmospheric air in the pump space during compression in order to maintain partial pressure of water vapor above its saturation value, thereby preventing condensation. The gas ballast valve is adjustable and can be shut off when not required.

Diffusion Pumps

A diffusion pump has a maximum pressure against which it can exhaust; this is usually in the milli torr region. (The maximum exhaust pressure is also known as the "tolerable fore pressure"). The mechanical pump provides and maintains this exhaust pressure for the diffusion pump. Fast pumping action is achieved through the use of high speed jets of oil vapor which collide with gas molecules and compress them in the direction of the mechanical pump. Refer Figure 19.5 for the construction of the diffusion pump. (The term "jet" is used to refer to both the vapor stream and to the nozzles from which the vapor issues.) The oil pool at the bottom of the pump is heated, causing oil vapor to be forced up the jet stack. The vapor strikes the umbrellas, and is projected downward and outward through the nozzles of the jet stack. In passing through the narrow jets, the oil vapor flows at a velocity near that of sound. The high speed vapor jet collides with gas molecules giving them a downward direction toward the fore line. The oil molecules condense on the walls of the pump which are cooled either by an air stream or by water, and flow back to the bottom pool. Thus, a continuous cycle of vaporization, condensation and revaporization takes place. The gas molecules are removed through the discharge line and out of the system by a mechanical backing pump. Oil of very low vapor pressure is used in these pumps.

19.8 Cryogenics

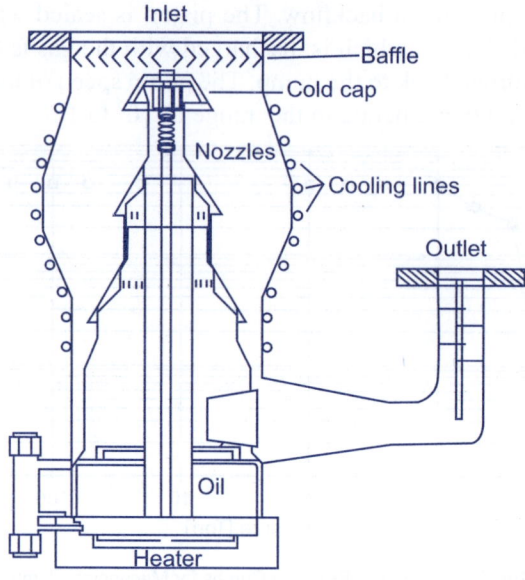

Fig. 19.5 Diffusion Pump

The diffusion pump usually operate in the free molecular regime. When the fore pressure i.e. outlet pressure of the pump is increased beyond a certain value, the vapor stream from the nozzles is terminated by a shock wave. The vapor molecules past the shock have randomly directed velocities and the pump becomes ineffective. Thus mechanical pump is a must to ensure reduction of fore pressure to 1 Pa or less thereby ensuring smooth operation of diffusion pump. The pump speed of diffusion pumps is nearly constant for most of the operating range and the ultimate pressure is around 1.5×10^{-8} torr. The working fluid for diffusion pump is silicon oil. The ratio of the actual pump speed to theoretical pump speed for a diffusion pump is known as H_o coefficient. Cold traps are used to condense stray oil vapors from the pump which can enter into the vacuum space.

Ion Pumps

An ion pump (also referred to as a sputter ion pump) is a type of vacuum pump capable of reaching up to 10^{-11} mbar under ideal conditions. An ion pump ionizes gases and employs a strong electrical potential, typically 3kV to 7kV, to accelerate them into a solid electrode. A swirling cloud of electrons produced in hollow Penning cells ionizes incoming gas atoms and molecules while they are trapped in a strong magnetic field. The swirling ions strike the chemically active cathode inducing sputter and are then pumped by chemisorption which effectively removes them from the vacuum chamber, resulting a net pumping action. Inert and lighter gases, such as He and H_2 do not effectively induce sputter and are absorbed by physisorption. Some fraction of the energetic gas ions (including gas that is not chemically active with the cathode material) that strike the metal cathode steal an electron from the surface and rebound as a neutral atom. These energetic neutrals are reflected back from the cathodes and buried as neutrals in exposed pump surfaces. There are three main types, the conventional or standard diode pump, the noble diode pump and the triode pump. Ion pumps are commonly used in ultra high vacuum (UHV) systems, as they can attain ultimate pressures less than 10^{-11} mbar. In contrast to other common vacuum pumps such as turbomolecular pumps and diffusion pumps, ion pumps have no moving parts and use no oil, and are therefore clean and low-maintenance, and produce no vibration, which is an important factor when working scanning probe microscopy. Refer Figure 19.6 for the construction and photograph of a typical ion pump.

The ion pump uses a combination of ionisation and chemiosorption to remove gas molecules from vacuum space. A large positive electric potential is applied to the anode and a large negative potential is applied to the sputter cathode. A collecter plate behind the anode is used to collect the ions. A magnetic field is applied to the entire unit. Electrons from the cathode travel to the anode and are deflected by the magnetic field. The electrons then travel in a cycloidal path until they collide with the gas molecules which wander in their path. The collisions produce additional electrons and ions. The electrons are attracted by the anode and ions move toward the sputter cathode. The ions pass through the anode and bombard the collector and are captured. Some ions bombard the sputter cathode and tear out bits of cathode (titanium) material. These bits shower on the collector plate and bury the ions collected on the surface, providing fresh surface for further ion capture by chemiosorption.

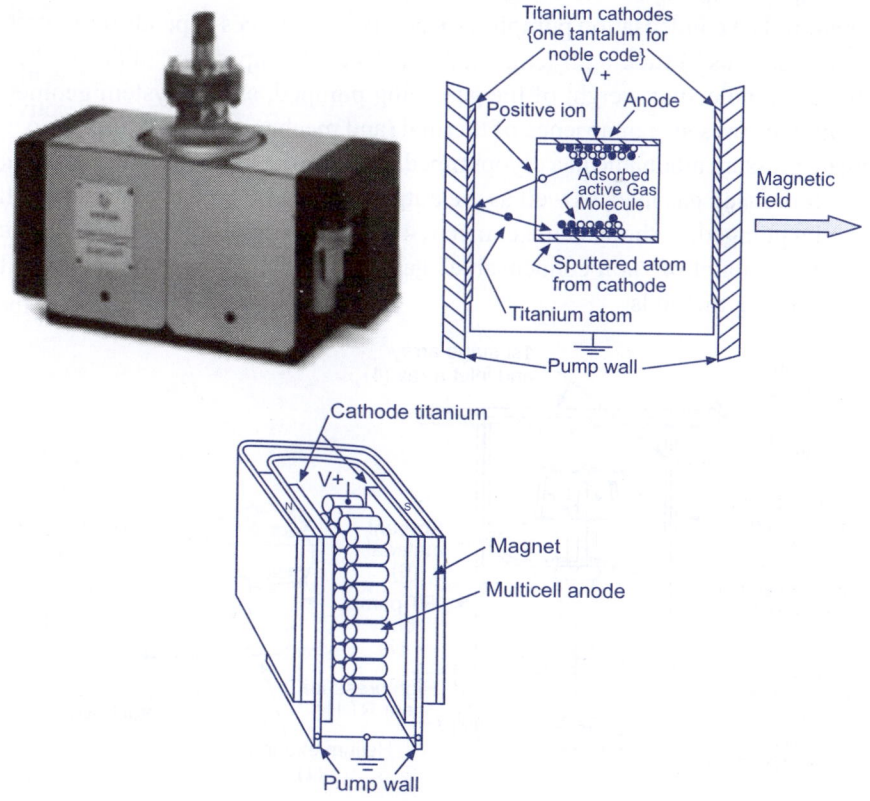

Fig. 19.6 Ion Pump

The ion pump is not suitable for applications above 1 Pa due to limitations of cathode life. The operating range of ion pump is 10^{-4} to 10^{-10} torr. The advantage of ion pump is its high pumping efficiency and its use as a vacuum gauge in high vacuum region. This is due to the fact that current drawn by the pump is proportional to the absolute pressure of the gas being pumped. The disadvantage of the ion pump is the high power requirement and high cost of the unit.

Cryopumping

The phenomenon of cryopumping is well known and has been used for many years to produce vacuum. The cryopumping is based upon the condensation and adsorption of gases on a cold panel. The most attractive features of cryopumps are their high pumping speed and clean contamination free operation. A cryopump is a vacuum pump that traps gases and vapours by condensing them on a cold surface. They are only effective on some gases, depending on the freezing and boiling points of the gas relative to the cryopump's temperature. They are sometimes used to block particular contaminents, for example in front of a diffusion pump to trap backstreaming oil, or in front of a McLeod gauge to keep out water. In this function, they are called a cryotrap or cold trap, even though the physical mechanism is the same as for a cryopump. The condensation pumping speed of cold surfaces depends on a number of factors such as 1) sticking fraction at the surface, 2) temperature of the gas being pumped, 3) molecular weight of the gas being pumped, and 4) system geometry, including the presence or absence of thermal (and mechanical) shielding. Also, the ultimate pressure attainable in a cryopumped chamber depends on 1) the magnitude of the introduced gas load, as well as the outgassing load, 2) the temperature of the gas being pumped, 3) the system geometry, 4) the vapor pressure of the condensate, and 5) the amount of "non condensable" gas present, which must be removed by other pumping methods.

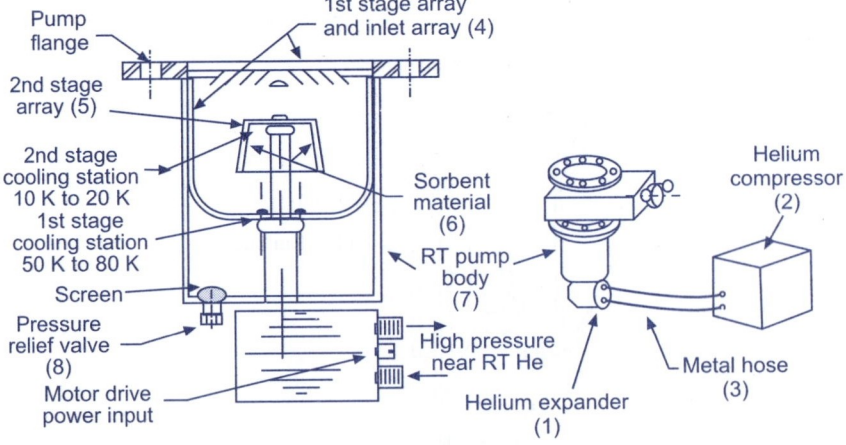

Fig. 19.7 Schematic Diagram of a Cryopump

The process of condensation of gas on cryogenically cooled surface to produce vacuum is called cryopumping. The cryopumping phenomenon involves phase change from gas to liquid at the cold surface as well as adsorption of the gas molecules. Very high pumping speeds of the order of 1000 m^3/s can be obtained. The pump is placed within the space to be evacuated thereby eliminating cost of piping. Refer Figure 19.7 for the schematic diagram of a cryopump and Figure 19.8 for a photograph of a cryopump and its cross-section with arrays. An electronic controller maintains the operating temperature of a cryopump by controlling a heater coupled to the cryopump. The heater is coupled to a cryopumping surface of the cryopump. The controller can control the operation of the heater in response to feedback from temperature sensors. The pump module consists of four major components: the refrigerator (expander), the first stage array (includes the baffle or sputtering/orifice plate), the second stage array (charcoal array) and the pump body/main flange. The expander in the pump module produces refrigeration power to cool the pump arrays to cryogenic temperatures. The first and second stage arrays provide pumping surfaces on which gases from the vacuum chamber are condensed or adsorbed. The cryo-arrays are mechanically attached to the cold head and cooled by the refrigeration power of the expander. The cryo-arrays pump gases in two ways: by cryosorption and by cryocondensation. The first stage pumps most of the easily condensed gas, such as water vapor, and thermally shields the second stage. The second stage pumps all other gases, such as argon, nitrogen, hydrogen, neon and helium.

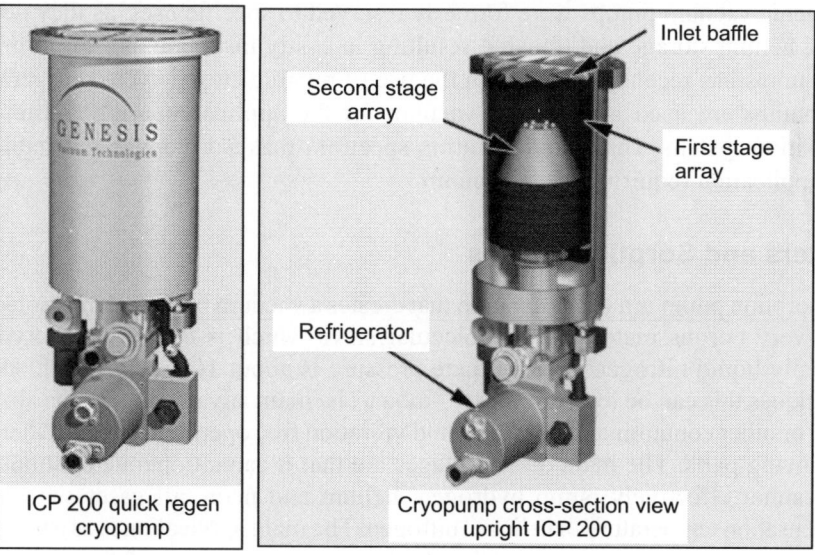

Fig. 19.8 Photograph and Cross Sectional View of a Cryopump

The surface of cryopump also known as the cryopanel is cooled to temperatures of 20 K internally using cold helium gas. The radiation losses are reduced by shielding the cryopanels. The gas removal rate of the gas molecules by cryopumping is given by the equation

$$S = n \times \alpha \times A \times S_m \times (1 - T_g/T_s \times P_{eq}/P_o)$$

where n = molecular density of gas in the tube, α = condensation coefficient, A = total cryopump area, S_m = theoretical maximum pumping speed of a unit area, T_s = surface temperature, P_{eq} = vapor pressure of the gas in equilibrium with a surface at temperature T_s, P_o = pressure of the gas at temperature T_g and T_g = Gas temperature.

The total mass flow rate leaving the cryopump surface is given by

$$m = m_1 + m_2 = (m_g + m_o)/f + [(1-f)m_g + m_o]/f$$

where m_g = mass of gas condensed, m_o = mass flow evaporated and f is the sticking coefficient which the ratio of gas molecules which stick to the surface to the number of gas molecules which strike the surface.

The pumping speed for the cryopump (S_p) is given by

$$S_p/A = (2gRT/\Pi M)^{1/2} + [1 - (p_u/p)]/[1 + (p_u/p)((T_1/T_2)^{1/2} - 1)],$$

where A = cross sectional area, g = gravitational acceleration, R = universal gas constant, M = Molecular weight of gas, p = gas pressure, p_u = ultimate pressure, T_1 and T_2 are hot and cold surface temperatures.

Sometimes the cryopump surface is protected by a baffle array. In order to improve efficiency of cryopanel, a layer of adsorbent such as charcoal is sometimes applied. Such a pump is called cryosorption pump. Despite number of advantages, cryogenic vacuum pumps were formerly reserved to specific uses, as they needed liquid helium storage and transfer, resulting in costly installations. A new era for cryopumps has recently begun with the use of reliable closed cycle refrigerators. cryopumps are used in ultrahigh vacuum for the application such as thin film deposition system. This wide operating spectrum makes cryopumps suitable for any application requiring clean vacuum.

Getters and Sorption Pumps

The sorption pump is a vacuum pump that creates a vacuum by adsorbing molecules on a very porous material like molecular sieve which is cooled by a cryogen, typically liquid nitrogen. The ultimate pressure is about 10^{-2} mbar. With special techniques this can be lowered till 10^{-7} mbar. The main advantages are the absence of oil or other contaminants, low cost and vibration free operation because there are no moving parts. The main disadvantages are that it cannot operate continuously and cannot effectively pump hydrogen, helium and neon, all gases with lower condensation temperature than liquid nitrogen. The main application is as a roughing pump for a sputter-ion pump in ultra-high vacuum experiments, for example in surface physics. Refer Figure 19.9 for the photograph and schematic diagram of the sorption pump. A sorption pump is usually constructed in stainless steel, aluminium or borosilicate glass. It can be a simple Pyrex flask filled with molecular sieve or an elaborate metal construction consisting of a metal flask containing perforated tubing and heat-conducting fins. A pressure relief valve can be installed. The design only influences the pumping speed and not the ultimate pressure that can be reached.

The design details are a trade-off between fast cooling using heat conducting fins and high gas conductance using perforated tubing. The typical molecular sieve used is a synthetic zeolite with a pore diameter around 0.4 nanometer and a surface area of about 500 m²/g. The sorption pump contains between 300 g and 1.2 kg of molecular sieve.

Getters are materials which remove residual gases in a vacuum system by one or more of the three sorption processes. 1) absorption 2) physical adsorption and 3) chemiosorption. Getters are used to maintain vacuum levels in annular spaces in dewars under static conditions by absorbing residual gases. Sorption pumping involves use of adsorbents to pump gases in a vacuum system. Sorption pumps can be single stage operating upto 10^{-3} torr or double stage operating upto 10^{-4} torr. The adsorbent usually charcoal or zeolite is cooled to lower temperatures using liquid nitrogen therby increasing its gas adsorbing capacity. The adsorbent gets saturated after continuous use and the gas must be desorbed by allowing the pump to warm to room temperature. The adsorbent is then heated to 200°C to remove any water vapor traces which are adsorbed.

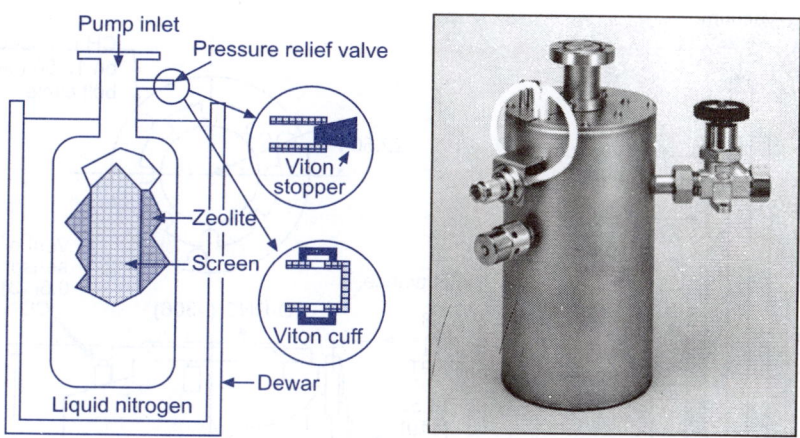

Fig. 19.9 Photograph and Cross Sectional View of a Sorption Pump

Baffles and Cold Traps

In vacuum applications, a cold trap is a device that condenses all vapors except the permanent gases into a liquid or solid. The most common objective is to prevent vapors from a vacuum pump from contaminating the experiment or sample of interest. Cold traps also refer to the application of cooled surfaces or baffles to prevent oil vapors from flowing from a pump and into a chamber. In such a case, a baffle or a section of pipe containing a number of cooled vanes, will be attached to the inlet of an existing pumping system. By cooling the baffle, either with a cryogen such as liquid nitrogen, or by use of an electrically driven Peltier element, oil vapor molecules that strike the baffle vanes will condense and thus be removed from the pumped cavity. Pumps that use oil either as their working fluid (diffusion pumps), or as their lubricant (mechanical rotary pumps), are

19.14 Cryogenics

often the sources of contamination in vacuum systems. Placing a cold trap at the mouth of such a pump greatly lowers the risk that oil vapours will backstream into the cavity. Cold traps can also be used for experiments involving vacuum lines such as small-scale very low temperature distillations/condensations. This is accomplished through the use of a coolant such as liquid nitrogen or a freezing mixture of dry ice in acetone or a similar solvent with a low melting point. Cold traps are also used in cryopump systems to generate hard vacuum by condensing the major constituents of air into their liquid forms. Refer Figure 19.10 for the construction of a cold trap and baffles with a photograph. The liquid nitrogen cold trap is essentially a dewar flask with the annular space connected to the piping between the vacuum pump and the space to be evacuated. The trap is constructed in such a way as to allow easy disassembly and reassembly after cleaning. The inner portion of the cold trap is constructed of brass or copper so as to maintain constant surface temperature. The liquid nitrogen level is maintained in the cold traps by automatic liquid level controllers.

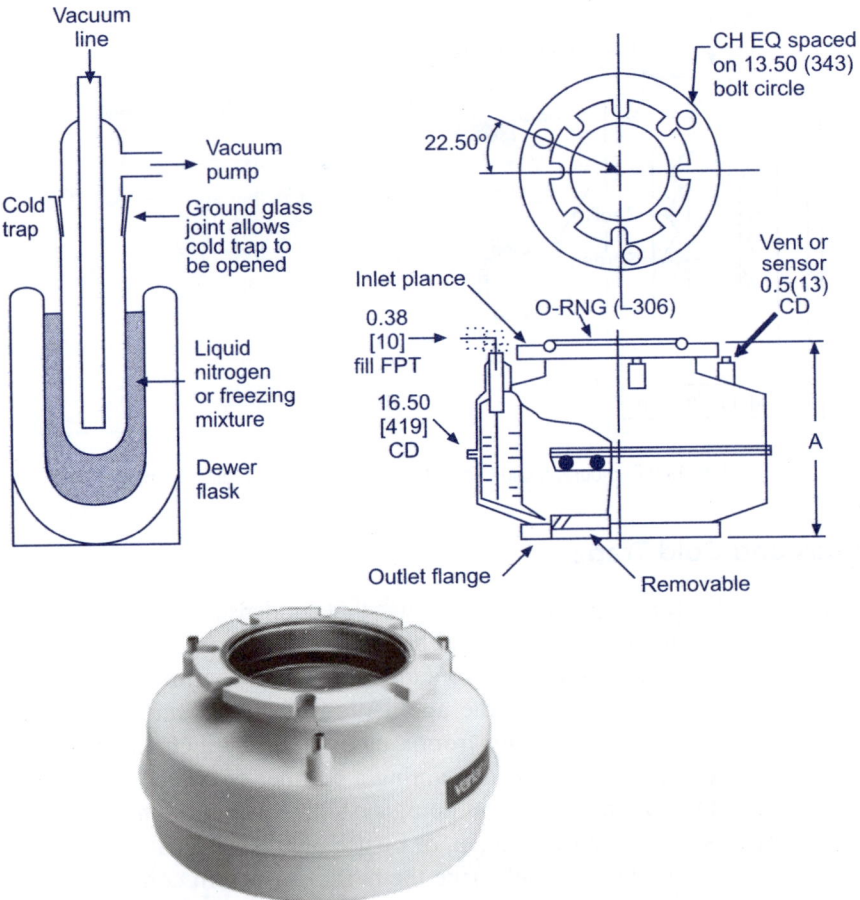

Fig. 19.10 Schematic of a Cold Trap and Baffles with Photograph

Baffles are the cooled surfaces placed near the inlet of the vacuum pump to prevent backstreaming of the pump fluid into the vacuum space. A cold trap is similar to a baffle, however the difference is that cold trap is always refrigerated by a cryogenic fluid or dry ice to condense and capture oil vapours and condensable gases. The design of the baffle is optically dense which means the gas molecules strike the baffle surface atleast once. The baffle is cooled by water circulating through tubes brazed to the baffle housing.

Vacuum Gauges

In vacuum systems pressures are several orders of magnitudes lower than atmospheric pressures, hence ordinary pressure guages like manometers do not work satisfactorily. A vacuum gauge determines the pressure in an evacuated apparatus by a measurement of some physical property of the residual gases, such as viscosity, heat conductivity, and so forth. The measurement of the response of a gauge to the residual gas naturally becomes more delicate as the gas becomes more and more tenuous. Finally, below a certain pressure limit (which is characteristic of a given gauge) the gauge does not behave measurably different from what it would if the vacuum were perfect. The variety of vacuum gauges available are 1) McLeod Gauge, 2) Pirani thermal conductivity Gauge, 3) Langmuir viscosity gauge, 4) Knudsen radiometer gauge, 5) Bavard Alpert ionization gauge. The characteristic limits for the above vacuum gauges are as follows:

- **Bavard Alpert gauge:** 10^{-9} mm of mercury
- **Knudsen gauge:** 10^{-6} mm of mercury
- **McLeod gauge:** 10^{-6} mm of mercury
- **Pirani gauge:** 10^{-5} mm of mercury
- **Langmuir's gauge:** 10^{-5} mm of mercury

A. **McLeod Gauge:** A McLeod gauge isolates a sample of gas and compresses it in a modified mercury manometer until the pressure is a few mmHg. The gas must be well-behaved during its compression (it must not condense). The technique is slow and unsuited to continual monitoring, but is capable of good accuracy. The useful range for the McLeod gauge is above 10^{-4} torr (roughly 10^{-2} Pa) as high as 10^{-6} Torr (0.1 mPa). 0.1 mPa is the lowest direct measurement of pressure that is possible with current McLeod gauge. Other vacuum gauges can measure lower pressures, but only indirectly by measurement of other pressure-controlled properties. These indirect measurements must be calibrated to SI units via a direct measurement, most commonly through a McLeod gauge. The operation of the McLeod gauge depends on a definite volume of residual gases being compressed, so that as the volume decreases, the pressure is increased to a value at which the hydrostatic head of mercury can be measured with an ordinary scale. Refer Figure 19.11 for a photograph and schematic of a McLeod vacuum gauge. The McLeod gauge is fragile. If it breaks, not only is the gauge lost but what is often more serious, mercury may get into the vacuum system.

19.16 Cryogenics

In glass vacuum systems using mercury pumps this is not as serious as it may be in kinetic vacuum systems. These systems, fabricated of brass with soft-soldered joints, are attacked by mercury and the joints are destroyed. Accidents with this gauge are usually caused by bringing the reservoir up too quickly.

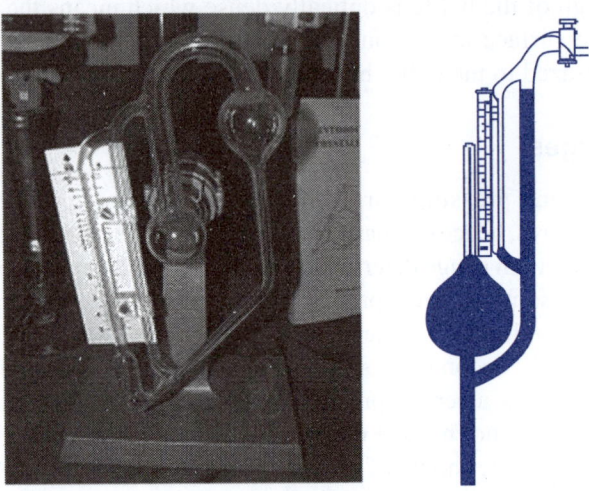

Fig. 19.11 Schematic and Photograph of a McLeod Vacuum Gauge

B. **Pirani Thermal Conductivity Gauge**: A Pirani thermal conductivity gauge consists of a metal wire open to the pressure being measured. The wire is heated by a current flowing through it and cooled by the gas surrounding it. If the gas pressure is reduced, the cooling effect will decrease, hence the equilibrium temperature of the wire will increase. The resistance of the wire is a function of its temperature: by measuring the voltage across the wire and the current flowing through it, the resistance (and so the gas pressure) can be determined. This type of gauge was invented by Marcello Pirani. Thermocouple gauges and thermistor gauges work in a similar manner, except a thermocouple or thermistor is used to measure the temperature of the wire. The Useful range for the Pirani gauge is 10^{-3} to 10 Torr (roughly 10^{-1} to 1000 Pa). Refer Figure 19.12 for a photograph and schematic of a Pirani vacuum gauge. The most satisfactory method is to connect the filament to one arm of a Wheatstone bridge and heat it by a constant current as shown in 19.12. If the bridge is balanced at one temperature of the filament, a change of its temperature caused by a change in the heat conductivity of the residual gases will unbalance it. Thus, the deflection of the bridge galvanometer indicates the pressure of the residual gases. Ordinarily, the filament is mounted in a bulb fitted with a connecting tube and is balanced with an identical compensating filament mounted in an adjacent arm of the bridge. This auxiliary bulb is evacuated and sealed off at a very low pressure. The use of an auxiliary bulb serves to makes the gauge insensitive to variations in room temperature. Changes in the over-all temperature of one bulb are the same as changes in the other, so that the galvanometer does not respond to these changes but only to the changes produced by the residual gas in the one bulb.

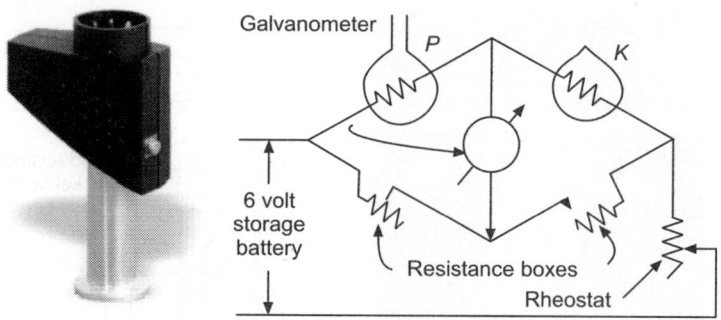

Fig. 19.12 Schematic and Photograph of a Pirani Vacuum Gauge

C. Bavard Alpert Ionisation Gauge: Ionization gauges are the most sensitive gauges for very low pressures. They sense pressure indirectly by measuring the electrical ions produced when the gas is bombarded with electrons. Fewer ions will be produced by lower density gases. The calibration of an ion gauge is unstable and dependent on the nature of the gases being measured, which is not always known. They can be calibrated against a McLeod gauge which is much more stable and independent of gas chemistry. Thermionic emission generate electrons, which collide with gas atoms and generate positive ions. The ions are attracted to a suitably biased electrode known as the collector. The current in the collector is proportional to the rate of ionization, which is a function of the pressure in the system. Hence, measuring the collector current gives the gas pressure. The useful pressure range is $10^{-10} - 10^{-3}$ torr (roughly $10^{-8} - 10^{-1}$ Pa). There are several sub-types of ionization gauge. Most ion gauges come in two types: hot cathode and cold cathode, a third type exists which is more sensitive and expensive known as a spinning rotor gauge, but is not discussed here. In the hot cathode version an electrically heated filament produces an electron beam. The electrons travel through the gauge and ionize gas molecules around them. The resulting ions are collected at a negative electrode. The current depends on the number of ions, which depends on the pressure in the gauge. Hot cathode gauges are accurate from 10^{-3} Torr to 10^{-10} Torr. The principle behind cold cathode version is the same, except that electrons are produced in a discharge created by a high voltage electrical discharge. Cold Cathode gauges are accurate from 10^{-2} Torr to 10^{-9} Torr. Ionization gauge calibration is very sensitive to construction geometry, chemical composition of gases being measured, corrosion and surface deposits. Their calibration can be invalidated by activation at atmospheric pressure or low vacuum. The composition of gases at high vacuums will usually be unpredictable, so a mass spectrometer must be used in conjunction with the ionization gauge for accurate measurement. Refer Figure 19.13 for a photograph and schematic of a Bavard Alpert vacuum gauge.

19.18 Cryogenics

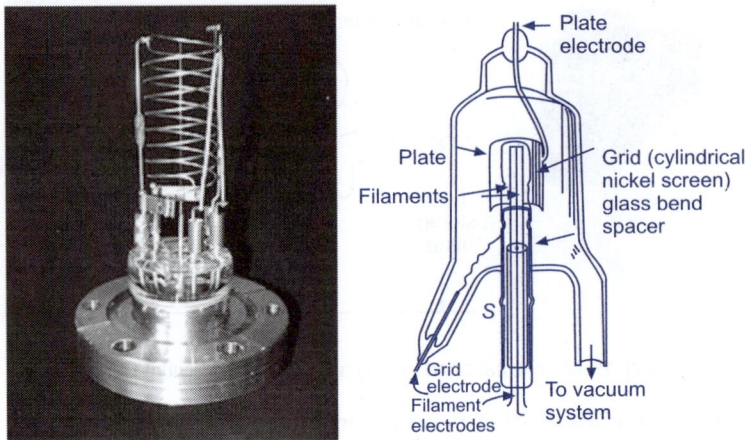

Fig. 19.13 Schematic and Photograph of a Bavard Alpert Vacuum Gauge

Bavard Alpert Ionization gauge is a triode mounted in a glass bulb connected to the apparatus in which the pressure is to be measured. They are electrically connected as shown in Figure 19.13. Electrons emitted from the filament are accelerated to the grid, and their momentum would carry them to the plate if an inverse field more than sufficient to prevent this were not impressed between the grid and the plate. They therefore return to the grid and are finally collected on it. However, while they arc between the grid and the plate, they bombard and ionize some of the molecules of the residual gas present there. These ions are collected on the plate and measured with a sensitive galvanometer. The ratio of this ion current to the current of bombarding electrons or grid current is proportional to the pressure at pressures below about 10^{-4} mm. An ionization gauge may be made from an ordinary three-element radio tube equipped with a glass connection to the vacuum system. Such gauges are useful for the pressure range from 10^{-3} to 10^{-6} mm of mercury.

D. Langmuir Viscosity Gauge: Langmuir's viscosity gauge is made with a flattened quartz fiber about 50m thick and from five to ten times as wide. This quartz ribbon is about 5 cm long and is mounted in one end of a glass tube about 25 mm in diameter as shown in Figure 19.14.

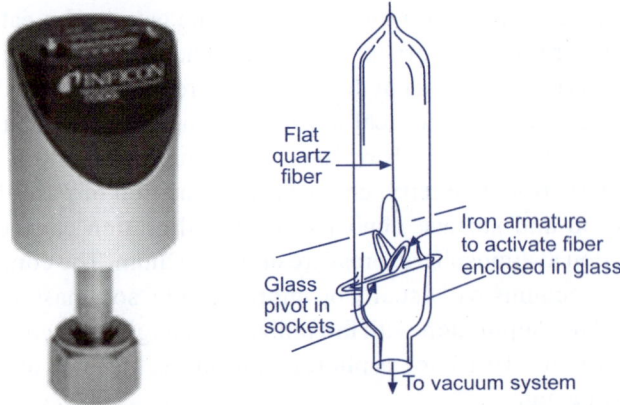

Fig. 19.14 Schematic and Photograph of a Langmuir Vacuum Gauge

The fiber is mounted together with a pivoted glass tube, which contains an iron armature operated by an external electromagnet, to start the fiber vibrating. When this ribbon is set vibrating in a high vacuum, the amplitude changes very slowly because the damping by the residual gas is almost negliglble, and, owing to the low internal viscosity of fused quartz, the loss of vibrational energy from this source is also low. From atmospheric pressure down to a few millimeters of mercury, the damping produced by the molecules of the residual gas is nearly independent of pressure. Over the transition range of pressure, where the damping varies from this constant value to zero, the time required for the amplitude of vibration to decrease to half value is an index of the pressure. A feature of this gauge is its small volume. Because there are no metal parts exposed, the gauge is suitable for measuring the pressure of corrosive gases like the halogens. This gauge, in conjunction with a McLeod gauge, may be used for measuring the molecular weight of an unknown gas at low pressures. Figure 19.13 shows the construction details and a photograph of a Langmuir Viscosity Gauge.

E. **Knudsen Radiometer Gauge**: Figure 19.15 shows the Knudsen gauge as designed by DuMond. When this gauge is constructed according to the specifications it has a definite sensitivity, so that no preliminary McLeod calibration is needed. The Knudsen gauge shown here is equipped with a permanent (Alnico) magnet for damping and has a special liquid air trap for determining what fraction of the pressure indication is produced by condensable vapors. The Knudsen gauge is preferred to the McLeod gauge where it is important to avoid contaminating a vacuum system with mercury. No expensive auxiliary instruments are required with the Knudsen gauge, as with the ionization gauge. Furthermore, the filaments will not burn out and the suspension is not delicate. All connections and supports fasten to one end plate which facilitates making repairs. The metal case thus becomes, in effect, a water-cooled covering "bell jar" fitted with a window.

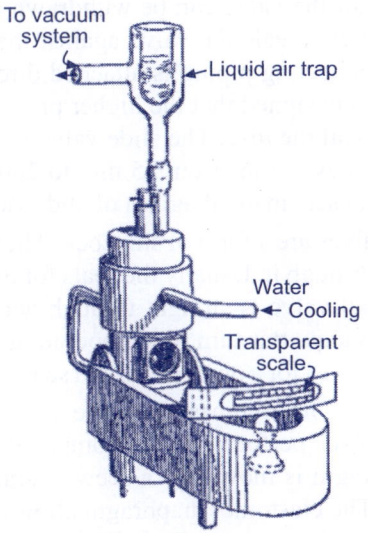

Fig. 19.15 Schematic of a Knudsen Vacuum Gauge

Vacuum Valves

Valves used in vacuum service have stringent design requirements such as 1) freedom from leakage in vacuum as any leak would destroy the vacuum, 2) maximum conductance when open so as to eliminate reduction of pumping speed and 3) absence of outgassing through seals. Four types of valves are used in vacuum service i.e. angle valve, slide valve, ball valve and diaphragm valve.

A. Angle Valve: Angle valves have relatively high conductance because the closing disc can be withdrawn completely from the flow passage. O-rings or bellows are used to seal the valve stem. The bellows seal provides a leakproof joint, however they are expensive and require a shorter stroke. Refer Figure 19.16 for a photograph and schematic diagram of angle valve.

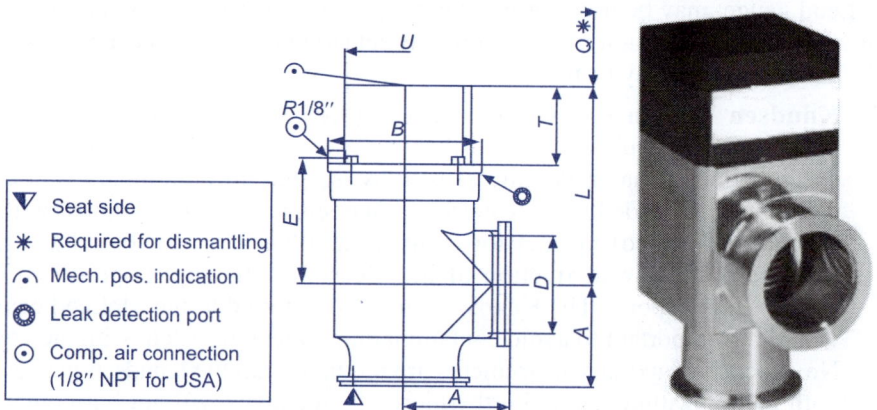

Fig. 19.16 Schematic and Photograph of an Angle Valve

B. Slide Valve: Slide valves have an inline design and high conductance because the closing portion of the valve can be withdrawn completely from the flow passage. The valve disc seals the valve against the valve flange through an O-ring. The interconnecting piping is attached directly to the valve flanges. The valve slide is so designed that the higher pressure acts on the side of the valve that helps to seal the disc. The slide valve is thin as compared to other valves and the thickness varies from 75 mm to 200 mm. Refer Figure 19.17 for a photograph and schematic diagram of slide valve.

C. Ball Valve: Ball valves are a form of stopcock. The valve consists of a sphere with a hole drilled through it. Usually the seals for the ball and operating shaft are O-rings. The ball valve is straight through and high conductance valve which only requires a quarter turn of the operating shaft to open or close the valve. Refer Figure 19.18 for a photograph and schematic diagram of ball valve.

D. Diaphragm Valve: The diaphragm valve is a vacuum valve of lower conductance and lower performance as compared to other valves. A metal or elastomer diaphragm is moved by a scew operator against the valve seat to close the valve. The elastomer diaphragm allows a larger possible motion, however is less permeable to gases. Refer Figure 19.19 for the photograph and schematic of diaphragm valve.

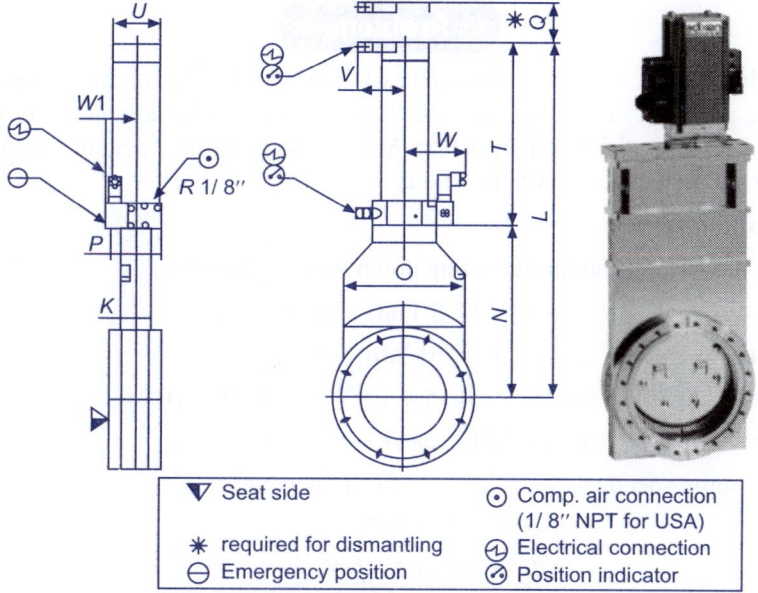

▼ Seat side
✱ required for dismantling
⊖ Emergency position
⊙ Comp. air connection (1/8" NPT for USA)
⌁ Electrical connection
⊘ Position indicator

Fig. 19.17 Schematic and Photograph of a Slide Valve

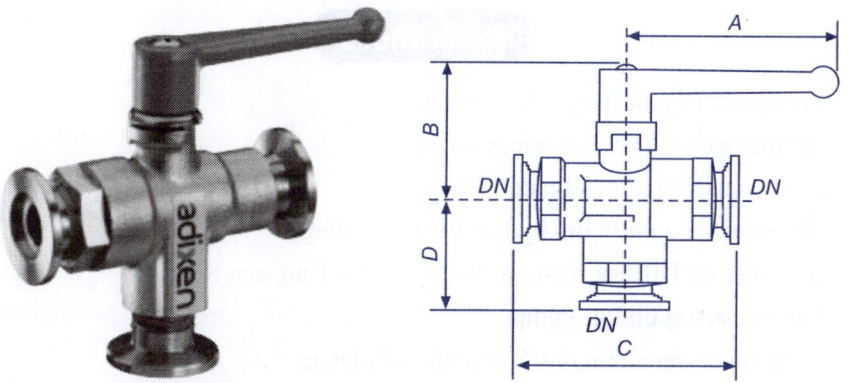

Fig. 19.18 Schematic and Photograph of a Ball Valve from Adixen

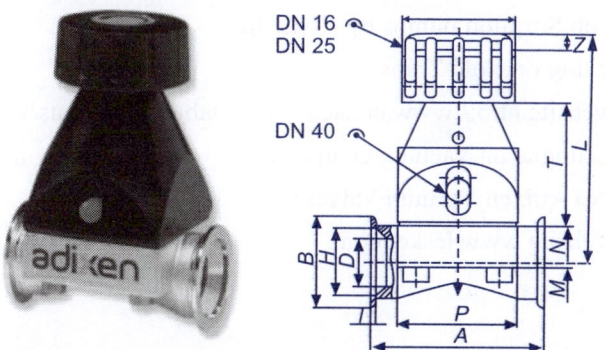

Fig. 19.19 Schematic and Photograph of a Diaphragm Valve from Adixen

Problem

Problem 1. A vacuum chamber has a volume of 1.25 m³ and conductance of piping is 12 dm³/s. The initial conditions of air are 101.3 kPa and 300 K. The ultimate pressure of the system is 0.15 MPa. Determine the vacuum pump speed to pump the system to 1.5 kPa in 1 hour.

Solution:

For the vacuum pump, the pump down time is given by

$$t_p = (V/S_s) \ln(p_1 - p_u / p_2 - p_u)$$

Thus $\quad S_s = (V/t_p) \ln(p_1 - p_u / p_2 - p_u)$

$S_s = (1.25/3600) \ln(101.3 - 0.15 \times 10^{-3} / 1.5 - 0.15 \times 10^{-3})$

$S_s = 3.861 \times 10^{-3} \, m^3/s = 3.861 \, dm^3/s$

The pump speed can be calculated from the equation

$$1/S_p = 1/S_s - 1/C_o$$
$$1/S_p = 1/3.861 - 1/12$$

Thus pump speed = S_p = 5.693 dm³/s ...(Ans)

References

1. Cryogenic Engineering – B.A. Hands.
2. Internet website www.wikipedia.com.
3. Cryogenic Systems – Randall Baron.
4. Notes on Vacuum Pump by Las Positas College.
5. Brochure on Diffusion pumps by Mid-west Tungsten Service.
6. Varian catalog on Ion Pump.
7. Paper on cryopumping by S. Gandhi of LDCE.
8. Journal paper by Welch et.al. (1999) American Vacuum Society, S0734-2101(99).
9. Brochure on Sorption pumps by Ulvac Inc.
10. Varian catalog on Cold Traps.
11. Internet website http://www.tau.ac.il/~phchlab/experiments/vacuum.
12. Inficon catalogue on Vacuum Gauges.
13. Brochure of Adixen Vacuum Valves.
14. Internet website www.lesker.com.

Questions

Q.1. Write a short note on cryopumping.

Q.2. What are the flow regimes in cryogenic vacuum systems?

Q.3. How is pump down time calculated for a vacuum system?

Q.4. Explain the different components in vacuum systems.

Q.5. Describe the ion pump and diffusion pump with a neat sketches.

Q.6. Describe some cryopump arrays.

Q.7. Describe the importance of baffles and cold traps.

Q.8. Describe a typical vacuum valve with neat sketch.

Q.9. A vacuum chamber has a volume of 1.3 m³ and conductance of piping is 15 dm³/s. The initial conditions of air are 101.3 kPa and 300 K. The ultimate pressure of the system is 0.18 MPa. Determine the vacuum pump size to pump the system to 1.8 kPa in 1 hour.

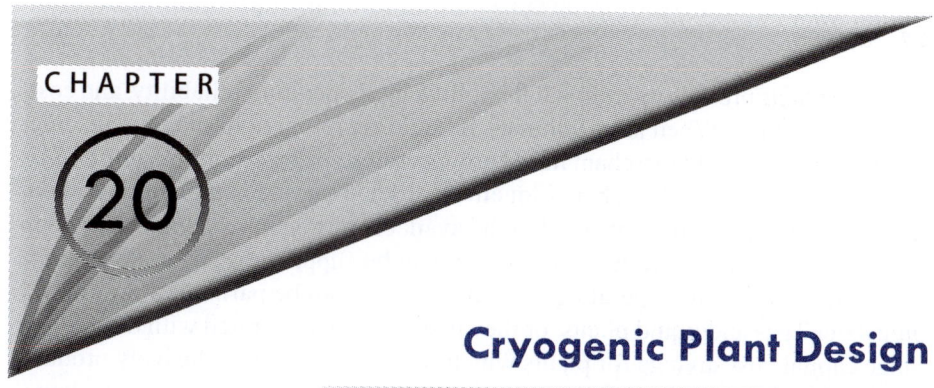

CHAPTER 20

Cryogenic Plant Design

INTRODUCTION

Cryogenic industry is now well established in the world. Today, liquid natural gas (LNG) represents one of the largest industrial domains of application of cryogenics together with the liquefaction and separation of air gases. The densification by condensation, and separation by distillation of gases remains the main driving force for the cryogenic industry. It is estimated that the world capacity of liquid oxygen is over 150 million tons per annum and that of LNG is over 60 million tons per annum. Gases like ethylene, helium, argon, nitrogen and hydrogen are also being consumed in large quantities through out the world and their demand is also increasing.

Cryogenic plants produce nitrogen, oxygen and argon products using very low temperature distillation to separate and purify components of air. Cryogenic plants are most commonly used to produce high purity products at medium to high production rates. Non-cryogenic plants produce gaseous nitrogen or oxygen products using near-ambient-temperature separation processes. There are two major types of non-cryogenic processes, using either selective adsorption or differential permutation through membranes to produce relatively pure oxygen or nitrogen. These processes use differences in properties such as molecular structure, size and mass to achieve the desired degree of separation. Non-cryogenic separation processes are most commonly used when high purify nitrogen or oxygen is not needed (e.g. 99 % purity nitrogen or oxygen at about 93% purity) and when product demand is relatively low; for example nitrogen at production rates less than about 500 Nm^3 or oxygen at production rates less than about 1500 Nm^3.

Cryogenic plant design is an extremely critical process. System optimization does not stop with specification of the cryogenic plant but a fully-optimal facility design will reflect all aspects of anticipated operating conditions at the site. Plant supply solutions which represent the best combination of cost and performance, as determined by customer priorities and criteria are desirable. The complete plant design integrates the air separation unit (and, if included, the liquefaction system) with appropriately configured and sized auxiliary equipment such as pipeline compressors, liquid storage tanks for backup and supplementation of plant production, liquid vaporizers (ambient or heated), elevated pressure gas storage, and bulk merchant liquid products storage and trailer loading systems. Well-optimized

20.2 Cryogenics

cryogenic plants provide operational flexibility, high operating reliability, and cost-effective operation. When large volumes of liquid are required to support customer operations and / or bulk merchant liquid product sales, the air separation unit (ASU) can be coupled with a liquefier. Liquefiers provide supplemental refrigeration, which is used to convert some or all of the available nitrogen, oxygen and argon to liquefied industrial gas products. Liquefiers can be supplied as stand-alone units, which allows maximum operating flexibility, which can be particularly useful for "piggyback" gas and liquid plants, or they may be fully integrated with the ASU to realize capital cost savings for plants which will primarily or exclusively produce bulk liquid products (liquid nitrogen, oxygen and argon). Figure 20.1 shows photograph of a typical cryogenic plant.

Fig. 20.1 Photograph of a Typical Cryogenic Plant

Cryogenic Plant Design : General Principles

The design of a process plant is a complex activity that will usually involve many different disciplines over a considerable period of time. The design may also go through many stages from the original research and development phases, through conceptual design, detailed process design and onto detailed engineering design and equipment selection. Many varied and complex factors including safety, health, the environment, economic and technical issues may have to be considered before the design is finalised. At each stage it is important that the personnel involved have the correct combination of technical competencies and experience in order to ensure that all aspects of the design process are being adequately addressed. The process design will often be an iterative process with many different options being investigated and tested before a process is selected. In many occasions a number of different options may be available and final selection may depend upon a range of factors. The process design should identify the various operational deviations that may occur and any impurities that may be present. In the mechanical design, the materials of construction chosen need to be compatible with the process materials at the standard operating conditions and under excursion conditions. The materials of construction also need to be compatible with each other in terms of corrosion properties. Impurities which may cause corrosion, and the possibility of erosion also need to be considered so that the detailed mechanical design can ensure that sufficient strength is available and suitable materials of construction are selected for fabrication.

Detailed mechanical, structural, civil and electrical design of equipment comes after the initial process design which covers the steps from the initial selection of the process to be used, through to the issuing of process flow sheets. Such flowsheets will include the selection, specification and chemical engineering design of the equipment. These are then used as the basis for the further detailed design. Design factors are an essential component in order to give a margin of safety in the design. Design factors may be appropriate in either the mechanical engineering design or in the process design where factors are often added to allow some flexibility in process operation. For mechanical and structural design the magnitude of design factors should allow for uncertainties in material properties, design methods, fabrication and operating loads. Plant design should take account of the relevant codes and standards. Conformity between projects can be achieved if standard designs are used whenever practicable.

Codes and Standards for Cryogenic Plants

Modern engineering codes and standards for cryogenics cover a wide range of areas including:

- Materials, properties and compositions;
- Testing procedures; for example for performance, compositions and quality;
- Preferred sizes; for example for tubes, plates and standard sections;
- Design methods and inspection and fabrication;
- Codes of practice for plant operation and safety.

Many companies have their own in-house standards which are primarily based on the published codes, such as BS5500, with added extras which cover either technical or contractual matters. In the safety report the base document for the in house codes should be clearly stated and the key safety related deviations or enhancements demonstrated so that the assessor can determine their adequacy. A safety report should demonstrate that consideration has been given to the appropriate standards and codes of practice developed by legislators, regulators, professional institutions and trade associations. It should also demonstrate that for any equipment that is installed, the operating procedures, testing regimes and maintenance strategies that are in place meet or exceed these requirements in terms of safety performance.

Safety Considerations in Design of Cryogenic Plants

The principles of inherently safer design are particularly important for major hazard plants and should be considered during the design stage. Some companies now have design procedures that require a review of designs and seek to ensure that inherently safer concepts have been addressed. Inherently safe design should be considered during the design stage in an effort to reduce the hazard potential of the plant. Protective equipment installed onto standard equipment to control accidents and protect people from their consequences is often complex, expensive and requires regular testing and maintenance. Attempts should be made to reduce the

20.4 Cryogenics

requirement for such protective equipment by designing simpler and safer processes in the first instance. A number of approaches can be considered but basically an inherently safer plant can be achieved by minimising the inventories of hazardous substances in storage and in process and hence the risk of a major accident can be significantly reduced.

Some of the techniques that can be considered are: 1) Intensification – this technique involves reducing the inventory of hazardous materials to a level whereby it poses a reduced hazard. This often means carrying out the reaction or unit operation in a smaller volume. 2) Substitution – this technique involves replacing a hazardous material or feature) with a safer one. For example, flammable solvents, refrigerants and heat transfer media can often be replaced by non-flammable or less flammable materials. 3) Attenuation – using a hazardous material under less hazardous conditions. For example, quantities of ammonia can be stored as refrigerated liquids under atmospheric pressure rather than under pressure at ambient temperature. 4) Limitation – affected by equipment design or changes to reaction conditions rather than by adding on protective equipment. For example, the selection of some types of gaskets can reduce leak rates from equipment in the event of a leak hence limiting the hazard. 5) Simplification – simpler plants are friendlier and safer than complex plants and therefore less likely to have a major accident caused by operator error. 6) Avoidance of incorrect assembly – for critical equipment plants can be designed so that incorrect assembly is difficult or impossible. Consideration should be given to installing different types of connections on inlet/outlet pipework to avoid the possibility of wrong connections being made.

Design Assessments for Cryogenic Plants

A design should be subject to a number of detailed assessments throughout its development. A number of different features can be examined and assessed. Examples are Value engineering assessment; Energy efficiency assessment; Reliability and availability assessment; Hazard identification and assessment; Occupational health assessment and Environmental assessment. A number of companies have developed detailed procedures for design studies that incorporate many of these assessments into a formalised structure. Evidence that Hazard identification and/or HAZOP studies have been carried out should be provided as evidence that a design has been evaluated and carefully considered before being installed on the cryogenic plant.

Thermal Design for Cryogenic Plants

The cryogenic system obeys the law of conservation of energy and the statement can be written as follows

$$\Sigma W + \Sigma Q + \Sigma mh = \Delta E$$

Here the ΣW stands for total work done, ΣQ stands for the total heat, m is the mass of the working fluid, h is the enthalpy of the working fluid and ΔE is the total energy change.

All cryogenic systems are related to cooling down of the working fluid and the rate of change of system energy must be negative or zero at constant temperature. Thus the derivative of the above change f energy term can be written as

$$d/dt(\Sigma W + \Sigma Q + \Sigma mh) \leq 0$$

In the cryogenic system, rate of cooling in relation to the temperature "T" depends on the thermal capacity of the system and difference between refrigeration power qr and total positive energy load. This can be written as

$$dT/dt = [qr - (\Sigma W + \Sigma Q)]/\Sigma m\, C_p$$

where C_p is the specific heat at constant pressure.

The thermal deign of a cryogenic system requires knowledge of thermal properties of materials along with heat transfer by three modes of conduction, convection and radiation.

At low temperatures, the specific heat is estimated from the energy distribution frequency of quantum particles. Debye model is used for this and can be written as

$$v_{max} = k\,\theta d/h$$

where k is the thermal conductivity, θd is the Debye temperature and h is the Planck's constant. The specific heat varies in proportion with the Debye temperature.

The cryogenic heat transfer by conduction can be estimated from the Fourier's law as given below.

$$q = Q/A = -k(dT/dx)$$

where q is the heat flux, Q is the total heat, A is the area of cross-section, k is the thermal conductivity, dT/dx is the temperature gradient over space.

The cryogenic heat transfer by convection can be estimated from the Maxwellian's law as given below.

$$q = nc\lambda m\, C_v(T_h - T_c)/L$$

where q is the heat flux, n is the molecular density, c is the molecular speed, λ is the mean free path of molecules, m is the molecular mass, C_v is the specific heat at constant volume, T_h and T_c are the hot and cold surface temperature and L is the distance between the two surfaces.

The cryogenic heat transfer by radiation can be estimated from the Stefan-Boltzmann law as given below.

$$q = \sigma A(T_h^4 - T_c^4)$$

where q is the heat flux, σ is the Stefan Boltzmann constant, A is the area of cross-section, T_h and T_c are the hot and cold surface temperatures.

Temperature Considerations for Cryogenic Plant Equipment

Temperature is a basic design parameter for cryogenic equipment. Any equipment that is to be installed should be designed to withstand the foreseeable temperature over the whole life of the plant. The combination of temperature and pressure should be considered since this affects the mechanical integrity of any equipment that is installed. In determining design temperatures a number of factors should be considered including: the temperature of the fluids to be handled, Joule-Thomson effect, which is the change in temperature that accompanies expansion of a gas without production of work or transfer of heat, ambient temperatures; solar radiation; and heating and cooling medium temperatures.

Consideration needs to be given to the temperature of the fluids that are to be handled and any excursions in temperature that could occur as a result of the failure of temperature control systems. The extremes of ambient temperature should be taken into account for plant situated outside buildings. Solar radiation on the exposed surface area of large storage tanks can significantly increase surface temperatures for storage vessels leading to significant thermal expansion of vessel contents. Likewise the low temperatures that can be achieved under conditions of snow, ice and wind, which can cause solidification of contents in vessels and pipelines, should also be considered. External facilities should be designed to accommodate the cycling of temperatures between extreme weather conditions. If secondary heating and cooling systems are employed then the maximum and minimum temperatures that can be achieved by these secondary systems should be assessed assuming failure of any control systems associated with these systems. Care should be taken to ensure that the maximum temperature that can be achieved by heating oil systems or the minimum temperature that can be achieved by cryogenic cooling systems does not compromise the design of the equipment. It should not adversely affect the mechanical strength and hence integrity, or result in additional process hazards as a result of overheating, decomposition or runaway reactions. The strength of materials decreases with increasing temperature and therefore the maximum design temperature should take into account the strength of material used for fabrication.

Pressure Considerations for Cryogenic Plant Equipment

A vessel should be designed to withstand the maximum pressure to which it is likely to be subjected in operation. For vessels under internal pressure the design pressure is usually taken at that which the relief valve is set. This is normally 5-10% above the normal working pressure to avoid inadvertent operation during minor process upsets. Vessels subjected to external pressure should be designed to resist the maximum differential pressure that is likely to occur. Vessels likely to be subjected to vacuum should be designed for full negative pressure of 1 bar unless fitted with an effective and reliable vacuum breaker device. Some form of pressure relief system may be appropriate in order to protect the equipment and prevent catastrophic failure of the equipment from occurring. Pressure vessels should be fitted with some form of pressure relief device set at the design pressure of the equipment to relieve over-pressure in a controlled. The set pressure of a relief valve should be such that the valve opens when the pressure rise threatens the integrity of the vessel but not when normal minor operating pressure deviations occur. The accumulation in the vessel is the permitted increase in the system pressure above the design pressure in an emergency overpressure situation. The maximum allowable accumulated pressure (MAAP) is specified within the various codes and this should be taken into account when the relief valve set point is selected. Normally the relief valve set point is set below or up to the maximum design pressure which allowing for the overpressure during a relief event ensures that the overall pressure is below the MAPP. Consideration should be given to the possibility of pressure cycling in equipment and subsequent failure of the equipment due to metal fatigue.

Materials of Construction for Cryogenic Plants

Selection of the material of construction for cryogenic plants is very critical. In some cases the available materials of construction may constrain the design temperatures and pressures that can be achieved and limit the design of the equipment. The most important characteristics that should be considered when selecting a material of construction are mechanical properties, tensile strength, stiffness, toughness, hardness, fatigue resistance, creep resistance, The effect of low and high temperatures on the mechanical properties, corrosion resistance, ease of fabrication, special properties – electrical resistance, magnetic properties, thermal conductivity, availability in standard sizes and cost. The selection of a suitable material of construction is often carried out by disciplines such as process engineers.

If materials to be used in the process are corrosive then this should be taken into account in the plant design and layout. Materials of construction should be carefully selected, protected where possible and regularly inspected if the presence of corrosive materials or a corrosive environment is anticipated. All possible forms of corrosion such as chemical attack, rusting, erosion and high temperature oxidation should be considered while selecting the appropriate material and that particular attention be paid to impurities and to fluid velocities. The life of equipment subjected to corrosive environments can be increased by proper consideration of design details. Equipment should be allowed to drain freely and completely and the internal surfaces should be smooth and free from locations where corrosion products can accumulate. Fluid velocities should be high enough to prevent deposition but not so high as to cause erosion. The corrosion allowance is the additional thickness of metal added to allow for material lost by corrosion and erosion or scaling. For carbon and low-alloy steels where severe corrosion is not expected a minimum allowance of 2 mm is often used, where more severe corrosion is anticipated an allowance of 4 mm is often used. Most design codes and standards specify a minimum allowance of 1 mm.

Erosion is often localised especially at areas of high velocity or impact. Occasionally corrosion and erosion combine to increase rates of deterioration. Erosion occurs primarily at sites where there is a flow restriction or change in direction including valves, elbows, tees and baffles. Erosion is promoted by the presence of solid particles, by drops in vapors, bubbles in liquids or two-phase flow. Conditions that can cause severe erosion include flashing flow and pump cavitation. If erosion is likely to occur then more resistant materials should be specified or the material surface protected in some way.

20.8 Cryogenics

Mechanical Design of Cryogenic Plant Equipment

The principles of machine design are applicable to cryogenic plant equipment.

The deflection of the pressure vessel's shell due to thermal expansion is calculated using the following equation

$$\delta_T = \alpha \Delta T L$$

where δ_T is the deflection, α is the thermal coefficient of expansion, ΔT is the temperature difference and L is the length.

The maximum stress due to temperature and torque should not exceed the minimum tensile strength using equation

$$\sigma = E\alpha\Delta T$$

where σ is the stress, α is the thermal coefficient of expansion, ΔT is the temperature difference and E is the modulus of elasticity.

The theoretical number of cycles before failure occurs due to fatigue is determined using equations

$$N_{Fail} = (\sigma a)^{1/b}$$
$$a = (0.9 S_{ut})^2 / S_e$$
$$b = -1/3 \log(-0.9 S_{ut}/S_e)$$

where N is the number of cycles, σ is the stress, a and b are constants, S_{ut} is the ultimate tensile strength and S_e is the actual tensile strength.

To determine the shear stress on each bolt the equation below is used.

$$\tau = \frac{F_B}{A_B}$$

where τ is the shear stress, F_B is the force per bolt, A_B is the cross sectional area per bolt.

The spring constant of the wall of the vessel can be approximated by,

$$k_{eq} = \frac{192 EI}{l^3}$$

where k_{eq} is the equivalent spring constant of a beam, E is modulus of elasticity of the material, I is the moment of inertia, and l is the length of the beam.

Design Considerations for Cryogenic Plant Specific Equipment

1. **Pressure Vessels:** Pressure vessels can be divided into 'simple vessels' and those that have more complex features. The relevant standards and codes provide comprehensive information about the design and manufacture of vessels and vessel design and fabrication is an area well covered by standards and codes. In general terms outright failure of a properly designed, constructed, operated and maintained pressure vessel is rare. Design and manufacture is normally carried out to meet the requirements of national and international

standards such BS 5500. The other most commonly used design code is ASME VIII. Generally pressure vessel design codes covers equipment such as distillation columns, storage drums, heaters, reboilers, vaporisers, condensers, heat exchangers, bullets, spheres etc. Basically any equipment with a "shell" that may experience some internal pressure is covered.

Factors that should be taken into account in the design process for pressure vessels include: internal and external static and dynamic pressures, ambient and operational temperatures, weight of vessel and contents, wind loading, residual stress, localised stress, thermal stress, stress concentrations; reaction forces, moments from attachments, fatigue, corrosion/erosion, creep and buckling. Pressure vessels are subject to a variety of loads and other conditions that cause stress and can result in failure and there are a number of design features associated with pressure vessels that need to be carefully considered. Consideration should also be given to other parts of the vessel not directly within the pressure envelope, but critical to vessel integrity i.e. any failure which could lead to breach of the pressure boundary e.g. vessel skirt or support legs. Other factors which require careful consideration include; a means of in-service periodic examination i.e. a means of determining the internal condition of the vessel by the provision of access openings; a means of draining and venting the vessel; and means by which the vessel can be safely filled and discharged.

Materials used for the manufacture of pressure vessels should have appropriate properties for all operating conditions that are reasonably foreseeable, and for all test conditions. They should be sufficiently chemically resistant to the fluid contained and not be significantly affected by ageing. The materials should be selected in order to avoid corrosion effects when the various materials are put together. Steel is the most common material of construction, including mild steel, low alloy steel, and stainless steel. It is often operating process temperature that determines the material used, but other equally important factors such as corrosion/erosion allowance, low temperature application etc. can determine selection. Pressure vessels are subject to a variety of loads and other conditions that cause stress and in certain cases may cause serious failure. Any design should take into account the most likely failure modes and causes of deterioration. Deterioration is possible on all vessel surfaces in contact with any range of organic or inorganic compounds, with contaminants, or fresh water, with steam or with the atmosphere. The form of deterioration may be electrochemical, chemical, mechanical or combinations of all.

2. **Storage Tanks:** Vertical storage tanks with flat bases and conical roofs are often used for the storage of liquids at atmospheric pressure and may vary in size considerably. The main load to be considered in the design of such tanks is the hydrostatic pressure of the liquid contained within the tank and the wind loading. The design of atmospheric storage tanks in general is governed by API Std 620. Tanks should be suitable for their operational duty and all reasonably expected forces such as tank contents, ground settlement, frost, wind and snow loadings, earthquake and others as appropriate. The

selection of the type of tank to be used for a particular duty will be influenced by considerations of safety, technical suitability and economy. The safety considerations are usually related to fire hazards which in turn are dependent on the physical properties of the stored material e.g. flash point, vapour pressure, electrical conductivity etc.

3. **Heat Exchangers:** The transfer of heat between two process streams is a common activity in a cryogenic plant. The most common form of equipment used to transfer heat is a heat exchanger which can be designed in many different shapes, sizes and configurations necessary to obtain the required heat transfer between one stream and another. A number of different heat transfer operations are possible with some involving a change of phase of one or more component. Heating, cooling, evaporation or condensation may all need to be considered and the equipment designed accordingly to account for the differing requirements. The basic design is commenced by an approximate sizing of the unit based on assumptions made concerning the heat transfer characteristics of the substances involved and the anticipated materials of construction. More detailed calculations are then required to confirm and refine the original design and to identify an optimum layout. Once the process design has been completed the mechanical design of the unit can then be carried out. The standards of the American Tubular Heat Exchanger Manufacturers Association (TEMA) are also widely used. The TEMA standards give the preferred shell and tube dimensions, the design and manufacturing tolerances, corrosion allowances and the recommended design stresses for materials of construction. Design temperatures and pressures for exchangers are usually specified with a margin of safety beyond the conditions normally anticipated. Major problems associated with heat exchanger design that may affect safety include fouling, polymerization, solidification, overheating, leakage, tube vibration and tube rupture. The shell of an exchanger is normally a pressure vessel and should be designed in accordance with the relevant pressure vessel design code.

4. **Pumps:** The basic requirements to define the application for pumps are usually the suction and delivery pressures, the flow rate required and the pressure loss in transmission. Special requirements for certain industrial sectors may also impose restrictions on the materials of construction to be used. Many designs have become standardised based on experience and numerous standards (API standards, ASME standards, ANSI standards) have become available. These standards often specify design, construction and testing details such as material selection, shop inspection and tests, drawings, clearances, construction procedures etc The choice of material of construction is dictated by consideration of corrosion, erosion, personnel safety and containment and contamination.

Many pumps are of the centrifugal type, although positive displacement types (such as reciprocating and screw types) are also used. Pumps are available throughout a vast range of sizes and capacities and are also available in a wide range of materials including various metals and plastics. Sealing of pumps is a very important consideration and is discussed later. The primary advantage of a centrifugal pump is its simplicity. Pumps are particularly vulnerable to mal-operation and poor installation practices. Proper installation and high quality maintenance is essential for safe operation. Problems associated with centrifugal pumps can include bearing and seal failure. Cavitation (the collapse of vapour bubbles in a flowing liquid leading to vibration, noise and erosion) can also result in damage to the pumping equipment. Key parameters for pump selection are the liquid to be handled, the total dynamic head, the suction and discharge heads, temperature, viscosity, vapour pressure, specific gravity, liquid corrosion characteristics, the presence of solids which may cause erosion etc.

5. **Compressors:** Both the positive displacement and centrifugal compressors are used in the cryogenic industry. They are complex machines and their reliability is crucial. It is very important that they are maintained to high operational standards. Centrifugal compressors are by far the most common although compression is generally lower than that given by reciprocating machines. On centrifugal compressors some of the principal malfunctions include rotor or shaft failure, bearing failure, vibration and surge. Reciprocating compressors are utilised for higher compression requirements. They may be either single or multi-stage units. Air compressors for dry air require special consideration and specific codes and standards exist.

6. **Fans:** The main applications for fans are for high flow, low pressure applications such as supplying air for drying, conveying material suspended in a gas stream, removing fumes, or in condensing towers. These units can be either centrifugal or axial flow type. They are simple machines but proper installation and maintenance is required to ensure high reliability and safe operation.

Maintenance, Inspection and Monitoring of Cryogenic plants

Plant equipment may be monitored during commissioning and throughout its operational life. This monitoring may be carried out on the basis of performance parameters such as flow, pressure, temperature, power etc. The alternative to performance monitoring is condition monitoring of which there are a number of techniques. The aim of such techniques is to identify deterioration and preempt imminent failures and so secure reliable/available plant, particularly for production and safety critical items. Some of these techniques are vibration monitoring, shock pulse monitoring, acoustic emission monitoring and oil analysis.

Construction of Cryogenic Plants

Structures are required to provide support for cryogenic plant and should be able to with stand all foreseeable loadings and operational extremes throughout the life of the plant. Failure of any structural component could lead to initiation of a major accident. Structural design should take into account natural events such as wind loadings, snow loadings and seismic activity and also plant excursions. It is critically important that construction phase is carried out according to the original specification and that no additional hazards are introduced to the plant during the construction phase. Poor construction can result in the integrity of the whole system being compromised resulting in an increased risk of a major accident. Building and construction are covered by a series of different building regulation. It is important to demonstrate that the correct materials of construction have been used and that appropriate construction techniques have been employed so as not to introduce construction faults and flaws into the plant. Commissioning of equipment should be carried out and records kept of the commissioning exercises. Following checks need to be carried out on the equipment: performance tests, pressure testing of systems, visual inspection checks, checking of pipework, checking of internal fittings installataion, checking the materials of construction, checking of rotating equipment for noise and vibration, checking plant against P and IDs and isometrics, pressure vessel and system tests, leak tests, protective devices tests and dynamic process fluid test. The management of the commissioning and verification stages should be identified under the safety management system. The system should focus on ensuring that the design intent is met, and that deviations are properly assessed and controlled. Systems should be in place to ensure that corrective action is taken on the identification of discrepancies between installed equipment and the design intent and to control any deviations from normal operation. Evidence of a number of pre-commissioning and commissioning checks should be presented to verify that the equipment as installed has been tested and is suitable for operation and meets the design intent.

References

1. Cryogenic Engineering – B.A. Hands.
2. Internet website www.wikipedia.com.
3. Cryogenic Systems – Randall Baron.
4. Nathan Hill – ESA Publication on Cryogenics – 2008.
5. Cryogenic Engineering – P.K. Bose.
6. Internet website www.hse.gov.uk.
7. Internet website www.uigi.com.
8. Design specifications for a CO_2 vessel – Brian Smith, Air Products Inc.

Questions

Q.1. Explain the importance of cryogenic plant design.

Q.2. Discuss the components of a typical cryogenic plant.

Q.3. Write a short note on materials for cryogenic plant equipment.

Q.4. What are the construction safety checks for a cryogenic plant?

Q.5. Explain briefly the thermal design considerations for a cryogenic plant.

CHAPTER 21

Safety for Cryogenic Systems

INTRODUCTION

Safety is critical for cryogenic installations due to hazards. The principal areas of hazard related to the use of cryogenic fluids are flammability, high pressure, freezing burns and asphyxiation. Hydrogen, methane and acetylene are obviously flammability hazards. However, oxygen greatly increases the flammability of ordinary combustibles and can cause non-combustible materials to burn readily. Liquefied inert gases (such as nitrogen and helium) or very cold metal surfaces can condense oxygen from the atmosphere. The high-pressure gas hazard is always present when cryogenic fluids are used because of the large expansion ratio from liquid to gas on evaporation. Equipment must be carefully selected for cryogenic service because of the changes in their properties at very low temperatures – normally ductile materials may become extremely brittle. Methods of joining materials must also receive careful consideration, because of different rates of contraction and embrittlement of sealant. Chemical reactivity between the fluid and apparatus must be studied also. Personnel hazards exist in several areas. Exposure of personnel to the above hazards must be avoided, but of prime concern is bodily contact with the extremely low temperatures involved. Brief contact with cryogenic materials can cause burns similar to thermal burns from high temperature contact. Prolonged contact with these temperatures will cause embrittlement of the affected parts because of the high water content of the body. The eyes are especially vulnerable, so eye protection is mandatory. While a number of gases in the cryogenic range are not toxic, they are capable for causing asphyxiation by displacing oxygen. Even oxygen may have harmful physiological effects on prolonged breathing of the pure gas.

Cryogenic Health Hazards

Some health hazards are directly associated with the cryogenic gas industry. Properties of certain gas products subject personnel to extreme cold temperatures, oxygen-deficient (asphyxiating) atmospheres, or oxygen-enriched (increased fire risk) atmospheres. Proper precautions, a basic knowledge of the behavior of these gases, and wearing proper protective equipment can minimize exposure to these hazards. Following are some hazards explained in detail.

21.2 Cryogenics

1. **Cryogenic Burns or Extreme Cold Hazard:** By definition, all cryogenic liquids are extremely cold. Cryogenic liquids and their vapors can rapidly freeze human tissue. Brief exposures that would not affect skin on the face or hands can damage delicate tissues such as the eyes. Prolonged exposure of the skin or contact with cold surfaces can cause frostbite. There is no initial pain but there is intense pain when frozen tissue thaws. Unprotected skin can stick to metal that is cooled by cryogenic liquids. The skin can then tear when pulled away. Even non-metallic materials are dangerous to touch at low temperature. Prolonged breathing of extremely cold air may damage the lungs. Cryogenic liquids can cause many common materials such as carbon steel; rubber and plastics to become brittle or even break under stress.

2. **Asphyxiation Hazard:** All cryogenic liquids produce large volumes of gas when they vaporize. For example, one volume of liquid nitrogen vaporizes to 694 volumes of nitrogen gas at 68 F and 1 atm. Air is normally 21% oxygen by volume. When this is reduced to 15-16% oxygen, symptoms of asphyxia (unconsciousness) will develop. At 12% oxygen, the individual will lose consciousness without warning and may be unaware of any danger. When there is not enough oxygen, asphyxiation and death can occur very quickly. When cryogenic liquids form a gas, that gas is very cold and usually heavier than air. This cold, heavy gas does not disperse very well and can accumulate near the floor. Even if the gas is non-toxic, it displaces the air. Oxygen deficiency is a serious hazard in enclosed or confined spaces. Signs of asphyxiation are giddiness, mental confusion, and loss of judgment, loss of coordination, weakness, nausea, fainting and death. Only a few breaths of oxygen-depleted air are required to cause a rapid drop in dissolved oxygen in the blood. Mental failure and coma follow within seconds. Symptoms or warnings are generally absent, but even if present, the loss of mental abilities, coordination and weakness may make it impossible for victims to help themselves or summon help from others. Most cryogenic liquids are odorless, colorless and tasteless when vaporized into the gaseous state. However, extremely cold liquids and their vapors have a built-in warning property that appears whenever they are exposed to the atmosphere. The cold "boil-off" gases condense the moisture in the surrounding air, creating a highly visible fog. Although fog clouds may be indicative of a release, they must never be used to define the leak area, which should not be entered.

3. **Flammability Hazard due to Oxygen Enriched Air:** Vaporization of liquid oxygen in an enclosed area can cause oxygen enrichment, which could saturate combustibles in the area such as workers' clothing. This can cause a fire if an ignition source is present. Although oxygen is not flammable it will support and vigorously accelerate the combustion of other materials. Liquids at or below the boiling point of liquefied air can actually condense the surrounding air causing a localized oxygen-enriched atmosphere. Extremely cold cryogens such Helium can even freeze or solidify the surrounding air.

4. **Explosion Hazard Due to High Pressure and Rapid Expansion:** Cryogenic liquids cannot be indefinitely maintained in the liquid state. If they are vaporized in sealed containers they can produce enormous pressures that could rupture the container. For this reason pressurized cryogenic containers are normally protected with multiple devices for over-pressure prevention. A pressure relief device must protect all selected equipment that may allow cryogenic liquid to be trapped.

5. **Cryogenic Frostbite and Hypothermia:** Cryogenic injuries or burns result from skin contact with very cold vapor, liquid, or surfaces. Effects are similar to those of a heat burn. Severity varies with the temperature and time of exposure. Exposed or insufficiently protected parts of the body can stick to cold surfaces due to the rapid freezing of available moisture, and skin and flesh can be torn on removal. Risk of frostbite or hypothermia (general body and brain cooling) in a cold environment is always present. There can be warning, in the case of frostbite, while the body sections freeze. As the body temperature drops, the first indications of hypothermia are bizarre or unusual behavior followed, often rapidly, by loss of consciousness.

6. **Cryogenic Respiratory Disorders:** Respiratory problems are caused by the inhalation of cold gas. Short-term exposure generally causes discomfort; however, prolonged inhalation can result in effects leading to serious illness such as pulmonary edema or pneumonia.

Personal Safety in Cryogenic Plants

The eyes are the most sensitive body part to the extreme cold of the liquid and vapors. The recommended personal protective equipment (PPE) for handling cryogens includes a full-face shield over safety glasses, loose-fitting thermal insulated or leather gloves, long sleeved shirts and trousers without cuffs. Gloves should be loose fitting to allow quick removal if liquid should be spilled inside. Gloves are not made to permit the hands to be immersed in a cryogenic liquid. They will only provide short-term protection from accidental contact with the liquid. No metal jewellery, rings, watches etc. should be worn on the hands or wrist while transferring cryogenic liquids.

Proper clothing and special equipment can serve to reduce fire hazards when working with cryogenic liquids or gases, but prevention of the hazard should be the primary objective. Insulated or leather gloves should be worn when handling anything that is cooled with cryogenic liquids or when participating in liquid loading and unloading activities. Gloves shall fit loosely so they can be removed easily if liquid splashes on or in them. A face shield or chemical splash goggles shall be worn at all times when handling cryogenic liquids. Clothing should have minimum nap. There are a number of flame retardant materials available such as Nomex for work clothing, but they can burn in high-oxygen atmospheres. There is some advantage in these materials as most of them would be self-extinguishing when removed to normal air atmospheres. All clothing should be clean and oil-free. No means of ignition should be carried. Footwear should not have nails or exposed metallic protectors that could cause sparking. If individuals inadvertently enter or are exposed to an oxygen-enriched atmosphere, they shall leave as quickly as possible. Avoid sources of ignition such as smoking. Opening the clothing and slapping it helps disperse trapped vapors.

General Safety Practices for Cryogenic Plants

1. Cryogenic liquids must be handled, stored and used only in containers or systems designed in accordance with applicable standards, procedures or proven safe practices.
2. All systems components piping, valves etc. must be of the appropriate materials to withstand the extreme temperatures.
3. Pressure relief valves must be placed in systems and piping to prevent pressure build up.
4. Any system section that could be isolated while containing cryogenic liquid must have a pressure relief valve. Pressure relief valve relief ports must be positioned to face toward a safe location.
5. Transfer operations involving open cryogenic containers, such as dewars must be done slowly, while wearing all required PPE. Care must be used not to contact non-insulated pipes and system components.
6. Open transfers of cryogenic fluids must be allowed only in well-ventilated areas.
7. Funnels should not be used while transferring cryogenic liquids.
8. Tongs or other similar devices should be used to immerse and remove objects from cryogenic liquids.
9. Hazard reviews are required on all newly purchased, built or modified tools using cryogenic materials.
10. Areas where it is possible to have low oxygen content shall be well ventilated. Inert gas vents should be piped outside of buildings or to a safe area.
11. Where an oxygen-deficient atmosphere is possible, special precautions such as installation of oxygen analyzers with alarms, ensuring a minimum number of air changes per hour, implementing special entry procedures should be taken.
12. Warning signs must be posted at all entrances to alert personnel to the potential hazard of an oxygen-deficient atmosphere.
13. Oxygen analyzer sensors should be located in positions most likely to experience an oxygen-deficient atmosphere and the alarm should be clearly visible, audible, or both at the point of personnel entry.
14. When there is any doubt of maintaining safe breathing atmosphere, self-contained breathing apparatus or approved masks should be used, particularly when personnel enter enclosed areas or vessels.
15. Breathing air should come from a qualified independent source; a plant instrument air system should not be used as a source of breathing air.

16. Personnel must be thoroughly instructed and trained in the nature of the hazards and precautions against them, including emergency procedures, operating equipment, safety devices, the properties of the materials used and personal protective equipment required.
17. If contact with a cryogenic fluid is possible, eye protection (or full-face protection is preferred), and an impervious or coat, trousers without turnups and high top shoes or boots must be worn.
18. Watches and jewellery should not be permitted in the plant i.e. anything capable of trapping a cryogenic fluid close to the skin.
19. Gloves may be worn as desired, but if they are necessary to handle containers or cold metal, they should be impervious and large enough to be tossed off the hand in case of a spill.
20. Care is required in the transport and storage of cryogenic substances. Slow evaporation is bound to take place leading to contamination of the atmosphere, unless there is good ventilation.
21. Cryogenic liquids should not be carried with passengers in a passenger lift.
22. Low boiling point gases must be protected in their vacuum vessel by a guard jacket of liquid nitrogen, which should be replenished regularly. If the supply of liquid nitrogen fails on a helium storage vessel, explosion may occur.
23. Keep equipment and systems scrupulously clean and avoid contaminants creating hazardous conditions on contact with the cryogenic fluids. This is particularly important when working with oxygen.
24. If the boiling point of a gas is below that of liquid gas, care must be taken in transferring the liquid e.g. into Dewar-type vessels or cryostat. A vacuum jacketed siphon must be used; if the gas is poured from the container, a plug of solid air may form in the outlet.
25. Liquid oxygen and nitrogen containers should not normally be emptied; they should be occasionally warmed to room temperature and purged with dry nitrogen. This avoids accumulation of hydrocarbon gases extracted from the atmosphere and the nuisance of water freezing out in the container.
26. It may be convenient to transfer liquid oxygen or nitrogen from its container by pressurization. This may be done safely if the appropriate dry gas is used and the container can withstand the pressure.
27. Instructions for handling liquid nitrogen, helium, argon and methane should be sought from the supplier and the experimental work should proceed based on their advice with the full knowledge and agreement of the safety department.

Cryogenic Plant Safety Considerations

A. **Site selection:** Cryogenic plant safety should begin with a safety evaluation of the proposed plant site. Generally, cryogenic plants are located in or near industrial areas as an adjunct to other industrial or chemical plants. A plant installation should conform to the applicable industry consensus standards as well as all applicable national regulations. The plant operation should be reviewed for compatibility with the surrounding area. The potential hazard of the cooling tower or cryogenic fog to nearby plants or vehicular traffic should be recognized. Adequate space should be provided for cryogenic liquid disposal.

B. **Safety factors in plant layouts:** The use of valve pits, trenches, or both for cryogenic gas or liquid piping systems is not recommended because oxygen-enriched or oxygen-deficient atmospheres can occur very easily with such installations. If gas and liquid piping systems are installed in enclosed spaces, precautionary measures such as forced ventilation and alarm systems are recommended. Appropriate warning signs shall be posted. Oxygen-rich liquid drain lines should not be installed in a trench. Over time, trenches can accumulate oil, grease, and trash or other debris. If a leak in the line develops, a fire could result. Caution should be taken to prevent liquid spills from entering floor drains or sewer systems.

C. **Materials of construction:** The materials used in a cryogenic plant are exposed to a wide range of temperatures, pressures, and purities during operation. Materials shall be selected that are compatible with the expected conditions including normal operation, startup, shutdown, and process upsets. For an oxygen system to operate safely, all parts of the system should be reviewed for compatibility with oxygen under all conditions they encounter. The system shall be designed to prevent oxygen combustion by selecting proper material, operating within the designed pressure, temperature, and flow limits, and obtaining/maintaining proper cleanliness. Substitution of materials should not be made without first consulting a qualified engineering source.

D. **Insulation:** Interconnecting process lines between components of a cryogenic plant require insulation to reduce process heat leak to an acceptable minimum and to prevent exposure of personnel to extremely low temperatures. The temperature and service of the line determine the type of insulation used. Insulation for LOX lines should be noncombustible to protect against a possible reaction in the event of a liquid leak. Other process lines operating at temperatures warmer than the liquefaction point of air, approximately $-192°C$, may be insulated with any commercially acceptable insulation that meets design requirements. Insulation that is non-combustible in air should be given preference. Oxygen-compatible binders, sealing compounds, and vapor barriers should be used on lines carrying oxygen or oxygen-enriched gases or liquids. Process lines operating at temperatures colder than the liquefaction point of air should be insulated with material compatible with oxygen. If the insulation cracks or deteriorates at these temperatures, air is diffused into the insulation, condenses against the surface of the pipe, and exposes the insulation material to oxygen-enriched liquid. Personnel should be protected from hot lines (greater than $60°C$) by either insulating the line or preventing access while the line is hot.

E. **Cleaning:** All materials for use in or interconnected with cryogenic liquids should be suitably cleaned before the system is put into service. Mill scale, rust, dirt, weld slag, oils, greases, and other organic material shall be removed. An improperly cleaned line in oxygen service can be hazardous because particulates, greases, oils, and other organic materials can ignite a fire. Fabrication and repair procedures should be controlled to minimize the presence of such contaminants and thereby simplify final cleaning procedures. Cryogenic process equipment and piping that handle inert fluids should be cleaned to prevent foreign material from reaching parts of the unit.

F. **Electrical requirements:** Applicable codes shall be followed. General purpose or weather proof types of electrical wiring and equipment are acceptable depending on whether the location of the cryogenic plant is indoors or outdoors. In areas where high oxygen concentrations could be expected, electrical equipment with open or unprotected make-and-break contacts should be avoided. The simple expedient of locating electrical equipment away from areas where high oxygen concentrations can occur eliminates potential hazards in these situations.

G. **Noise:** The noise produced by compressors and their drives; by expansion turbines; by high gas velocities through piping and valves; and by pressure relief valves, vents, or bypasses shall be considered from the standpoint of potential hazard of hearing damage to employees. To assess the hazard, noise surveys should be performed after initial inspection or when modifications are made that could change the noise emitted. Noise abatement and use of personnel ear protection shall follow national government guidelines available. New equipment and varying operating conditions require a continuing program of noise level surveillance. Periodic audiometric checks of personnel might be necessary depending on exposure times and noise levels.

H. **Intake air quality:** Air quality can have an impact on the cryogenic plant site selection and should be carefully evaluated. The cryogenic plant typically is located in an industrial area and thus some degree of contamination released from industrial and/or chemical plant operations can be expected to be present in the air. Trace contaminants in the atmospheric air, particularly hydrocarbons, have a direct bearing on the safe operation of a cryogenic plant. It is important to identify these contaminants and their levels of concentration in the atmospheric air. Short-term air quality analysis might not be representative of long-term air contaminant levels. Changing site conditions can have an impact on air quality and should be evaluated periodically or when the surrounding industries change. Trace contaminants can be put into three main categories i.e. plugging, reactive and corrosive based on the potential problems they cause in the plant.

I. **Control systems:** Instrumented systems are required to perform safety-related functions as well as traditional control functions of cryogenic plants. System architecture ranges from simple pneumatic control loops with electrical relay logic to sophisticated computer-based systems allowing automated start and shutdown as well as unattended and remote operation based on complex control algorithms. Critical safety systems are used to prevent an uncontrolled release of a toxic substances or fires or explosions or any other unplanned incident that could cause death or life-threatening injury to employees of the plant. Operational safety systems should be used to prevent an unplanned incident that could cause non life-threatening personnel injury, limited equipment damage, or minor off-site impact; and routine plant operation control should be used for routine plant operation and equipment protection.

Cryogenic Plant Critical Safety Systems

Critical safety systems must be provided in cryogenic plants and should be failsafe. The failure of any critical component shall result in the shutdown and isolation of the system in a predetermined manner. Critical safety systems shall be protected from accidental change by use of passwords, key locks, or other methods. Critical safety systems may be separate from controls necessary for routine plant operation. These systems also can require redundancy through duplication of critical components or functions. The critical safety system may share components with the routine plant control system if it can be shown that failure of the routine plant control system does not compromise the critical safety system. The proper operation of critical safety systems shall be verified and documented as follows:

- during initial control system commissioning and startup;
- after maintenance is performed on the system;
- at periodic intervals; and
- after an extended outage.

Modification of any critical safety system including bypassing functionality for temporary operation shall require a documented management of change (MOC) procedure including review by technically competent personnel and authorization by the appropriate personnel. An external override (a plant emergency shutdown that is independent of the plant control system) must be provided to immediately shut down part or all of a facility to safeguard personnel and mitigate the potential consequences of a major operational safety event. The external override shall require manual reset by a separate and secure means to prevent unintentional restart. Any external override shall be clearly identified and plant personnel made aware of its location.

Procedures should be developed to cover the proper response to anticipated emergency conditions for which the plant operators may have to contend. Potential emergency conditions should include mechanical malfunctions, and power failures, as well as environmental and civil disturbances that can affect plant safety. Emergency conditions that should be considered are: energy release, cryogenic liquid spill, fog cloud from a cryogenic release, site security threat, severe weather conditions such as tornado or flood and adjacent industry incidents such as explosions or toxic gas releases.

References

1. Cryogenic Safety Manual by AIGA.
2. Cryogenic Systems – Randall Baron.
3. Internet website http://www.gla.ac.uk/.
4. Cryogenic Engineering – P.K. Bose.
5. Internet website http://www.ncsu.edu.

Questions

Q.1. What are Cryogenic Hazards ? Explain any two in detail.

Q.2. Enumerate some basic safety precautions when dealing with cryogenic fluids.

Q.3. Write a short note on safety considerations for cryogenic plants.

CHAPTER 22

Advances in Cryogenics

INTRODUCTION

Cryogenics, the science and technology of producing temperatures below 120 K has entered its second century of existence. It is the result of a historical conjunction of progress in science – the gradual construction of thermodynamics throughout the Nineteenth century, from the macroscopic theory of energy of J. Joule and S. Carnot, to the quest for liquefying the so-called "non-condensable" gases of the atmosphere, calling for the ingenuity of engineers to design compressor machinery, heat exchangers, as well as thermal insulation techniques.

Cryogenic industry is now well established in the world. Today, liquid natural gas (LNG) represents one of the largest industrial domains of application of cryogenics together with the liquefaction and separation of air gases. The densification by condensation, and separation by distillation of gases remains the main driving force for the cryogenic industry, exemplified not only by liquid oxygen and nitrogen used in chemical and metallurgical processes, but also by the cryogenic liquid propellants of rocket engines and the proposed use of hydrogen as a "clean" energy vector in transportation. It is estimated that the world capacity of liquid oxygen is over 150 million tons per annum and that of LNG is over 60 million tons per annum. Gases like ethylene, helium, argon, nitrogen and hydrogen are also being consumed in large quantities through out the world and their demand is also increasing.

The UK and France appear to be the two leading European countries in cryogenics field. There are a number of companies worldwide that develop and manufacture cryogenic systems. In the main, they are not focused on space as a market, with the majority manufacturing cryogenic systems for analytical and scientific instrumentation including superconducting magnets, together with a mostly separate industry producing cryogenic systems for other purposes. There is a considerable related supply chain, with companies producing vacuum vessels and components, electronics and accessories.

Cryogenics has shown a sustained development towards lower temperatures attaining values down to about 0.1 nK today in specialized laboratories through a combination of helium dilution and adiabatic demagnetization techniques. This chapter takes a review of recent advances in cryogenics.

Advances in Superconductivity

The low absolute temperatures encountered in cryogenics can be used to reveal and study physical phenomena with low characteristic energy e.g. the intrinsic transport properties of metals and alloys and to create new forms of condensed matter. Although the application potential of these phenomena was identified very early, it took an unusually long time span to bring them out from the laboratory into practical devices. It is only with the development of practical type II superconductors and the progress in power refrigeration technology down to helium II temperature, that superconductivity and superfluidity have become industrial techniques, applied on a grand scale. Superconductivity has been applied in nuclear magnetic resonance (NMR) systems, magnetic-confinement fusion device and high-energy particle accelerators. With some 15000 units operating worldwide and a continuously growing demand, magnetic resonance imaging (MRI) devices constitute the first market application of superconductivity. Figure 22.1 shows the applications of superconductivity.

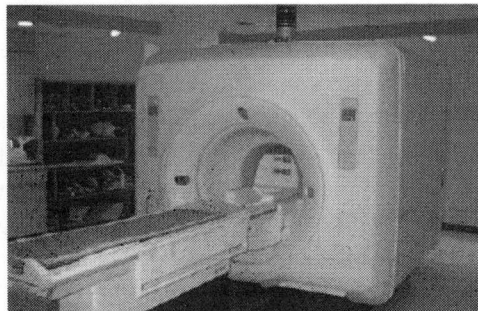 

Fig. 22.1 Applications of Superconductivity: MRI and Particle Accelerator

Practical "high-temperature" superconductors (HTS) — a class of materials still in the development phase — will eventually supersede conventional technology in specific niches of the electro-technical machine and power line markets. The main benefit of HTS would be to greatly simplify cryogenics by operating at liquid nitrogen temperatures of 63 K. Some electro-technical applications of superconductivity, however have no conventional equivalent. The absence of electrical resistance has permitted to plan and build super conducting magnetic energy storage (SMES) devices, used for peak shaving of demand, network stabilization or uninterrupted power supply. The particular property of superconductors to switch rapidly from the super conductive to the resistive state when their critical current is exceeded, has already been applied to fault current limiters protecting networks from destructive over-intensities.

The small value of the magnetic flux quantum in superconductors is used for precision measurement standards and sensitive detection of magnetic fields. Superconducting quantum interference device (SQUID) detectors have become irreplaceable tools in geophysical surveying, underwater detection, and magnetoencephalography. With the operation of sensitive detectors and reduction of thermal noise, cryogenics has found its place to cool embarked semiconductor detectors looking at the cosmic microwave background of the universe or at astronomical objects in different ranges of the electromagnetic spectrum, from space probes in Earth orbit or beyond.

Advances in Cryogenic Propulsion

Gas liquefaction is second to none when the final product must be kept at maximum density, for liquid rocket propellants. In liquid-hydrogen fuelled rockets, which have developed to very powerful launch vehicles operating with industrial reliability subcooled liquid and even solid-liquid mixtures ("slush") could further increase the mass of propellant per unit volume of tank, while reducing loss of propellant ("zero boil-off") for long missions. Current research at the Cryogenic Engineering Group covers linear alternators for cryogenic power sources in space applications, new generation linear compressors with greater efficiency and low mass, valved linear compressors for mechanical coolers for sub 12 K temperatures – to replace stored cryogen systems, linear motors at low cost and compatible with new clearance cell technology,

Fig. 22.2 Space Shuttle using Liquid Hydrogen

Advances in Cryocoolers

Cryocoolers are small machines ranging from a few mW to a few hundred W cooling power, requiring no manipulation of cryogens and often integrated in the piece of equipment they serve. The issue of efficiency is less critical, but that of reliability

22.4 Cryogenics

is essential for aerospace applications with no possibility of servicing or repair. An important element of reliability is design simplicity, exemplified by the sorption compressors and non-contact pressure oscillators of space cryocoolers, as well as the absence of cold moving parts in pulse tube refrigerators. Noteworthy developments in this domain concern the use of selected refrigerant mixtures in Joule-Thomson or pulse-tube coolers, as well as of regenerator materials showing magnetic phase transition to beat the decrease in specific heat of solids at low temperature and permit operation of pulse-tube coolers down to below 4 K. While the domain of large-scale cryogenic refrigeration is in the hands of a few companies world wide, the large variety of specific applications and the ingenuity of the researchers and developers render the field of cryocoolers very lively and industrially attractive for small and medium-size industry. Figure 22.3 shows a typical cryocooler.

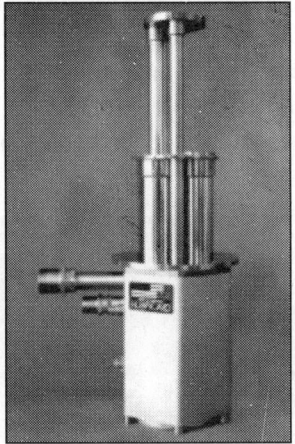

Fig. 22.3 A Typical Cryocooler

Cryocoolers are required for a wide variety of applications, and the number of applications keeps expanding as improvements to cryocoolers are made. One of the common applications is for cooling infrared sensors to about 80 K for night vision capability of the military. Over 125,000 Stirling cryocoolers for this tactical military application have been produced to date. In the last decade or so the desire for night vision surveillance and missile detection from satellites has prompted much research on improved cryocoolers to meet the very stringent requirements for space use. Each application has a particular set of requirements that have led to many recent improvements in cryocoolers. Lifetimes of 10 years are now standard requirements for most space applications. Until recently efficiencies of 3 to 10% of Carnot were typical of these small cryocoolers, whereas now efficiencies of 15 to 25% of Carnot are now possible. New commercial applications, particularly the use of high temperature superconductors as microwave filters in base stations for wireless communication systems, have stimulated much research on ways to reduce the cost of cryocoolers while still maintaining a lifetime goal of 3 to 5 years.

Adiabatic Demagnetization Refrigerators (ADR)

Adiabatic demagnetization refrigerators (ADR) are ultra low temperature systems, which offer a simple method of achieving temperatures of 50 mK. These systems make use of two built in pills that operate at temperatures of 1 K and 50 mK, supported below a 4 K surface that is cooled with either liquid helium or a mechanical (pulse tube) cooler. These systems offer a typical hold time of two to three days below 100 mK, and require no active pumping on the cryostat. A recycling period of two to three hours is needed at the end of the three-day period, to reduce the temperature back to 50 mK. These systems include a conductively cooled superconducting magnet, which is used to magnetize the two pills, and a mechanical heat switch that links and isolates the two stages from the 4 K surface. After magnetizing the pills at 4 K, the pills are isolated from the 4 K surface and demagnetized to achieve the lowest temperature. The cold stage may then be allowed to drift up in temperature or controlled to operate at any higher temperature between 50 mK and 150 mK. Refer Figure 22.4 for an ADR developed by Janis Corporation

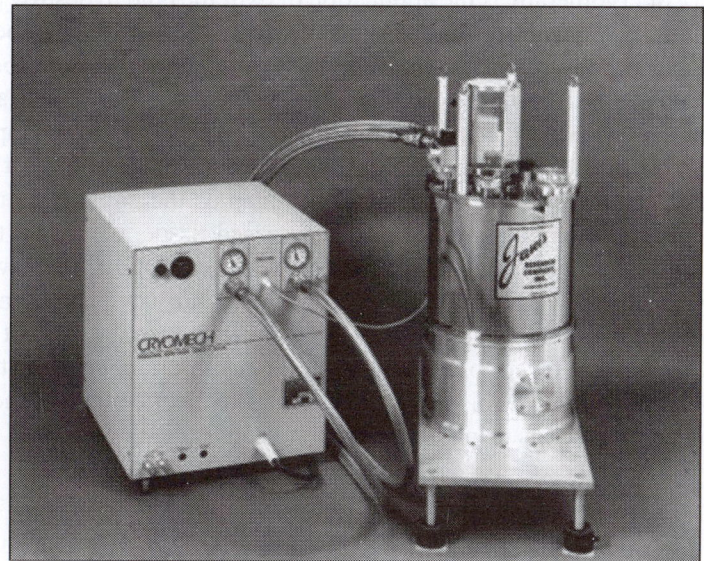

Fig. 22.4 Adiabatic Demagnetization Refrigerator

The adiabatic demagnetization refrigerator (ADR) system includes a Pulse Tube refrigerator with internal highly conductive flexible thermal linkage to cool two stages, the ADR support stages, a superconducting magnet and associated leads and thermometry. It also cools two sets of radiation shields that protect the cold assembly from the room temperature radiational heat load. The two-stage ADR system contains a gadolinium-gallium garnet (GGG) crystal and a ferric ammonium alum (FAA) salt pill. Each pill has its own ultra low thermal conducting support structure isolating it from the 4 K flange and the intermediate stage. The two stages operate at a temperature of ~1 K (GGG) and 50 – 100 mK (FAA), with a typical hold time of three days below 100 mK. A mechanical thermal heat switch assembly is provided for attaching and detaching the two stages of the ADR during initial cool down, and during its magnetization and demagnetization cycles. Two ruthenium oxide thermometers with appropriate heat sunk wiring at the various intermediate stages are provided to monitor the temperature of the two stages of the ADR. The two thermometers are supplied with a standard calibration data between 50 mK and 20 K. A 4-Tesla conductively cooled multi-filamentary Niobium-Titanium superconducting magnet surrounds the GGG and FAA pills. The magnet is wired with an optimized arrangement of low and high temperature superconducting leads, and is provided with voltage taps and two standard thermometers to monitor its cool down and operating temperature. The magnet reaches its rated field of 4-Tesla at a nominal 10 amperes. A high permeability magnetic shield surrounds the magnet and the two pills, to reduce the stray magnetic field in the vicinity of the cold stages.

Advances in Cryosurgery

Cryosurgery or cryotherapy is the application of extreme cold to destroy abnormal or diseased tissue. Cryosurgery has been historically used to treat a number of skin diseases and disorders as well as cancers. Generally, all tumors that can be reached by the cryoprobes used during the operation. Cryosurgery works by taking advantage of the destructive force of freezing temperatures on cells. At low temperatures, ice crystals form inside the cells, which can tear them apart. The most common method of freezing lesions is using liquid nitrogen as the cooling solution. The super-cooled liquid may be sprayed on the diseased tissue, circulated through a tube called a cryoprobe, or simply dabbed on with a cotton or foam swab. Less frequently, doctors use carbon dioxide "snow" formed into a cylinder or mixed with acetone to form a slush that is applied directly to the treated tissue. Recent advances in technology have allowed for the use of argon gas to drive ice formation using the Joule-Thomson effect. This gives physicians excellent control of the ice, and minimizing complications using ultra-thin cryoneedles.

References

1. Cryogenic Engineering – B.A. Hands.
2. Cryogenic Systems – Randall Baron.
3. Internet Wikipedia website on Cryogenics.
4. Refrigeration and Air Conditioning – Dr. S.S.Thipse.
5. Paper on Cryogenics by Phillip Liberian, CERN.
6. Nathan Hill – ESA publication on Cryogenics – 2008.
7. Internet website www.janis.com.
8. Internet website www.wikipedia.com.

Questions

Q.1. Enumerate some recent advances in Cryogenics.

Q.2. Write a short note on Advances in Superconductivity.

Q.3. Discuss the benefits of advances in Cryocoolers.

Q.4. What are ADRs? Explain their operation.

Q.5. What are the advances in Cryosurgery?

CHAPTER 23

Applications of Cryogenics - I
Industrial Liquefied Gases

INTRODUCTION

The term "Industrial Gases" refers to gases, which are produced in relatively large quantities by gas companies for use in a variety of industrial manufacturing processes. They are produced and supplied in both gas and liquid form. They are transported to customers in cylinders, as bulk liquid, or as pipeline gases. Industrial gases include common atmospheric gases like nitrogen, oxygen and argon, air derived rare gases like neon, krypton and xenon, natural gas-derived gases such as methane, helium, hydrogen, carbon monoxide and carbon dioxide along with welding gases such as acetylene and nitrous oxide. Sometimes industrial gas mixtures are also used, such as acetylene-oxygen etc. Gas mixture classes are characterized by the degree of accuracy, level of traceability, blend tolerance, process accuracy, analytical accuracy and other critical parameters.

Industrial gases are used for welding, medical and numerous process applications. Gases are used in almost every product chain. For instance when the local workshop fine tunes the car ignition, the result is checked with the help of instruments calibrated with specialty gas mixtures. When diving really deep specialty gas mixtures are required to avoid caisson disease. Purity of industrial gases is commonly expressed as a two digit number. For example, Helium with a purity of 99.9995% would be described as Helium 5.5 with the first 5 representing the number of nines, while the second 5 represents the first digit following the number of nines. Often it may not just be a question of the gas purity, but the degree of impurity that makes the real difference. Two gases of nominally identical purity may not necessarily have the same effect. This difference in purity is important with many new techniques that are sensitive to all kinds of contamination. Every liquefied gas has to be transported and fractionated to be used. Special pumps are needed for this kind of work since liquefied gases have a very special behavior depending on how they are used.

Industrial gases are valued for their physical and chemical properties such as reactivity, inertness and coldness. These properties are utilized to produce specialty products, protect and maintain product quality, and lower operating costs in steel making, metals manufacturing and fabrication, petroleum refining, chemicals and pharmaceuticals manufacture, production of electronic equipment and components, the rubber and plastics industries, food and beverage processing, glass manufacture, healthcare, pulp and paper and environmental protection operations. Oxygen, in particular, is valued for its reactivity. Oxygen enrichment of air is used to increase the amount of oxygen available for combustion or biological activity. This can increase reaction rates and lead to greater throughput in existing equipment and smaller sizes for new equipment. Oxygen's reactivity is used in metals processing (steel, copper, lead, zinc), glass furnaces, cement kilns, chemical manufacture, sewage treatment, pulp and paper manufacture, welding and cutting of metals. Hydrogen, methane and carbon monoxide are gases that react with oxygen and other materials. These gases are often used as raw materials for chemical manufacturing processes. Hydrogen is used in refineries to remove sulfur and to chemically restructure (reform) hydrocarbons. It is used to hydrogenate fatty acids in animal and vegetable oils. It is also used as a reducing agent in steel and zinc manufacture. Methane generated by biological activity, is the primary component of "natural gas". It is used as a fuel and as a chemical raw material. Carbon monoxide is co-produced with hydrogen by steam reforming plants using methane or other hydrocarbons as feedstock. It is a raw material for making monomers and other chemical products. Carbon dioxide does not react with oxygen, but will combine with other elements and compounds. Thus its commercial uses include raw material for various chemical processes and neutralizing agent for alkaline materials.

Inertness is somewhat relative. Some industrial gases (helium, neon, argon, krypton and xenon) are almost totally inert. Helium and argon are commercially available in relatively large quantities; neon, krypton and xenon have much more limited availability. Many applications requiring an "inert gas" rely on nitrogen or carbon dioxide, because they have very little reactivity under normal pressure and temperature conditions and are much less expensive than the other "inert" gases. The "noble" gases are all monatomic. Nitrogen and argon are commonly used in the gaseous form to shield potentially reactive materials from contact with oxygen. Applications based on inertness include blanketing of storage tanks and vessels that contain flammable liquids or powders; blanketing of materials that would degrade in air, such as vegetable oil, spices, and fragrances; maintaining controlled atmospheres for industrial activities such as growing silicon and germanium crystals, manufacturing precision electronic devices, welding and soldering; preventing light bulb filaments from burning; retarding evaporation of filaments with high molecular weight inert gases; filling insulating spaces between multi-pane windows; and creating non-flammable lighter-than-air devices such as balloons.

Liquid nitrogen and solid carbon dioxide are valued because they combine intense coldness with inertness. This combination is employed to rapidly chill and freeze food items (meat, fruit, vegetables, baked goods, and dairy products). Rapid freezing results in very small ice crystals, less cellular damage, and better-quality products after thawing. The intense cold produced by these products can also be used to make normally soft and flexible materials hard and rigid, allowing them to be ground, machined or fractured. Carbon dioxide is also used in some applications for its inertness, in particular for fire fighting. Both portable fire extinguishers and total room fire extinguisher systems use carbon dioxide to extinguish flames without damage to materials and without the risk of short circuiting electrical systems or damaging electronic components. Liquefied industrial gases are supplied in ELF, dewar and ISO containers. ELF tank has a capacity ranging from 105 m^3 to 132 m^3, depending on the type of gas. Dewar tank contains liquid nitrogen and varies in capacity. ISO container is a liquid gas tank that is permanently mounted on a frame and is used in multimode transportation. Material safety data sheets provide information about the composition, characteristics and potential hazards of industrial gases plus precautions that should be taken and tips for safe handling.

History of Industrial Liquefied Gas Industry

Various researchers living in several different countries beginning in the later half of the eighteenth century discovered the gases that make up the multi-billion-dollar industrial gas industry. Nitrogen was isolated in 1772 by Daniel Rutherford (1749-1819), a British physician; in 1776, it was identified as an elemental gas by the great French chemist Antoine-Laurent Lavoisier (1743-1794). Also about 1776, oxygen was discovered by two chemists working independently in Europe; the English scientist Joseph Priestley (1733-1804) and Swedish chemist Carl Wilhelm Scheele (1742-1786) share credit for the discovery. During the late 1800s, it was used for medical purposes and put to commercial use in welding. Oxygen was also used to generate limelight for theaters and music halls. Acetylene was discovered in 1863 and first produced commercially in 1892. In 1897, Georges Claude (1870-1960), a French researcher, developed a method of dissolving acetylene in acetone at low pressures. Claude's process enabled the development of methods that allowed the movement of the gas via transportation cylinders. The first acetylene-burning torches were developed around 1900. In 1877, two researchers, Louis-Paul Cailletet (1832-1913) in France and Raoul Pierre-Pictet (1846-1929) in Switzerland, developed similar processes for the fractional distillation of liquid air. This procedure made it possible to produce large volumes of oxygen economically. In 1903 the Linde Air Products Company constructed the first commercial oxygen plant in the United States. Events of the early twentieth century demanded increasing amounts of industrial gases. World War I required large amounts of oxygen and acetylene for welding; during World War II pilots of high-altitude aircraft needed oxygen for their flights. Following the wars, researchers used inert gases such as argon and helium in electric arc welding. Growing industrialization in the western world brought rapid expansion to the gas industry. Oxygen demand continued growing through the 1950s as steel manufacturers turned to the gas to improve production methods. Maturing uses for nitrogen, previously considered a waste material, developed during the 1960s, along with advances in the uses of helium and argon. The 1970s brought large-scale expansions in the capacity to produce industrial gases. The decade also saw growth in the use of specialty gases by the electronics industry. By the mid-1980s, the electronics industry used an estimated 15% of the industry nitrogen output.

23.4 Cryogenics

Industrial Liquefied Gases: Industry Scenario

The global market for industrial gases experienced $ 29 billion in sales 1996 and is expected to reach more than $ 60 billion by 2010. Because of problems related to the transportation and storage of gas products, most production occurred close to its point of use. There was, therefore, very little international trade in industrial gases. Instead of transporting products, large international corporations functioned by operating production facilities in many countries.

Industrial gases touch virtually every facet of life. The three major atmospheric gases—oxygen, nitrogen, and argon—are used in steel production. Oxygen enhances kiln firing to reduce brick-making costs. Liquid oxygen and liquid hydrogen fuel rockets. Nitrogen is used in brewing beer, recycling tires, and applying metallic finishes on toys. Ammonia is synthesized from nitrogen for use in fertilizers and urea. It is also important in the production of nitrous oxide (also known as laughing gas) that is used as an anesthetic in some types of surgery. Liquid nitrogen and liquid carbon dioxide are used to make plastic fittings for moldings, enhance oil recovery from wells, and enable solvent recycling. Argon contributes to stainless steel manufacturing and serves as a component in fluorescent lighting. The industrial gas industry differs from many other types of manufacturing because its raw materials are primarily extracted from the atmosphere. The two principal gases produced by the industry are nitrogen and oxygen. Dry air is composed of 78.1% nitrogen, 20.9% oxygen, and just less than 1% argon. All other atmospheric gases, often called rare gases, make up the remaining one-tenth of 1%. Additional industrial gases such as hydrogen, acetylene, and carbon dioxide are obtained as co-products or by-products of other operations. Production costs within the industry are divided among labor, energy, and distribution.

The industry uses three different techniques to separate gases from the atmosphere. Cryogenic methods are the oldest and most widely used. Cryogenic separation relies on cooling and pressurizing the air until it becomes liquid. Oxygen, when held at a pressure of 80 pounds per square inch, liquefies at minus 274 degrees Fahrenheit; nitrogen liquefies at a colder temperature. As the atmospheric gases liquefy, they are extracted by means of a distillation process. Additional distillation steps are necessary to produce argon and other rare gases such as krypton and xenon. Helium liquefies only at temperatures approaching absolute zero. As a result, cryogenic production is not economically feasible for helium. Most commercially available helium is derived from natural gas rather than from the atmosphere. Two non-cryogenic gas production methods are membrane separation and pressure swing absorption (PSA). Membrane separation uses hollow fibers, most frequently made of organic polymers, to recover gases such as hydrogen from oil refineries or carbon dioxide from natural gas supplies. Pressure swing absorption (PSA) relies on a molecular sieve material that selectively absorbs atmospheric components at specific temperatures and pressures. Figure 23.1 shows a typical industrial gas plant.

Fig. 23.1 Industrial Gas Plant

The industrial gas industry is divided into two major segments. The first, called the "tonnage" or "supply scheme" market, is composed of large-volume users who usually receive gas via a direct pipeline from an on-site production facility. Under typical on-site contracts, a gas supplier constructs a production plant at or adjacent to a gas user's facility. The gas supplier owns and operates the plant for the benefit of the gas customer. Within this market segment, gas sold is measured in terms of tons per day. Examples of customers who routinely purchase industrial gases on the tonnage market include chemical, petroleum, electronics, and steel manufacturers. The other major market segment is known as the "merchant" or "bulk liquid" market. Customers within this market generally have fluctuating demand rates or operate multiple facilities in scattered locations. They often purchase gas products under short-term contracts of less than five years in duration. Suppliers deliver liquid gas in cryogenic tanker trucks or by rail. Gases are shipped and stored in liquid form because of volume constraints. For example, liquid oxygen takes up less than 1 percent of the space required to contain the same amount in a gaseous state. Examples of customers in this category include the metal, food processing, electronics, chemical, aerospace, plastics, medical, glass, and paper industries. A third, but much smaller market segment, consists of cylinder gas deliveries. Cylinder gas shipments are generally limited to expensive specialty gases and mixtures. A typical tanker truck carries the equivalent of 1,600 large cylinders. A train of ten cars carries the equivalent of 57,000 cylinders.

The industrial gas market is dominated by several suppliers. In the mid-1980s, Union Carbide was the largest industrial gas supplier in the United States. It provided approximately one-third of the nation's merchant gas. In 1985 the company opened six new nitrogen plants, with most of its production capacity aimed at the fast-growing high-tech market. In 1992, Union Carbide's industrial gas unit was spun off to become an independent entity, Praxair, Inc., which had revenues of $4.4 billion in 1996. Another major producer was Air Products and Chemicals, Inc. Founded in 1940, Air Products pioneered on-site industrial gas manufacturing. In 1991, the company introduced small volume, low cost, non-cryogenic nitrogen for use by metal heat-treating firms. The firm's industrial gas segment was growing less rapidly than its chemical and related businesses. Liquid Air Corporation, a subsidiary of the French company L'Air Liquide, entered the U.S. market in 1968. By the mid-1980s L'Air Liquide operated in 66 countries. Liquid Air is the company's headquarters for its operations in North and South America. It supplies products including oxygen, nitrous oxide, hydrogen, nitrogen, specialty gases, chemical gases, and rare gases to a wide variety of industrial users. Another international gas producer with a strong presence in the U.S. market is the BOC Group, which derived more than half of its worldwide sales in 1996 from industrial gases. The BOC Group originated in England with the incorporation of the Brins Oxygen Company Limited in 1886. BOC acquired the American company Airco in 1978. By the mid-1980s Airco provided 20% of the U.S. domestic merchant gas. BOC's expansion continued; by the late 1990s, the company operated units on all continents except Antarctica.

During the 1990s, pollution abatement was one of the most rapidly developing areas of study within the industrial gases industry. Waste water treatment was successfully been improved by oxygen injection, and oxygen is also used in hazardous waste incineration. Large quantities of oxygen and hydrogen are consumed in the production of directly reduced iron, which replaces scrap metal resources in producing steel for electric-arc furnaces. Recovery systems using nitrogen to condense and recapture solvents and chemical vapors helped manufacturers come into compliance with the Clean Air Act Amendments of 1990. An innovative technology based on carbon dioxide offered promise for reducing the environmental impact of solvent use within the paint and coatings industry. Additionally, carbon dioxide-based refrigeration systems were introduced to replace systems that relied on chlorofluorocarbons (CFCs). Research into new or refined uses for industrial gases also continued. Liquid nitrogen was being considered as a possible aid in reducing problems associated with cracking in structural concrete. Xenon provided sun-like brightness to meet the special lighting needs of airports, stadiums, the motion picture industry, and copying machine manufacturers. Other rare gases were also being developed for use in diagnostic technologies and pharmaceutical applications.

Types of Industrial Gases

Industrial gases can be classified as 1) Purified Gases 2) Mixture Gases 3) Specialty Gases.

1. **Purified Gases:** A wide range of purified and ultra purified gases area available to meet customer needs and requirements. The high purity gases available are Nitrogen, Helium, Oxygen, Carbon dioxide and Hydrogen. All the purified and ultra purified gases can be supplied in individual high pressure cylinder ranging from 10 liters to 50 liters size. Each pure gas is distinguished by stringent high-purity specifications, from leading-edge research and development through routine quality assurance and production control. Pure gases are guaranteed to be free of critical contaminants that can cause instrument interference and

chromatographic capillary column degradation. Consistent high-purity is achieved through extensive gas cylinder preparation, dedication of gas cylinders by gas service, and statistical quality control. Use of pure gases optimizes analytical results, regardless of application, and significantly extends instrument life. Applications of purified gases are:

- Carrier and detector pure gases for GC and GC-MS
- Pure gas and liquid hydrocarbon feedstock
- VOC-free pure gases for ppb measurements
- Mobile phases for supercritical fluid chromatography and extraction
- Pure gases and liquids for reaction chemistry and chemical processing
- Pure gases for elemental analysis
- Instrument zero gases
- Inert pure gases for blanketing
- Flame ionization fuel gas mixtures
- Thermal conductivity pure gas mixtures.

Pure gases can be divided into five categories 1) Research grade with purity: 99.9995%, 2) Ultra high purity grade with purity 99.999 %, 3) Zero grade with purity 99.998 %, 4) Pre-purified grade with purity 99.996 %, 5) High purity grade with purity 99.98 %.

The common practice of specifying purity as 99.999% (5 nines), 99.9999% (6 nines), and so on. The problem with 9's is that there is no standard way of deriving these numbers The procedure varies from product to product and from vendor to vendor. Gases must be pure enough so that the remaining impurities don't interfere with analysis or damage the equipment. The nature of the remaining impurity matters. There are several ways to manipulate gas purity levels–one of the easiest is to simply limit the number of analyzed contaminants to arrive at a target purity.

2. **Mixture Gases:** To meet the requirements of the customer, gas mixtures can be blended by gravimetric and manometric (pressure) method in one cylinder according to the tolerance and accuracies required. All the mixtures gases are documented in quality assurance manual and are traceable. With stringent quality control production and supply various types of mixtures gas can be done for following applications:
    - Medical and pharmaceutical application
    - Calibration gas mixture for an analytical application
    - Environmental compliance mixtures
    - Laser gas mixtures
    - Electronics gas mixtures
    - Leak detection mixtures
    - Petrochemical Engineering and Calibration mixtures
    - Chemical and Fertilizer Industry standard mixtures.

Three grades of mixture gases are available covering most specification requirements. Primary standard, certified standard, and unanalyzed mixture are the three grades.

Primary Standards: These should be used when the application demands the highest mixture accuracy and reliability. Primary Standards are filled gravimetrically on a high-load, high-sensitivity scale, calibrated with NIST Class S weights. Gravimetric blending offers the closest tolerance available, sometimes better than some methods of analysis. A dual verification of mixture accuracy is also performed by routine quality control analysis.

Certified Standards: These, sometimes referred to as working standards, are analyzed calibration mixtures used routinely in science and industry. For the majority of applications, the tolerance of a Certified Standard is acceptable. These standards are generally prepared either by partial pressure or gravimetrically. Certification of the standard is usually done through quality control analysis.

Unanalyzed Mixtures: Although prepared by the same techniques as Primary and Certified Standards, Unanalyzed Mixtures are not verified or checked by analysis. These mixtures should only be used in applications where the accuracy is not an issue.

Two tolerances are associated with all primary and certified standards. First is the blend or preparation tolerance. This is the minimum acceptable uncertainty associated with the actual production of the blend. These uncertainties are accumulated during the manufacturing process because of the inherent inaccuracies of equipment used in production, such as pressure gauges, and the inherent inaccuracies due to the physical properties of the gases. Second is the analytical or certification tolerance, which is the minimum acceptable uncertainty associated with the analysis of the blend. This uncertainty is accumulated throughout the analytical procedure and includes instrument and calibration errors. For most applications, the certification tolerance is of greater importance than the preparation tolerance because it represents the range in which the true or actual concentration may be in relation to the analyzed concentration. For some applications, such as those that require an upper or lower range of concentration that cannot be exceeded, the preparation tolerance becomes equally if not more important.

3. **Specialty Gases:** Specialty gases are used for specific applications and tailor made to suit customer requirements. Specialty gases are used for all analytical, laboratory, medical, pharmaceutical or biotechnology applications with highest levels of purity and consistency. These gases are available in gas cylinder ranging from 2L to 133L cylinder or in bulk liquid tanks. The commonly used specialty gases are Xenon, Neon, Propane, Propylene, Ethylene, Ethane and Butane.

Let us now study some specific industrial gases in detail.

Industrial Gas: Liquid Oxygen

Oxygen (O_2) constitutes approximately 21% of air, has a gaseous specific gravity of 1.105, that makes it slightly heavier than air, and has a boiling point of $-183°C$.

Upon liquefying it becomes a transparent, pale blue liquid that is slightly heavier than water. Oxygen is also colorless, odorless and tasteless and has poor solubility in water. The principal uses of oxygen are indicative of its strong oxidizing and life-sustaining properties. It's used in medicine for therapeutic purposes and used in treatment of patients with respiratory disorders. In the chemical and petroleum industries, oxygen can be used as a chemical reactant and in combustion systems to have increased reaction rates and reduced air emissions. Furthermore, in acid sulfuric mill, oxygen is used to alter the structure of feed stocks through oxidation, producing nitric acid and other chemicals. Oxygen is used in the pulp and paper industry for a variety of applications, including pulp bleaching, black liquor oxidation and lime kiln enrichment. Oxygen is frequently employed in iron, copper and lead blast furnaces, iron melting foundry cupolas to increase productivity by increasing combustion temperatures and reduce the specific coke consumption. Oxygen is also used to support oxyfuel cutting operations. Oxygen/fuel combustion is used in the glass industry to reduce particulate and Nitrogen Oxide (NOx) emissions in melting operations. In waste water/water treatment, oxygen can be used in place of air in the activated sludge process to maintain a higher population of microorganisms and reducing odor.

Industrial Gas: Liquid Nitrogen

Nitrogen (N_2) makes up 78.03% of air, has a gaseous specific gravity of 0.967 and a boiling point of –195°C at atmospheric pressure. It is colorless, odorless, non-toxic and chemically inert. Nitrogen is non-flammable and capable of suppressing combustion processes. In addition, as an inert gas it has an asphyxiating effect, as it displaces the oxygen required for breathing. Nitrogen has numerous applications in the industrial and research sectors because of its non-reactive nature with many materials. In the chemical and petroleum industries, gaseous nitrogen applications include storage tanks, purging vessels and pipelines of flammable or toxic gases and vapors, and the sparging and pressure transfer of liquids. While in electronics field, it is used in the inerting of printed circuit board reflow and wave ovens, to prevent oxidation. In metals industry, nitrogen is used in iron and steelworks applications for degassing and metal-stirring, purging, cooling, slag splashing, gas knives and as a carrier gas for injecting de-sulphurising compounds. Nitrogen is also used as an assist gas for laser cutting of stainless steel, aluminum and non-metallic materials. Food industry uses liquid nitrogen for freezing food. In addition, Nitrogen is used in brewing, soft drinks and wine-making industries to exclude air from the product. Medical industry uses liquid nitrogen in cryosurgery, where nitrogen is used to safely remove skin lesions such as warts, by the use of spray tips or probes. Furthermore, liquid nitrogen can be used to store medical sample such as blood, plasma and semen. Another interesting nitrogen applications is shrink-fitting where liquid nitrogen is used to shrink components so they are small enough to be inserted into another component.

Industrial Gas: Liquid Argon

Argon is chemically inactive gas and belongs to the family of inert gases. It is 1.38 times heavier than air. It is odorless, tasteless, colorless, non-toxic, non-combustible, non-corrosive and does not react with other elements or chemical compounds. Argon is widely used in incandescent lamp industry or the filling of light bulbs and fluorescent lamps. Furthermore, it also used with other rare gases in filling of special bulbs and display tubes for lighted signs. It is used to provide an inert gas shield for arc welding to prevent oxidation of the metal being welded. In plasma gas cutting, argon is heated to high temperatures used for cutting operation and for coating metals with refractory materials. In addition, argon also used in refining of stainless steel, refining of melted aluminum, silicon ingot production, semiconductor fabrication, powdered metal fabrication and as a carrier gas for various analytical instruments.

Industrial Gas: Liquid Acetylene

Acetylene is a compound of carbon and hydrogen. It is a colorless flammable gas with a garlic-like odor. It is slightly lighter than air. Its combustion with oxygen produces a flame with temperature of approximately 3150°C, allowing it to be used in welding, cutting, brazing, heating and soldering metals. The oxy-acetylene torch can be used to repair ships underwater, to construct bridges, pipelines, dams, tunnels, buildings and to reinforce concrete. When used with either oxygen or air, acetylene can produce a thin layer of black carbon particles. This can prevent components sticking in their moulds and help on mould release. In laboratory, acetylene is used as an instrumentation gas in optical spectrometry and in atomic absorption for elemental analysis.

Industrial Gas: Solid Carbon Dioxide

Carbon dioxide (CO_2) is a nonflammable, colorless, odorless and liquefied gas. Carbon dioxide gas is relatively nonreactive and nontoxic. It exists simultaneously as a solid, liquid and gas at a temperature of –56.6°C and a pressure of 416 kPa. At the temperature of –79°C and 1 atmospheric pressure, carbon dioxide is solidified and forms the "dry ice" at a density of 97.4 pounds per cubic foot. Carbon dioxide is widely used in food and beverages industry. It is the source of the bubbles in soft drinks and other carbonated beverages. As a natural anti-microbial, carbon dioxide is also used to increase the shelf life of juice and dairy products, protecting taste and texture, and reducing the need for preservatives. In addition, as a "dry ice" it also used in food freezing and chilling and packaging, mixer and blender cooling, and for transit refrigeration. Carbon dioxide is commonly used as a shielding gas during welding because it prevents atmospheric contamination of molten weld metal. It is also used for CO_2 lasers in welding and cutting. Carbon dioxide also plays an important role in many industrial process such as grinding sensitive material, rubber tumbling, cold-treating metals, shrink fitting of machinery parts and industrial cleaning by blasting and polishing. It is used to fill certain types of fire extinguishers that rely on its inert properties, density, and low temperature when released from high-pressure storage. Injection of carbon dioxide allows controlling the acidity of liquid effluents.

Industrial Gas: Liquid Hydrogen

Hydrogen (H_2) is a colorless, odorless, tasteless, flammable gas and the lightest gas ever known. Hydrogen finds use in diverse applications covering many industries. Large quantities of hydrogen is used in chemical syntheses mainly to manufacture ammonia and methanol, furthermore it also used to hydrogenate non-edible oils for soaps, insulation, plastics, ointments and other special chemicals. In food industry hydrogen is used to hydrogenate edible oils such as soybean, fish, cottonseed and corn, converting them to semisolid materials such as shortenings, margarine and peanut butter. Hydrogen is used with oxygen in oxy-hydrogen welding and cutting in certain brazing operation, welding aluminum and magnesium especially thin sections. In addition, it commonly mixed with argon for welding austenitic stainless and also used to support plasma welding and cutting operations. Hydrogen is a carbon-free energy source used in the fuel cells that create electricity through an electrochemical process in combination with oxygen. Hydrogen is used as carrier gas in semiconductor processes, especially for silicon deposition. Hydrogen is used for desulfurization of gasoline from crude oil in petroleum recovery and refinery. For power generation, large electrical generators are sometimes run in a hydrogen atmosphere to reduce windage losses and remove heat.

Industrial Gas: Liquid Helium

Helium (He) is a colorless, odorless, tasteless inert gas at room temperature and atmospheric pressure. It is the second-lightest element found in earth with the gaseous specific gravity of 0.138. Its boiling point is –268.9°C at atmospheric pressure. It is a rare gas in the atmosphere. Helium is widely used in the welding industry as an ideal shielding gas for arc welding of stainless steel, aluminum, copper, nickel, titanium and many alloys. It is also extensively used in filling of balloons for upper atmosphere and cosmic studies, weather forecast and advertisement purpose. Helium is the most commonly gas used as carrier in gas chromatography. Helium's molecules are extremely small, making it a perfect leak detection gas to ensure no leaks in pressure and vacuum systems. Liquid helium is also used as a cooling fluid in superconducting magnet applications for the MRI. A mixture of helium and oxygen finds use as a breathing gas in deep-sea diving. Besides, it also used for anesthesia of patients suffering from asthma.

References

1. Internet website http://www.sig.net.my.
2. Cryogenic Systems – Randall Baron.
3. Internet website http://www.uigi.com.
4. Cryogenic Engineering – P.K. Bose.
5. Internet website http://www.answers.com/topic/industrial-gases.
6. Internet website http://hiq.linde-gas.com.
7. Internet website www.mathesontrigas.com.
8. Internet website www.xpresscryogenics.com.

Questions

Q.1. What are industrial gases? Give at least three examples.

Q.2. Explain any five applications of liquefied industrial gases.

Q.3. Write a short note on types of industrial gases.

Q.4. Discuss the current scenario of industrial gas industry

Q.5. Write a short note on history of industrial gases.

CHAPTER 24

Applications of Cryogenics - II
Biological and Medical Applications

INTRODUCTION

Cryobiology studies the effects of low temperatures on living things. The word cryobiology is derived from the Greek words "cryo" = cold, "bios" = life, and "logos" = science. In practice, cryobiology is the study of biological material or systems at temperatures below normal. Materials or systems studied may include proteins, cells, tissues, organs, or whole organisms. Temperatures may range from moderately hypothermic conditions to cryogenic temperatures. Cryonics is the low temperature preservation of humans and animals in a damaged state with the intention of future revival. Cryonics is not part of mainstream cryobiology because cryonics still depends heavily on speculative future technology.

The use of cryogenics to produce low-temperature environments in the study of living plants and animals is common. The principal effects of cold on living tissue are destruction of life and preservation of life at a reduced level of activity. Both of these effects are demonstrated in nature. Death by freezing is a relatively common occurrence in severe winter storms. Among cold-blooded animals winter weather usually results in a coma like sleep that may last for a considerable length of time. In cryobiological applications much lower temperatures are used than are present in natural environments. The extreme cold of liquid nitrogen (boiling at −320°F or −196°C) can cause living tissue to be destroyed in a matter of seconds or to be preserved for years and possibly for centuries with essentially no detectable biochemical activity. The result achieved when heat is withdrawn from living tissue depends on processes occurring in the individual cells. Basic knowledge of the causes of cell death, especially during the process of freezing, and the discovery of methods which circumvent these causes have led to practical applications both for long-term storage of living cells or tissue (cryopreservation) and for calculated and selective destruction of tissue (cryosurgery).

The major areas of study in cryobiology are as follows:
1. Study of cold-adaptation of microorganisms, plants and animals.
2. Cryopreservation of cells, tissues, gametes, and embryos of animal and human origin for long-term storage by addition of cryoprotectants which protect the cells during freezing and thawing.

3. Preservation of organs under hypothermic conditions for transplantation or dead bodies also known as Cryonics.
4. Freeze-drying of pharmaceuticals.
5. Cryosurgery, a (minimally) invasive approach for the destruction of unhealthy tissue using cryogenic gases/fluids.

History of Cryobiology

As early as in 2500 BC low temperatures were used in Egypt in medicine. The use of cold was recommended by Hippocrates to stop bleeding and swelling. Robert Boyle studied the effects of low temperatures on animals. In 1949 sperm was cryopreserved for the first time by Polge. This led to a much wider use of cryopreservation, with many organs, tissues and cells routinely stored in low temperatures. Human sperm, eggs and embryos were routinely stored in fertility research and treatments. In the early part of twenty first century a baby was born from a cryopreserved egg fertilized by a cryopreserved sperm.

Human Cryobiology

Cryobiology in humans with regards to infertility is also called vitrification which is a process in which a woman's eggs are flash frozen in a matter of seconds. Freezing the eggs at such a rapid rate greatly improves their survival rate and prevents ice crystals from forming in the egg that could damage the DNA. Once the couple is ready to conceive the eggs are thawed and through the process of Invitiro Fertilization (IVF), the sperm is placed with the eggs and together they are placed back into the uterus. Vitrification has its glitches and is not as reliable as freezing fertilized eggs or embryos because eggs alone are extremely sensitive to temperature. Many researchers are also freezing ovarian tissue in conjunction with the eggs in hopes that the ovarian tissue can be transplanted back into the uterus, stimulating normal ovulation cycles.

Cryopreservation

Cryopreservation as an applied science is primarily concerned with low temperature preservation in the −80°C to −196°C temperature range. Organs, and tissues and single cells have been the most common objects cryopreserved. The earliest commercial application of cryopreservation was in the storage of animal sperm cells for use in artificial insemination. The microorganisms used in cheese production can be frozen, stored, and transported without loss of lactic acid–producing activity. Pollen from various plants can be frozen for storage and transport, facilitating plant-breeding experiments. Among the most valuable applications of cryopreservation is the storage of whole blood or separated blood cells.

Cryopreservation of cells is guided by the "Two-Factor Hypothesis" of American cryobiologist Peter Mazur, which states that excessively rapid cooling kills cells by intracellular ice formation and excessively slow cooling kills cells by either electrolyte toxicity or mechanical crushing. During slow cooling ice forms extracellularly, causing water to osmotically leave cells, thereby dehydrating them. Intracellular ice can be much more damaging than extracellular ice. For red blood cells the optimum cooling rate is very rapid (nearly 100°C per second), whereas for stem cells the optimum cooling rate is very slow (1°C per minute). Cryoprotectants, such as DMSO (dimethyl sulfoxide) and glycerol, are used to protect cells from freezing. Cryobiologists attempt to optimize cryoprotectant concentration (minimizing both ice formation and toxicity) as well as cooling rate. Cells may be cooled at an optimum cooling rate to a temperature between −30°C and −40°C before being plunged into liquid nitrogen. Cryobiologists are also increasingly using mixtures of cryoprotectants for full vitrification (zero ice formation) in preservation of cells, tissues and organs. Vitrification methods pose a challenge in the requirement to search for cryoprotectant mixtures that can minimize toxicity.

Cryonics

Cryonics is the preservation of legally dead humans or pets at very low temperature (below −200°F, −130°C) in the hope that future science can restore them to life, youth and health. Cryonicists are people who use or advocate cryonics to greatly extending life and youth. Most diseases, including the progressive deterioration known as "getting old", are the result of damage to organs, tissues, cells and cellular components. With enough progress of medicine and molecular repair capability, all diseases should eventually be curable, including aging. Medicine in the future should be able to restore and maintain people in a condition of youth and health. Cryonics could be a lifeboat (or "first aid") to future medicine. Because it is based on speculation about the capabilities of future science cryonics is not a science. Figure 24.1 shows a typical cryonics facility.

Fig. 24.1 Cryonics Facility

24.4 Cryogenics

Legally, cryonics can only be applied to a person who has been pronounced dead by an authorized health professional. But the criterion for death is nearly always the cessation of heartbeat. Almost all cells of the body, including those in the brain are still alive when death is pronounced. The main damage within the first hour is due to clotting and blood vessel injury. It usually takes many hours for all cells to die at room temperature, including those in the brain. Cryonicists do not want to wait hours or days before starting the process of cooling. If all of the tissues in the body can be preserved close to the condition they were in immediately following cessation of heartbeat, deterioration is minimized. Cryonics procedures involve replacing body water with anti-freeze mixtures called "cryoprotectants". By perfusing these biologic anti-freezes through blood vessels for a few hours most body water can be removed and replaced by cryoprotectant. At very low temperatures these cryoprotectants harden like glass, without forming damaging ice crystals. This glass formation is known as vitrification and is an important part of the efforts of cryonicists to preserve people and pets in the best possible condition to maximize the chance of future youthful life. Although cooling a human or animal to below $-200°F$ can potentially preserve them unaltered for thousands of years, this process can cause additional damage to that caused by aging and disease, such as damage due to thermal stress, cryoprotectant toxicity and even freezing damage when cryoprotectant perfusion is poor.

Cryosurgery

Cryosurgery (cryotherapy) is the application of extreme cold to destroy abnormal or diseased tissue. Cryosurgery has been historically used to treat a number of diseases and disorders, especially a variety of benign and malignant skin conditions. Cryosurgery was first carried out by James Arnott in 1845 in an operation on a patient with cancer. Although not very widespread, cryosurgery has its benefits. Cryosurgery works by taking advantage of the destructive force of freezing temperatures on cells. At low temperatures, ice crystals form inside the cells, which can tear them apart. More damage occurs when blood vessels supplying the diseased tissue freeze

Cryosurgery is a minimally invasive procedure, and is often preferred to more traditional kinds of surgery because of its minimal pain, scarring, and cost; however, as with any medical treatment, there are risks involved, primarily that of damage to nearby healthy tissue. Damage to nerve tissue is of particular concern. Warts, moles, skin tags and small skin cancers are candidates for cryosurgical treatment. Several internal disorders are also treated with cryosurgery, including liver cancer, prostate cancer, lung cancer, oral cancers, cervical disorders and, more commonly in the past, hemorrhoids. Refer Figure 24.2 for a typical skin treatment with cryosurgery.

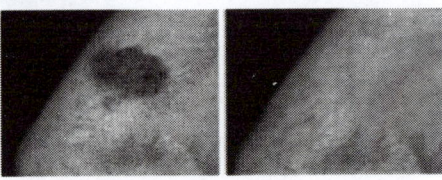

Fig. 24.2 Skin Treatment by Cryosurgery

The most common method of freezing lesions is using liquid nitrogen as the cooling solution. The super-cooled liquid may be sprayed on the diseased tissue, circulated through a tube called a cryoprobe, or simply dabbed on with a cotton or foam swab. Less frequently, doctors use carbon dioxide "snow" formed into a cylinder or mixed with acetone to form a slush that is applied directly to the treated tissue. Recent advances in technology have allowed for the use of argon gas to drive ice formation using the Joule-Thomson effect.

One of the significant advantages of cryosurgery is that the apparatus can be employed to cool the tissue to the extent that the normal or the aberrant function is suppressed; yet at this stage the procedure can be reversed without permanent effect. A second major advantage of cryosurgery is that the advancing front of reduced temperatures tends to cause the removal of blood and the constriction of blood vessels in the affected area. This means that little or no bleeding results from cryosurgical procedures. A third major advantage of cryosurgery is that cryosurgery equipment currently employs a freezing apparatus that can be placed in contact with area to be destroyed with a minimum incision to expose the affected area.

The various cryogens used for cryosurgery are Nitrous oxide (–75°C), Carbon-dioxide snow (–79°C) and liquid Nitrogen (–196°C). Cryocan is a container that stores liquid nitrogen and cryojet is a hand held unit that is used to spray the liquid nitrogen over the diseased part. Liquid nitrogen has a temperature of (–196°C) and freezes tissues to (–60°C). Refer Figure 24.3 for a photograph of a cryocan and a cryojet. The procedure involves spraying liquid nitrogen over the diseased part leading to freezing of the tissues. This causes a stinging sensation and numbness initially. Later on there may be varying degree of pain, which may need analgesics.

Fig. 24.3 Cryocan and a Cryojet

Cryogenic Medical Gases

Modern anesthetic practice is almost totally dependent on medical gases like oxygen which provide a dimension of safety that makes possible some of the complicated medical maneuvers required by modern surgery, and without which the risks would be too great to justify the method. Of equal value is the use of liquefied oxygen gas in intensive care units where it provides the main life-saving support, particularly for patients whose lungs need to be artificially ventilated if they are to survive. The use of oxygen in chronic respiratory and cardiac conditions is very often essential and it is needed for the relief of all forms of hypoxia. In short, modern medicine could not be practiced without the support of cryogenic medical gases. Liquefied medical gases are widely used in clinical practice to provide a basis for virtually all modern anesthetic techniques as well as pre and post-operative management.

Liquefied Medical Oxygen

Oxygen is an essential ingredient for aerobic life. It is present in the atmosphere at 21% and is an absolute necessity for life. Supplementary oxygen is often required for clinical purposes – and is supplied in gaseous or liquid form. Medical oxygen is used to restore the tissue oxygen tension towards normal by improving oxygen availability in

- Cardiac and respiratory arrest;
- Resuscitation of critically ill when circulation is impaired
- Cyanosis of recent origin as a result of cardio-pulmonary disease
- Surgical trauma, chest wounds and rib fractures
- Hyperpyrexia
- Carbon Monoxide poisoning
- Shock, severe haemorrhage and coronary occlusion
- Neonatal resuscitation

Now-a-days oxygen concentrators are available which are designed to meet the most comprehensive long-term oxygen therapy requirements. Their advantages are low operation cost; easy to operate, safe, low power consumption; quiet operation and low maintenance cost. Figure 24.4 shows an oxygen concentration unit from BOC.

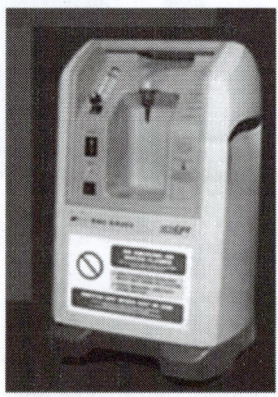

Fig. 24.4 Oxygen Concentration Unit

Liquefied Medical Oxygen Plant

These plants work on the Pressure Swing Adsorption technique. This technology has been proven all over the world for its working efficiency and trouble free operation. Figure 24.5 shows a typical PSA medical oxygen plant. The operation of the plant consists of the following steps.

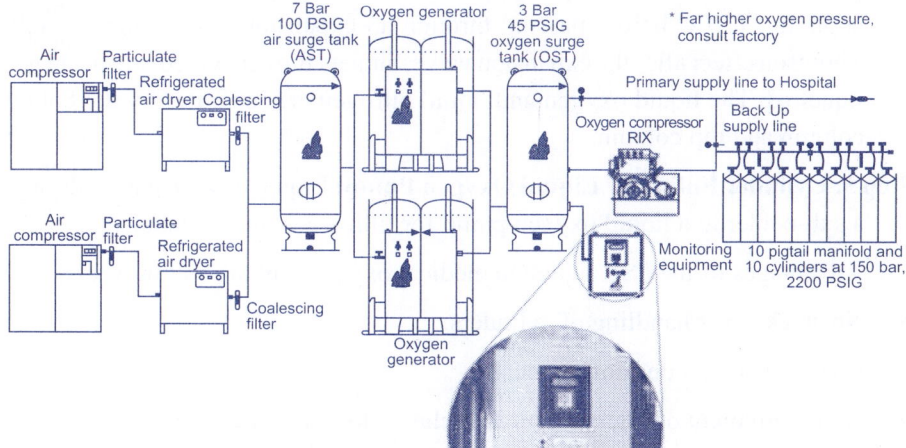

Fig. 24.5 PSA Medical Liquefied Oxygen Plant

Step 1: Compression of Air: The free saturated air is sucked from atmosphere through a highly efficient dry-type suction filter into the first stage of the horizontally balanced opposed, lubricated reciprocating air compressor.

Step 2: Purification of Air: This consists of purification of the air by removing moisture, oil traces and carbon dioxide in the process air. Compressed air is chilled to 12°C in a chilling unit and evaporation cooler, and then passes to a moisture separator, where the condensed moisture gets removed before entering into a molecular sieve. Before sending the air to the sieve, air is passed through an oil absorber where air becomes oil free. Chilled air passes through the Molecular Sieve consisting of twin towers to remove moisture and carbon dioxide present in the air. Any dust particle gets filtered in the dust filter before air enters the air separation column.

Step 3: Expansion of Air: The process air before liquefaction in the air separation unit needs to be cooled to cryogenic temperatures. The main portion of the air after the purification stage enters the expansion engine through the heat exchanger after pre-cooling. The temperature of the air drops to around –165°C by the expander, which is an efficient design with Teflon piston rings and hydraulic mechanism with leak proof ball valves. Rest of air at –80°C from heat exchanger enters into an expansion engine, where the air further gets cooled down to –150°C before entering into the bottom column. The liquefied air from both these streams, which is collected at the bottom column is termed as rich liquid.

Step 4: Separation of Air: After the expansion, the air enters the air separation unit (cold box), where the air converts into liquid air by deep cooling at cryogenic temperatures and is separated into liquid oxygen and nitrogen. Chilled, oil-free and moisture-free air enters into multi-pass heat exchanger where it gets cooled to –80°C by cold gained from outgoing waste nitrogen and oxygen. A part of air, enters a multi-pass liquefier made of special alloy tubes. This air cools to –170°C before passing through an expansion valve. Due to Joule-Thomson effect after the expansion valve, air gets further cooled down and gets liquefied. The liquid oxygen and liquid nitrogen are separated by the bottom column and top column.

Step 5: Cylinder Filling by Liquid Oxygen Pump: Liquid oxygen passes through a sub-cooler to a liquid oxygen pump for filling gas into cylinders.

Advantages of the PSA liquefied medical oxygen unit are as follows:

- No stocking or handling of cylinders.
- Gas delivery at constant pressure.
- No requirement of high pressure regulators to reduce cylinder pressure.
- Loss of gas due to residual pressure of cylinders is eliminated.
- Higher stock capacity.
- High purity is ensured since the gas is in liquefied form.

References

1. Cryogenic Engineering – B.A. Hands.
2. Cryogenic Systems – Randall Baron.
3. Internet website http://www.ogsi.com.
4. Nathan Hill – ESA publication on Cryogenics - 2008.
5. Internet website http://www.psaoxygenplants.com.
6. Internet website http://www.wikipedia.com/.
7. Cryogenic Engineering – P.K. Bose.
8. Internet website of BOC gases.

Questions

Q.1. What is Cryobiology ? Enumerate at least two applications.

Q.2. Distinguish between cryonics and cryo-preservation.

Q.3. Write a short note on PSA liquefied medical oxygen plant.

Q.4. Explain the need for cryosurgery. How is it beneficial?

Q.5. Describe the benefits of cryobiology for humans.

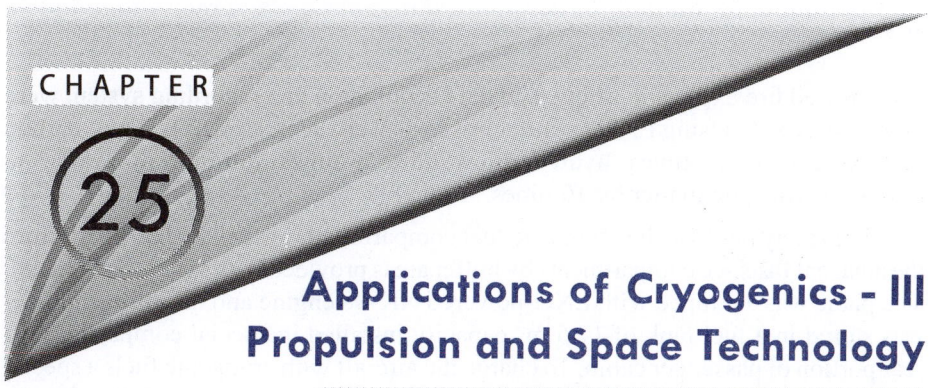

CHAPTER 25

Applications of Cryogenics - III Propulsion and Space Technology

INTRODUCTION

From the first rockets built by the Chinese over a millennium ago to the precision engines used by modern missiles, rocket engines work in accordance with Newton's Third Law of Motion: For every action there is an opposite and equal reaction. In a rocket engine, hot gas expelled at high velocity generates thrust in the opposite direction. The most common means of doing so uses chemical reactions to produce the hot gas. The first rockets used solid propellants, such as black powder, but they were very inefficient. Liquid-propellant rocket engines, first developed in 1926 by Robert H. Goddard, are much more powerful and opened the way to space flight. Thus the cryogenic era of propulsion and space technology was born.

Further a family of rocket engines were developed that burned kerosene and liquid oxygen (LOX) based on German V-2 rocket technology obtained after World War II. As these rockets were adapted to their new role as launch vehicles in the 1960s, still larger versions of their engines were built for the Saturn rockets that sent Apollo missions to the Moon. The newer designs included a distinctive stage-and-a-half design, which allows it to jettison a pair of booster engines when they were no longer needed, leaving a smaller sustainer engine to power the stage. Cryogenic fuels, mainly liquid hydrogen, have been used as rocket fuels. Liquid oxygen is used as an oxidizer of hydrogen, but oxygen is not, strictly speaking, a fuel. For example, NASA's workhorse space shuttle uses cryogenic hydrogen fuel as its primary means of getting into orbit, as did all of the rockets built for the Soviet space program.

Cryogenic Aircraft Development

Russian aircraft manufacturer Tupolev is currently developing a cryogenic plane TU-155 with a cryogenic liquefied natural gas or LNG fuel system. Initial efforts to use liquid hydrogen failed due to safety concerns. Natural gas turned to be of paramount importance in the energy program of Russia. The content of natural gas exceeded 50% of energy balance and hence LNG was used. This is how the first in the world Cryogenic Aircraft was built. The TU-155 a/c was built on the basis of serial TU-154B a/c. To use cryogenic fuel airframe and some standard systems were modified, cryogenic fuel charging, storage and feeding systems were installed that ensured fire/explosion safety, and data acquisition and recording system were modified as well. Using LNG aircraft emissions were decreased as follows: carbon monoxide about 10 times, hydrocarbons 2.5 to 3 times, nitrogen oxides 1.5 to 2 times, particulate matter by 10 times.

For safety purpose the cryogenic fuel compartment was isolated on board from the adjacent fuselage compartments by buffer areas provided with ventilation system. The plane was equipped with LNG-powered NK-88 engine and the cryogenic fuel was stored in a fuel tank of 17.5 m^3 capacity installed in special compartment in rear portion of passenger cabin. To charge the aircraft with cryogenic fuels a special charging module was developed which was located on rear side of the aircraft. Cryogenic fuel was delivered to the site by filling trucks. Some issues concerning components and systems of the cryogenic plane TU-155 a/c were studied on ground rigs. Creation of the aircraft was accompanied by serious scientific and research works and elaboration of large amount of regulatory documentation. Figure 25.1 shows the aircraft. After flight development and initial testing, on 18th January, 1989, the TU-155 a/c cryogenic plane performed its first flight on LNG fuel. The cryogenic engine, its components, fuel pumps, pressure maintenance system and cryogenic fuel tanks – all these were developed from scratch and were utilized in future developments. Appearance of TU-155 a/c changed dramatically the scope of cryogenic aviation. It was demonstrated in reality that LNG-powered aircraft can be operated with the same safety level than those working on aviation gasoline.

Fig. 25.1 The Cryogenic Aircraft TU-155 of Tupolev

Cryogenic Propulsion

Cryogenic propulsion era started with the use of liquid rockets. A liquid-fuel rocket is a rocket with an engine that uses propellants in liquid form. Liquids are desirable because their reasonably high density allows the volume and hence the mass of the tanks to be relatively low, resulting in a high mass ratio. Liquid rockets have been built as monopropellant rockets using a single type of propellant, bipropellant rockets using two types of propellant, or more exotic tripropellant rockets using three types of propellant. Bipropellant liquid rockets generally use one liquid fuel and one liquid oxidizer, such as liquid hydrogen and liquid oxygen.

The only known claim to liquid propellant rocket engine experiments in the nineteenth century was made by a Peruvian scientist named Pedro Paulet. Paulet described laboratory tests of liquid rocket engines, but did not claim to have flown a liquid rocket. The first flight of a vehicle powered by a liquid-rocket took place on March 16, 1926 at Auburn, Massachusetts, when American professor Robert H. Goddard launched a rocket which used liquid oxygen and gasoline as propellants. The rocket, which was dubbed "Nell", rose just 41 feet during a 2.5-second flight that ended in a cabbage field, but it was an important demonstration that liquid rockets were possible.

Unlike gases, a typical liquid propellant has a density similar to water, approximately 0.7-1.4g/cm^3 while requiring only relatively modest pressure to prevent vaporisation. This combination of density and low pressure permits very lightweight tankage. For injection into the combustion chamber the propellant pressure needs to be greater than the chamber pressure at the injectors; this can be achieved with a pump. Turbopumps are usually extremely lightweight and can give excellent performance. Indeed, overall rocket engine thrust to weight ratios including a turbopump have been as high as 133:1 with the Soviet NK-33 rocket engine. Liquid propellant rockets can be throttled in realtime, and have good control of mixture ratio; they can also be shutdown without any great drama, and, with a suitable ignition system, restarted. They can also employ regenerative cooling which uses the fuel to cool the chamber prior to injection.

Disadvantages of Cryogenic Propulsion

Use of liquid propellants can be associated with a number of issues:
1. As the propellant is a very large proportion of the mass of the vehicle, the center of mass shifts significantly rearward as the propellant is used causing loss of control.
2. Liquid propellants are subject to *sloshing*, which is undesirable movement.
3. Liquid propellants often need ullage motors in zero-gravity or during staging to avoid sucking gas into engines at startup.
4. Liquid propellants can leak, possibly leading to the formation of an explosive mixture.
5. Turbo pumps to pump liquid propellants are complex to design, and can suffer serious failure modes if they run dry.
6. Propellants are subject to vortexing within the tank, particularly towards the end of the burn, which can result in gas being sucked into the engine or pump.
7. Cryogenic propellants, such as liquid oxygen freezes atmospheric water vapour into very hard crystals. This can damage or block seals and valves and can cause leaks and other failures.

25.4 Cryogenics

8. Liquid rockets tend to be very complex, which increases the opportunities for malfunctions.
9. Liquid rockets require considerable preparation immediately before launch. This makes them less practical than solid rockets for most military applications.

Cryogenic Propellants

Thousands of combinations of fuels and oxidizers have been tried over the years as propellants. Some of the more common propellants are:

- Liquid oxygen (LOX) and liquid hydrogen (LH_2)
- Liquid oxygen (LOX) and kerosene
- Liquid oxygen (LOX) and ethanol (C_2H_5OH)
- Liquid oxygen (LOX) and gasoline
- Nitric acid (HNO_3) and kerosene
- Nitric acid (HNO_3) and dimethyl hydrazine $(CH_3)_2N_2H_2$
- Hydrogen peroxide and kerosene
- Hydrazine (N_2H_4) and red fuming nitric acid
- Aerozine 50 and dinitrogen tetroxide

One of the most efficient mixtures, oxygen and hydrogen, suffers from the extremely low temperatures required for storing hydrogen and oxygen as liquids (around 20 K) and low fuel density (70 kg/m³), necessitating large and heavy tanks. For storable ICBMs and interplanetary spacecraft, storing cryogenic propellants over extended periods is awkward and expensive. Because of this, mixtures of hydrazine and its derivatives in combination with nitrogen oxides are generally used for such rockets.

Cryogenic Injectors

Cryogenic injectors can be as simple as a number of small diameter holes arranged in carefully constructed patterns through which the fuel and oxidiser travel. The speed of the flow is determined by the square root of the pressure drop across the injectors, the shape of the hole and other details such as the density of the propellant. The first injectors used on the V-2 created parallel jets of fuel and oxidiser which then combusted in the chamber. This gave quite poor efficiency. Injectors today classically consist of a number of small holes which aim jets of fuel and oxidiser so that they collide at a point in space a short distance away from the injector plate. This helps to break the flow up into small droplets that burn more easily. The main type of injectors are

- Shower head type
- Self impinging doublet type
- Cross impinging triplet type
- Centrifugal or Swirling type

Other injector types include the pintle injector, which potentially permits good mixture control over a wide range of flow rates. To avoid instabilities such as chugging which is a relatively low speed oscillation the engine must be designed with enough pressure drop across the injectors to render the flow largely independent of the chamber pressure. This is normally achieved by using at least 20% of the chamber pressure across the injectors. Injectors are commonly laid out so that a fuel rich layer is created at the combustion chamber wall. This reduces the temperature there, and downstream to the throat and even into the nozzle and permits the combustion chamber to be run at higher pressure, which permits a higher expansion ratio nozzle to be used which gives a higher specific impulse and better system performance. Regenerative cooling is also often used in liquid rocket engines.

Ignition can be performed in many ways, but perhaps more so with liquid propellants than other rockets a consistent and significant ignitions source is required; failure to ignite within a few tens of milliseconds can cause overpressure of the chamber due to excess propellant. Generally, ignition systems try to apply flames across the injector surface, with a mass flow of approximately 1% of the full mass flow of the chamber. Safety interlocks are sometimes used to ensure the presence of an ignition source before the main valves open; however reliability of the interlocks can in some cases be lower than the ignition system. Methods of ignition include pyrotechnic, electrical and chemical methods.

Cryogenic Engine

Cryogenic engines are rocket motors designed for liquid fuels that have to be held at very low "cryogenic" temperatures to be liquid – they would otherwise be gas at normal temperatures. Typically hydrogen and oxygen are used which need to be held below 20°K (–423°F) and 90°K (–297°F) to remain liquid. The components of the engine have to be cooled to similar low temperatures so that the fuel doesn't boil to a gas in the lines that feed the engine. The thrust comes from the rapid expansion from liquid to gas with the gas emerging from the motor at very high speed. The energy needed to heat the fuels comes from burning them, once they are gasses. The engines on the NASA shuttles are cryogenic engines and they have a very high thrust to weight ratio. Currently only the US, Russia, China, France, Japan and India have mastered cryogenic engine technology.

Fig. 25.2 Cryogenic Rocket Engine

25.6 Cryogenics

The use of cryogenic engines was first considered by the German, American and Soviet engineers independently, and all discovered that rocket engines needed high mass flow rate for both liquid oxidizer and fuel, for generating the necessary thrust. Higher thrust levels were achieved when liquid oxygen (LOX) and liquid hydrocarbon were used as fuel. At atmospheric conditions, LOX and low molecular weight hydrocarbons are in gaseous state and to get required mass flow rate, the only option is to feed them to the engine in liquid form. For that, these are stored in liquid form by cooling them down and hence the name Cryogenic Rocket Engines. Various cryogenic fuels combination were tried and the liquid oxygen (LOX) oxidizer and the liquid hydrogen (LH_2) fuel combination caught special attention of engineers as both are easily available, bio-friendly, non-corrosive and have the highest entropy release by combustion. All cryogenic rocket engines work on expander cycle or gas-generator cycle or staged combustion cycle depending on thrust requirement, since the oxidizer and fuel are at sub-zero temperatures.

The major components of a cryogenic rocket engine are the Thrust chamber or combustion chamber, pyrotechnic igniter, fuel injector, fuel turbo pumps, gas turbine, cryo valves, regulators, the Fuel tanks and rocket engine nozzle. The fuel flow can be differentiated into a main flow or a bypass flow configuration. In the main flow design, the entire fuel is fed through the gas turbines, which intern drive the cryopump for fuel and oxidizer, and then injected to the combustion chamber. In the bypass configuration, the fuel flow is split, the main part is goes to combustion chamber to generate thrust, while a small amount of the fuel goes to turbine, to drive the cryopumps for fuel and oxidizer and is subsequently injected to combustion chamber.

The Indian Space Research Organisation (ISRO) has successfully conducted a test of its indigenous cryogenic engine to be used in the next geo-synchronous launch vehicle mission. The flight acceptance hot test of the cryogenic engine was carried out at the liquid propulsion systems centre at Mahendragiri in Tamil Nadu.

Cryogenics for Space Applications

Space cryogenics is the application of cryogenics to space missions. These applications fall into two broad areas, supporting space science missions and supporting the space transportation infrastructure. Many of these science missions use infrared, gamma ray, and x-ray detectors that operate at cryogenic temperatures. The detectors are cooled to increase their sensitivity. Astronomy missions often use cryogenic telescopes to reduce the thermal emissions of the telescope, permitting very faint objects to be seen. A broad range of cryogenic technology is needed to support these missions. For instance, materials change their properties (strength, dimensions etc). These changes need to be considered when building an instrument for space. It is a challenge to design a telescope that is assembled at room temperature and then cooled to 20 K (–253°C) or so and launched into space. After surviving the high vibration environment of launch and the dimensional changes of cooling down, the instrument must be in focus and provide an undistorted image. For lower temperatures, instruments have used stored solid cryogens (such as nitrogen, neon, or hydrogen). For lower temperatures, liquid helium can be used in the 1-2 Kelvin range. Containing liquids while venting the effluent vapor has been a challenge. The disadvantage of using a stored cryogen is that it is converted to vapor by heat dissipated in the instrument or that comes in through the supports and insulation. Eventually, the cryogen is consumed, ending the mission. Recently there have been many advances in building closed cycle refrigerators for space applications. These coolers have extended mission durations and extended the range of temperatures available to 0.05 Kelvin. These coolers are required to be long lived, 5-10 years, have a low system mass and vibration levels. Another area of space science, which makes use of cryogenics, is sample preservation. This includes the preservation of biological samples from experiments on the shuttle and the station and the preservation of material gathered from comets, asteroids, and other planets. These applications have used phase change materials (solid to liquid transition) or liquid nitrogen absorbed in fine pore as coolants. Closed cycle coolers are now being developed for these applications.

Three areas of cryogenic technology pertinent to space applications are discussed in detail 1) Cryogenic tanks for long-term storage, 2) Solid cryogens for cooling space borne sensors and equipment and 3) Cryosorption technology for space simulation chambers.

Cryogenic Tanks

Cryogenic tanks are vessels provided for containment of cryogenic fluids. The wall of the vessel may be of a sandwich construction comprising a first layer of fiber reinforced polymer, a layer of gas expanded polymer foam and a second layer of fiber reinforced polymer similar to the first layer. Variations of the basic sandwich construction may be provided. The method of construction comprises a sandwich construction having a predetermined configuration to allow construction of a vessel having a cylindrical configuration, for example. The sandwich construction of the present invention allows suitable bonding of sections with a mixture of reinforcing fiber and polymer. Refer Figure 25.3 for the photograph of cryogenic tanks.

Fig. 25.3 Cryogenic Tanks

25.8 Cryogenics

Cryogenic tanks used to store liquid oxygen, liquid hydrogen and liquid helium. On some scientific satellites, cryogenic tanks are also used to store the cryogen that provides the cooling power required by the scientific instruments.

Solid Cryogens

Stored-cryogen expendable-coolant systems use either cryogenic liquids in the subcritical or supercritical state, cryogenic solids, or high-pressure gas combined with a Joule-Thomson (JT) expansion valve system to provide cooling of spacecraft components. These technologies can provide cooling to temperatures ranging from near absolute zero to more than 300 K. The advantages of these systems are simplicity, reliability, relative economy, and negligible power requirements. The disadvantages are the systems' limited life, resulting from parasitic heat leakage, and the high weight and volume penalty for extended durations of use. The primary limitations of the liquid-storage-type systems are the complex tank design (needed to minimize boil-off), the need for phase separation of subcritical fluids in the space environment, and the large weight and volume penalty for extended mission time. An alternative to these systems is the use of stored cryogens in the solid state, which provides a higher heat content, higher density, and simpler system design than the liquid systems. Limitations of solid systems include restrictions on detector mounting, specialized filling procedures, the need for a metal form or mesh to reduce temperature gradients as the solid dissipates, and more complex ground-hold considerations. The operating pressure of the cryogen must also be tightly controlled.

Solid cryogen cooling systems are mostly used for spacecraft instrument cooling. The solid cryogen coolers, are open-cycle systems that provide temperature stability in the range of a few degrees over the 7 to 12-month orbital lifetime. The expendable coolant is stored as a solid during orbital operation by maintenance of the vapor pressure below the triple point, so that sublimation occurs, and heat is removed by the subliming solid whose vapor is vented directly into space. Recent developments in solid cooler technology include: 1) lower operating temperature systems: hydrogen (8 K) and neon (18 K); 2) hybrid systems: combination of solid cryogens with radiative coolers or mechanical refrigerators; 3) multimission cooler approach: cooler designed to satisfy wide range of instrument requirements with interchangeable cryogens and 4) shielding for long-life helium systems by solid cryogens.

Cryosorption

The first generation of space simulators were capable of operating down to vacuum levels of 8 torr. This level of vacuum limits the testing and evaluation of spacecraft to problems dealing with corona and thermal balance. Current missions require more sophisticated and complex systems for simulation of spacecraft environment. In order to study these phenomena and to define the duration of space simulation tests an ultra-high vacuum test chamber is used which requires incorporation of a pumping system to provide a high pumping speed. The space simulator consists of a series of concentric chambers, providing a guard vacuum for the inner chamber, which forms the working test volume. Cryosorption can be used for the space simulators by use of a cryosorption pump having an improved construction for obtaining a high vacuum condition in a pumping system.

Cryosorption refers to the adsorption of a gas on a surface maintained at cryogenic temperatures. The advantages of cryosorption pumping have been recognized in the past, but two engineering difficulties have prevented its direct use in space simulation chambers. One was maintaining the surface of a bed of adsorbent at 20 K, the other was simply retaining the adsorbent in the chamber so as not to contaminate the test space, Both problems have been solved by the recent successful bonding of an adsorbent material to a metallic plate, the bonding being unaffected by temperature cycling from room temperature down to 20 K. Theoretical isotherms were used to predict the performance of these cryosorption panels.

A usual cryosorption pump is provided with a pumping chamber to be evacuated to obtain a high vacuum condition and with a cryosorption member having an extremely low temperature surface, i.e. cryosurface, disposed at substantially the central portion of the pumping chamber so as to be cooled to a temperature of about 4.2 K (absolute temperature) by means of liquid helium. However, helium is not condensed and evacuated on the cryosurface at the temperature of about 4.2 K for the reason that helium has a high equilibrium vapor pressure. In order to evacuate the helium, the cryosurface of the cryosorption member is coated with an adsorbent such as active carbon and molecular sieves, or a condensed gas layer composed of a gas such as Ar or CO_2 gas.

References

1. Cryogenic Engineering – B.A. Hands.
2. Internet website of NASA.
3. Cryogenic Systems – Randall Baron.
4. Alternative Fuels – Dr. S.S. Thipse.
5. Internet website of Cryogenic Society of America.
6. Air Liquide Brochure on Cryogenic Tanks.
7. NASA Report on Cryosorption.
8. Internet website of Wikipedia.
9. Internet website of ISRO.

Questions

Q.1. What are the applications of cryogenics in propulsion?

Q.2. Explain the benefits and drawbacks of using cryogenic propellants in rockets.

Q.3. Write a short note on cryogenic engine.

Q.4. What is cryosorption? Where is it used?

Q.5. How is cryogenics used in space applications?

CHAPTER 26

Applications of Cryogenics - IV
Food Freezing and Storage

INTRODUCTION

The term food preservation refers to any one of a number of techniques used to prevent food from spoiling. It includes methods such as canning, pickling, drying and freeze-drying, irradiation, pasteurization, smoking, and the addition of chemical additives. Food preservation has become an increasingly important component of the food industry as fewer people eat foods produced on their own lands, and as consumers expect to be able to purchase and consume foods that are "out of season." Freezing is an effective form of food preservation because the pathogens that cause food spoilage are killed or do not grow very rapidly at reduced temperatures. The process is less effective in food preservation than are thermal techniques such as boiling because pathogens are more likely to be able to survive cold temperatures than hot temperatures. In fact, one of the problems surrounding the use of freezing as a method of food preservation is the danger that pathogens deactivated (but not killed) by the process will once again become active when the frozen food thaws.

Freezing food is a common method of food preservation which slows both food decay and, by turning water to ice, makes it unavailable for most bacterial growth and slows down most chemical reactions. Freezing only slows the deterioration of food and while it may stop the growth of microorganisms, it does not necessarily kill them. Many enzyme reactions are only slowed by freezing. Therefore it is common to stop enzyme activity before freezing, either by blanching or by adding chemicals. Foods may be preserved for several months by freezing. Long-term freezing requires a constant temperature of –18 degrees Celsius (0 degrees Fahrenheit) or less. The time food can be kept in the freezer is reduced considerably if the temperature in a freezer fluctuates. Fluctuations could occur by a small gap in the freezer door or adding a large amount of unfrozen food. Cryogens, like liquid nitrogen, are used for specialty chilling and freezing applications. The freezing of foods like vegetables, fish, fruits, flowers and biotechnology products, like vaccines, requires nitrogen in blast freezing or immersion freezing systems. Special cryogenic equipments are used to remove heat from the organic matter and provide a low temperature environment.

A number of factors are involved in the selection of the best approach to the freezing of foods, including the temperature to be used, the rate at which freezing is to take place, and the actual method used to freeze the food. Because of differences in cellular composition, foods actually begin to freeze at different temperatures ranging from about 31°F (–0.6°C) for some kinds of fish to 19°F (–7°C) for some kinds of fruits. The rate at which food is frozen is also a factor, primarily because of aesthetic reasons. The more slowly food is frozen, the larger the ice crystals that are formed. Large ice crystals have the tendency to cause rupture of cells and the destruction of texture in meats, fish, vegetables, and fruits. In order to deal with this problem, the technique of quick-freezing has been developed. In quick-freezing, a food is cooled to or below its freezing point as quickly as possible. The product thus obtained, when thawed, tends to have a firm, more natural texture than is the case with most slow-frozen foods. About a half dozen methods for the freezing of foods have been developed. One, described as the plate, or contact, freezing technique, was invented by the American inventor Charles Birdseye in 1929. In this method, food to be frozen is placed on a refrigerated plate and cooled to a temperature less than its freezing point. Or, the food may be placed between two parallel refrigerated plates and frozen.

Another technique for freezing foods is by immersion in cryogenic liquids. At one time, sodium chloride brine solutions were widely used for this purpose. A 10% brine solution, for example, has a freezing point of about 21°F (–6°C), well within the desired freezing range for many foods. More recently, liquid nitrogen has been used for immersion freezing. The temperature of liquid nitrogen is about –320°F (–195.5°C), so that foods immersed in this substance freeze very quickly. As with most methods of food preservation, freezing works better with some foods than with others. Fish, meat, poultry, and citrus fruit juices are among the foods most commonly preserved by this method.

Need for Food Freezing

For too long the catering industry has not chilled food properly, simply leaving it out to cool in the kitchen, or placing it in a general purpose refrigerator which is too slow. These are highly dangerous practices, especially in summer, and could very easily lead to food poisoning, which can be fatal. According to the US Food Standards Agency, an estimated 5.5 million people a year are affected by food poisoning, the vast majority as a result of eating in a restaurant, café, takeaway or fast food outlet. The US Food Safety Act. 1990 makes it an offence to "render food injurious to health" or to sell food that does not comply with safety requirements. The Guidelines for Cook–Chill and Cook–Freeze Catering Systems, 1989 state that "food should be chilled to between 0°C and +3°C within a period of 90 minutes. A specially designed rapid chilling apparatus is required if rapid reduction of temperature is to be achieved.

Freezing is the use of storage temperatures below –18°C to accomplish the followings:

1. **Preservation**

    Pathogen growth is halted below –4°C

    Spoilage microorganisms don't grow below –10°C

    Chemical reaction rates are significantly reduced.

2. **Processing aid**

   Freezing changes the texture and viscosity for further processing, e.g. slicing meat products.

3. **Product definition**

   Freezing defines some food products, e.g. ice cream and frozen deserts.

History of Food Freezing

According to trade history, it was Clarence Birdseye who invented, developed, and commercialized a method for quick-freezing food products in convenient packages and without altering the original taste. His invention was based on his observation of the Eskimos of the Arctic, who preserved fresh fish and meat in barrels of seawater quickly. frozen by the arctic temperatures. This finding led him to conclude that it was the rapid freezing in the extremely low temperatures that made food retain freshness when thawed and cooked months later. And thus came into being the technique of quick freezing, which is a process through which items are frozen at such a speed that only small ice crystals are able to form. In the process the cell walls are not damaged, and the frozen food, when thawed, keeps its maximum flavor, texture, and color. Birdseye experimented with quick-freezing other foods, including fruits and vegetables, and soon became convinced that he had a viable commercial venture. In 1923, with an investment of $7 for an electric fan, buckets of brine, and cakes of ice, Clarence Birdseye invented and later perfected a system of packing fresh food into waxed cardboard boxes and flash freezing under high pressure. Birdseye refined and perfected a machine called a "Quick Freeze Machine" that he unveiled in 1925.

Advantages of Food Freezing

In general, boiling foods can cause them to lose important nutrients, including vitamins. In particular, Vitamins C and Folic acid are susceptible to loss during the commercial process. There has always been controversy as to whether frozen food is better or worse than fresh food. However, reports show that frozen foods are not so much nutritionally different from fresh ones. According to a 1998 report of the Food and Drug Administration, freezing is a very efficient method of preserving the nutritional value, texture and flavor of many foods. Most vitamins will keep well in frozen foods. Carotene (a compound that is converted to vitamin A in the body) may actually be better preserved in frozen agricultural produce because packaging keeps the food away from light (which destroys carotene). For example, frozen peas typically have about 60% more carotene than 'fresh' peas (that have been exposed to light during their trip to the market and while awaiting sale). The vitamin losses associated with blanching and the thawing/cooking process are similar to those that occur during normal cooking of fresh vegetables. This means that, provided they have been stored and then cooked properly, frozen vegetables provide similar levels of nutrition to fresh vegetables. It is also worth noting that for cooking both frozen and fresh vegetables, microwave cooking and steaming are both superior (in terms of retaining nutritional value) than boiling in a large volume of water.

Freezing of Vegetables

Frozen vegetables are omnipresent in the markets, and are increasingly becoming an important part of the food industry worldwide. Frozen vegetables are packaged vegetables that are sold in the frozen section of the store, usually packaged in either rectangular boxes or plastic bags. Ordinarily frozen vegetables found in the markets include spinach, broccoli, cauliflower, peas, corn, yam and mixtures of these and other vegetables. Vegetables are usually frozen within hours of being picked, and during the thawing process, they are very close to fresh in taste and texture. The frozen vegetables are becoming increasingly popular in time-starved nuclear households. If the meal can be heated in a microwave, total time from freezer to table can be less than five minutes. Besides offering fresh taste and convenience, freezing is also a safe method of preservation, as most pathogens are inactivated at low temperatures. Apart from this nutritional benefit, frozen vegetables allow enjoyment of any vegetable at any time of year, no matter what the season.

Fig. 26.1 Frozen Vegetables

But in the recent times, despite the inflation, changing lifestyles and economic development are creating a shift in buying patterns. Consumers are discovering they have less time to shop and prepare meals at home. This is especially true for contemporary women, who are entering the workforce in record numbers. Consumers are learning that frozen vegetables are convenient and easy to prepare, and when it comes to storage, they have a longer shelf life. As a result, frozen vegetables purchases are up, creating excellent business opportunities for food freezing firms for frozen potatoes, corn, broccoli, cauliflower, and other vegetables.

Freezing of Fish

Fish are cold-blooded, aquatic vertebrates. Important saltwater fish include herring, cod, rosefish, mackerel and tuna. Fish is even more perishable than meat. This is attributable not only to the protein-decomposing enzymes which are still active at low temperatures and the large proportion of microorganisms associated with fish, but also to the low connective tissue content of fish. In comparison with lean fish (trout, codling, rose fish or Pollack), oily types of fish, such as eel, herring or salmon are particularly at risk of spoilage. Fish are frozen by the flash or rapid freezing process, sometimes while still on board the fishing vessels. The best quality fish is that frozen by the flash or rapid freezing process. Rapid freezing results in the formation of only very small ice crystals, which, unlike large crystals, do not rupture the cell walls and thus do not result in the loss of cell fluids (drip) on thawing.

The following criteria may be used to assess fresh fish: unobtrusive odor, firm and resilient flesh, brilliant red color of gills, bright, glossy color, glossy black pupils. The fish must be properly deep frozen on loading. If it is not at the required core temperature, it will spoil during a long voyage. Checks must accordingly be carried out during loading. Properly deep frozen fish sounds like wood when struck. The core temperature should be measured for each batch by drilling a hole into the middle of the fish and measuring the temperature with a meat thermometer. In the case of (unpack aged) tuna, a spike is used to make holes 3–5 cm in depth in the fish so that the measurement can be made. Occasionally, fish is delivered which, after freezing, has been exposed to higher temperatures. Such incorrect storage results in depreciation and may be recognized by the formation of frost on the cartons. Fish covered with a thick layer of ice or with brown discoloration or freezer burn must also be rejected prior to loading. The duration of storage for various types of fish is given as follows in Table 26.1.

Table 26.1 Storage Periods for Fish

Designation	Temperature	Rel. humidity	Max. duration of storage
Frozen oily fish	-28 to -18°C	90 - 95%	8 months
Frozen lean fish	-20°C	90 - 95%	12 months
Frozen filleted fish	-28 to -23°C	90%	6 - 9 months

Frozen fish, which has been stored for an excessively long period, has a dry, straw-like texture and poor flavor and is described as freezer damaged. Figure 26.2 shows the freezer-damaged fish. At temperatures of –62°C, the "eutectic point" (EP) is reached. Only once the EP is reached is all the water in the cells of the product completely frozen and all microbial decomposition brought to a standstill, i.e. at temperatures of below –62°C it is possible to transport or store foodstuffs for an "infinite" period without loss of quality

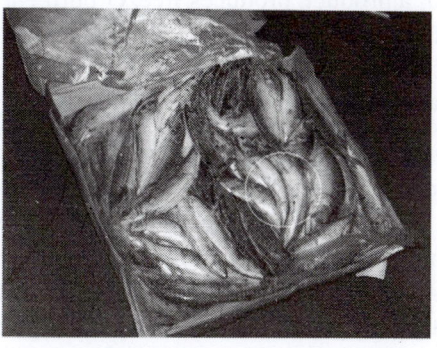

Fig. 26.2 Freezer Damaged Fish

Hazards of Improper Food Freezing

1. **Chemical changes during freezing:** Fresh produce contains chemical compounds called enzymes, which cause the loss of color, loss of nutrients, flavor changes, and color changes in frozen fruits and vegetables. These enzymes must be inactivated to prevent such reactions from taking place. Another group of chemical changes that can take place in frozen products is the development of rancid oxidative flavors through contact of the frozen product with air. Using a wrapping material, which does not permit air to pass into the product, can control this problem.

2. **Textural changes during freezing:** Freezing fruits and vegetables actually consist of freezing the water contained in the plant cells. When the water freezes, it expands and the ice crystals cause the cell walls to rupture. Consequently, the texture of the produce, when thawed, will be much softer than it was when raw. This textural difference is especially noticeable in products, which are usually consumed raw. Textural changes due to freezing are not as apparent in products which are cooked before eating because cooking also softens cell walls. These changes are also less noticeable in high starch vegetables, such as peas, corn, and lima beans.

3. **Changes caused by fluctuating temperatures:** Fluctuating temperatures in the freezer can cause the migration of water vapor from the product to the surface of the container. This defect is sometimes found in commercially frozen foods, which have been improperly handled.

4. **Moisture loss during freezing:** Dehydration is of particular interest because it is less obvious, harder to quantify and often has a large economic impact. It is the result of the inevitable loss of water vapor that occurs when a product is exposed to air or another gaseous medium. Frost accumulating on the coil surfaces provides a gross indicator of the rate of dehydration moisture loss. Fast cooling and freezing greatly reduce dehydration.

5. **Freezer burn due to freezing:** Moisture loss, or ice crystals evaporating from the surface area of a product, produces freezer burn—a grainy, brownish spot where the tissues become dry and tough in frozen storage. This surface freeze-dried area is very likely to develop off flavors. Packaging in heavyweight, moisture proof wrap will prevent freezer burn.

6. **Microbial growth in the freezer:** The freezing process does not actually destroy the microorganisms, which may be present on fruits and vegetables. While blanching destroys some microorganisms and there is a gradual decline in the number of these microorganisms during freezer storage, sufficient populations are still present to multiply in numbers and cause spoilage of the product when it thaws. For this reason it is necessary to carefully inspect any frozen products which have accidentally thawed by the freezer going off or the freezer door being left open.

7. **Loss of nutrient value of frozen foods:** Freezing, when properly done, is the method of food preservation, which may potentially preserve the greatest quantity of nutrients. However if improperly done can cause loss of nutrient value. It is important to store the frozen product at 0°F and to use it within suggested storage times.

Food Freezing Techniques

There are many different types of freezing techniques available for food preservation and consumers are often uncertain about which type is best suited to their needs. Three factors may be initially considered when selecting a freezing technique namely financial, functional and feasibility. Financial considerations will take into account both the capital and running cost of the equipment and also projected losses such as product damage and dehydration. Expensive freezers should therefore justify their purchase by giving special benefits and if these benefits are not worthwhile, they need not be considered. Functional considerations will take into account such things as whether the freezer is required for continuous or batch operation and also whether the freezer is physically able to freeze the product. For instance, a horizontal plate freezer would be inappropriate for freezing large whole tuna. Feasibility will take into account whether it is possible to operate the freezer in tile plant location. A liquid nitrogen freezer (LNF), for instance, may be suitable in every respect for freezing the product and the high costs of using this method of freezing may be justified. However, if the location of the plant is such that there can be no guaranteed supply of liquid nitrogen, the freezer should not be considered.

The main techniques for food freezing are
A. Blast Freezing
B. IQF or Quick Freezing
C. Spiral Freezing
D. Immersion Freezing
F. Fluidised Bed Freezing
G. Plate Freezing
H. Flash Freezing

Let us study each technique in detail.

(A) Blast Freezing

In Blast freezing technique, vigorous circulation of cold air enables freezing to proceed at a moderately rapid rate. Products are placed on trays, either loose or in packages and the trays are placed on freezing coils in a low temperature room with cold air blowing over the product. In some installations of this system, the cold air that is in the low temperature room is circulated by means of large fans, whereas in other installations the air is blown through refrigerated coils located either inside the room or in an adjoining blower room. Tunnel freezing is possibly the most commonly used blast freezing system. In this system a long, slow moving mesh belt passes through a tunnel or enclosure containing very cold air in motion. The speed of the belt is variable according to the time necessary to freeze the product. Usually the cold air is introduced into the tunnel at the opposite end from the one where the product to be frozen enters, that is, the airflow is usually counter to the direction of the flow of the product. The temperature of the air is usually between −18° and −34°. The air velocity varies, however if rapid freezing is to be had, it is necessary to recirculate a rather large volume of the air in order to obtain a relatively small rise in the temperature of the air as it touches and leaves the product. Air has a very low specific heat and for that reason a large volume must be carefully distributed through the system. Air velocities ranging all the way from 100 ft. per min. up to 3500 ft. per min have been reported, and it is difficult to establish any speed as having more or less common usage. Possibly 2500 ft. per min may be considered a practical and economical air velocity at −29°C. Air blast freezing is economical and is capable of accommodating foods of a variety of sizes and shapes. It can however result in excessive dehydration of unpack aged foods if conditions are not carefully controlled, and this in turn necessitates frequent defrosting of equipment. Undesirable bulging of packaged foods, which are not confined between rigid surfaces during freezing, is another drawback of this method.

Blast Freezer

Blast Freezers are fully assembled using only USDA accepted materials, and are ready for hook-up to customer's electricity, water, and refrigeration equipment. Some Design features are as follows:

Galvanized steel evaporator coil
Available in Ammonia or Freon
Stainless steel interior
Insulated, galvanized floor with drains
Power available in 60 cycle or 50 cycle
FIFO (First In First Out) system
Doors on both ends provide straight line production flow

Figure 26.3 shows a typical blast freezer.

Fig. 26.3 Typical Blast Freezer

Benefits of Blast Freezing

1. **Improved food quality:** Damaging bacteria are rendered dormant, minimizing food spoilage. Color, texture, flavor, structure and nutritional value are locked in the frozen food. Delicate food surfaces such as pasta and fruit are protected, as rapid chilling stops an "ice skin" forming which otherwise dehydrates and damages the products' appearance. Blast freezing also helps to keep food looking good. The slower food freezes, the larger the ice crystals formed; and large ice crystals can damage food, dry it out, and break down the physical structure. Air chill/freeze temperatures are fully adjustable, depending on the food to be chilled, to ensure the end product is of the best quality.

2. **Reduction in kitchen wastage:** Kitchens with blast chillers and freezers throw away less food. Food left over, for example, can quickly and safely be chilled for later use, with complete confidence in its quality and safety. Once prepared and chilled, only the correct numbers need be reheated, as legislation states that chilled food can be safely kept below +3°C for up to 5 days after production. Blast chillers can even be used to chill wine, cans or other drinks, buffet dishes, salads and serving dishes.

3. **Improved kitchen efficiency:** Blast chilling allows advance food preparation of larger batches as per requirement and then chill for later use. This makes for fewer batches, more efficient use of ovens, more predictable working hours and optimum use of the chef's schedule, and gives time to improve the presentation. Blast chilling can increase the potential profits by increasing production capacity, and reducing food wastage.

(B) IQF or Quick Freezing

When product is individually frozen, it is commonly known as Individual Quick Freezing or I.Q.F. This value-added process is more labor intensive and requires costlier freezing equipment. Freezing in homogeneous mass or "block freezing," such as in two-kilo or five-pound boxes of shrimp, is relatively low in cost. Cryogenic freezing systems, which usually employ liquid nitrogen are also very effective, but due to the high cost of the gas, are usually limited to high cost products. I.Q.F. in mechanical freezers, of items such as cut broccoli and cauliflower, peas, diced vegetables, berries, etc. presents a variety of problems. The quantity to be frozen in pound per hour of different varieties is dependent upon the specific heat of the product, the size and shape, the temperature when entering the freezer and the free water content. In order to attain some degree of efficiency in mechanical freezers, a combination of low temperature, airflow and rotation of product is carries out so that the articles do not stick together. These three parameters are augmented and rendered more efficient by use of proper variable speed conveyers, with shakers or other means, and de-watering devices to control free water content.

The purpose of these processes is to instantly crust-free the product as it enters in the initial freezing phase in a single layer. As this is accomplished (so that the articles do not stick), they are then conveyed to the second phase where they may be several inches thick, while still in a low temperature airflow. The product is then in a state known a "deep fluidization." This final phase freezes the product totally, where it may be packaged, stick-free like "marbles in a sack." The variety of products to be frozen requires selection of the proper freezing system for each category of products. The selection must be based upon the specific heat of the product, the size and shape, the water content and free water, as well as the entering temperature. Freezing systems for particular operations require consultation to be sure that there is a balance between amount of product to be frozen and the cost of the equipment. Some flexibility should be considered so that the bulk of the product can be frozen with a minimum equipment investment and consideration given to the growth of production capability. Two types of IQF chilling cycles are available as shown in Figure 26.4.

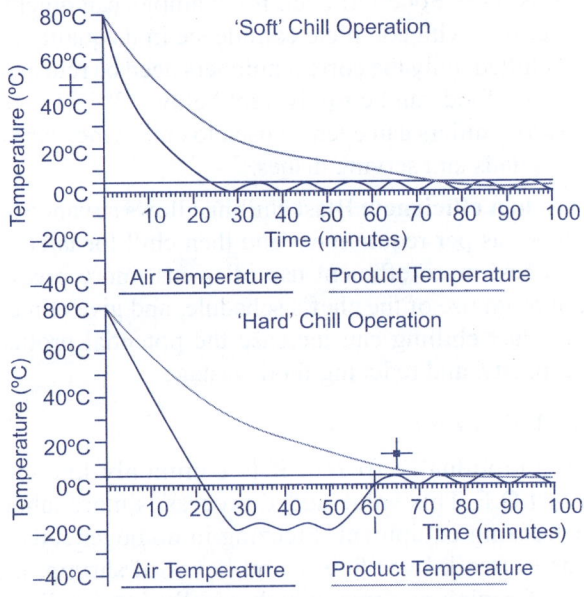

Fig. 26.4 IQF Chilling Cycles

Soft Chill– Is used for the safe chilling of delicate products. The Soft Chill cycle is ideal for the rapid but gentle chilling of any dish. This cycle brings the temperature of the food down whilst retaining a positive air temperature. This prevents large ice crystals forming which can damage the structure of delicate foods such as gateaux and patisserie items, and high water content items such as vegetables, rice and pasta. Using this cycle maintains the texture, consistency and appearance of these foods, with no dehydration or cell damage.

Hard Chill– Is used for general purpose chilling. The Hard Chill cycle is ideal for chilling 'standard products' such as cooked meat, pies, individually portioned meals etc. The air temperature is kept at around –15°C for the first 70% of the cycle, to reduce temperature quickly. Air temperature is then increased for the last 30% of the cycle to stop surface damage and to ensure quality.

(C) Spiral Freezing

Depending on the upstream process and capacity required, this type of freezer is available in a range of models with different belt widths and may be completely factory assembled or partially assembled in modules for quick installation and future portability. Spiral belt freezers use a product belt that can be bent laterally. The original spiral belt design uses a spiraling rail system to carry the belt, with a central drum that drives the belt through friction at the belt edge. The latest spiral belt design uses a self-stacking, self-enclosing stainless steel belt for compactness, greater reliability and improved airflow. This design eliminates the traditional rail system and friction drive. The number of tiers in the belt stack can be varied to accommodate different capacities. In feeds and out feeds can be located to suit most line layouts. Spiral freezers are good systems for products requiring a long freezing time (generally 10 minutes to 2 hours) and for products that require careful handling. Typical products frozen in spiral belt freezers include raw and cooked meat patties, fish fillets, chicken parts, pizza and a variety of packaged products.

(D) Immersion Freezing

Immersion freezing is a method of commercially preparing frozen foods so that the product remains suitable for consumption over a long period of time. The process helps to lock in moisture as well as maintain the flavor and taste of the processed food. Immersion freezing is very successful in the food industry. One of the chief applications of immersion freezing has to do with the preparation of meats and poultry for use in frozen foods such as frozen entrees and cuts of meat. Essentially, the meat is exposed to liquid nitrogen. Generally, the time for exposure is somewhere between six and ten seconds. This type of freezing helps to lock in important vitamins and nutrients as well as delay the process of decomposition. Meats and poultry that have been prepared for use in frozen dinners are normally cooked to specifications, then subjected to immersion freezing. However, such items as uncooked chicken tenders and beef patties may also be preserved using immersion freezing.

Both raw and cooked fruits and vegetables are often preserved with immersion freezing. Because of the rapid freeze that occurs, immersion freezing allows the foods to retain their color and texture for long periods of time. They also help to maintain the nutrient content of the vegetables, making them similar to fresh fruits and vegetables when it comes to steaming, braising or broiling the vegetables. Combinations of cooked vegetables, such as in frozen soup mixes, are also often preserved using the process of immersion freezing.

Another advantage of immersion freezing is that the process helps to prevent the escape of steam after the food has been cooked. This is important when a food manufacturer is preparing packaged dinners, as the presence of steam could cause the container to burst or tear after the product has been packaged. With immersion freezing, there is no chance for pressure from within the package to burst the container and thus expose the product to the open air. Immersion freezing has made it possible to mass-produce a wide variety of frozen foods that can serve as an alternative to canned and fresh foods. Because the process makes it possible for food to retain color, texture, and nutritional value, many people prefer to use foods prepared by immersion freezing. This is especially true with seasonal foods that are not widely available at all times of the year.

Immersion Freezers

Equipped with an accurate level control system, the immersion freezer allows storage of high quality products such as seafood (shrimps, shellfish), soft fruits (strawberries, raspberries) or vegetables. Easy to open and to clean, this freezer can also be combined with other mechanical freezers. Most of the times it is a two-stage freezer for space efficient individual quick-freezing of food products using liquid nitrogen (LN_2) refrigerant. It consists of a freezing tank with a long immersion belt. In this first stage, the outer layer of food product is quick-frozen. The product then transfers to the freezer tunnel consisting of a stack of long stainless steel mesh belts one over the other, with a height clearance between them. There is an LN_2 spray manifold at the top of the insulated tunnel (i.e. over the topmost belt), along with circulation fans. Each belt has a variable speed drive. As the product passes through the second stage, the liquid nitrogen residual from the immersion bath evaporates, effectively pulling out the remaining residual heat such that, in proper operation, the product is frozen all the way through, while leaving the product in discrete, individual pieces. Immersion freezers are manufacturer rated for 4,000 lbs per hour throughput and are made of stainless steel. Refer Figure 26.5 for a typical immersion freezer in the USA.

Fig. 26.5 Typical Immersion Freezer

Benefits of Immersion Freezing

- High and consistent quality frozen products.
- High capacity at minimum floor space.
- Instantaneous freezing of the surface prevents moisture and quality losses.
- Maximum efficiency.

(E) Fluidized Bed Freezing

Solid food particles ranging in size from peas to strawberries can be fluidized by forcing air upward through a bed of particles 1-5 inch deep on a mesh belt to partially lift or suspend the particles. If the air used for fluidization is appropriately cooled, freezing can be accomplished at a rapid rate. An air velocity of at least 375 ft/min is necessary for fluidization to be accomplished, and this in turn depends on size, shape, and the uniformity of particles. A depth of slightly more than 1 in. is suitable for easily fluidized particles, such as peas and whole kernel corn; a depth of 3-5 in. is used for partially fluidizable particles, such as green beans; and a depth of 8-10 in. can be used for non-fluidizable products, such as fish fillets. Fluidized bed freezing has proved successful for many kinds and sizes of unpackaged food tissues, although the best results are obtained with products that are relatively small and uniform in size (e.g., peas, limas, cut green beans, strawberries, whole kernel corn, brussel sprouts). The advantages of fluidized bed freezing as compared to air-blast freezing are: 1) More efficient heat transfer and more rapid rates of freezing. 2) Less product dehydration and less frequent defrosting of equipment. 3) Short freezing time is apparently responsible for the small loss of moisture. A major disadvantage of fluidized bed freezing is that large or non-uniform products cannot be fluidized at reasonable air velocities.

(F) Plate Freezing

Food products can be frozen by placing them in contact with a metal surface cooled by refrigerants such as ammonia. Packaged food products may rest on, slide against, or be pressed between cold metal plates. Plate freezers consist of a series of flat hollow refrigerated metal plates. The plates are mounted parallel to each other and may be either horizontal or vertical. The spaces between the plates are variable, the plates being opened out for loading and, prior to the freezing operation, closed so that the surface of the plates is in intimate contact with the packaged or unpackaged food. Clearly the frozen product is in the form of blocks and, during the freezing process, heat flow is perpendicular to the faces of the plates. A moderate pressure is maintained between the plates and the package surfaces during freezing to promote good face-to-face contact.

Double contact plate freezers are commonly used for freezing foods in retail packages. This equipment, which may be batch, semi automatic, or automatic, consists of a stack of horizontal cold plates with intervening spaces to accommodate single layers of packaged product. The filled unit appears much like a multilayered sandwich containing cold plates and product in alternating layers. The plates are arranged so that they can be opened or closed in a vertical manner, enabling product to be added or removed. When closed, the plates make firm contact with the two major surfaces of the packages, thereby facilitating heat transfer and assuring that the major surfaces of the packages do not bulge during freezing. Contact plate freezing is an economical method that minimizes problems of product dehydration, defrosting of equipment and package bulging. Disadvantages of this method are 1)

Packages must be uniform of thickness. 2) Freezing occurs at a moderately slow rate as compared to other modern methods. 3) Freezing times are extended considerably when the package contains a significant volume of void space.

(G) Flash Freezing

The present freezing methods, including blast freezing and immersion freezing techniques, are subject to many disadvantages including inefficiency of operation, length of time required to satisfactorily freeze the products, and the cost factor. Flash freezing is a revolutionary process of quick freezing (usually within seconds). This locks in all the flavors, juices, vitamins and minerals and allows the products to keep perfectly for long periods. Each product is held in this condition until it is thawed, ensuring that it will be just as fresh as when it was frozen. Flash freezing is achieved with sophisticated, ultra-low temperature freezers to freeze food products like meats immediately after they are individually vacuum packed. This quick freeze process is absolutely essential to maintain the flavor and texture of food products. Flash frozen foods have a much longer shelf life and they do not suffer from freezer burn. Rapid freezing has nutritional benefits because it maintains the quality of the product without artificial preservatives. Taste is also superior because these foods are selected at the peak of quality and then flash frozen to seal in flavor, aroma and nutrition.

This technique involves flash freezing of articles in a substantially thermally isolated chamber. The method employs utilization of the latent heat of vaporization during the "flashing" of liquid nitrogen to a gas in the immediate vicinity of the product to be frozen. The ultra-cold nitrogen gas is effectively utilized to aid in the freezing process by recirculating the gas over the product at high velocities to bombard and completely encompass the product to take advantage of the chilling effect of the fast moving gas. In order to effectively utilize the latent and specific heat of the liquid and gaseous nitrogen the freezing process takes place in a chamber that is fully insulated by highly efficient insulation techniques involving drawing and maintaining a vacuum between spaced walls that surround the process chamber.

A particularly effective arrangement for insulating the main process chamber is using Concentric shells or sleeves having suitably spaced walls to encircle the product. Vacuum is maintained in the annular region between these walls and all connections of the external components required for the liquid nitrogen system and the gaseous nitrogen recirculation system are made through insulated end spools so that the vacuum section is not broken or otherwise interrupted by external connections. The temperature within the process chamber is sensed by a temperature transmitting means that serves to control the temperature within the chamber by regulating the flow of liquid nitrogen into the chamber. It has been found that for certain applications, exposure of the food product for 2 to 6 minutes within the chamber is sufficient to lower the temperature of the product the desired amount. For bakery goods, the system utilizes about 1 pound of liquid nitrogen for each pound of product.

The temperature of the entering cryogenic liquid is about −320°F. The flow volume of the gas recirculation stream between the intermediate set of plenum chambers is about 1,000 cubic feet per minute and the gas temperature is about −200°F. The total mass flow rate of the gas in these transverse flow streams is substantially greater than the mass rate of liquid flashing to gas in the chamber. The high volume, high velocity per recirculation creates turbulence at the surface of the product or article and effects better heat transfer. Transverse flow is more efficient as it is easier to achieve high volume and high velocity and it affords a better angle of impingement upon the product. The spray technique used progressively and repeatedly wipes the surface of the product with liquid nitrogen droplets to promote rapid heat transfer. The gas generated on the surface of the product in the freezing process is penetrated by the liquid nitrogen droplets which leave the spray nozzles with adequate velocity for this purpose. The progressively deposited, full coverage high velocity spray technique results in very rapid heat transfer between the product and the liquid nitrogen. This results in high production for a relatively small unit.

Benefits of Flash Freezing

- Can be fitted into existing systems
- Easily expanded
- Easy access for cleaning
- Efficient and flexible freezing
- High productivity
- Low capital/rental cost
- Premium cryogenic quality product
- Simple installation and start-up

This completes the review of some major food preservation techniques using cryogenics.

References

1. Cryogenic Engineering – B.A. Hands.
2. Internet website http://science.jrank.org/pages/2814/Food-Preservation.html.
3. Cryogenic Systems – Randall Baron.
4. Internet website www.alard-equipment.com.
5. Internet website of Air Products, www.airproducts.com.
6. Foster Refrigerator Blue Paper on food freezing, UK.
7. Internet website www.wisegeek.com.
8. Paper by UNIDO on food freezing methods.
9. Internet website http://www.tis-gdv.de/tis_e/ware/fisch/gefroren/gefroren.htm.

Questions

Q.1. Why is food preservation necessary? How does food freezing help in this regard?

Q.2. Enumerate the important methods of food freezing.

Q.3. Write a short notes on
   (a) Blast freezing
   (b) Immersion freezing
   (c) Flash freezing.

Q.4. Discuss the importance of Liquid Nitrogen for food freezing.

Q.5. Describe any cryogenic food storage technique.

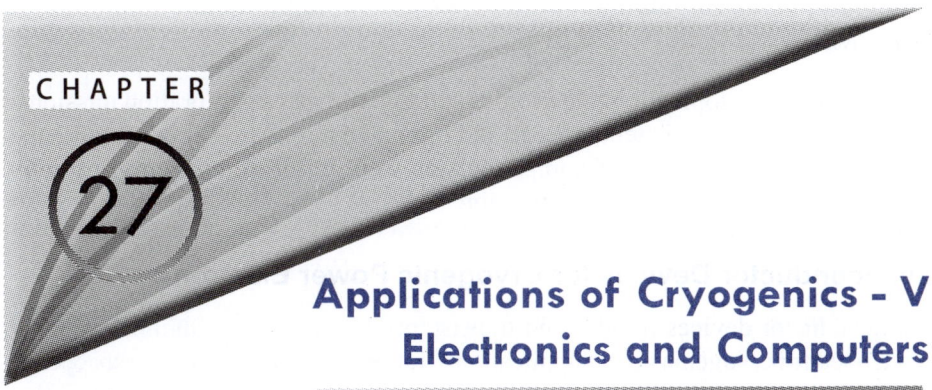

CHAPTER 27

Applications of Cryogenics - V Electronics and Computers

INTRODUCTION

Cryo-electronics is the branch of electronics that deals with the design, construction, and use of cryogenic devices. Cryotronics is the production of electronics that utilize superconductivity, and is not to be confused with cryo electronics. The simplest use of cryotronics is the cryotron, which is a switch. Additionally, the sensitive amplifiers used in radio telescopes are cooled using liquid helium to reduce thermal noise. Cryotronics includes creating, supersensitive miniature receivers for detecting weak radio signals. And crytron switches that depends on the effect of inducing persistent current in a closed superconducting circuit. The electronics industry is using cryogenics in increasing proportion. The electronic industry depends heavily on the use of liquid nitrogen and electronics production relies on cheap quality control techniques for mass production systems. Extreme cold provided by liquid nitrogen is used to 'shock' test assemblies to ensure there are no electrical connection faults and no contamination by dirt or oxidisation during the manufacturing process where cleanliness is vital.

Cryogenic Power Electronics (CPE) provides promising benefits for power conditioning system compared to their room-temperature counterparts in terms of reduced size and weight (increased power density), improved efficiency, improved switching speed, and improved reliability. Active devices such as semiconductor switches can exhibit performance improvements such as reduced conduction losses, higher switching speed, and reduced diode reverse recovery, greater device gain, higher over-current capability, and increased power levels. Passive devices (inductors, capacitors, interconnects) will also improve by the lowered resistance of their constituent metal conductors or the use of superconductors. Advanced power electronic packaging methods that adapt well to cryogenics and help further improve system performance are being developed. System designers can develop the best circuit topologies for maximum CPE system performance.

Space exploration missions require electronics capable of efficient and reliable operation at low temperatures. Presently, spacecraft on-board electronics are maintained at approximately 20°C through the use of radioisotopes. Cryogenic electronics enhance efficiency of space systems, improve reliability, and simplify their design. A Low Temperature Electronics Program at NASA focuses on research and development of electronics suitable for space exploration missions. The effects of cryogenic temperature and thermal cycling are being investigated for commercial-off-the-shelf components as well as for components specially developed for low temperature operation.

Semiconductor Devices for Cryogenic Power Electronics

Semiconductor devices (diodes and transistors), based on germanium, are being developed for cryogenic power applications. Target applications include spacecraft for cold environments as well as commercial, industrial, and defense systems that incorporate cryogenics. The primary reason for basing these devices on Germanium is that it enables good performance down to deep cryogenic temperatures (down to 20 K and lower), although there are other advantages to using Germanium such as a low p-n junction forward voltage and high mobility. Following devices have been designed, fabricated and characterized using germanium as a base. Germanium power diodes (10-A) that operate from room temperature to approximately 20 K and germanium field effect transistors with a current of more than 0.5 A at liquid-nitrogen and liquid-helium temperatures, and a transconductance of approximately 200-300 mS at cryogenic temperatures. Germanium power devices are practical and offer important advantages for cryogenic space and ground applications.

Further steps in developing power semiconductor devices based on the silicon-germanium (SiGe) materials are being taken. The applications and motivation are similar to those for development of Ge devices for spacecraft for cold environments as well as commercial, industrial, and defense systems that incorporate cryogenics. The SiGe materials system has proved its benefits in devices for telecommunications. It also has valuable features for power electronics and cryogenic operation. The objective is to take advantage of the features of SiGe in combination with those of Si and Ge to develop diodes and transistors for cryogenic power operation. These features include: Si: an extensive technology base, high breakdown voltage, an excellent grown oxide. Ge: low p-n junction forward voltage, low freeze-out temperature, high mobility at low temperature. SiGe: bandgap engineering, selective placement, a developing technology base and compatibility with Si processing. The first device that is developed for cryogenic power is the hetero-junction bipolar transistor (HBT). These follow a standard design, using SiGe for the base region to maintain high gain over a wide temperature range from room temperature to deep cryogenic temperatures. However, we are designing the structure for high current and voltage. The current gain of HBT increases upon cooling from room to liquid-nitrogen temperature, which is an important outcome. The charge carriers must be placed in a channel region, separated from the ionized dopants. The source of the carriers, i.e. the supply layer, must be highly doped in order to prevent carrier freeze-out at low temperatures. Using more appropriate materials and designs it is expected to improve the HBT characteristics considerably. Successful development of bipolar structures could form the basis for fabrication of more complex power devices for cryogenic operation, such as the insulated-gate bipolar transistor (IGBT) and MOS-controlled thyristor (MCT).

Cryogenic Electronics for Space Operation

Planetary exploration missions and deep space probes require electrical power management and control systems that can operate efficiently and reliably in very low temperature environments. Presently, spacecraft operating in the cold environment of deep space carry a large number of radioisotope heating units to maintain the surrounding temperature of the onboard electronics at approximately 20°C. Electronics capable of operation at cryogenic temperatures would not only tolerate the hostile environment of deep space but also reduce system size and weight by eliminating or reducing the radioisotope heating units and their associate structures. Thereby, such electronics would reduce system development as well as launch costs. In addition, power electronic circuits designed for operation at low temperatures are expected to result in more efficient systems than those at room temperature. This improvement results because semiconductor and dielectric materials have better behavior and tolerance in their electrical and thermal properties at low temperatures.

The Low Temperature Electronics Program at the NASA is focusing on the research and development of electrical components, circuits, and systems suitable for applications in the aerospace environment and in deep space exploration missions. Research is being conducted on devices and systems for reliable use down to cryogenic temperatures. Some of the commercial off-the-shelf as well as developed components that are being characterized include semiconductor switching devices, resistors, magnetics, and capacitors. Semiconductor devices and integrated circuits including digital-to-analog and analog-to-digital converters, dc-dc converters, operational amplifiers, and oscillators are also being investigated for potential use in low-temperature applications. For example, Figure 27.1 shows the output response of an advanced oscillator at room temperature and at –190°C. Most oscillators can operate at temperatures down to only –55°C. It can be seen that, for this oscillator, the low temperature of –196°C changed the leading and trailing edges of the oscillator pulses by producing overshoot.

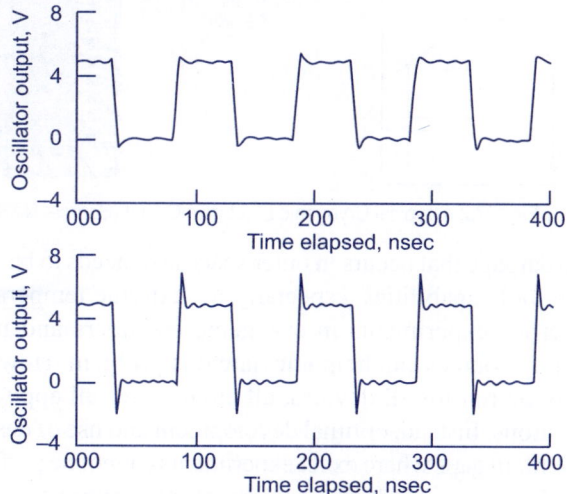

Fig. 27.1 Output waveforms of an advanced oscillator. Top: Operating temperature, 25°C. Bottom: Operating temperature, –196°C. (Source : NASA)

27.4 Cryogenics

CAD tools, models and methodologies for electronics design for circuit operation in extreme environments with a focus on very low temperature and radiation effects are being developed. These new tools and methodologies will help enable NASA to design next generation electronics. Such capabilities will significantly improve reliability, performance and lifetime of electronics that are used for space applications, including satellites and space travel. This will be achieved through the development of novel physics-based modeling techniques and verified by experiment. The new cryogenic design tools will greatly reduce the chances of error during actual circuit implementation, and thus reduce the number of design cycles, thereby substantially decreasing fabrication times and expenses. Models and CAD tools are relatively inexpensive as compared to fabrication costs, thus the results of this project should provide a very large return on investment. There has not been a significant effort to design electronics that operate reliably in outer space. Most of the electronics design software currently in use do not even give results for extreme temperature conditions. The details of the semiconductor physics that occur at cryogenic temperatures simply has not played a sufficiently large role in electronics design development to provide the existing knowledge base necessary for robust cryogenic development. The main difficulties with cryogenic design arise from changes in carrier mobilities, as well as the carrier freeze-out phenomena that occur at near absolute zero. These effects must be fully understood for modern devices, and incorporated accurately into electronic design tools.

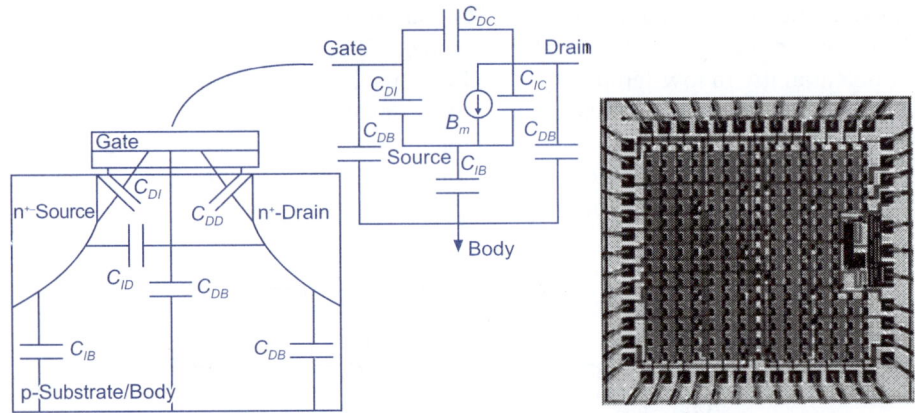

Fig. 27.2 A Typical Cryogenic Electronic Circuit (Source: NASA)

Radiation damage that occurs in outer space also needs to be more fully worked into the design tool capabilities, especially for extreme temperature applications. Running numerous experiments in cryogenic chambers and in the presence of radiation emitting sources can help alleviate this problem. However, experiments cannot possibly be run for all devices, all circuits, for all applications and for all operating conditions. Instead, optimal development and use of design software must be employed to fill in gaps where exact experiments cannot be performed for extreme environments. Each semiconductor process of interest must be characterized and modeled for low temperature applications. Figure 27.2 shows a typical cryogenic electronic circuit.

Cryogenic Electronics for Superconductivity

The most impressive application of cryogenic electronics is for superconductivity. When electrical current passes through conventional copper wire, it encounters resistance, impeding its flow. Overcoming this resistance requires energy, which is lost as heat. Superconducting wires and cables, on the other hand, offer no resistance to direct currents. Currents flow with zero losses, allowing for astonishing electrical efficiencies in superconducting systems. These cables are potentially much lighter than their copper counterparts. Refer Figure 27.3 for an illustration of superconductivity.

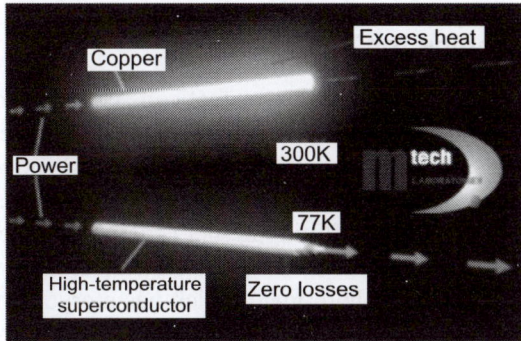

Fig. 27.3 Cryogenic Superconductivity Illustration (Source: MTech labs)

Superconducting motors, transformers, transmission lines, and circuits are capable of exceptional performance. The illustration below shows a high-temperature superconducting (HTS) motor designed by American Superconductor Corporation. HTS components are typically operated at around 77 K (–196°C), the temperature of liquid nitrogen. Achieving such low temperatures requires energy. Further losses are incurred when electrical leads and interfaces transition from these low-temperature elements to room-temperature motor controller power circuits. Moreover, the power circuits themselves can be relatively inefficient.

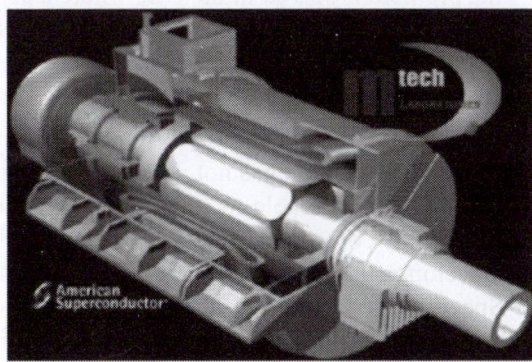

Fig. 27.3 a Cryogenic Superconductivity Motor (Source: American Superconductor)

27.6 Cryogenics

Replacing conventional semiconductors with cryogenic power electronics can greatly reduce these losses. Cryogenic Power Electronics can optimize the advantages of the super conducting elements in a system, and requires little additional hardware to implement, especially when the cryogenic platform is already present for super conducting elements. Cryogenic superconductivity is based on the discovery that the loss-producing on-state resistance of many power semiconductor devices drops drastically when these devices are cooled to 77 degrees Kelvin, the temperature of liquid nitrogen. At room temperature, power semiconductors carrying large currents often reach temperatures up to 400 K, further increasing their resistance. When cooled to 77 K, this resistance is reduced by up to 37 times.

Cryogenic Electronics for Transport

A cryogenic power conditioning system for fuel cells is proposed which is cooled by liquid hydrogen or liquid natural gas (methane) used to power these fuel cells, or by liquid nitrogen supplied by high-temperature super conducting cables. The main applications are in large vehicles such as transit buses. The result is a combined motor and motor-drive system exhibiting higher efficiency, lower weight, smaller size and lower cost. General Motors developed a car called "Zafira" which operates on fuel cells powered by liquid hydrogen, which has a cryogenic temperature of 20.27 K. Other fuel cells use a reformer which extracts hydrogen gas from natural gas (methane) or gasoline. Alternatively, liquid natural gas (LNG) at a temperature of 112 K presents interesting possibilities.

Fuel cells deliver DC power, which is then converted by so-called inverters into AC power required for general use and for the efficient operation of electrical motors. These inverters are often larger, heavier, and more expensive than the motors they control. The new concept of Cryogenic Energy Conversion (CEC) has achieved drastic reductions in size, weight, and cost in the field of power electronics. Such size and weight reductions are important in vehicles, where energy savings are crucial. However, when applied to motor drives for transportation (or other) applications, CEC presents a major problem: the cooling, which adds further weight and requires an additional tank. This problem is solved in the case where a cryogenic fuel such as liquid hydrogen or liquid natural gas are already available, opening up interesting possibilities. Wireless interconnection provides many advantages such as higher switching speeds, higher frequencies, and improved efficiency and reliability.

The great potential of silicon applications in power and energy conversion (solid-state transformers, inverters, etc.) has not yet been adequately addressed by the semiconductor industry. It is desirable to change this situation by promoting the concept of cryogenic energy conversion. The widespread application of electric vehicles requires the development of a sufficiently small, light, and efficient motor drive or Adjustable Speed Drive (ASD) to couple the fuel cell or battery output to the motor. Therefore multichip modules are proposed for an efficient motor drive system. Sooner or later, High-Temperature Superconductors (HTS) will be commercially available for applications in the power and energy generation and distribution fields at competitive prices: HTS cables, HTS transformers, HTS motors, and HTS generators. HTS components require Cryogenic Cooling. In most cases, such as in HTS cables and transformers, liquid nitrogen (LN_2), at a temperature of 77 K, will be used. The availability of HTS components requires a rethinking and redesigning of many energy systems. HTS Technology can best be supported by the new concept of Cryogenic Energy Conversion (CEC) based on Low Temperature Electronics (LTE) and Cryo-MOSFEts, Cryo-IGBTs, or other cryogenically operated devices. The efficiency of electrical motor operation can be drastically enhanced by applying CEC to Motor Drives or ASDs. CEC would miniaturize these drives, which can be 2-3 times more expensive in prior art technologies, and are also much larger and heavier than the motors they control. Size and weight reduction, along with improved conversion efficiency, is nowhere more important than in transportation vehicles. Every kilogram of weight reduction translates into a considerable energy saving for vehicles traveling hundreds of thousands of miles in a lifetime.

Motor drives using Cryo-Multichip Modules (CMCM) are intended for application in vehicles (buses, trucks, trains, ships, airplanes) as one important component in the coming age of Cryogenics which will combine High-Temperature Super conductors with Low-Temperature Semiconductors. Such Adjustable Speed Drives (ASDs) will, of course, find applications in stationary systems as well. Refer Figure 27.4 for a schematic of the cryobus fitted with ASD drive. Many manufacturing plants requiring ASDs already use liquid nitrogen for other purposes. The proposed CMD will provide smaller size, reduced weight and increased efficiency due to its application of Cryogenic Energy Conversion (CEC). Every kilogram of weight reduction translates into a considerable fuel saving over the lifetime of a vehicle running hundreds of thousands of miles. Such a development should be desirable in view of the fact that the federal government now mandates that cities of certain sizes must provide alternatively-fueled methods of public transportation.

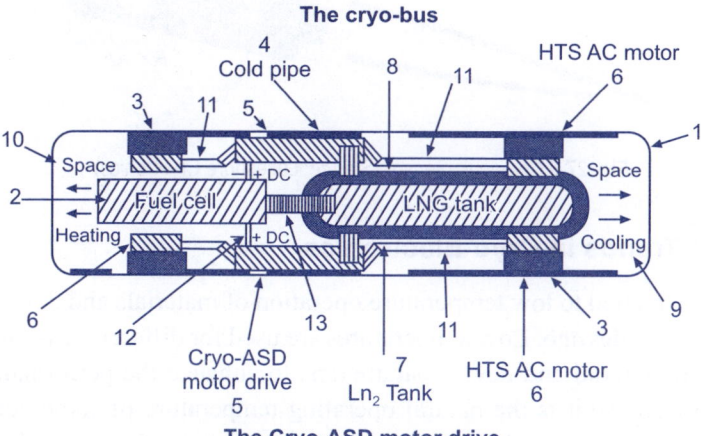

Fig. 27.4 Cryogenic Bus with ASD Drive

27.8 Cryogenics

Cryogenic Electronics in Advanced Sensor Systems

Cryogenic Electronics technologies are important from the standpoint of meeting sensor performance requirements. Radar and Electronic Warfare (EW) Systems requirements are based on the need to detect and track targets in the presence of clutter and jamming signals. Radars must transmit high fidelity signals, and receive returns with high fidelity and wide dynamic range for Doppler Processing, to discriminate targets from clutter and jamming. Requirements are translated into radar specifications: transmitter, receiver, antenna array, and other components must produce low phase noise. The receiver and analog to digital converter (ADC) need wide linear dynamic range. HTS technology provides key devices to achieve phase noise suppression and large signal dynamic range. Examples include: HTS resonators generate stable reference signals in the radar; switchable preselectors remove out of band jamming; HTS circuits enable digital to analog converters (DACs) and ADCs to provide direct synthesis of complex waveforms and digitization of received signals. EW System requirements are high dynamic range reception, and real-time signal processing; they must be wide-band and recognize a diverse set of waveforms. An HTS switchable preselector and delay line provide the front end for an advanced EW receiver and processor. HTS is cooled by cryogenic liquids. Special cryogenic sensors such as Ruthenium oxide temperature sensors have been developed. These are thick-film resistors used in applications involving magnetic fields. These composite sensors consist of bismuth ruthenate, ruthenium oxides, binders, and other compounds that allow them to obtain the necessary temperature and resistance characteristics. Each sensor model adheres to a single resistance versus temperature curve. Figure 27.5 shows a cryogenic sensor.

Fig. 27.5 Cryogenic Sensor (Source: Lakeshore Cryotronics)

Research Trends in Cryo-Electronics

The research related to low temperature operation of materials and devices started more than 5 decades ago. Low temperatures are used for different reasons: to study fundamental material and device parameters, to enhance the performance of the devices, or because it is the natural operating temperature of some components (e.g. superconductors, single electron transistors, etc.). Application fields are very broad including astrophysical, detectors and sensors, medical diagnostics, space, magnetic levitation transport systems, cryogenic instrumentation, etc. This paper aims to review the present status of silicon-based cryogenic electronics and to give an outlook related to the so-called emerging device technologies that are gaining in interest and are looking very promising for extending the ITRS roadmap. Special attention is given to dedicated cryogenic phenomena and to the increasing importance of quantization effects. With the very recent advent of nanotechnologies, multiple-level-wiring interconnections, hybrid superconductor-semiconductor systems, the rapid evolution of mini and micro-coolers, and the appearance of physical limitations for further downscaling, it is worth to explore the option of cryo-operation of electronic devices/systems, and identify possible limitations, possible applications, and establish future guidelines.

Introduction to CryoComputing

CryoComputing can be broken into two parts, Cryo and Computing. Cryo means cold, and computing refers to the use of computer devices. In general, CryoComputing refers to the use of cold in computers. Other commercial applications of the cryogenic heat pipes and phase change materials include cooling of supercomputers.

When the temperature of a conductor is lowered, the molecules of the conductor gather together. The denser material is easier for flowing electrons to move through. Lowering the temperature of a conductor reduces the energy of the conductor, and therefore reduces the number of phonons, This decreased concentration of phonons reduces the electron scattering, increasing conductivity. Because of reduced phonon collisions, electrical resistance is proportional to temperature T at lower temperatures, while at higher temperatures, the resistance is proportional to temperature T. A classical physical explanation of the relationship between temperature and resistance also exists. In a metal the atoms are arranged in a crystal lattice. As temperature increases, the atoms vibrate more rapidly and more randomly about their given positions in the lattice. The opposite is true at lower temperatures. The atoms vibrate less, and are more closely confined to their positions. The apparent random motion of particles at high temperatures causes more collisions When the resistance in microchips is reduced, electrons are allowed to move faster and without loss through materials. The computer on-off (high-low) signals it sends from one part of the computer to another are expedited when the conductors are cooled. Cooling the CPU is one of the most efficient ways to increase computer speed as CPU speeds are a bottleneck in the computer environment.

Both Intel and DEC are working on CryoComputing technologies. This is exciting new technology and Intel recently announced breaking the one GHz CPU speed barrier through the use of CryoComputing.

Cryogenic Processor

A Cryogenic processor is a unit designed to reach ultra-low temperatures (usually around the temp of $-150°C$) at a slow rate in order to prevent thermal shock to the components being treated. The first commercial unit was developed by Ed Busch in the late 1960s. The development of programmable microprocessor controls allowed the machines to follow temperature profiles that greatly increased the effectiveness of the process. Some manufacturers make cryoprocessors with home computers to define the temperature profile. The reliability of these computers in an industrial environment is greatly in doubt. Before programmable controls were added to control cryogenic processors, the "treatment" process of an object was previously done manually by immersing the object in liquid nitrogen. This normally caused thermal shock to occur within an object, resulting in cracks to the structure. Modern cryogenic processors measure changes in temperature and adjust the input of liquid nitrogen accordingly to ensure that only small fractional changes in temperature occur over a long period of time. Their temperature measurements and adjustments are condensed into "profiles" that are used to repeat the process in a certain way when treating for similarly grouped objects. The general processing cycle for modern cryogenic processors occurs within a three day time window, with 24 hours to reach the optimal bottom temperature for a product, 24 hours to hold at the bottom temperature, and 24 hours to return to room temperature. Depending on the product, some items will be heated in an oven to even higher temperatures. Some processors are capable of providing both the negative and positive extreme temperatures, separate units (a cryogenic processor and a dedicated oven) can sometimes produce better results depending upon the application. Cryogenic processors now-a-days are ensuring the accurate and consistent results for all products. As the technology sector improves, cryogenic processors will get better as they benefit from new computer systems. Ongoing research in the future will also improve their temperature treatment profiles.

Cooling of Computers by Cryocoolers or Cryogenic Liquids

An energy efficient cryocooler microprocessors and computer components using liquid and/or gaseous refrigerants in phase change. A cryogenic cooling system provides simultaneous steady stream (or pulse) cryocooling of a computer components with each part of the computer system receiving cooling at its optimum temperature. The system comprises an evaporator with liquid and gaseous refrigerant which evaporator is placed onto the microprocessor to disperse heat from said microprocessor and a enclosure around the cooled system that cools all other components in one stage. The use of the invented cryocooling system resulted in the increase 2.4 GHz "Celeron D" microprocessor frequency to 4.78 GHz. The RAM was also over clocked to 60 MHz above normal on a test system. The cryogenic cooling system is cost effective and energy effective in power and refrigerant consumption. On earlier computers the processors and memory are submerged in cryogenics fluids to lower the temperature at which they work fastest. Electrical resistance in metal alloys reduces dramatically at cryogenic temperatures sometimes behaving as superconductors.

Fig. 27.6 Liquid Nitrogen Cooling of Computers

An uncommon practice is to submerse the computer's components in a cryogenic liquid as shown in Figure 27.6. Personal computers that are cooled in this manner do not generally require any fans or pumps, and may be cooled exclusively by passive heat exchange between the computer's parts, the cooling fluid and the ambient air. Evaporation can pose a problem, and the liquid may require either to be regularly refilled or sealed inside the computer's enclosure. Liquid may also slowly seep into and damage components, particularly capacitors, causing an initially functional computer to fail after hours or days immersed. Liquid nitrogen may be used to cool an over clocked PC. As liquid nitrogen evaporates at −196 °C, far below the freezing point of water, it is valuable as an extreme coolant for short over clocking sessions. In a typical installation of liquid nitrogen cooling, a copper or aluminum pipe is mounted on top of the processor or graphics card. After being heavily insulated against condensation, the liquid nitrogen is poured into the pipe, resulting in temperatures well below −100°C.

Cryogenic Cooling of Supercomputers

Supercomputers are a class of extremely powerful computers. The term is commonly applied to the fastest high-performance systems available at any given time. Such computers have been used primarily for scientific and engineering work requiring exceedingly high-speed computations. Common applications for supercomputers include testing mathematical models for complex physical phenomena or designs, such as climate and weather, evolution of the cosmos, nuclear weapons and reactors, new chemical compounds (especially for pharmaceutical purposes), and cryptology. As the cost of supercomputing declined in the 1990s, more businesses began to use supercomputers for market research and other business-related models.

Supercomputers have certain distinguishing features. Unlike conventional computers, they usually have more than one CPU (central processing unit), which contains circuits for interpreting program instructions and executing arithmetic and logic operations in proper sequence. The use of several CPUs to achieve high computational rates is necessitated by the physical limits of circuit technology. Electronic signals cannot travel faster than the speed of light, which thus constitutes a fundamental speed limit for signal transmission and circuit switching. This limit has almost been reached, owing to miniaturization of circuit components, dramatic reduction in the length of wires connecting circuit boards, and innovation in cooling techniques (e.g., in various supercomputer systems, processor and memory circuits are immersed in a cryogenic fluid to achieve the low temperatures at which they operate fastest). Rapid retrieval of stored data and instructions is required to support the extremely high computational speed of CPUs. Therefore, most supercomputers have a very large storage capacity, as well as a very fast input/output capability. Figure 27.7 shows a Cray supercomputer. Cray used expensive state-of-the-art custom processors and cryogenic cooling systems with liquid nitrogen to achieve speed records.

Fig. 27.7 The Cray Supercomputer

Cryogenic Cooling of CPU Chips

The cooling of CPU chips such as the Pentium^{TM} and ALPHA^{TM} chips to cryogenic temperatures demonstrated increase in their speeds. Different technologies have been proposed to implement cryogenic cooling such as pool boiling of MCM modules where high power chips are housed close together using liquid cryogens, direct cooling of chips using micro-sterling based cryocoolers such as pulse-tube refrigerators, thermoelectric coolers such as those used in the cooling of infrared sensors, and others. All of these technologies have in common complex cooling delivery systems in addition to the modification of the target chip in order to be properly cooled.

One further challenge to the thermal management of high power electronics to cryogenic temperatures is the density or number of heat generation sources to be cooled. When the designer is faced with cooling one or two 50W chips to 80 K, the equipment available, despite its costs and complexity, is being manufactured by a few companies around the world and are specifically tailored to low heat removal rates. In cases when the power dissipations are approaching 500 to 1000 W, cryogenic systems to generate the cooling load are approaching industrial size systems such as gas liquefaction systems for 70-80 K temperature range and industrial vapor-compression refrigeration systems for the −30 to −50°C temperature range. The limitations are not confined to heat removal capacities. Volume restrictions are a very important consideration. The push for MCM packaging for cryogenically cooled computer chips is just one manifestation of the problem. MCM packaging reduces the distance between the CPU chip and its peripherals, thus increasing the speed of the system. Unfortunately, reduced volume packaging imposes very stringent requirements on the cooling system at room temperatures, and extraordinary requirements at cryogenic temperatures. Cryocooling systems take up space when existing technology is applied to single chips.

Cryogenics for Telecommunication Applications

In many telecommunications applications, switching/signal processing equipment is commonly placed in outdoor cabinets. The housed equipment generates heat that must dissipated while keeping the air temperature inside the cabinets within prescribed limits for optimum performance. Furthermore, the enclosure, being outdoors, receives full solar irradiation, creating an extra heat load that must be handled. In many cases the temperature inside the cabinets can remain higher than ambient, but in many applications, such as cable TV and cellular phones, the temperature inside the cabinets must be kept between 20 to 30°C to maintain high reliability. This clearly demands refrigeration units to be installed as part the equipment. However, air conditioning equipment is bulky, expensive to maintain and, must be ozone friendly. Thus, the push for higher processing speeds and capabilities in electronics and telecommunications systems with increasing clarity and reduced distortion and noise are making equipment designers look for highly innovative technologies. Higher speeds mean higher power dissipation densities; and cooling equipment (especially superconductors) to cryogenic temperatures appears to be one of the promising technologies. However, achieving cryogenic cooling is not simple and requires sophisticated, special equipment. Novel refrigeration systems derived from currently used low-temperature cryogenic cycles are being proposed and developed. These new systems will in all likelihood occupy less space than currently installed units, and some may use air as the working fluid —which will eliminate all environmental concerns. These systems are brought about by a push in installing equipment that works best at low temperatures down to cryogenic levels (0°C to −100/200°C) since reliability and power consumption reductions are enhanced.

References

1. Internet website www.electrochem.org.
2. Internet website www.grc.nasa.gov.
3. Internet website http://www.cryocircuits.com.
4. US patent 6798083 Cryogenic power conversion for fuel cell systems especially for vehicles.
5. Internet website www.lakeshore.com.
6. Internet website www.thermal-engineering.com.
7. Internet website of Wikipedia on cryogenics.
8. Cryogenic Engineering – Barron.

Questions

Q.1. What is cryo-electronics? What are its applications?

Q.2. Write a short note on cryogenic electronics for space systems.

Q.3. What is cryo computing? What are its applications?

Q.4. What is a supercomputer? How is it cooled?

CHAPTER 28

Applications of Cryogenics - VI Transport

INTRODUCTION

The principal use of cryogenics in transport applications is use of cryogenic fuels. Cryogenic fuels require storage at extremely low temperatures in order to maintain them in a liquid state. Cryogenic fuels most often constitute liquefied gases such as liquid hydrogen, liquid natural gas and liquid nitrogen. Cryogenic fuels have been used as propellants as well as fuels for IC engine vehicles. Russian aircraft manufacturer Tupolev is currently researching a version of its popular design Tu-154 with an LNG cryogenic fuel system. Cryogenic fuels are the fuel of choice for some car and vehicle manufacturers. These vehicles such as buses require longer range capabilities and longer hours of operation than other vehicles and cryogenic fuels meet these requirements. Even though cryogenic liquid fuel tanks can store more in a given volume than compressed gas tanks and liquid fuel carries more energy density than gaseous fuel, there are some downsides to using cryogenic liquid fuel to power large vehicles. Issues with fuel boil-off, liquefaction, weight, volume and tank cost need to be addressed for economical use of cryogenic liquid fuels. Safety concerns are also more of an issue with fuels that is in a cryogenic liquid state. Cryogenic liquid fuels require special handling, methods of storage and transport need to be strictly adhered to in order to avoid critical accidents. Elimination of sparks from electrical equipment, static electricity, open flames or extremely hot objects need to be employed to insure one's safety. These safety measures for dealing with cryogenic liquid fuel will be required for safe working of the vehicle as well as fueling stations and other parts of manufacturing and distribution chain.

BMW has a production ready Mini vehicle powered by cryogenic liquid hydrogen, called the BMW Mini Hydrogen rolled out in 2001. In 1999, the BMW 750hl was introduced as the first liquid hydrogen powered car. The BMW 750hl was powered by a 12-cylinder combustion engine with two independent fuel induction systems, which are electronically controlled to run off of either gasoline or liquid hydrogen tanks. The BMW 750hl also contains a solar-powered voltaic array, which generates electricity to split water into hydrogen and oxygen. The oxygen is released into the atmosphere and the hydrogen is liquefied and stored at extremely low temperatures (–253°C). During the internal combustion phase, the hydrogen combines with oxygen to power the vehicle and water is released as steam.

28.2 Cryogenics

Another exciting development in cryogenic liquid hydrogen fuel technology is the world's first Unmanned Aerial Vehicle (UAV) powered by liquid hydrogen has completed flight tests. California-based AeroVironment developed the UAV for purposes such as hurricane tracking from heights such as 65,000 - 98,000 feet above sea level. The unmanned plane can stay in the air for approximately 24-hours before it needs refueling. This kind of cryogenic liquid fuel technology could eventually lead to breakthroughs in the airline industry, future reducing dependence upon foreign oil and greenhouse gases.

Liquefied Natural Gas (LNG) as a Transportation Fuel

Natural gas was first liquefied in the 19th century by Michael Faraday and Karl Von Linde. LNG became viable in 1917, when the first LNG plant went into operation in West Virginia. The first commercial liquefaction plant was built in Cleveland, Ohio, in 1941. In January 1959, the world's first LNG tanker carried LNG cargo from Lake Charles (Louisiana) to Canvey Island (United Kingdom). This event demonstrated that large quantities of LNG could be transported safely across the ocean. The first liquefaction plant in the world was commissioned in Algeria to supply this contract with gas production coming from huge gas reserves found in the Sahara. In 1969, Alaska's Kenai plant (which currently has a capacity of 1.3 mtpa) began LNG deliveries to Japan's Tokyo Electric Power Company (TEPCO). In 1972, Brunei became Asia's first producer of LNG, bringing on stream an LNG plant at Lumut, which now has a capacity of 6.5 mtpa, and supplies LNG to Korea and Japan. The demand for LNG in Asia continued to rise and Malaysia entered the LNG market in 1983 followed by Australia in 1989.

For decades, natural gas has provided clean power to thousands of households and businesses nationwide. Today, more than 1,000 vehicles traveling on U.S. roads are powered by natural gas that is cooled to form liquefied natural gas (LNG). LNG is formed when natural gas is cooled to temperatures of 262°C below zero, thus producing a viable vehicle fuel used mainly in heavy-duty trucks and buses. LNG is odorless, colorless, non-corrosive, and non-toxic. When extracted from underground reserves, natural gas is composed of approximately 90% methane. During the liquefaction process, oxygen, carbon dioxide, sulfur compounds, and water are removed, purifying the fuel and increasing its methane content to almost 100%. As a result, LNG-fuelled vehicles can offer significant emissions benefits compared with older gasoline or diesel vehicles, and can significantly reduce carbon monoxide and particulate emissions as well as nitrogen oxide emissions. The number of LNG-fueled transit buses is expanding rapidly as many cities try to reduce air pollution levels. Bus fleets in Orange County (California) and Phoenix (Arizona) are fueled entirely by LNG. Its cleaner burning characteristics can result in longer engine life and reduced maintenance costs. Using LNG eliminates the need for periodic tank inspections. In addition, when compared with gasoline-powered vehicles, some maintenance savings are anticipated for vehicles using LNG because of the reduced frequency of oil changes. Because of the fuel's low temperatures, only trained personnel should maintain LNG vehicles. An LNG chain refers to the combined activities of exploration, shipping, storage and dispensing to vehicles or industries. Overall LNG has a bright future as a transportation fuel.

Production of LNG

LNG is created when natural gas is cooled down to $-162°C$ using a variety of techniques. Even when liquefied, LNG contains only approximately 60% of the energy of gasoline per unit volume. While the technology of insulating containers has advanced significantly, loss of LNG due to heat transfer into the containers eventually occurs. The natural gas is purified or conditioned by removing water and CO_2 impurities from it since these may clog the processing equipment due to the extremely low temperatures. For instance, water would convert to ice and clog valves or it could enter the storage tank and be dispensed in fuel that would contaminate the vehicles fuel delivery system equipment.

The natural gas is then compressed to approximately 3,000 pounds per square inch gauge (psig) and cooled by an industrial refrigeration unit consisting of a compressor, evaporator and a chiller heat exchanger. The gas is then cooled to $-60°F$ and then routed to the main heat exchanger, which has a gas to gas counter current design, which reduces the natural gas temperature to $-100°F$. The gas is then expanded through a microprocessor-controlled expansion valve (a Joule-Thomson valve) to a pressure of 90–125 psig. This process uses a closed loop system for identifying temperatures and pressures, properly controlling the JT orifice to provide relatively high liquefaction efficiencies over the fluctuating range of temperatures and pressures. Approximately 50% of the flow across the orifice can be liquefied, and the remaining 50%, still a cold gas in the range of $-180°F$, is still not cold enough to be a liquid. This vapor must be withdrawn from the storage vessels in order to maintain the operating pressure of the storage vessels. Therefore, the cold gas is removed and passes through the counter current side of the main heat exchanger. This cold gas is the medium that reduces the temperature of the feedstock natural gas from $-60°F$ to $-100°F$. Although liquefaction can be achieved at pressures as low as 681 psig, the most effective pressure to liquefy natural gas for small-scale on-site liquefaction, appears to be between 2700 psig and 3000 psig. There is a lower efficiency beyond 3100 psig, which means that the energy spent for compression over 3000 psig yields very little if any increase to the liquefaction rate. This is apparent from the entropy chart for natural gas.

Properties of LNG

Formula = CH_4

Molecular weight = 16

Temperature = $-162°C$

Density = -26.5 Lb/Cu.Ft

Boiling point = $-259°F$

28.4 Cryogenics

Flammability limits = 5 to 15

Stoichiometric air to fuel ratio = 17.2

Flame visibility - Visible in all conditions

Octane number = 120

Autoignition temperature = 540°C

Lower heating value = 88100 Btu/gallon

Economics of LNG

Depending on the quantity of vehicles purchased and the equipment used, LNG heavy-duty trucks or buses can cost an additional $30,000 to $50,000 over base CNG vehicles. Industry experts expect these costs to drop as the LNG market development and vehicle production rises. Fuel dispensing and fuel storage required for LNG typically costs $15,000 to $22,000 per vehicle. In addition, LNG's price is highly dependent on geographic location, purity of feedstock, transportation costs, and quantity of fuel purchased, but LNG's cost per mile is generally less than or equal to the price of gasoline.

Advantages of LNG

1. LNG produces less greenhouse gas and toxic emissions than low-sulfur diesel and biodiesel.
2. Due to the molecular structure of gaseous fuels, LNG will always result in lower particulate emissions than diesel.
3. The supply of LNG is dependable due to its vast reserves around the world.
4. LNG engines can achieve a reduction of up to 50% in noise levels compared to diesel engines.
5. LNG prices are always stable.
6. LNG storage pressure is 700 kPa, which is lower than the CNG pressure of 2000 kPa.
7. LNG has around 3.5 times the fuel density of CNG. Vehicle operators can therefore achieve much greater haulage ranges using LNG.
8. Fill times for LNG vehicles are typically at the same rate as diesel or faster.
9. LNG can safely be stored in stainless steel vessels.
10. OEM engines with warranty and service are available.

Disadvantages of LNG

1. Higher cost
2. Inadequate distribution infrastructure

3. Hazards of frost bite in case of leakage
4. Need for monitoring of tanks to prevent fuel vent off
5. Absence of vehicle standards

Hazards of LNG

An LNG vehicle parked indoors and unmoved for a week or more will vent a flammable gas mixture that could catch fire in the vicinity of an ignition source. To address this safety issue, LNG use should be restricted to frequently driven fleet vehicles or to vehicles stored outdoors. Only trained personnel should service the vehicles. In addition, refueling vehicles with LNG requires training because of the fuel's ultra-low temperature. It can cause frostbite if it comes in contact with skin. Since LNG is almost 100% methane—a greenhouse gas—it can also contribute to global climate change if accidentally released into the air. Methane is slightly soluble in water and, under certain environmental conditions (anaerobic), it does not biodegrade easily. If excess amounts accumulate, the gas can bubble from the water, possibly creating a risk of a fire or an explosion.

Material Compatibility of LNG

Being very clean and pure minimizes the materials compatibility concerns for LNG. For LNG fuel tanks, stainless steel is the preferred material and instances of material compatibility problems are rare. Aluminum also has been used as a tank material without such problems. Carbon steels are not used since their performance at low temperatures is questionable, that is, they become susceptible to brittle factures. While tanks are usually made from stainless steel or aluminum, LNG fittings may use some nickel alloys, brass and copper, in addition to stainless steel aluminum. Due to the cryogenic temperatures, few elastomers are relied on in LNG fuel systems. For sealing purposes, welded fittings, compression fittings, and flanged joints are commonly used with little need for elastomers. For places where elastomers must be used, Teflon is the preferred choice since it retains its strength unlike typical synthetic elastomers that become brittle at low temperatures.

Transportation of LNG

Transportation and supply is an important aspect in the gas sector, since LNG reserves are normally quite distant from consumer markets. LNG is usually transported through pipelines. There is a pipeline network in Europe and North America. Apart from pipelines, LNG is also transported using tanker trucks, railway tankers, and purpose-built ships known as LNG carriers. LNG is sometimes taken to crygenic temperatures to reduce the tanker capacity. Figure 28.1 shows an LNG carrier.

Fig. 28.1 LNG Carrier

Storage of LNG

On land and at offshore terminals, LNG is stored in specially engineered and constructed double-walled storage tanks. The LNG storage tanks have approximately 3-ft thick exterior concrete walls. An inner tank is made of a special steel/nickel alloy to accommodate the very cold LNG. The space between these walls is filled with insulation to maintain a cold environment for the LNG. Should a leak occur in the inner wall, all the LNG would be contained in the space by the outer wall. Large tanks have a low-aspect ratio (height to width) and are cylindrical in design with a domed roof. Storage pressures in these tanks are very low at less than 50 kPa. Figure 28.2 shows the LNG storage site. Sometimes more expensive frozen-earth, underground storage is used. Pre-stressed concrete backed up with suitable thermal insulation is designed to be under and above ground to suit sites conditions and local safety regulations and requirements. Sophisticated monitoring systems provide constant surveillance for internal leaks. The tanks are not pressurized. The liquid is taken from the tank and regasified for delivery through pipelines to the industry. Smaller quantities (190,000 U.S. gallons and less) may be stored in horizontal or vertical, vacuum-jacketed pressure vessels. These tanks may be at pressures anywhere from less than 50 kPa to over 1700 kPa. LNG tanks are all double-walled because an evacuated space is required to achieve sufficient insulation to keep the gas from vaporizing too quickly. A "hold time" (time from when the tank is filled to when the pressure build-up need the vapour to be vented) of at least five days is required for practical LNG tanks. In addition to the evacuated interstitial space, several different types of insulation may be used. The goal is to make the conductive, convective, and radiant heat transfer into the LNG as low as possible. LNG tanks are made from stainless steel or aluminum according to ASME pressure vessel codes. The primary concern for LNG tanks is the thermal cycling they must endure from ambient temperature to 162°C. These tanks are usually cylindrical, but they may also be spherical. The center section of cylindrical tanks is made from rolled sheet welded together. The ends are usually formed under hydraulic pressure, and then welded to the center section. Stand-offs are welded between the inner and outer tanks to connect them and provide stability to the operation. To protect the tank from rupture due to overpressure, vent valves are included that will release some vapor when the maximum set pressure is reached.

Fig. 28.2 LNG Terminal

Piping for LNG

Piping for LNG comes in two types: single-wall and vacuum-jacketed. All the piping is made from stainless steel to provide durability at cryogenic temperatures. The inner section of vacuum-jacketed piping is also stainless steel for the same reasons. Single-wall piping may also be insulated through the addition of external polyurethane or multi-layered insulation. Piping for LNG service is connected via threaded joints, flanged connections, swaged fittings, or welded connections. Long runs of LNG piping will expand and contract significant amounts when LNG is present. Bellows-type expansion joints are recommended to prevent stresses from building up in piping systems due to temperature cycling. These expansion joints can be welded into the piping or attached using flanged connections. Pressure relief valves play an important role in LNG piping systems. Any section of LNG piping that can trap LNG must have a pressure relief valve to prevent failure of that section of pipe should all the LNG vaporize.

LNG Dispensers

Primary purpose of LNG dispensers is to meter and control the flow of LNG to the vehicle. LNG dispensers do not include a pump. The meters inside LNG dispensers have low-flow restriction and use stainless steel internal flow surface. Typically, vacuum-jacketed piping inside the dispenser is made of stainless steel. LNG filters in LNG dispensers are not common, though most will have at least stainless steel wire mesh strainers that will prevent metal shavings and large foreign matter from going into the vehicle fuel tank. LNG dispenser hoses should have breakaway fittings to prevent a vehicle from pulling away from the dispenser during refueling.

LNG Vehicles

Researchers at West Virginia University developed LNG trucks are using the Cummins L-10 (natural gas engine) with advanced electronic fueling control. The L-10 engine was equipped with an oxygen sensor feedback. The Kenworth Truck Company in U.S. has developed T 800 range of LNG trucks at its manufacturing facility in Renton, Washington. Under an exclusive agreement with Westport Innovations Inc. in Vancouver, Kenworth will use Westport's LNG fuel system technology adapted for the Cummins ISX 15-litre engine. Pacific Gas and Electric Company in San Francisco recently became the first utility in the nation to operate Kenworth T 800 LNG-powered trucks. Figure 28.3 shows the Kenworth LNG Truck in the U.S.A.

Fig. 28.3 LNG Truck

The Westport LNG system is available with 400 and 450 horsepower ratings and up to 1,750 lb-ft torque for heavy-duty port, freight, and vocational applications. LNG fuel tanks can be configured to suit customers' requirements. The Westport LNG system comprises LNG fuel tanks, proprietary Westport fuel injectors, cryogenic fuel pumps and associated electronic components to facilitate robust performance and reliable operation. The Westport LNG system is 2007 EPA and CARB certified to 0.8g/bhp-hr NOx and 0.01g/bhp-hr. In September 1997, China's first commercial buses powered by LNG took to the streets of Haikou, capital of China's island province of Hainan in the South China Sea. In early 1997, LNG storage and fuel-handling equipment made by Denver, Colorado-based Cryenco Sciences, Inc., was licensed for sale in China for the first time, further signaling the country's growing interest in LNG. An earlier LNG demonstration project took place in 1990, when Kaifeng Cryogenic Devices Manufacturing Company converted a bus to LNG in the eastern interior province of Henan.

Since LNG engines are derivatives of heavy-duty diesel engines, it is possible for them to have the same or more power than the equivalent displacement diesel version. Limiting factors to power output include oxides of nitrogen emissions (increased power means richer operation and generation of more oxides of nitrogen for a given displacement engine) and exhaust valve life (spark-ignition engines experience higher exhaust valve temperatures than diesel engines). Rather than acceleration, heavy-duty engine performance is more a function of maximum horsepower and "torque rise." Both characteristics can be made equivalent or better for LNG versions of heavy-duty diesel engines. LNG vehicles should have the same exhaust emissions as CNG vehicles except that these vehicles might also occasionally vent methane from the fuel storage system. If LNG vehicles are used regularly, no venting of methane is necessary. However, if an LNG vehicle is left idle for a week or more, it will need to start venting to prevent excessive pressure in the fuel tanks. Actual LNG emissions will vary with engine design; however, the potential emission reductions offered by LNG relative to diesel are as follows: Production of half the particulate matter of average diesel vehicles with significant reduction in CO, nitrogen oxide and volatile organic hydrocarbon emissions.

Liquefied Hydrogen (LH$_2$) as a Transportation Fuel

Liquid hydrogen is the liquid state of the element hydrogen. It is a common liquid propellant for rocket applications. In the aerospace industry, its name is often abbreviated to LH$_2$. To exist as a liquid, H$_2$ must be pressurized and cooled to a very low temperature, 20.27 K (−252.87°C). One common method of obtaining liquid hydrogen involves a compressor resembling a jet engine in both appearance and principle. Liquid hydrogen is typically used as a concentrated form of hydrogen storage. As in any gas, storing it as liquid takes less space than storing it as a gas at normal temperature and pressure. Once liquified it can be maintained as a liquid in pressurized and thermally insulated containers also known as Dewar vessels as shown in Figure 28.4

Fig. 28.4 Liquid Hydrogen Tank

In rocket engines, liquid hydrogen is frequently used to cool the engine nozzle and other parts before being mixed with the oxidizer (often liquid oxygen) and burned. The resulting exhaust of such LH_2 engines is water with traces of ozone and hydrogen peroxide. Liquefied hydrogen can be used as a fuel in an internal combustion engine or fuel cell. Various concept hydrogen vehicles have been built using this form of hydrogen like the BMW H_2R. LH_2 injection is preferable as compared to gas injection as it reduces the gas-displacement effect and also improves breathing and lowers pumping losses.

History of Liquid Hydrogen

When the young James Dewar was appointed Fullerian Professor of Chemistry at the Royal Institution, London, in 1877, he set about understanding of heat transfer processes and of how to achieve thermal insulation; the second was the basic data on the properties of fluids for producing low temperatures. After fifteen years in 1892, he was able to employ his invention of the silvered, double walled, glass vacuum vessel to contain cryogenic liquids for the first time, for relatively long periods, before they evaporated. His discovery of silvering as an effective means of reducing the radiated heat flux component was a breakthrough. From 1802, the Dewar flask quickly became the standard container for cryogenic liquids, leading to the successful liquefaction of hydrogen in later years. Although liquid oxygen was first used in rockets from 1926 by Goddard and in the German V2 rockets of 1944, the combination of liquid hydrogen and liquid oxygen was first used in the U.S. rocket "Atlas Centaur" in 1961. This achievement represents the breakthrough in space rocketry which led to the large-scale production of liquid hydrogen and the Saturn series from 1963, the ESA Ariane series from 1975, the NASA Space Shuttles from 1983 and others. Many early rocket theorists believed that hydrogen would be a marvellous propellant, since it gives the highest specific impulse. As hydrogen in any state is very bulky, for space vehicles it made sense to store it as a cryogenic liquid.

Production of Liquid Hydrogen

The path to liquid hydrogen involves the liquefaction process, which is expensive, energy consuming and has a low thermal efficiency. The process involves cooling gaseous hydrogen to below $-253°C$ using liquid nitrogen and series of compression and expansion steps. The liquid hydrogen so produced is then stored in large insulated tanks and then filled into delivery trucks and transported to dispensing sites, where it is dispensed to the vehicles. Countries like Japan have set up liquid hydrogen production facilities that obtain hydrogen from producer gas. These facilities represent the world's first approach to mass-producing liquid hydrogen from a by product gas. Figure 28.5 shows the photograph of a typical liquid hydrogen facility. These facilities produce liquid hydrogen, which is efficiently transportable and storable, from large quantities of by-product gas generated in the steel making process. They can produce 0.2 ton of high-purity liquid hydrogen everyday. The manufacture of liquid hydrogen ensures supply of hydrogen to fuel cell vehicles. The Helium brayton cycle is used for hydrogen liquefaction and the output of the plant is about 200 kg/day which is suffcient for filling approximate 50 fuel cell vehicles per day. The temperature of liquid hydrogen is –250°C and the pressure is 10 bar with a purity level of 99.99%.

Fig. 28.5 LH_2 Production Facility in Japan

Properties of Liquid Hydrogen

Formula = H_2

Molecular Weight: 2.016

Boiling Point @ 1 atm: –423.0°F (–252.8°C, 20°K)

Freezing Point @ 1 atm: –434.5°F (–259.2°C, 14°K)

Critical Temperature: –399.8°F (–239.9°C)

Critical Pressure: 188 psi (12.9 atm)

Density, Liquid @ B.P., 1 atm: 4.23 lb./cu.ft.

Specific Gravity, Liquid @ B.P., 1 atm: 0.0710

Latent Heat of Vaporization: 389 Btu/lb. mole

Flammable Limits @ 1 atm in air 4.00%: –74.2% (by Volume)

Detonable Limits @ 1 atm in air 18.2%: –58.9% (by Volume)

Autoignition Temperature @ 1 atm: 1060°F (571°C)

Expansion Ratio, Liquid to Gas, B.P. to 68°F (20°C): 1 to 848.

Hazards of Liquid Hydrogen

The hazards associated with handling liquid hydrogen are fire, explosion, asphyxiation, and exposure to extremely low temperatures. The potential for forming and igniting flammable mixtures containing hydrogen may be higher due to migration of hydrogen through cracks. The minimum ignition energy for flammable mixtures containing hydrogen is extremely low. Burns may result from unknowingly walking into a hydrogen fire. The fire and explosion hazards can be controlled by appropriate design and operating procedures. Adequate ventilation will help reduce the possible formation of flammable mixtures in the event of a hydrogen leak or spill and will also eliminate the potential hazard of asphyxiation. Protective clothing should be worn to prevent exposure to extremely cold liquid and cold hydrogen vapors. Cold burns may occur from short contact with frosted lines, liquid air that may be dripping from cold lines or vent stacks, vaporizer fins, and vapor leaks. Air will condense at liquid hydrogen temperatures and can become an oxygen-enriched liquid, which increases the combustion rate of flammable and combustible materials. Full face shield, safety glasses, insulated or leather gloves, long-sleeved shirts, and pants without cuffs should be worn when working on liquid hydrogen systems.

Advantages of Liquid Hydrogen

1. High energy density
2. Liquid tankers are cheap and efficient
3. Can be converted to gaseous hydrogen
4. Eliminates need for compression.

Disadvantages of Liquid Hydrogen

1. Thermodynamics limits energy efficiency
2. High energy consumption during liquefaction
3. High costs
4. Complexity of handling cryogenic liquids.

Storage of Liquid Hydrogen

State-of-the-art mobile liquid hydrogen storage systems, as shown in Figure 28.6, consist of double-wall vessels of cylindrical shape. Preferred shell materials are stainless steel or aluminum alloy, since they are very resistant to hydrogen brittleness and show negligible hydrogen permeation. For stainless steel, the minimum wall thickness for cryogenic vessels is at least 2 mm. The weight of the whole tank system including valves and heat exchanger is above 1000 kg. The space between the inner and outer vessel is mainly used for the thermal insulation. Several layers of insulation foils reduce the heat entry by thermal radiation. The high vacuum with a lower storage pressure reduces the thermal convection. The insulation consists of reflective foils (aluminum or aluminized polymer) separated by spacers (glass fibre or polymeric). Support structures, which keep the inner tank in position to the outer tank, are made of materials with low thermal conductivity (e.g. glass- or carbon-fibre reinforced plastics). The storage tanks are designed as per ASME (section VIII), ISO TC 220/CD 21009 and CGA H-3 Codes. To protect the tank from rupture due to over-pressure, vent valves are included that will release some vapor when the maximum set pressure is reached. Vent valves may be made from stainless steel, brass, and aluminum.

Fig. 28.6 Liquid Hydrogen Storage Tank from Air Products

Most of the liquid hydrogen tanks are above ground, however underground tanks are also available. Underground tanks must be protected from corrosion, and actions must be taken to prevent the surrounding ground from freezing which could put external stress on the tank. Underground storage tanks provide higher capacity per unit volume and increase safety. Furthermore an underground tank can better maintain the cryogenic temperatures and pressures. Liquid hydrogen tanks can be spherical or cylindrical. Spherical tanks reduce the surface area and thereby reduce the evaporative losses. Tank capacities range from 5700 litres to 95000 litres. Sometimes geological storage can also be used for hydrogen if suitable rocky terrain is available.

Transportation of Liquid Hydrogen

Liquid hydrogen can be transported in pressurized tanks by truck, rail, barge, or ship. Insulation of the storage tanks is of utmost importance. Due to the very low boiling point of hydrogen, losses resulting from boil-off can be considerable. Cryogenic liquid hydrogen trucks have capacity of 3000 to 4000 kg of liquid hydrogen. The tankers have similar construction as stationary tanks, but have to meet DOT requirements as well. Figure 28.7 shows a liquid hydrogen tanker.

Fig. 28.7 Liquid Hydrogen Tanker

Piping for Liquid Hydrogen

Piping for LH_2 comes in two types: single-wall and vacuum-jacketed. All the single wall piping is made from stainless steel for durability purposes at cryogenic temperatures. The inner section of vacuum-jacketed piping is also stainless steel for the same purpose. Single-wall piping may also be insulated through the addition of external polyurethane or multi-layered insulation. Piping for LH_2 service is connected via threaded joints, flanged connections, or welded connections. Welding provides the most secure connection and should be used whenever possible. Flanged connections should be used where welded connections cannot. Flanged fittings tend to loosen if they are subjected to repeat temperature cycling. Threaded connections should be used only sparingly, unless welded after installation. If threaded connections are used in several consecutive 90-degree turns, obtaining a leak-tight system becomes very difficult.

Dispensers for Liquid Hydrogen

LH_2 dispensers are very similar to LNG dispensers in function as their primary purpose is to meter and control the flow of liquid hydrogen to the vehicle. The meters inside LH_2 dispensers have low flow restriction and use stainless steel internal flow surface. Piping inside the dispenser is typically vacuum-jacketed stainless steel. Recent development in the liquid hydrogen filling of the cars in Germany is the automatic filling by a robot. ARAL and BMW have developed the robotic systems and Linde is supplying the liquid hydrogen lines and the LH_2 coupling. Refer Figure 28.8 for the robotic liquid hydrogen dispenser. The safety authority TÜV has certified the safety-related interfaces between the robotics, the liquid hydrogen lines and the LH_2 coupling. The robotics system is installed above the carriageway on an island in an aboveground level version. The movement mechanism can be moved in four axes, so that the end effector (end point of the robot arm) can be moved freely in a limited working area (1000 × 800 × 200 mm). The LH_2 hose runs externally and the electric cables and other control lines run inside the robot arm to the end effector.

Fig. 28.8 Robotic Liquid Hydrogen Dispenser

Mechanical claddings, which activate safety-oriented switch elements in the event of a collision, meet the requirements for the protection and safety of users. In addition, gas sensors on the robot and underneath the filling station roof check that the robotics system is operating safely. In the first project phase, the filling robot carried out 500 filling operations. The time taken to fill a tank with a volume of 125 Litre is approximately three minutes. The use of the cold drawable Linde coupling means that it is possible to fill several vehicles consecutively without any problem.

Vehicle Emissions from Liquid Hydrogen

Hydrogen is the cleanest fuel available to man. Hydrogen-fueled ICE's and gas turbine engines have negligible emissions of air pollutants. Hydrogen-powered-fuel-cell vehicles have zero emissions. Burning hydrogen with air under appropriate conditions in engines or gas turbines results in very low emissions. Trace hydrocarbon and carbon monoxide emissions, if generated at all, can result only from the combustion of motor oil in the combustion chamber of ICE's. Nitrogen oxides (NOx) emissions are formed and they increase exponentially with the combustion temperature. Therefore, these can be influenced through appropriate process control. Particulate and sulfur emissions are limited to small quantities of lubricant remnants. Engines fueled with hydrogen produce no carbon dioxide emissions.

BMW Liquid Hydrogen Cars

BMW has a Mini vehicle powered by liquid hydrogen, called the BMW 750hl as shown in Figure 28.9. A 12-cylinder engine with two independent fuel induction systems, which are electronically controlled to run off either gasoline or liquid hydrogen, powers the BMW 750hl. The BMW 750hl also contains a solar-powered voltaic array, which generates electricity to split water into hydrogen and oxygen. The oxygen is released into the atmosphere and the hydrogen is liquefied and stored at extremely low temperatures ($-253°C$). During the internal combustion phase, the cryogenic hydrogen combines with oxygen to power the vehicle and water is released as steam. The fuel tank of the BMW 750hl vehicle holds approximately 125 litres of liquid hydrogen corresponding to a driving range of over 300 km. The hydrogen combustion engine accelerates the vehicles to over 200 km/h.

Fig. 28.9 BMW Liquid Hydrogen Cars

The hydrogen is burned in the engine with excess air. This concept lowers the flame temperature in the combustion area to below the critical limit above which nitrogen oxides are created. Even without additional waste gas processing by catalytic converters, hydrogen engines thus run practically without emissions. BMW has recently unveiled its series 7 luxury Sedan on liquid hydrogen as shown in Figure 28.9. On board this BMW hydrogen vehicle, a PEM fuel cell is provided for the on-board power supply, which is used as a battery substitute (Auxiliary Power Unit, APU). It provides an output of up to 5 kW. As a further development, the integration of a new larger tank system with a capacity of 170 liters is used in this vehicle.

References

1. Cryogenic Engineering – B.A. Hands.
2. Cryogenic Systems – Randall Baron.
3. Alternative Fuels – Dr. S.S. Thipse.
4. Internet website of Petronet LNG.
5. Nathan Hill – ESA Publication on Cryogenics - 2008.
6. Internet website of BMW Corporation.
7. Heat Transfer in LNG Engineering - Adorjan.

Questions

Q.1. What are the applications of cryogenics in Transport?

Q.2. Explain the benefits and drawbacks of using cryogenic alternative fuels.

Q.3. Write a short note on LNG as a fuel.

Q.4. What are the advantages and disadvantages of LNG?

Q.5. Write a short note on Liquid Hydrogen as a fuel.

Q.6. Discuss the economics of a Liquid Hydrogen vis-à-vis Gaseous Hydrogen.

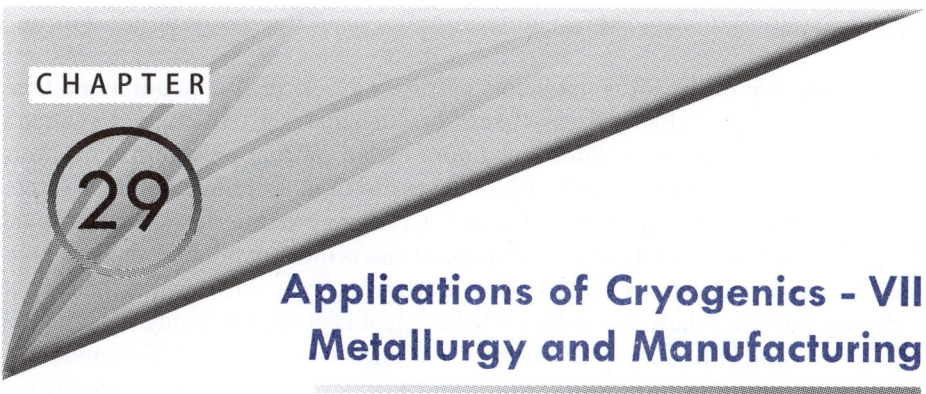

CHAPTER 29

Applications of Cryogenics - VII
Metallurgy and Manufacturing

INTRODUCTION

Metallurgy is a domain of materials science that studies the physical and chemical behavior of metallic elements, their intermetallic compounds, and their mixtures, which are called alloys. Application of cryogenics to metallurgy applications is termed as cryo-mettalurgy. Cryogenic treatment of metals can significantly reduce retained internal stresses which cause an in strength and durability. The history behind cryogenic processing can be dated back probably 50 years to the occasional use of low temperature treatments in the hardening of steels. Over the past 40 years or so reports have been made of substantial benefits of treating steel at low temperatures. Usually the temperature was around that of liquid nitrogen, which is –196 degrees Celsius. Recently in the United States claims of improvement have been expanded to copper, high temp alloys, carbides, plastics, and nylon.

Cryogenic treatment avoids formation of stress boundary areas, which are more susceptible to micro cracking and can lead to premature fatigue and eventual failure of the stressed part. Residual stresses exist in all types of parts from engines to tooling. The stress is introduced into the parts at the time of forging, casting, heat treating and final machining. These stresses create a complex, invisible random pattern in the alloy. As parts expand from the heat generated during operation, the retained stresses cause uneven expansion, increased dimensional instability, and increased wear as well as decreased performance. Stress relief takes place when the entire mass of the part is at an equal temperature (surface and core), and then slowly cycled (less than one degree per minute) through a wide temperature range. By exposing parts to ultra low temperatures using cryogenic techniques for a prolonged period a very dense molecular state is created. Absolute zero (–459.67 Fahrenheit) is known as the zero motion molecular state of mass. The slow rate of temperature change allows thermal compression and thermal expansion to occur, which causes the release of stress. The result is a dimensionally stabilized part, which will resist distortion and warpage and significantly increase performance and durability. Fine carbides and tight lattice structures are precipitated from cryogenic treatment. The wide distribution of these fine carbide particles increase wear resistance. These particles are responsible for the exceptional wear characteristics imparted to materials by the process, due to a denser structure and resulting larger surface area of contact, reducing friction, heat and wear.

29.2 Cryogenics

Cryogenic treatment is not a coating or a surface treatment, but a one-time, permanent, irreversible process that alters the material structure. Cryogenic temperatures are required to effect a complete molecular change in most alloy steels, converting the retained austenite into Martensite (a more refined grain structure, which is more uniform than Austenite). Cryogenic tempering transforms the microstructure into a more uniform structure that is more durable, stronger, longer lasting, and more dimensionally stable. Cryogenic treatment ensures there is no retained Austenite during quenching. When steel is at the hardening temperature, there is a solid solution of carbon and iron, known as austenite. The amount of Martensite formed at quenching is a function of the lowest temperature encountered. At any given temperature of quenching there is a certain amount of Martensite and the balance is untransformed Austenite. This untransformed Austenite is very brittle and can cause loss of strength or hardness, dimensional instability, or cracking.

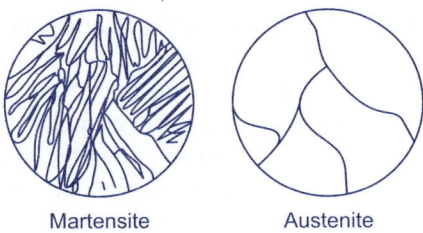

Martensite   Austenite

Fig. 29.1 Microstructure Changes in Steel due to Cryogenic treatment

Quenches are usually done to room temperature. Most medium carbon steels and low alloy steels undergo transformation to 100% Martensite at room temperature. Refer Figure 29.1 for the changes in microstructure of steel due to cryogenic treatment.

Deep Cryogenic Treatment (DCT)

Machining companies are using Deep Cryogenic Treatment (DCT) for treating steels cost effectively. This type of treatment cools the steel to –190°C, which enhances the steels mechanical properties much like heat treatment. The main difference between the two is that DCT is much more cost effective in delivering same type of results. DCT is used to enhance the mechanical properties of steel by increasing wear resistance, hardness, and dimensional stability. Many manufacturing companies waste money each year on worn out steel tools. DCT increases tool life and reduces the overall cost for the company and the end user. The problem with conventional heat treatment is that when steel is heated up to a phase called Austenite and then quenched with oil or water it does not form the desired 100% Martensite as the steel must reach a much lower temperature then room temperature to successfully transform. Another problem associated with heat treatment is the internal stresses and the dimensional instability. The DCT process cools the steel down to –196°C avoiding internal stresses. The cooling must not be done to fast or it will induce thermal shock into the steel. The steel is inserted in a freezer chamber for 24 hours. The freezer has liquid nitrogen floating in the atmosphere. After the steel has been in the chamber for 24 hours it is slowly heated back up to room temperature and then conventionally tempered. After the DCT process has been completed it usually increases tool life by 200-600 percent by relieving internal stresses that is from retained Austenite. Wear resistance is also increased because the cold temperatures squeeze the carbon atoms out; the new uniform microstructure formed reduces friction and in return reduces the heat involved with friction.

Auto-racing teams use DCT on their engine parts because it increases the longevity of the engine. It is also used for gun barrels because the gun shoots much more accurately when the stresses are relieved. DCT does not replace the standard heat treatment, it simply adds another step to the process of treatment of steels. The main reason that steel is subjected to this low temperature treating is to refine the molecular structure, which creates a stronger and more durable product. This treatment basically changes the whole molecular structure of the material; this change lasts throughout the entire life of the part because it is not a surface treatment. Even after a number of sharpening the material will not lose strength, durability and wear resistance. Cryogenic treatment also increases the tensile strength of the material, which increases the life of the tool. Another thing that cryogenic treatment does is that it takes the brittleness out of materials, which makes them less apt to break if they happen to get dropped. Table 29.1 below shows the percent increase of wear resistance in these materials.

Table 29.1 Improvement in Strength of Steels

AISI#	Name of Alloy	Increase in Strength
D-2	High carbon/chromium die steel	817 %
52100	Standard steel	420 %
A-10	Graphite tool steel	264 %
T-1	Tungsten high-speed tool steel	176 %
P-20	Mold Steel	130 %
430	Ferritic stainless	119 %
8620	Nickel-chromium-moly alloy	104 %
AQS	Graphitic cast iron	97 %
S-7	Silicon tool steel	503 %
0-1	Oil hardening cold work die steel	418 %
M-1	Molybdenum high-speed steel	225 %
CPM-10V	Alloy steel	131 %
440	Martensitic stainless	121 %
303	Austenitic stainless	110 %
C-1020	Carbon steel	98 %
T-2	Tungsten high-speed steel	92 %

However, high carbon and high alloy steels have retained Austenite at room temperature. To eliminate retained Austenite, the temperature has to be lowered. In Cryogenic treatment the material is subject to deep freeze temperatures of as low as −185°C, but usually −75°C is sufficient. The Austenite is unstable at this temperature, and the whole structure becomes Martensite. This is the reason to use cryogenic treatment.

Cryogenic Tempering of Tools

Cryogenic processing methods such as tempering dramatically improved wear resistance for perishable tooling and dies. Longer tool or die life and reductions in machine downtime and maintenance can significantly cut operating costs. Simply chilling metals to sub-zero temperatures for stress relief and stabilization is a very old technique. In fact, Swiss watchmakers have reportedly used cold mountain caves to condition and stabilize parts for more than 100 years. Extreme temperatures and computer based controls are key elements that set cryogenic processing apart from this sort of traditional cold treatment. Cryogenic processing to increase wear resistance is a fairly recent development that has spawned a new industry. Cryogenic tempering of tools is accomplished in various ways. The most popular approach in the United States is a deep, dry, controlled cryogenic process. Deep cryogenic tempering takes place at around −320F (196°C), near the temperature of liquid nitrogen. Shallow cryogenic tempering takes place around −120°F (−84°C), near the temperature of dry ice. When material is immersed in liquid nitrogen, the process is considered "wet." A dry process is one during which the material is not immersed in a liquid.

Tools sent to a commercial cryogenic processor are generally batched with items from other companies until enough are obtained to fill a specially constructed freezer that uses liquid nitrogen as the refrigerant. The temperature is gradually lowered or ramped down to −320°F. Items remain at that temperature for 20 to 60 hr (commonly referred to as the soak). Then, the temperature is gradually raised to room temperature or beyond. If the material requires additional tempering (to stabilize freshly transformed martensite), the system slowly raises the temperature (to around 375°F or 191°C for most tool steels) before gradually reducing it to room temperature. Insufficient soak time, quick cooling or warming, and skipping the post-soak temper can hamper the effectiveness of cryogenic tempering. Any one of these factors can cause inconsistent results. Fortunately, today's cryogenic processors are able to provide more consistent results than older equipment.

Tools for myriad applications can benefit from cryogenic treatment. Property enhancements are claimed for steel, aluminum, brass, copper, nylon, plastics, and carbides. Applications include but aren't limited to aerospace, manufacturing, sports, music, firearms, motorsports. When it comes to steel tools, cryogenic tempering is widely accepted technology. Metal tools and parts that may benefit from cryogenic processing include drill bits, end mills, cutters, dies, punches, bearings, cams, crankshafts, blocks, and pistons. Improved wear resistance of tool steels after cryogenic tempering, is due to three factors: retained austenite (RA) conversion, carbide precipitation, and thermal-mechanical stabilization. Changes from the last two factors, however, can only be seen under a microscope.

Plastics, aluminum, copper, brass, and other materials that don't have retained Austenite to convert may be enhanced by thermal-mechanical stabilization-expansion and contraction of the crystalline structure. Cryogenic tempering decreases latent stress in materials however avoiding thermal shock is the key. In steel, a differential in the rate of expansion from core to surface results in thermal shock, which makes the material brittle and very susceptible to cracking.

Introduction to Cryogenic Manufacturing

Cryogenics is now widely used for manufacturing applications. The major cryogenic manufacturing processes are (1) cryogenic deburring (2) cryogenic milling (3) cryogenic shrink fitting (4) cryogenic deflashing (5) cryogenic grinding. The benefits of cryogenic manufacturing are:

- Improved productivity due to increased throughput
- No caking of the product during manufacture, reducing cleaning downtime
- Less wear of grinding tools
- Superior oxidation resistance of product

The five cryogenic manufacturing techniques are explained in greater detail.

Cryogenic Deburring

Cryogenic deburring is a process to remove machine burrs from machined parts made from plastics, polymers, alloys like stainless steels and metals. Depending on the machined material and the size of the part variables such as throw wheel speed, temperature, and time are set for the cryogenic deburring equipment. The parts are then placed into the chamber, exposed to sub-zero temperatures (achieved with the use of liquid nitrogen) and then tumbled and blasted with a cryogenic-grade polycarbonate media. Because the burrs have a high surface area relative to their mass, they "freeze" quickly and become brittle. The tumbling action combined with the impact of the media removes the burrs without damaging the part itself. Cryogenic deburring is used by industries as their machine tools create burrs on their parts in blind and recessed holes, as well as other edges of the part from drilling, milling, grinding, and turning. The benefits of cryogenic deburring include its cost-effectiveness when compared to other types of deburring, repeatable automated results with each part, non-abrasive media which will not attack the structural integrity of the parts, and eco-friendly nature.

Cryogenic Milling

Cryogenic milling or cryomilling is a variation of mechanical milling, in which metallic powders or temperature sensitive samples are milled in a cryogen (liquid nitrogen or liquid argon) slurry at cryogenic temperatures. This leads to formation of nanostructured microstructure. Cryomilling takes advantage of both the cryogenic temperatures and conventional mechanical milling. The extremely low milling temperature suppresses recovery and recrystallization and leads to finer grain structures and more rapid grain refinement. The embrittlement of the sample makes even elastic and soft samples grindable. Final finenesses of below 5 μm can be achieved. The ground material can be analyzed by a laboratory analyzer. Freezer milling is a type of cryogenic milling that uses a solenoid to mill samples. The solenoid moves the grinding media back and forth inside the vial, grinding the sample down to analytical fineness. This type of milling is especially useful in milling temperature sensitive samples, as samples are milled at liquid nitrogen temperatures. The idea behind using a solenoid is that the only "moving part" in the system is the grinding media inside the vial. The reason for this is that at liquid nitrogen temperatures (–198°C) any moving part will come under huge stress leading to potentially poor realiability. Cryogenic milling using a solenoid has been used for over 50 years and has been proved to be a very reliable method of processing temperature sensitive samples in the laboratory.

Cryogenic Shrink Fitting

Shrink fitting, (or "compression fitting" as it is sometimes called), is a method used to insert a pin or bushing into a housing or other assembly requiring an extremely tight tolerance fit. It can be used as an alternative to conventional press fitting, or more likely, to permit a mechanical fit that otherwise could not be accomplished via the mechanical force of press fitting. In application, the insert is cooled via exposure to a cryogen, typically carbon dioxide (solid or liquid) or liquid nitrogen in order to reduce its size through the contraction usually associated with reduced temperatures. A companion operation of heating the housing that receives the insert is conducted to enlarge the opening by taking advantage of the expansion usually associated with increased temperatures. While it is not always necessary to use both heating and cooling in combination, the most demanding applications, including those with the tightest tolerances, often require this multi-step approach. Care should be taken when using any cryogen and consideration of the material and relative masses needs to be considered carefully. In addition, because steels may be subject to additional transformation when exposed to cryogenic temperatures, additional processing either before or after may be warranted. Heating of metals (for expansion) should also be controlled and not induce unnecessary thermal stress on the component. For these reasons, many companies seek expert advice for their compression shrink fitting requirements.

Cryogenic Deflashing

Cryogenic Deflashing is employed to remove undesired residual mold flash that remains on molded parts after they are removed or ejected from the mold cavity. Typically, this flash occurs in areas where different sections of the mold come together (and apart) and is known as "parting line flash". The remaining mold flash typically traces around where the different mold sections "mate" and is created when the liquid mold material escapes out of the mold cavity into the tight area where the mold sections press against one another. Molding flash can be caused from old or worn mold cavities that no longer fit tightly together. Most often, the type of material being molded, and its attendant viscosity in its liquid form, is the primary factor that leads to the creation of the unwanted mold flash. To remove flash, manufacturers use a cryogenic deflashing process. By processing the parts in a frozen state, the flash becomes stiff or brittle and breaks away cleanly during its processing.

The process typically works as follows: Parts are loaded into a basket that is placed into an insulated chamber. Using liquid nitrogen, the temperature of the chamber (and the parts) is lowered to approx minus 200°F. The parts are tumbled at a predetermined rate (approx. 50 RPM) and blasted with a cryogenic grade polycarbonate media that is sized at between 0.015 to 0.060 inches. The temperature, tumble rate, size of the blast media and process time varies widely depending on the part size and geometry as well as the type of material being processed. Cryogenic deflashing offers the advantage of not degrading or otherwise affecting the finish of the parts. In addition, only the undesired flash is removed and the integrity of the part shape is fully maintained. Sharp edges are not rounded from the process and the media can penetrate into recessed sections and clean blind and through holes with remarkable precision. Typical products that use cryogenic deflashing include O-rings and gaskets, medical catheters, insulators, valve stems, washers, fittings, tubes, flexible boots, face masks and goggles.

Cryogenic Grinding

Cryogenic grinding or cryogrinding is the act of cooling or chilling a material and then reducing it into a small particle size. Often plastic, elastic and heat-sensitive materials cannot be finely ground. In cryogenic grinding, nitrogen or carbon dioxide is used to lower the temperature of these materials to their precise embrittlement point, prior to grinding. This process has proven to be a cost and energy-efficient compliment to traditional size-reduction methods. Cryogenic grinding has been successfully used to produce powder coatings, food products and pharmaceuticals as well as in recycling plastics materials and tires. Thermoplastics are difficult to grind to small particle sizes at ambient temperatures because they soften, adhere in lumpy masses and clog screens. When chilled by dry ice, liquid carbon dioxide or liquid nitrogen, the thermoplastics can be finely ground to powders suitable for electrostatic spraying and other powder processes. Cryogenic grinding of plant and animal tissue is a technique used by microbiologists. Samples that require extraction of nucleic acids must be kept at −80°C or lower during the entire extraction process. Figure 29.2 shows a cryogenic grinding apparatus.

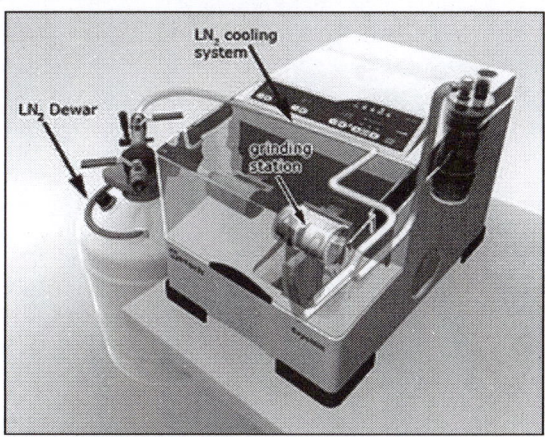

Fig. 29.2 Cryogenic Grinding Apparatus

References

1. Cryogenic Engineering – B.A. Hands.
2. Cryogenic Systems – Randall Baron.
3. Internet website http://web.vtc.edu/mt/114/Research/Russ/DCT.htm.
4. Nathan Hill – ESA publication on Cryogenics - 2008.
5. Internet website http://www.300below.com/site/manufacturingengineering.html.
6. Internet website of Air products.
7. Internet website of Wikipedia.
8. Internet website www.cryocentral.com.

Questions

Q.1. What are the applications of Cryogenics in Metallurgy?

Q.2. Explain the benefits and drawbacks of using cryogenic tempering.

Q.3. Write a short note on deep cryogenic treatment (DCT) process.

Q.4. List the applications of Cryogenics in Manufacture. Describe one process in detail.

Q.5. What are the advantages of cryogenic manufacturing process?

Q.6. Describe the cryo-milling process. Where is it used?

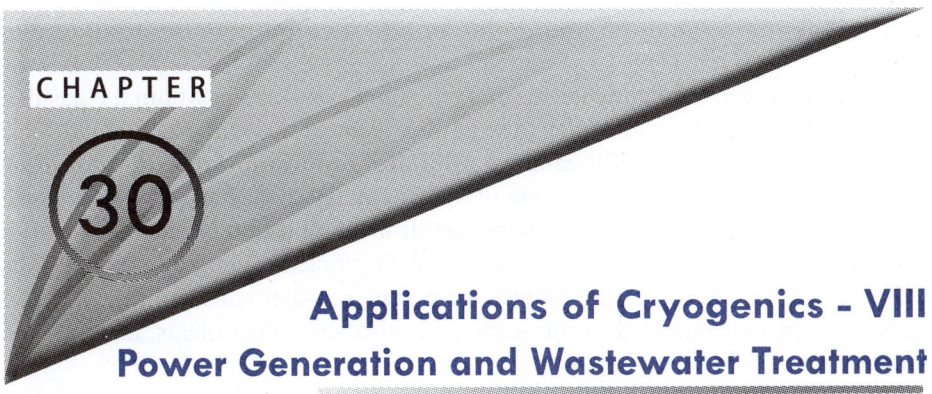

CHAPTER 30

Applications of Cryogenics - VIII
Power Generation and Wastewater Treatment

INTRODUCTION

Cryogenics has been used in industrial sectors such as energy and power generation. Low pressure elevated cryogenic air separation units (ASUs) have been successfully applied to support gasification projects worldwide. ASU technology has ranged from traditional low pressure standalone facilities supplying liquefied gases for gasification to highly integrated, elevated pressure facilities that obtain air feed and inject excess nitrogen in gas turbines. The current trends for ASUs are increased capacity units for generation of higher-pressure ratios for larger size turbines. ASUs with capacities ranging from 10,000 to 20,000 metric ton per day are being developed. Cryogenics also plays an important role in the Integrated Gasification Combined Cycle (ICGC) power generation plant. Air separation units are used in IGCC to improve efficiency of cycle.

Cryogenics plays an important role in oil recovery operations as well. Very high-pressure nitrogen is used to increase recovery of oil from wells. This nitrogen is produced from multiple cryogenic ASUs on site. These ASUs are favoured due to their low operating costs. Furthermore cryogenics is playing an active role in coal gasification plants such as those owned by SASOL in South Africa. Coal is gasified with oxygen to produce synthetic natural gas (SNG) or Syngas and is further converted to liquid fuels using the Fischer-Tropsch Process. Liquid oxygen purity levels of 98.5% are desirable and a typical plant consumes 3500 tonnes of LOX per day. Most oxygen plants for gasification use the liquid pump principle. In this process liquid oxygen is pumped to the required pressure and vaporised by heat exchange with nitrogen from air.

IGCC Power Generation Plant

An Integrated Gasification Combined Cycle, or IGCC, is a power plant using synthesis gas (syngas). This gas is often used to power a gas turbine whose waste heat is passed to a steam turbine system (combined cycle gas turbine). IGCC is a technology that aims to extract the maximum energy out of coal that is burned. This is achieved by gasification, which converts coal into synthetic gas or syngas. Syngas is a mixture containing mainly carbon monoxide (CO) and hydrogen (H_2) and some carbon dioxide as well. Coal can be gasified in various ways – by controlling the mix of

30.2 Cryogenics

coal, air or oxygen (O_2)-enriched air or even pure O_2 and steam within the gasifier. In an IGCC power generation plant, the gasifier is integrated with a combined cycle power plant (CCPP). A CCPP employs more than one thermodynamic cycle. Heat engines are able to use only a portion of the energy that their fuel generates, which is usually less than 50%. The remaining heat from combustion is generally wasted. But by combining two or more "cycles", overall efficiency can be improved. The IGCC technology substantially reduces air pollution, water consumption and solid waste production compared to conventional coal-based power plants, making it an important technology for the future.

The gasification process can produce syngas from high-sulfur coal, heavy petroleum residues and biomass. The impurities from the coal gas in IGCC are removed before combustion which results in lower emissions of sulfur dioxide, particulates and mercury. It also results in improved efficiency compared to conventional pulverized coal. Thus the lower emissions that IGCC technology allows may be important in the future as emission regulations tighten due to growing concern for the impacts of pollutants on the environment and the globe. Figure 30.1 shows a schematic flow diagram of an IGCC plant.

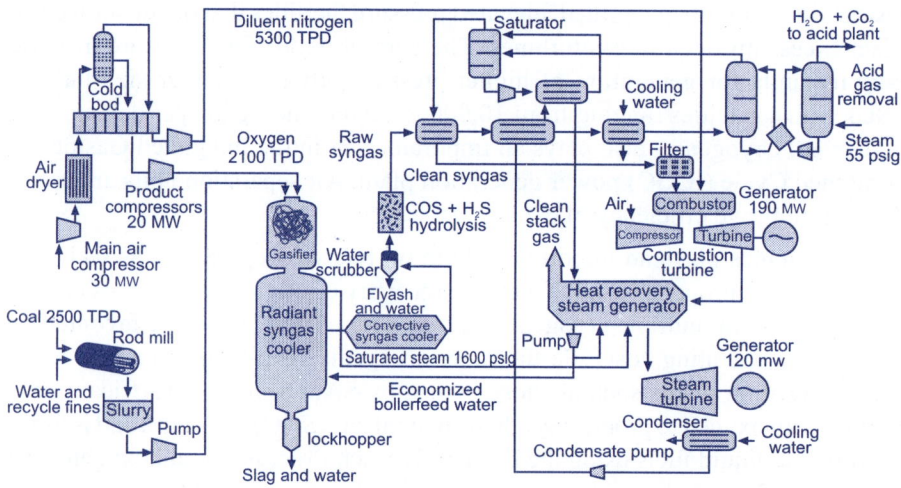

Fig. 30.1 IGCC Power Plant using Cryogenic Technology

The plant is called "integrated" because its syngas is produced in a gasification unit in the plant which has been optimized for the plant's combined cycle. In this example the syngas produced is used as fuel in a gas turbine which produces electrical power. To improve the overall process efficiency heat is recovered from both the gasification process and also the gas turbine exhaust in 'Waste Heat Boilers' producing steam. This steam is then used in steam turbines to produce additional electrical power. The cryogenic air seperation unit produces the liquid oxygen which is fed to the gasifier as well as the dilunt nitrogen which is injected in the turbine.

IGCC is now touted as "capture ready" and could potentially capture and store carbon dioxide. Next generation IGCC plants with CO_2 capture technology will be expected to have higher thermal efficiency and to hold the cost down because of simplified systems compared to conventional IGCC. The main feature is that instead of using oxygen and nitrogen to gasify coal, they use oxygen and CO_2. The main advantage is that it is possible to improve the performance of cold gas efficiency and to reduce the unburned carbon (char). With a 1300°C class gas turbine it is possible to achieve 42% net thermal efficiency, rising to 45% with a 1500 degree class gas turbine, with CO_2 capture. In case of conventional IGCC systems, it is only possible to achieve just over 30% efficiency with a 1300° gas turbine. The CO_2 extracted from gas turbine exhaust gas is utilized in this system. Using a closed gas turbine system capable of capturing the CO_2 by direct compression and liquefaction obviates the need for a separation and capture system.

Introduction to Cryogenics for Waste Water Treatment

Waste water treatment covers the mechanisms and processes used to treat waters that have been contaminated in some way by anthropogenic industrial or commercial activities prior to its release into the environment or its re-use. Most industries produce some wet waste although recent trends in the developed world have been to minimise such production or recycle such waste within the production process. However, many industries remain dependent on processes that produce wastewaters. Many industries have a need to treat water to obtain very high quality water for demanding purposes. Water treatment produces organic and mineral sludges from filtration and sedimentation. Ion exchange using natural or synthetic resins removes calcium, magnesium and carbonate ions from water, replacing them with hydrogen and hydroxyl ions.

Conventional wastewater treatment solutions are based on biological processes. Often, these solutions are unable to handle large variations in the input load. These plants require large footprint, constant supervision and consume a large amount of electricity. Some also require periodic, expensive degrading to digest the accumulated sludge. The performance of these plants is severely affected by the absence of power. New cryogenics based innovative processes for treatment of wastewater are being developed using physico-chemical treatment methodology. This system is an excellent and robust solution specifically designed for applications with irregular and intermittent loads. The Table 30.1 shows the improvement in Water Quality after treatment.

Table 30.1 Waste Water Quality Improvement

Parameter	Typical Waste Water	Treated Water
BOD: Biological Oxygen Demand	< 250 ppm	< 30 ppm
COD: Chemical Oxygen Demand	500 ppm	< 100 ppm
TSS: Total Suspended Solids	500 ppm	< 100 ppm
TKN: Total Kjeldahi Nitrogen	30-70 ppm	< 20 ppm
TP: Total Phosphorus	10-30 ppm	< 0-2 ppm

Cryogenic Waste Water Treatment Plant

The typical cryogenic wastewater facility is shown in Figure 30.2. Raw waste water enters the plant through mechanically cleaned bar screens with each screen followed by a grit collection system. The screenings and grit removed from the flow are moved on belt conveyors to trucks for hauling to a commercial landfill.

The partially treated wastewater flows through an aerated channel to rectangular sedimentation tanks. The settled solids are removed from the bottom of the sedimentation tanks by scrapers supported from carriages moving on rails. The monorake mechanism also skims floating material (scum) from the surface of the wastewater in the sedimentation tanks. The raw primary sludge is pumped from hoppers in the ends of the sedimentation tank floors to four anaerobic digesters. The scum is also pumped into the storage tanks. Settled waste water from the sedimentation tanks flows to the wet well of a pump station, where it is pumped over ultra-high rate biological trickling filters referred to as Bio-towers. A plastic "honeycomb" media in the Bio-towers provides sites for bacteria to grow. These bacteria consume some of the organic pollutants in the settled wastewater, reducing organic loads by about 20% as the waste water trickles down through the honeycomb.

Fig. 30.2 Cryogenic Waste Water Treatment Facility

After passing through the Bio-towers, the wastewater flows through a parallel arrangement of five oxygenation trains, each of which contains four successive aeration chambers or stages. Oxygen gas and return sludge are added to the wastewater in the first chamber of each train and this mixture (mixed liquor) flows through the four successive chambers. Blowers and agitators in each chamber will mix the oxygen-enriched gas with the mixed liquor, thus keeping the solids in suspension. Three cryogenic oxygen-generating plants produce maximum high purity oxygen gas for the aeration process. The mixed liquor effluent from the oxygenation trains flow into twenty final settling tanks. Flow controllers distribute the flow equally between the tanks. Settled sludge from the tanks is recycled to the oxygenation trains or wasted to sludge thickeners. The secondary system, consisting of the oxygenation trains and the final settling tanks, is designed to provide rated performance. Disinfection of the flow is accomplished by injecting a solution of sodium hypochlorite (bleach) into the secondary effluent. Under normal flows, the bleach contact time is 30-45 minutes. Dechlorination is accomplished by the addition of sodium bisulfite prior to discharging the effluent to the nearby river.

A typical cryogenic waste water treatment plant handling 800,000 m^3/day of waste water costs in excess of $200 million including the construction of oxygen reactors, final clarifiers, cryogenic facilities, disinfection and service water facilities and a pumping plant and requires 3-4 years for commissioning.

References

1. Cryogenic Engineering – B.A. Hands.
2. Internet website of Wikipedia.
3. Cryogenic Systems – Randall Baron.
4. Internet website www.airproducts.com.
5. Review article by Walter Castle on Cryogenics.
6. Nathan Hill – ESA publication on Cryogenics - 2008.
7. Paper on Cryogenic Oxygen Supply for Gasification by Allam et.al, Air Products.

Questions

Q.1. What are the applications of Cryogenics in Power Generation?

Q.2. Explain the working of an IGCC power plant with a sketch.

Q.3. Write a short note on use of cryogenics for waste water treatment.

Q.4. Write a short note on Liquid Hydrogen as a fuel.

Appendix – I

Properties of Cryogens

Cryogen	Property	Value	Units	Comments
Nitrogen N_2	Boiling Point	77.348	K	1 Atm
	Melting Point	63.2	K	1 Atm
	Triple Point	63.2	K	triple point temperature
		0.127	Atm	triple point pressure
	Critical Point	126.19	K	critical point temperature
		33.534	Atm	critical point pressure
		0.31311	g/ml	critical point density
	Density	0.80663	g/ml	liquid, 1 Atm, @ boiling point
	Dielectric Constant	1.434		liquid, 1 Atm, @ boiling point
	Latent Heat	198.3	kJ/kg	vaporization @ 1 Atm
Ammonia NH_3	Boiling Point	239.82	K	1 Atm
	Melting Point	195.5	K	1 Atm
	Triple Point	195.49	K	triple point temperature
		6.06	kPa	triple point pressure
	Critical Point	405.40	K	critical point temperature
		111.85	Atm	critical point pressure
		0.22500	g/ml	critical point density
	Density	0.68197	g/ml	liquid, 1 Atm, @ boiling point
	Dielectric Constant	22		liquid, 1 Atm, @ boiling point
	Latent Heat	1369	kJ/kg	vaporization @ 1 Atm
Argon Ar	Boiling Point	87.3	K	1 Atm
	Melting Point	83.8	K	1 Atm
	Triple Point	83.8	K	triple point temperature
		0.679	Atm	triple point pressure
	Critical Point	150.7	K	critical point temperature
		48.3	Atm	critical point pressure
		0.544	g/ml	critical point density
	Density	1.403	g/ml	liquid, 1 Atm, @ boiling point
	Dielectric Constant	1.520		liquid, 1 Atm, @ boiling point
	Latent Heat	161.6	kJ/kg	vaporization @ 1 Atm

A.2 Appendix — I

Substance	Property	Value	Unit	Description
Carbon Dioxide CO_2	Sublimation Point	194.7	K	1 Atm
	Triple Point	216.6	K	triple point temperature
		5.12	Atm	triple point pressure
	Critical Point	304.21	K	critical point temperature
		72.878	Atm	critical point pressure
		0.46650	g/ml	critical point density
	Density	1.563	g/ml	subliming solid @ 1 Atm
	Latent Heat	573	kJ/kg	subliming solid @ 1 Atm
Helium 4 ^{4}He	Boiling Point	4.2304	K	1 Atm
	Lambda Point	2.18	K	lambda point temperature
		0.050	Atm	lambda point pressure
	Critical Point	5.1953	K	critical point temperature
		2.2449	Atm	critical point pressure
		0.069641	g/ml	critical point density
	Density	0.12473	g/ml	liquid, 1 Atm, @ boiling point
	Dielectric Constant	1.049		liquid, 1 Atm, @ boiling point
	Latent Heat	20.73	kJ/kg	vaporization @ 1 Atm
Helium 3 ^{3}He	Boiling Point	3.19	K	1 Atm
	Critical Point	3.35	K	critical point temperature
		1.15	Atm	critical point pressure
	Density	0.059	g/ml	liquid, 1 Atm, @ boiling point
	Latent Heat	8.48	kJ/kg	vaporization @ 1 Atm
Hydrogen e-H_2	Boiling Point	20.27	K	1 Atm
	Melting Point	14.0	K	1 Atm
	Triple Point	13.81	K	triple point temperature
		0.0695	Atm	triple point pressure
	Critical Point	32.98	K	critical point temperature
		12.76	Atm	critical point pressure
		0.0314	g/ml	critical point density
	Density	0.0708	g/ml	liquid, 1 Atm, @ boiling point
	Dielectric Constant	1.230		liquid, 1 Atm, @ boiling point
	Latent Heat	446	kJ/kg	vaporization @ 1 Atm
Methane CH_4	Boiling Point	111.67	K	1 Atm
	Melting Point	90.7	K	1 Atm
	Triple Point	88.7	K	triple point temperature
		0.099	Atm	triple point pressure
	Critical Point	190.56	K	critical point temperature
		45.391	Atm	critical point pressure
		0.16266	g/ml	critical point density
	Density	0.42236	g/ml	liquid, 1 Atm, @ boiling point
	Dielectric Constant	1.676		liquid, 1 Atm, @ boiling point
	Latent Heat	511.5	kJ/kg	vaporization @ 1 Atm

Appendix – I **A.3**

Neon Ne	Boiling Point	27.10	K	1 Atm
	Melting Point	24.58	K	1 Atm
	Triple Point	24.56	K	triple point temperature
		0.427	Atm	triple point pressure
	Critical Point	44.49	K	critical point temperature
		26.44	Atm	critical point pressure
		0.482	g/ml	critical point density
	Density	1.204	g/ml	liquid, 1 Atm, @ boiling point
	Dielectric Constant	1.188		liquid, 1 Atm, @ boiling point
	Latent Heat	86.2	kJ/kg	vaporization @ 1 Atm
Oxygen O_2	Boiling Point	90.188	K	1 Atm
	Melting Point	54.4	K	1 Atm
	Triple Point	54.4	K	triple point temperature
		0.0015	Atm	triple point pressure
	Critical Point	154.58	K	critical point temperature
		49.771	Atm	critical point pressure
		0.43614	g/ml	critical point density
	Density	1.1412	g/ml	liquid, 1 Atm, @ boiling point
	Dielectric Constant	1.487		liquid, 1 Atm, @ boiling point
	Latent Heat	212.9	kJ/kg	vaporization

Appendix – II

List of International Standards for Cryogenic Vessels – ISO TC 220

Standard Name	Number	Year
Liquefied natural gas -- Transportable tanks for use on board vehicles	ISO/AWI 12991	2004
Cryogenic vessels -- Large transportable vacuum-insulated vessels -- Part 1: Design, fabrication, inspection and testing	ISO 20421-1	2006
Cryogenic vessels -- Large transportable vacuum-insulated vessels -- Part 2: Operational requirements	ISO 20421-2	2005
Cryogenic vessels -- Static vacuum-insulated vessels -- Part 1: Design, fabrication, inspection and tests	ISO 21009-1	2008
Cryogenic vessels -- Static vacuum insulated vessels -- Part 2: Operational requirements	ISO 21009-2	2006
Cryogenic vessels -- Gas/materials compatibility	ISO 21010	2004
Cryogenic vessels -- Valves for cryogenic service	ISO 21011	2008
Cryogenic vessels -- Hoses	ISO 21012	2006
Cryogenic vessels -- Pressure-relief accessories for cryogenic service -- Part 1: Reclosable pressure-relief valves	ISO 21013-1	2008
Cryogenic vessels -- Pressure-relief accessories for cryogenic service -- Part 2: Non-reclosable pressure-relief devices	ISO 21013-2	2007
Cryogenic vessels -- Pressure-relief accessories for cryogenic service -- Part 3: Sizing and capacity determination	ISO 21013-3	2006
Cryogenic vessels -- Pilot operated pressure relief devices -- Part 4: Pressure-relief accessories for cryogenic service	ISO/AWI 21013-4	2004
Cryogenic vessels -- Cryogenic insulation performance	ISO 21014	2006
Cryogenic vessels -- Toughness requirements for materials at cryogenic temperature -- Part 1: Temperatures below -80 degrees C	ISO 21028-1	2004
Cryogenic vessels -- Toughness requirements for materials at cryogenic temperature -- Part 2: Temperatures between -80 degrees C and -20 degrees C	ISO 21028-2	2004
Cryogenic vessels -- Transportable vacuum insulated vessels of not more than 1000 litres volume -- Part 1: Design, fabrication, inspection and tests	ISO 21029-1	2004
Cryogenic vessels -- Transportable vacuum insulated vessels of not more than 1000 litres volume -- Part 2: Operational requirements	ISO 21029-2	2004
Cryogenic vessels -- Cleanliness for cryogenic service	ISO 23208	2005
Cryogenic vessels -- Pumps for cryogenic service	ISO 24490	2005

Appendix – III

University Papers - Question Bank

Unit – I

Q.1. Define the following terms: Figure of merit, coefficient of performance, liquid yield, Joule-Thomson coefficient, cryogenic engineering.

Q.2. Determine the refrigeration effect, COP and figure of merit for a Linde-Hampson refrigerator operating from 300 K and 1 atm to 100 atm. The overall efficiency of compressor is 75% and heat exchanger effectiveness is 0.96. Working fluid used is nitrogen.

Q.3. Write short notes on: Applications of cryogenics, Cryogenic flow rate measurement, Liquid nitrogen freezing of food, Walker's chart of classification, Insulation in cryogenic systems.

Q.4. What is cryogenic temperature? How is it measured?

Q.5. Describe the single stage Stirling cycle cryocooler with a neat sketch.

Q.6. Explain in detail five low temperature properties of engineering materials.

Q.7. Discuss the limitations of Brayton cycle for cryogenic applications and effect of variation in refrigeration temperature on cycle performance.

Q.8. Sketch the Heylandt, Claude and Kapitza cycles on T-S diagram and compare them.

Q.9. Derive the expression for work requirement in an ideal separation system to separate a mixture of "n" different gases.

Q.10. Discuss the operation cycle of double volume Gifford-McMahon Cryorefrigerator.

Q.11. What is cryopumping? Discuss various cryopump configurations used in practice.

Q.12. Derive an expression for COP of a V-M refrigerator. Compare it with COP of Phillips refrigerator.

Q.13. Describe the Simon Liquefaction System for Helium. What are its advantages?

Unit – II

Q.1. In a reversed Brayton cycle, derive an expression for COP in terms of ambient and refrigeration temperature, Δt in cooler and rise in temperature during isobaric heat absorption.

Q.2. In a reversed Brayton cycle, $T_a = 300K$, $T_r = 200\ K$, $p_a = 1$ atm, $\Delta tr = \Delta tc = 5\ K$ and overall $\Delta t = 50\ K$. Assume cycle efficiency as 20%. The isotropic efficiency of compressor is 90%. Find the highest temperature, pressure, and efficiency of expander and COP for the system.

Q.3. In an ideal refrigeration system for liquefaction, derive an expression for work required/unit mass compressed.

Q.4. Stating assumptions, find an expression for liquid yield and FOM for simple L-H cycle.

Q.5. In a dual pressure Linde system working between pressures 1 atm and 200 atm, determine the intermediate pressure giving a liquid yield of 0.065 for $i =$ 0.7. Also calculate the FOM.

Q.6. The equation $y = ax + b$ is the equation of operating line in the upper part of the rectification tower. Comment on this statement.

Q.7. To separate a mixture of "n" different gases using ideal separation system, determine the work required.

Q.8. Explain with a neat sketch Heylandt's and Kapitza system. Also mention similarities between them.

Q.9. In an ideal Claude system, air enters the compressor at 1 atm and 300 K and is compressed to 40 atm. Determine the expander flow rate ratio for liquid yield 0.19. Air enters the isentropic expander at 40 atm and 240 K.

Q.10. Write short notes on: Applications of cryogenics, Pulse tube refrigerator, Simon liquefaction system, and Murphree efficiency.

Q.11. Describe the two stage and single stage Stirling cycle cryocooler with a neat sketch.

Q.12. Discuss the limitations posed by regenerator with positive refrigerating effect on operation of Stirling system.

Q.13. Derive an expression for COP of a V-M cryocooler. Compare it with COP of Phillips refrigerator.

Q.14. Explain V-M cryocooler with a neat sketch.

Q.15. Write a note on cryogenic insulations.

Q.16. What is cryopumping? Explain with a neat sketch.

Q.17. Measurement of cryogenic temperature is critical – comment. Describe various methods used.

Q.18. What are cryogens? Suggest methods to store them.

Q.19. Write short notes on: Walker's chart of classification, Superfluidity and conductivity, Liquid nitrogen for food preservation, Solvay refrigerator, Mechanical properties at cryogenic temperature.

Unit – III

Q.1. For a Brayton cycle, derive an expression for COP in terms of temperature at inlet and outlet of compressor and expander.

Q.2. In a Brayton cycle, $T_a = 300K$, $Tr = 250$ K, $p_a = 1$ atm, $\Delta tr = \Delta tc = 10$ K and inlet temperature of compressor is 240 K. Relative efficiency is 5% for refrigerating effect of 20 kJ/kg. The isentropic efficiency of compressor is 80%. Find the highest temperature, highest pressure, efficiency of expander. Assume working fluid as air.

Q.3. Describe thermodynamically ideal system of liquefaction with a neat sketch. In an ideal refrigeration system for liquefaction, derive an expression for work required for liquefaction of nitrogen.

Q.4. Explain the simple L-H cycle and Claude cycle used in liquefaction with a schematic diagram and representation on T-S diagram.

Q.5. In a Linde dual pressure system, find the change in yield and FOM if intermediate pressure is increased from 25 atm to 75 atm. Assume $P_L = 1$ atm, $P_H = 200$ atm; $x = 0.8$, $\epsilon = 1$, compressor efficiency $= 0.9$. Derive equations used.

Q.6. Rectification is the cascading of several evaporations and condensations carried out in counter flow. Explain with a neat sketch.

Q.7. Derive an equation of operating line in upper portion of rectification column if the flow at the bottom contains 95% oxygen at 45 kg moles/hr. The top product is 98% nitrogen and 1000 kW heat is removed from the top of the column. Determine the number of plates assuming air feed stream with 21% O_2 and 79% N_2.

Q.8. Write short notes on: Kapitza system, Otho-Para Hydrogen conversion, Simon Helium liquefaction system, Critical components in liquefaction systems.

Q.9. Explain the classification of cryocoolers using Walker's chart.

Q.10. Describe ideal thermodynamic cryogenic system assuming isothermal and isobaric source and derive expressions for COP in each case.

Q.11. Explain flow rate measurement systems in cryogenic fluid transfer.

Q.12. Describe with a neat sketch, a cryogenic fluid storage vessel.

Q.13. Discuss a Phillip's refrigerator and explain importance of regenerator effectiveness.

Q.14. Compare Solvay and GM refrigerators with representation on T-S chart and schematic diagrams.

Q.15. Explain the importance of vacuum technology in cryogenics and describe mechanical and diffusion vacuum pumps.

Q.16. Write short notes on: Applications of Cryogenics, V-M Cryocooler, Mechanical properties at cryogenic temperature.

Unit – IV

Q.1. For a Brayton Cycle, show that COP = $T_1/(T_2 - T_1) = T_4/(T_3 - T_4)$, where T_1 = Temperature at suction of compressor, T_2 = Temperature at discharge of compressor, T_3 = Temperature at inlet of expander and T_4 = Temperature at outlet of expander.

Q.2. Discuss the limitations of Brayton cycle for cryogenic applications and effect of variation in refrigeration temperature on performance of cycle.

Q.3. Explain the term "Figure of Merit" and derive an expression for yield and FOM for a simple Linde-Hampson cycle.

Q.4. A Linde Hampson cycle uses air as a working fluid. 90 atm and 1 atm are the higher and lower pressures and isothermal compression occurs at 300 K. Find the effectiveness of heat exchanger, when yield is 70% of the maximum possible yield.

Q.5. Sketch on T-S diagrams the Heylandt, Claude and Kapitza cycles and compare them.

Q.6. Explain with a neat diagram, air separation column.

Q.7. Discuss the cycle of operation for a double volume Gifford-McMahon Cryorefrigerator.

Q.8. A cryogenic refrigeration system operates between T_o = 300 K and T_r = 100 K using air as a working medium. Find (a) Work input for refrigerating effect of 100 W for ideal isothermal source. (b) For ideal isobaric source of refrigeration with Δt = 10°C, find the work input.

Q.9. Write short notes on: Applications of Cryogenic temperature, Hydrogen liquefaction, Pulse tube refrigerator, Dual pressure liquefaction systems, Effect of variable specific heat in heat exchanger.

Q.10. Describe the single stage Stirling cycle cryocooler with a neat sketch.

Q.11. In case of Stirling cycle cryocooler for 1% deviation from ideal value of effectiveness of regenerator, calculate% loss of RE when volume expansion ratio is 1.5 and helium is working fluid. Assume suitable ambient temperature.

Q.12. Explain V-M cryocooler with a neat sketch.

Q.13. Define COP for V-M cryocooler and compare it to COP of Phillips Cryocooler.

Q.14. What is cryogenic temperature? How is it measured?

Q.15. Describe essential elements of a Dewar vessel.

Q.16. Compare various insulations used in cryogenic systems.

Q.17. What is cryopumping ? Discuss various cryopump configurations used in practice.

Q.18. Write short notes on: Walker's chart, Superconductivity, Liquid nitrogen for food freezing, Flow rate measurement in cryogenic systems, Properties of materials at cryogenic temperature.

Unit – V

Q.1. For a reversed Brayton Cycle with compression and 100% efficiencies, show that the COP = $(Tr - \Delta tr - \Delta t)/((T_o - T_r) + (\Delta to + \Delta tr + \Delta t))$, where T_o = Ambient Temperature, Tr = Refrigeration Temperature, Δto = Temperature difference in cooler, Δtr = Temperature difference in refrigerator, Δt = Rise in temperature during isobaric refrigeration.

Q.2. In a reversed Brayton cycle, T_o = 300K, T_r = 200 K, p_a = 1 atm, Δtr = Δtc = 5 K and overall Δt = 50 K. Assume cycle efficiency as 20%. The isentropic efficiency of compressor is 90%. Find the highest temperature, highest pressure, efficiency of expander and system COP.

Q.3. Derive an expression for work input per unit mass liquefied for an ideal system and an expression for FOM for a simple Linde-Hampson system.

Q.4. A liquefaction system for air operates on a simple Linde-Hampson cycle with P_L = 1atm and P_H = 200 atm. Temperature of isothermal compression is 308 K. Heat leakage into storage vessel is 3.5 kJ/kg of air. Heat exchanger effectiveness is 0.97. Find the power required to produce 50 kg of liquid air per hour.

Q.5. Show that the FOM of a pre-cooled L-H cycle is better than that of a simple L-H cycle. What are the limitations of the pre-cooled cycle?

Q.6. What are the advantages of using the dual pressure cycle for liquefaction?

Q.7. Compare the Heylandt, Claude and Kapitza cycles.

Q.8. Show that the equation of operating line in the upper part of the rectification tower is of the form $y = ax + b$. State the assumptions made in deriving the equation.

Q.9. Derive an expression for work requirement in an ideal separation system to separate a mixture of "n" different gases.

Q.10. Write short notes on: Inversion Curve, Simon liquefaction system, Linde double column separation system, Murphree efficiency concept.

Q.11. Classify properties of engineering materials at low temperature and explain them in detail.

Q.12. What is superfluidity? Explain type I & II super conductors.

Q.13. Discuss ideal Stirling cycle used in Phillip's cryocooler with the help of P-V diagram.

Q.14. Explain importance of regeneration and determine the minimum effectiveness for a cryocooler necessary for operation when T_a = 300 K, T_e = 75 K and v_1/v_2 = 1.5 for helium as a working fluid.

Q.15. What is cryopumping? Explain with a neat sketch.

Q.16. "Diffusion pump cannot operate if pump exhaust pressure is above 10^{-1} torr". Explain.

Q.17. Describe with a neat sketch an Ion pump.

Q.18. Explain measurement of temperature at cryogenic range.

Q.19. Discuss various cryogenic insulations in order of increasing cost.

Q.20. Write short notes on: Applications of Cryogenics in engineering industry, V-M Cryocooler, Solvay refrigerator, Properties of cryogenic liquids at normal boiling points, Cryogen storage vessel.

Unit – VI

Q.1. For a Brayton Cycle, derive an expression in terms of following temperatures. T_1 = Temperature at suction of compressor, T_2 = Temperature at discharge of compressor, T_3 = Temperature at inlet of expander and T_4 = Temperature at outlet of expander.

Q.2. A Carnot cycle works between T_o = 300 K and T_r = 210 K using 30 KJ/kg of air. A Brayton cycle has same refrigeration effect as the Carnot cycle and P_L =1 atm, P_H = 40 atm, $\Delta tr = \Delta tc$ = 5 K and compression and expansion are irreversible. The isentropic efficiency of compressor is 85%. Find the highest temperature, heat rejected, efficiency of expander and work input.

Q.3. Define Murphree efficiency and discuss the factors affecting its highest value.

Q.4. Explain the way to determine number of plates in a rectification column for O_2/N_2 separation.

Q.5. Find Murphree efficiency if pressure is 1 atm for (a) liquid leaves at 105 K and mole fraction = 0.46, (b) vapor leaves with mole fraction = 0.66, (c) vapor rising with mole fraction = 0.52.

Q.6. Derive an expression for W/mass compressed for ideal liquefaction cycle and Carnot cycle. Draw neat sketches.

Q.7. Determine Δt due to throttling from 200 atm to 1 atm for temperatures (a) 140 K, (b)180 K, (c) 205 K, (d) 300 K.

Q.8. A mixture of 76% Nitrogen, 20% Oxygen and 4% Argon by volume is separated into pure components. Work required is 130 kJ/kg. Find FOM and derive the formula used.

Q.9. Explain Claude system for liquefaction of air. Derive the expression for yield and FOM, if expander output is utilized for the compressor.

Q.10. Write short notes on: Magnetic refrigeration, Otho-Para Hydrogen conversion, Simon liquefaction system, Heat exchangers in liquefaction.

Q.11. Explain in detail Phillip's refrigerator.

Q.12. Determine refrigerating effect, COP and FOM for Phillips's refrigerator to maintain 80 K. Heat is rejected at 305 K. P_{min} = 1 atm and P_{max} = 10 atm. Assume all processes as ideal.

Q.13. Derive an expression for COP for V-M cryocooler and compare it to COP of Phillips Cryocooler.

Q.14. Explain the working of V-M cycle refrigerator.

Q.15. Classify cryocoolers using Walker's chart.

Q.16. What are cryogens ? Describe how they are stored.

Q.17. Explain the importance of insulation in cryogenics with their types.

Q.18. Cryogenic temperature measurement is critical. Comment.

Q.19. Write short notes on: Applications of Cryogenics, G-M refrigerator, Mechanical properties at cryogenic temperature, Cryopumping.

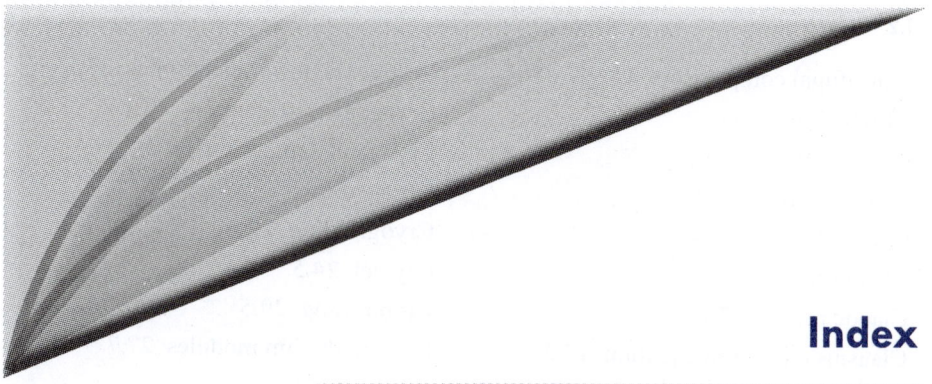

Index

A

Absolute zero 29.1
Activated sludge 23.9
Active magnetic regenerator 9.17
Adiabatic 13.6
Adiabatic demagnetization 4.14, 22.1
Adiabatic demagnetization refrigerators 14.12
Adiabatic expansion 4.8
Adjustable speed drive 27.6
Agitators 30.4
Air separation 12.1
Air separation unit (asu) 12.6, 20.2, 30.1
Anesthetic 24.5
ANSI 6.6
ASME 6.6
Asphyxiant 6.2
Asphyxiating) atmospheres 21.1
Asphyxiation 28.12
Attenuation 20.4
Austenite 29.3
Avogadro's number 5.9
Azeotropic mixture 11.4

B

Baffles 19.5
Ball valve 19.20
Bavard Alpert ionization gauge 19.18
Bayonet joint 18.20
Bell coleman cycle 4.11
Bellow tube 18.12
Bipropellant liquid 25.3
Black body 1.11
Block freezing 26.9
Boil off losses 9.17
Booster engines 25.1
Bosons 6.12
Bourdon gauge 17.11
Brayton-cycle cryocoolers 15.8
Brazing 23.10
Brine solutions 26.2
Bubble flow 18.24

C

Capacitance level probes 17.3
Capacitance meter 17.10
Capacitors 27.11
Capillary column 23.7
Capillary tube 17.5
Carbide precipitation 29.4
Carnot demagnetization process 9.18
Carnot efficiency 13.2
Carotene 26.3
Cascade process 7.12
Catalytic converters 28.16
Caustic scrubbers 12.1
Central processing unit 27.11

Centrifugal compressors 15.9
CERN 3.4
CGA 6.10
Charcoal array 19.11
Charles's law 17.21
Chemiosorption 19.9
Claude process 7.13
Clausius clapeyron equation 17.23
Coefficient of performance (COP) 13.3
Colburn factor 10.9
Cold box 12.13
Cold trap 19.5
Combined cycle power plant 30.2
Composite cryogenic tanks 18.9
Composite layer 16.1
Compressors 2.2
Condensation 11.5
Conductance 19.3
Contaminant 6.11
Continuous-cycle dilution refrigerators 14.19
Control algorithms 21.7
COP 2.3
Critical pressure 18.5
Crosshead 12.11
Cryobiology 24.1
Cryocomputing 27.9
Cryocoolers 8.5, 15.1
Cryo-electronics 27.1
Cryo-focusing 18.22
Cryogen 25.7
Cryogen 3.6, 6.1, 25.7
Cryogenic adsorption 11.32
Cryogenic deburring 29.5
Cryogenic deflashing 29.6
Cryogenic distillation 9.16
Cryogenic fuels 28.1
Cryogenic liquid spill 21.8
Cryogenic power electronics 27.1
Cryogenic refrigerators 13.1
Cryogenic tempering 29.2
Cryogenics 3.1
Cryogrinding 29.7
Cryojet 24.5
Cryomilling 29.5
Cryo-multichip modules 27.7
Cryonics 24.3
Cryoprotectants 24.4
Cryopumping 19.10
Cryosorption 25.9
Cryostat 15.5
Cryostats 8.8
Cryosurface 25.9
Cryosurgery 22.6
Cryotronics 27.1

D

Darcy's law 12.17
Dead space 13.13
Debye model 20.5
Debye temperatures 5.8
Dechlorination 30.5
Deep cryogenic treatment 29.2
Deep fluidization 26.10
Defrosting 26.13
Dehydration 26.6
Dense liquid 18.24
Desiccant 9.2, 16.8
Dewar flask 3.3, 9.12, 18.1
Dialysis 12.16
Dielectric constant 17.4
Diffusion pumps 19.9
Digital to analog converters 27.8
Dilution refrigerator 8.8
Dimensionally 29.2

Diodes 27.2
Displacer 14.9, 15.17
Distillation columns 20.9
DOT 6.6
Double rectification columns 12.10
Dry air 23.4
Dry ice 23.10
Dynamic gas 14.7

E

EER 2.3
Effluent 25.7
Elastomer diaphragm 19.20
Elastomers 28.5
Electrical conductivity 5.11
Electrical resistance 17.3
Electrolysis 9.12
Electromagnetic flowmeter 17.10
Electromagnetic interference (EMI) 15.2
Electromagnetic noise 15.22
ELF tank 23.3
Embrittlement 21.1
EMF 17.17
Enthalpy composition diagram 11.6
Entropy 1.1, 28.3
Epoxy 18.6
Equilibrium 1.2
Erosion 20.7
Eutectic point 26.5
Evacuation 6.7
Evaporators 2.2
Exergy analysis 12.3
Expanded foam 16.3
Expander 7.14
Expander mass flow rate ratio 13.12
Expanders 8.10
Expansion engines 11.28

Expansion ratio 6.3
Extended outage 21.8

F

Fatigue strength 5.2
Fatigue-failure models 18.10
Fermi energy 5.15
Ferric ammonium alum (FAA) salt 22.5
Fiberglass 16.6
Figure of merit (FOM) 4.2
Fin effectiveness 10.10
Flanged connections 28.7
Flash tank 9.10
Flashing 26.14
Fluidized bed 26.13
Fluorescent lamps 23.10
Fog cloud 21.8
Food preservation 26.1
Forecooler 11.23
Freezing point 8.2
Frostbite 21.2, 28.5
Fuel dispensing 28.4

G

Gadolinium-gallium garnet 14.12
Gas separation 11.1
Geo-synchronous 25.6
Getters 19.13
Giaque-hampson coil tube 10.5
Gibbs phase rule 11.3
G-m cryocooler 15.13
G-m refrigerator 14.8
Gravimetric blending 23.8

H

Heat exchangers 7.12, 8.5
Helium dilution 22.1

Hermetic 10.13
Hetero-junction bipolar transistor 27.2
Heylandt air-separation system 11.23
Heylandt system 7.16
High-temperature" superconductors 22.2
Hold time 28.6
Honeycomb 30.4
Hydrazine 25.4
Hydrogen – deuterium separation 11.28
Hydrogenate edible oils 23.11
Hydrostatic gauges 17.1
Hypothermia 21.3
Hypothermic 24.2
Hypoxia 24.6

I

Ice skin 26.9
Ideal gas 1.3
Immersion freezers 26.12
Inert 6.13
Inertness 23.2
Infrared sensors 15.7
Insulated-gate bipolar transistor 27.2
Insulation foils 28.12
Integrated gasification combined cycle 30.1
Intensification 20.4
Intermolecular forces 8.1
Inversion curve 4.6, 7.4, 15.7
Inverted annular 18.25
Invitiro fertilization 24.2
Ion exchange 30.3
Ion pump 19.8
Ionisation 19.9
Irreversibilities 12.3, 13.9
Isentropic expansion coeffcient 7.6

J

JET 19.7
Joule-thomson coefficient 4.3
Joule-thomson effect 3.3
Jt-nozzle 10.5

K

Kapitza system 7.15
Kapton 18.10
Kelvin scale 3.2
Kennard equation 19.2
Kettle liquid 11.21
Knudsen gauge 19.19
Knudsen number 19.1

L

Labyrinth seal 10.19
Langmuir's viscosity gauge 19.18
Latent heat 6.3
Linde cycle 7.7
Linde-bronn hydrogen separation system. 11.26
Liquefaction 7.9, 28.3
Liquefier cold box 9.15
LMTD 10.2
LNG 18.1

M

Magnetic bond 17.5
Magnetic permeability 5.17
Magnetic resonance imaging 3.6
Magnetic thermometers 17.20
Magnetocaloric effect 14.16
Magneto-caloric materials 14.14
Manometer 17.2

Martensite 29.3
Matrix capacity ratio 13.14
Maximum allowable accumulated pressure 20.7
Mccabe and thiele method 11.11
Mean free path 19.1
Mean-time-to-failure 15.4
Mechanical claddings 28.15
Mechanical cryocoolers 15.3
Meissner effect 5.15
Membrane 11.2
Membrane technology 8.2
Mercaptan 7.2
Microorganisms 26.6
Microspheres 16.13
Mist flow 18.24
Mixed fluid cascade 9.8
Modulus of elasticity 20.8
Molecular sieve 12.13
Molecular sieve 7.1
Monotonic 17.15
MR cocktail 9.6
MRI systems 15.13
Multi layer insulation 16.2, 16.10
Multi-stage compressor 12.5
Murphree efficiency 11.9
Mylar tube 18.18

N

Nausea 6.11
Neon lamps 8.4
NIST 15.16
Non-condensable 19.10
Non-combustible materials 21.1
NPSH 10.19
NTU 10.2
Number of theoretical plates 11.17
Nusselt number 10.9

O

Ohm's law 19.3
Oil scrapper rings 12.11
Opacified powders 16.10
Orbital lifetime 25.8
Orbital wrapping 16.11
Orifice plate 17.6
Orifice pulse tube refrigerator 15.16
Ortho-hydrogen 6.8
Oscillator 27.3
Outgassing rate 19.4
Oxygen-compatible binders 21.6
Oxygen-enriched atmosphere 21.2

P

Para-hydrogen 6.8
Paramagnetic pill 14.13
Paramagnetic salt 8.12
Partial pressure 11.5
Pathogen 26.2
Peltier effect 4.17
Permeation 12.17
Personal protective equipment 21.3
Personnel hazards 21.1
Phase equilibrium curve 11.4
Phillips refrigerator 14.2
Phonons 5.8
Physisorption 15.10
Piggyback 12.4, 20.2
Pirani thermal conductivity gauge 19.16
Planck's constant 20.5
Plank sorption cryocoolers 15.11
Plate freezer 26.7
PLC 12.12
Pointing stability 15.23
Poisseuille equation 19.2

Polyurethane 16.4
Positive gas displacement pumps 19.6
Power circuits 27.5
Power dissipation 15.22
Prandtl number 10.9
Pre-cooling 8.4
Preparation tolerance 23.8
Pressure swing absorption (PSA) 23.4
Propellants 25.1
Protocol requirements 14.1
PSP 17.13
Pulmonary edema 21.3
Pulse tube cryocoolers 15.2
Pump speed 19.3
Purge 6.7
Pyrex flask 19.12

Q

Quantum considerations 14.18
Quick freeze machine 26.3

R

Radiant barrier 16.1
Raoult's law 11.5
Receiver 7.8
Rectification 11.8
Recuperative 10.6
Recycling 13.13
Reflux ratio 11.13
Refrigerant separator 9.5
Refrigeration 2.1
Refrigeration effect 13.3
Refrigerator optimization 13.9
Regenerative 10.6
Regenerator 3.6, 15.19
Regenerators 11.22
Reinforced polymer 25.7

Relied on chlorofluorocarbons (CFCS) 23.6
Resistance temperature detector 17.17
Reynolds number 10.9, 19.1
Rupture disks 6.10

S

Safety interlocks 25.5
Scavenger system 9.1
Sedan 28.16
Selectivity 12.18
Semiconductor 27.4
Shear forces 18.24
Shear stress 20.8
Shielding gas 23.11
Shrink fitting 29.6
Shrinkage 10.4
Silicon-on-sapphire 17.13
Single-shot 14.18
Slush 22.3
Solvay refrigerator 14.6
Space simulators 25.8
Spatial resolution 15.23
Specialty gas mixtures 23.1
Specific gravity 6.5
Specific heat 5.7
Specific impulse 28.10
Specific power 4.1
Spring constant 20.8
Stefan boltzman's equation 16.7
Stirling cycle cryocooler 15.19
Stirling cycle refrigerators 14.2
    Stodola equation 10.18
Sub-cooling 9.7
Submersible cryogenic pumps 10.19
Substitution 20.4
Super critical expansion turbine 9.14
Supercomputer 27.12

Superconducting quantum interference device 22.2
Superconductivity 3.6
Superconductivity 5.11
Supercritical 6.14
Superfluidity 22.2
Superfluidity 6.16
System optimization 20.1

T

Teflon 24.7
Thermal conductivity 5.5, 16.1
Thermal diffusivity 1.7
Thermal valves 14.18
Thermistor 17.19
Thermodynamic work of separation 12.2
Thermoelectric coolers 27.12
Thermoelectric difference 17.16
Thermos bottle 18.3
Thermo-siphon 8.9
Thin film pressure transducer 17.12
Throttle valve 13.8
Throttling 4.16
Throttling valve 7.7
Throughput 19.2
Tolerable fore pressure 19.7
Ton 2.2
Tonnage 23.5
Torque rise 28.9
Torr 19.2
Transducer 17.11
Transistors 27.2
Turbine flowmeters 17.8
Turbo expanders 10.15
Two-factor hypothesis 24.2
Two-phase cryogenic flow 18.23

U

Ultra high vacuum 19.9
Ultrasonic flow equipment 17.10
Unmanned aerial vehicle 28.2

V

Vacuum gauge 19.15
Vacuum insulation 16.8
Vacuum jacketed pipe 18.13
Vacuum-jacketed piping 28.7
Van der walls forces 15.10
Vapor cooled shield 16.13
Vaporizers 18.8
Venturi tube 17.7
Voltaic array 28.1
Vortexing 25.3
Vuilleumier cycle 14.5
Vuilleumier refrigerator 14.5

W

Walker's chart 15.2
Waste heat boilers 30.2

Y

Yield 7.3
Yield strength 5.2
Young's modulus 5.4

Z

Zeolite 19.13
Zeolitic material 12.15
Zero-gravity 15.12

NHTI Library
Concord's Community College
Concord, NH 03301